FUNDAMENTALS OF
CHEMICAL REACTION
ENGINEERING

SECOND EDITION

FUNDAMENTALS OF CHEMICAL REACTION ENGINEERING

CHARLES D. HOLLAND
RAYFORD G. ANTHONY

Department of Chemical Engineering
Texas A & M University

PRENTICE HALL
Englewood Cliffs, New Jersey 07632

Library of Congress Cataloging-in-Publication Data

Holland, Charles Donald.
Fundamentals of chemical reaction engineering / Charles D.
Holland, Rayford G. Anthony.— 2nd ed.
  p. cm.
  Includes bibliographical references and index.
  ISBN 0-13-335639-6
  1. Chemical engineering. 2. Chemical reactions. I. Anthony,
Rayford G. (Rayford Gaines) II. Title.
TP149.H64 1989   88-6489
660.2′99 — dc 19   CIP

Editorial / production supervision: Evalyn Schoppet
Manufacturing buyer: Mary Anne Gloriande

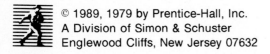 © 1989, 1979 by Prentice-Hall, Inc.
A Division of Simon & Schuster
Englewood Cliffs, New Jersey 07632

Printed in the United States of America

10 9 8 7 6 5 4 3 2 1

ISBN 0-13-335639-6

Prentice-Hall International (UK) Limited, *London*
Prentice-Hall of Australia Pty. Limited, *Sydney*
Prentice-Hall Canada Inc., *Toronto*
Prentice-Hall Hispanoamericana, S.A., *Mexico*
Prentice-Hall of India Private Limited, *New Delhi*
Prentice-Hall of Japan, Inc., *Tokyo*
Simon & Schuster Asia Pte. Ltd., *Singapore*
Editora Prentice-Hall do Brasil, Ltda., *Rio de Janeiro*

Dedicated to

FRED J. BENSON

Dean Emeritus of the College of Engineering

and in memory of

JAMES DONALD LINDSAY

Professor Emeritus and former Head of the
Department of Chemical Engineering
Texas A & M University

Contents

Preface

This book is written for the beginning student and practicing engineer in chemical kinetics and chemical reaction engineering. All topics are initiated by commencing with first principles. Derivations are complete, rigorous, and easily followed. The chapters are ordered with respect to the increasing difficulty of the subject matter. For example, Chapter 12, Theory of Reaction Rates, could have been included in Chapter 1 but was deferred because the level of that subject matter is higher than that needed by the student in order to master the material presented in subsequent chapters.

In this edition, the first four chapters present the design and analysis of data for isothermal reactors for irreversible, reversible, parallel, series, and complex reactions. New problems have been introduced which in many cases require use of numerical methods to obtain the solutions. Chapter 5 presents a rigorous review of thermodynamics required in the analysis and design of chemical reactions for ideal and nonideal systems. These principles are used in Chapters 6, 7, 8, and 9 in the design and analysis of tubular reactors, thermal and catalytic, with plug flow and for convective or dispersive fluxes in three dimensions, batch and perfectly mixed flow reactors. The terms for ideal and non-ideal contributions to the enthalpy of the solution are clearly presented. Several new design problems representing actual case studies were added to the problems.

The book is divided into three sections, fundamental principles of chemical kinetics and reactor engineering, design of thermal and catalytic reactors, and advanced topics on polymerization reactions, mixing effects, and theory of reaction rates. Since tracer techniques are important in the diagnosis of mixing and flow problems in reactors, Chapter 11, Use of Tracer Techniques to Determine Mixing Effects, was added in this edition. Chapters 1 through 7 and part of Chapter 8 are covered in an undergraduate course, and Chapters 8 through 12 serve as a basis for a graduate course at Texas A & M University.

Helpful suggestions from Professors A. Akgerman, J. Bullin, P. T. Eubank, A. Gadalla, R. White, and D. T. Hanson of the Department of Chemical Engineering are appreciated. The authors also wish to thank D. L. Tanner of Dow Chemical Company for his assistance in the solution of examples in Chapter 7, and T. Helton, H. Ghaeli, P. Moore, Y-T. Chen, and L. Czarnecki for their assistance on some of the problems requiring computer solutions.

FUNDAMENTALS OF CHEMICAL REACTION ENGINEERING

Section I:
Fundamentals

Reaction Rates and Batch Reactors

1

I. INTRODUCTION

Chemical kinetics is the branch of physical chemistry that is concerned with the rates and mechanisms of chemical reactions. The first quantitative data on rates of reaction were reported in 1777 by Wenzel [1], who investigated the effect of acid concentration on the rate of dissolution of copper and zinc metals in acid solutions. The law of mass action for homogeneous reactions was formulated by Guldberg and Waage in 1867, and Van't Hoff extended it for heterogeneous reactions in 1877. The concept of the reaction velocity constant and the classification of reactions according to molecularity were introduced in 1884 by Van't Hoff [2]. In 1889, Arrhenius [3] suggested that only those molecules having an energy equal to or greater than a given level had the potential for reaction.

Van Narum appears to have been among the first to recognize catalysis when he observed in 1796 the effect of the surface catalysis in the dehydrogenation of alcohol. In 1902, Ostwald [4] defined a catalyst as "any substance which alters the velocity of a chemical reaction without appearing in the end products."

The beginning of the use of reaction kinetics in the design of reactors is marked by a paper published in 1923 by Lewis and Ries [5] on the catalytic oxidation of sulfur dioxide. Hougen [6] has given major credit for the early development of the theory and applications of chemical kinetics to process design, particularly in catalytic processes, to Kenneth M. Watson. A text written by Hougen and Watson [7] led to the further development and acceptance of the subject of reactor design.

Reactions may be classified in a variety of useful ways. Some of the methods of classification are enumerated. Reactions may be classified according to the number of phases involved in the reactions:

1. Homogeneous
2. Heterogeneous or multiphase

They may be classified according to the manner in which the reaction is carried out:

3. Thermal or uncatalyzed
4. Catalyzed

Sometimes reactions are classified according to the nature of the reaction, such as:

5. Unidirectional or irreversible
6. Reversible
7. Isothermal or nonisothermal
8. Biological

Sometimes reactions are classified according to the type of equipment used to carry out the reactions:

9. Batch reactors
10. Flow reactors
    (a) tubular   (c) trickle bed   (e) slurry
    (b) continuous stirred tank   (d) bubble column   (f) ebullating flow
11. Catalytic reactors with fixed or fluidized beds

Batch reactors were no doubt the first type of reactor to be devised. They consist of well-stirred pots or kettles in which the reactants are mixed by stirring and allowed to react over a period of time. Provisions are generally made for heating or cooling the reacting mixture. Batch reactors are still used in the modern petrochemical industry. Figure 1-1 shows a reactor that may be used in either a batch or a continuous process.

The use of batch reactors in the manufacture of soap has been known for centuries and was recorded by early historians. During the excavation of Pompeii, a soap factory was discovered, complete with boiling pans and a quantity of soap which had been buried under volcanic ash for nearly 2000 years [8]. Batch-type reactors are still in use in the manufacture of soap by the "kettle or full-boiled method." This method consists of the saponification of

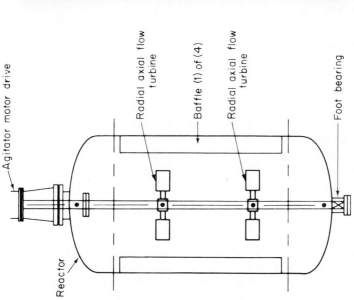

Figure 1-1. Reactor used in the production of synthetic rubber (Courtesy of Goodyear Tire and Rubber Company).

Agitator motor drive

Radial axial flow turbine

Baffle (1) of (4)

Radial axial flow turbine

Foot bearing

Reactor

neutral fats and oils (triglycerides) with caustic soda as follows:

$$
\begin{array}{l}
\text{H} \\
| \\
\text{H}-\text{C}-\text{OOC}-\text{R}_1 \\
| \\
\text{H}-\text{C}-\text{OOC}-\text{R}_2 + 3\text{NaOH} \longrightarrow \\
| \\
\text{H}-\text{C}-\text{OOC}-\text{R}_3 \\
| \\
\text{H}
\end{array}
\qquad
\begin{array}{l}
\text{H} \\
| \\
\text{H}-\text{C}-\text{OH} \quad \text{Na}-\text{OOC}-\text{R}_1 \\
| \\
\text{H}-\text{C}-\text{OH} + \text{Na}-\text{OOC}-\text{R}_2 \\
| \\
\text{H}-\text{C}-\text{OH} \quad \text{Na}-\text{OOC}-\text{R}_3 \\
| \\
\text{H}
\end{array}
$$

    Triglyceride   Caustic soda   Glycerol   Sodium soap

High-grade toilet soaps as well as low-grade industrial soaps with a high glycerol recovery efficiency may be made by this method [9].

    The fundamental principles upon which the field of chemical reaction engineering is based are introduced in Parts II and III of this chapter. The rate of reaction, introduced in Part II, is shown to depend on the manner in which the molecules participate in a chemical reaction. After the development of the material-balance equations for batch reactors has been presented in Part III, expressions are developed for the variation of the moles reacted as a function of time.

II. FUNDAMENTAL KINETIC AND THERMODYNAMIC RELATIONSHIPS

Rate expressions for various types of reactions are formulated in this section. A summary of the fundamental principles of the theory of reaction rates which are needed in these formulations are also presented. (A comprehensive treatment of the theory of reaction rates is presented for the benefit of the advanced student in Chapter 12.) Also presented in this section is the relationship between reaction kinetics and thermodynamics, as well as other relationships which are needed in subsequent chapters.

A. Definition of the Rate of Reaction

    The following definition for the rate of reaction leads to rate expressions, based on experimental results, which are valid over wide ranges of operating conditions. The net rate of disappearance (or appearance) of component A in a reacting mixture is defined as follows:

$$
r_A = \frac{\text{moles of } A \text{ disappearing (or appearing)}}{(\text{unit time})(\text{unit of reactor volume})} \tag{1-1}
$$

Suppose that the reacting mixture consists of components A, B, C, and D at a given temperature and pressure. Then the rate of reaction may be expected to depend on the temperature, pressure, and concentrations of the reacting species; that is,

$$r_A = f(T, P, C_A, C_B, C_C, C_D) \tag{1-2}$$

The precise form of this functional relationship has been deduced for molecular or elementary reactions by use of the kinetic theory of gases and the transition state theory, but for most complex reactions encountered in chemical engineering practice, the functional form must be determined experimentally. Hereafter, this functional relationship is referred to as simply the *rate expression*. The form of the rate expression depends on *how* the molecules participate in the reaction, and if properly determined it is independent of the type of reactor used to carry out the reaction.

B. Principles of the Kinetic Theory of Bimolecular Reactions

While the values of the rate constants predicted by the kinetic theory are generally highly inaccurate, many concepts of the kinetic theory carry over to subsequent theories used to describe chemical reactions. Thus, a summary of the results of the theory developed in Chapter 12 is presented next. The two principal postulates of this theory are (1) before two molecules A and B can react, they must collide, and (2) of the molecules which collide, only those react which have a kinetic energy equal to or greater than ϵ along the line of centers of the molecules at the moment of contact, provided that they are properly orientated.

On the basis of these postulates, it is shown in Chapter 12 that the rate of reaction, based on the kinetic theory of gases, for two unlike molecules A and B is given by

$$r_A \left(\frac{\text{g mol}}{\text{s cm}^3} \right) = r_B = k_c C_A C_B \tag{1-3}$$

and the rate constant k_c is given by

$$k_c = p\sigma_{AB}^2 \sqrt{\frac{8\pi RT}{\mu}} \, e^{-E_A/RT} \tag{1-4}$$

where C_A, C_B = concentrations of A and B, g mol/cm^3

   E_A = energy of activation, cal/g mol

   p = fraction of the molecules that collide that have the proper orientation

   R = gas constant, cal/(g mol)(K)

M_A, M_B = molecular weights of A and B, g/g mol

$$\sigma_{AB} = \frac{\sigma_A + \sigma_B}{2} = \text{collision diameter of molecules } A \text{ and } B \text{ having}$$
diameters of σ_A and σ_B

$$\mu = \frac{M_A M_B}{M_A + M_B} = \text{reduced mass of } A \text{ and } B$$

The rate expression given by Equations (1-3) and (1-4) for the bimolecular reaction $A + B \rightarrow$ products is in agreement with the experimental evidence that the rate is proportional to the product of the concentration, $C_A C_B$, and that the rate should vary exponentially with temperature. Thus, the kinetic theory predicts the correct form of the rate expression even though the numerical value of the rate constant calculated from this theory is generally too large.

C. Transition State Theory

The development of the modern transition state theory, which is also called the activated complex theory, as well as the theory of absolute reaction rates, is presented in Chapter 12. The postulates on which this theory is based are as follows:

1. For a reaction to occur, the reactants must form an activated complex.
2. A state of equilibrium exists between the reactants and the activated complex.
3. The rate of reaction is equal to the concentration of the activated complex that passes from reactants to products divided by the time required for an activated complex to traverse the distance across the energy barrier from the reactant to the product side.

On the basis of these postulates and by use of Maxwell–Boltzmann statistics, the following expression for the rate of reaction of $A + B \rightarrow$ products was deduced.

$$r_A = r_B = k_c C_A C_B \tag{1-5}$$

and

$$k_c = \kappa \frac{kT}{h} e^{-\Delta G^\dagger / RT} = \kappa \left(\frac{kT}{h} e^{\Delta S^\dagger / R} \right) e^{-\Delta H^\dagger / RT} \tag{1-6}$$

where k = Boltzmann's constant = 1.3805×10^{-16} erg/(K)(molecule)

h = Planck's constant = 6.624×10^{-27} erg s/molecule

$\Delta G^\dagger$, $\Delta S^\dagger$, $\Delta H^\dagger$ = standard free energy change, standard entropy change, and standard enthalpy change, respectively, for the equilibrium reaction between the reactants and the activated complex

κ = probability that an activated complex will react to products

The transition state theory is also in agreement with experimental observations in that it predicts that the rate of reaction should be proportional to the product of concentrations $C_A C_B$ and that the reaction rate constant should exhibit exponential behavior with temperature. Like the kinetic theory model, the values of k_c computed by this theory are generally unreliable, and thus k_c must be determined experimentally.

D. Relationships of the Rates of Reaction of the Members of a Reacting Mixture

Suppose that the stoichiometry of a reaction is given by

$$aA + bB = cC + dD$$

In the definition of the rates of reaction, it is convenient to define the rates for the reactants in terms of disappearance and the rates for the products in terms of appearance. Let the rates of reaction r_A and r_B for A and B, respectively, be defined in terms of disappearance and the rates r_C and r_D for C and D, respectively, be defined in terms of appearance. From the *law of definite proportions* and the stoichiometric equation, b moles of B react every time a moles of A react. Then B disappears b/a times as fast as A, or

$$r_B = \frac{b}{a} r_A \tag{1-7}$$

For every a moles of A which react, c moles of C and d moles of D are formed. Thus

$$r_C = \frac{c}{a} r_A, \qquad r_D = \frac{d}{a} r_A \tag{1-8}$$

EXAMPLE 1-1

Phosphine decomposes by the following stoichiometric equation:

$$4PH_3 \rightarrow P_4 + 6H_2$$

At a given instant, phosphine decomposes at the rate of 10×10^{-5} g mol/s/liter. Let the rate of reaction for phosphine be defined in terms of disappearance; that is,

$$r_{PH_3} = \frac{\text{g mol of } PH_3 \text{ disappearing}}{\text{s liter}} = 10 \times 10^{-5}$$

(a) Compute the rates of appearance of phosphorus and hydrogen.
(b) Compute the rates of disappearance of phosphorus and hydrogen and the rate of appearance of phosphine.

SOLUTION

(a) Let the rates of reaction for phosphorus and hydrogen be defined in terms of appearance.

$$\frac{\text{g mol of P}_4 \text{ appearing}}{\text{s liter}} = r_{P_4} = \frac{1}{4} r_{PH_3} = 2.5 \times 10^{-5}$$

$$\frac{\text{g mol of H}_2 \text{ appearing}}{\text{s liter}} = r_{H_2} = \frac{6}{4} r_{PH_3} = 15 \times 10^{-5}$$

(b)

$$\frac{\text{g mol of P}_4 \text{ disappearing}}{\text{s liter}} = -r_{P_4} = -2.5 \times 10^{-5}$$

$$\frac{\text{g mol of H}_2 \text{ disappearing}}{\text{s liter}} = -r_{H_2} = -15 \times 10^{-5}$$

$$\frac{\text{g mol of PH}_3 \text{ appearing}}{\text{s liter}} = -r_{PH_3} = -10 \times 10^{-5}$$

E. Order and Molecularity

In general, it is not possible to predict the mechanism from a knowledge of the stoichiometry. The mechanism and functional form of the rate expression [see Equation (1-2)] must be determined experimentally. If it is found by experiment that the expression for the rate of reaction is given by

$$r_A = k_c C_A^{n_1} C_B^{n_2} C_C^{n_3} \tag{1-9}$$

then the *order N* of the reaction is defined by

$$N = n_1 + n_2 + n_3 \tag{1-10}$$

Even though k_c is a function of temperature, it is called the *rate constant* or the *specific reaction rate*. The order of a reaction is sometimes expressed relative to a component. For example, the rate of reaction given by Equation (1-9) is of *order* n_1 with respect to component A, where n_1 may be zero, negative, or positive integers or fractions.

The molecularity of a single chemical reaction is defined as the minimum number of molecules which must combine or interact when the reaction goes one time in a given direction. The molecularity and the *law of mass action* are closely related. In 1867, Guldberg and Waage gave the first general statement of this law. The present accepted statement of this law is that the rate of a molecular reaction is proportional to the product of the concentrations of the reactants, where the concentration of each reactant is raised to the power given by the stoichiometric coefficient of that component. The rate expressions for unimolecular, bimolecular, and trimolecular reactions are as follows:

Unimolecular: $A \rightarrow$ products

$$r_A = k_c C_A \tag{1-11}$$

Bimolecular: $2A \rightarrow$ products

$$r_A = k_c C_A^2 \tag{1-12}$$

Bimolecular: $A + B \rightarrow$ products

$$r_A = k_c C_A C_B \tag{1-13}$$

Trimolecular: $3A \rightarrow$ products

$$r_A = k_c C_A^3 \tag{1-14}$$

Trimolecular: $2A + B \rightarrow$ products

$$r_A = k_c C_A^2 C_B \tag{1-15}$$

Trimolecular: $A + B + C \rightarrow$ products

$$r_A = k_c C_A C_B C_C \tag{1-16}$$

Although the same symbol k_c is used in each of the preceding cases, it must be realized, of course, that it differs from reaction to reaction. Also, k_c in these rate expressions is defined with respect to component A. For example, for the reaction $2A + B \rightarrow$ products,

$$r_B = \frac{1}{2} r_A$$

and, consequently,

$$r_B = \frac{1}{2} k_c C_A^2 C_B \tag{1-17}$$

where k_c is the rate constant with respect to component A.

Molecularity as well as order is always stated for a single chemical reaction in a single direction. The molecularity of the reverse reaction of a reversible reaction may differ from that of the forward reaction. For example, in the reaction

$$A \underset{k_c'}{\overset{k_c}{\rightleftarrows}} R + S$$

the forward reaction is unimolecular and the reverse reaction is bimolecular. Therefore, the net rate of disappearance of A is given by

$$r_A = r_{Af} - r_{Ar} = k_c C_A - k_c' C_R C_S \tag{1-18}$$

If component A reacts simultaneously by each of two reactions, then the net rate of reaction of A is given by the algebraic sum of the rates for each of the reactions. For example, consider the reactions,

$$A \xrightarrow{k_1} R, \qquad r_{A1} = k_1 C_A$$

$$A \xrightarrow{k_2} 2S, \qquad r_{A2} = k_2 C_A$$

Then

$$r_A = r_{A1} + r_{A2} = k_1 C_A + k_2 C_A \tag{1-19}$$

Also, first-order isomerization reactions of the following type may be encountered.

$$A \underset{k_1'}{\overset{k_1}{\rightleftharpoons}} R \underset{k_2'}{\overset{k_2}{\rightleftharpoons}} S \underset{k_3'}{\overset{k_3}{\rightleftharpoons}} A$$

In this case, the net rate of reaction of A is given by

$$r_A = k_1 C_A - k_1' C_R + k_3' C_A - k_3 C_S \tag{1-20}$$

Multiple reactions are discussed in greater detail in Chapter 3.

If a molecular reaction goes precisely as written, the order of the reaction is equal to the molecularity. However, the converse of this statement is not necessarily true; that is, a reaction may be first order, but it is not necessarily unimolecular. The reaction may in fact have a complex reaction mechanism, whose rate may be represented by an empirical expression that depends on the first power of the concentration. For example, to explain the decomposition of nitrous oxide in Example 1-2 by a first-order rate expression, a reaction mechanism is required.

The rate expressions presented previously apply regardless of whether the reaction occurs in the gas phase or the liquid phase.

Gas Phase Reactions. The rate expressions for reactions which occur in the gas phase may be stated in terms of either concentrations or partial pressures. For example, the rate of reaction for the unimolecular reaction,

$$A \rightarrow \text{products}$$

may be stated in terms of p_A, the partial pressure of A; that is,

$$r_A = k_p p_A \tag{1-21}$$

where k_p is the rate constant when partial pressures are used. Since the rate of reaction r_A is the same regardless of whether concentrations or partial pressures are used, it follows that

$$r_A = k_c C_A = k_p p_A \tag{1-22}$$

Similarly, for any of the previous reactions which occur in the gas phase, the rates may be expressed as a product of k_p and the appropriate partial pressures.

F. Dimensions of the Rate Constants k_c and k_p

Since r_A has the same units regardless of whether concentrations or partial pressures are used, it follows that k_c and k_p have precisely those dimensions required for the dimensions on the left side of each equation to equal the dimensions on the right side. For example, consider Equation (1-22):

$$r_A \left[\frac{\text{g mol}}{\text{s m}^3} \right] = k_c \left[\frac{1}{\text{s}} \right] C_A \left[\frac{\text{g mol}}{\text{m}^3} \right] = k_p \left[\frac{\text{g mol}}{\text{s m}^3 \text{ atm}} \right] p_A \text{ (atm)} \qquad (1\text{-}23)$$

Similarly, for the second-order reaction $2A \rightarrow$ products,

$$r_A \left[\frac{\text{g mol}}{\text{s m}^3} \right] = k_c \left[\frac{\text{m}^3}{\text{s g mol}} \right] C_A^2 \left[\frac{(\text{g mol})^2}{(\text{m}^3)^2} \right]$$

$$= k_p \left[\frac{\text{g mol}}{\text{s m}^3 \text{ atm}^2} \right] p_A^2 [\text{atm}^2] \qquad (1\text{-}24)$$

G. Relationships of the Rate Constants

If the reacting mixture is a perfect gas mixture, then

$$p_A = C_A RT \qquad (1\text{-}25)$$

and it follows from Equation (1-24) that

$$k_c C_A^2 = k_p C_A^2 (RT)^2$$

or
$$k_c = k_p (RT)^2$$

Thus, in general, for an Nth-order reaction which occurs in the gas phase,

$$k_c = k_p (RT)^N \qquad (1\text{-}26)$$

where the volumetric behavior of the gas phase approximates that of a perfect gas.

EXAMPLE 1-2

The decomposition of nitrous oxide in glass or silica vessels is generally a homogeneous second-order reaction. In the presence of a gold wire at a high temperature, the decomposition of nitrous oxide was found by Hinshelwood and Prichard [10] to be first order. That is, the stoichiometry is given by

$$2N_2O \rightarrow 2N_2 + O_2$$

and the rate of reaction in the presence of a gold wire catalyst is given by

$$r_{N_2O} = \left(\frac{\text{g mol of } N_2O \text{ disappearing}}{\text{s cm}^3} \right) = k_c C_{N_2O}$$

At 900°C, Hinshelwood and Prichard found that

$$k_c = 0.013 \text{ s}^{-1}$$

Calculate the corresponding value of

$$k_p \left(\frac{\text{g mol}}{\text{s cm}^3 \text{ atm}} \right)$$

SOLUTION

For a first-order reaction, Equation (1-26) becomes

$$k_p = k_c (RT)^{-1}$$

Thus

$$k_p = (0.013 \text{ s}^{-1}) \left[\left(82.05 \frac{\text{atm cm}^3}{\text{g mol K}} \right) (1173.1 \text{ K}) \right]^{-1}$$

$$= 1.3504 \times 10^{-7} \frac{\text{g mol}}{\text{s cm}^3 \text{ atm}}$$

H. Reversible Reactions and Thermodynamic Equilibrium

The rate of reaction r_A as defined by Equation (1-1) is equal to the net rate of disappearance of A by all reactions that occur. When a single reversible reaction such as

$$2A \underset{k_c'}{\overset{k_c}{\rightleftarrows}} R + S$$

occurs, the net rate of reaction is equal to the forward rate of reaction minus the reverse rate of reaction. That is, A disappears by the forward reaction at the rate r_{Af},

$$r_{Af} = k_c C_A^2 \tag{1-27}$$

and A appears by the reverse reaction,

$$R + S \xrightarrow{k_c'} 2A$$

at the rate r_{Ar},

$$r_{Ar} = k_c' C_R C_S \tag{1-28}$$

Then r_A, the net rate of disappearance of A, is given by

$$r_A = r_{Af} - r_{Ar} = k_c C_A^2 - k_c' C_R C_S \tag{1-29}$$

At thermodynamic equilibrium,

$$r_A = 0 \tag{1-30}$$

and Equation (1-29) may be arranged in the well-known form

$$K_c = \frac{k_c}{k'_c} = \frac{C_R C_S}{C_A^2} \tag{1-31}$$

where K_c = equilibrium constant where concentrations are employed. For ideal solutions, the equilibrium constant K_c remains fixed (at a given temperature) for all possible combinations of the concentration of A, R, and S that satisfy the equality given by Equation (1-31). If the particular reaction also occurs in the gas phase, then, in a manner similar to that described previously,

$$K_p = \frac{p_R p_S}{p_A^2} \tag{1-32}$$

where K_p = equilibrium constant where partial pressures are employed. If the mixture obeys the perfect gas law, the equilibrium constant K_p is independent of the total pressure as well as the partial pressure of A, R, and S at a given temperature.

In the literature, rate expressions can be found which are thermodynamically inconsistent. For example, for the reaction

$$A + 3B \rightleftarrows 2R$$

suppose that the experimental rate of formation of R is represented by $r_R = kC_A C_B^{1/2} - k'C_R^2$. At $r_R = 0$,

$$\frac{k}{k'} = \frac{C_R^2}{C_A C_B^{1/2}}$$

The equilibrium constant K_c for the reaction as written is given by

$$K_c = \frac{C_R^2}{C_A C_B^3}$$

Comparison of the expressions for k/k' and K_c shows a disagreement in the power of C_B. Thus, the rate expression for r_R is said to be thermodynamically inconsistent. Since k/k' is unequal to K_c, one should guard against committing the fallacy of setting them equal in order to determine the reverse rate constant k'. Inconsistent thermodynamic rate expressions result from the fact that the reaction is not a molecular reaction and the order for a component is not equal to the stoichiometric coefficients. Instead of a single reaction being involved, a mechanism consisting of several reactions is needed to describe the experimental results that are discussed in subsequent chapters.

The use of thermodynamics in the analysis of reaction systems and in reactor design is presented in Chapter 5 with applications in Chapters 6 and 7.

EXAMPLE 1-3

One method for producing hydrogen is to treat water gas (carbon monoxide and hydrogen) with steam. Water gas is made by reacting the carbon in a fossil fuel such as coal or lignite with steam at temperatures in the neighborhood of 1500 K [11]. The *water gas* reaction is

$$C + H_2O \rightarrow CO + H_2$$
(water gas)

The water gas produced can be further reacted with steam at temperatures around 400°C in the presence of a catalyst to produce additional hydrogen by the *water gas shift* reaction [12].

$$CO + H_2O \rightarrow CO_2 + H_2$$

At 400°C, $K_p = 11.772$ for the water gas shift reaction [13]. Calculate the moles of hydrogen produced when an equimolar mixture of carbon monoxide and hydrogen is treated with 100% excess of steam at a pressure of 1 atm (absolute).

SOLUTION

As a basis, suppose that the feed consists of 1 mol of CO, 1 mol of H_2, and 2 mol of H_2O. Let x be the moles of CO reacted at equilibrium. Then the moles of CO, H_2O, CO_2, and H_2 at equilibrium may be stated in terms of x as follows:

$$\text{mol of CO} = 1 - x$$

$$\text{mol of } H_2O = 2 - x$$

$$\text{mol of } CO_2 = x$$

$$\underline{\text{mol of } H_2 = 1 + x}$$

$$\text{total moles} = 4$$

$$p_{CO} = \left(\frac{1-x}{4}\right)P, \qquad p_{H_2O} = \left(\frac{2-x}{4}\right)P$$

$$p_{CO_2} = \left(\frac{x}{4}\right)P, \qquad p_{H_2} = \left(\frac{1+x}{4}\right)P$$

Then

$$K_p = \frac{p_{CO_2}\,p_{H_2}}{p_{CO}\,p_{H_2O}} = \frac{x(1+x)}{(1-x)(2-x)}$$

This equation can be solved for x in terms of K_p by use of the quadratic formula to give

$$x = \frac{(3K_p + 1) - \sqrt{(3K_p + 1)^2 - 8K_p(K_p - 1)}}{2(K_p - 1)}$$

After $K_p = 11.772$ has been substituted, one obtains

$$x = \frac{36.316 - \sqrt{(36.316)^2 - 8(11.772)(10.772)}}{21.544} = 0.876$$

which is equal to the moles of hydrogen produced.

I. Effect of Temperature on the Rate Constant, k_c

The rate constant normally varies in an exponential manner with temperature as proposed by Arrhenius [3].

$$k_c = A \exp\left(\frac{-E_A}{RT}\right) \tag{1-33}$$

where A = frequency factor or Arrhenius constant
E_A = activation energy
R = gas law constant in consistent units

This general form of the expression for k_c is suggested by the kinetic theory of gases (see Chapter 12 and Equation (1-4)). By use of this theory, k_c is found to be proportional to

$$\sqrt{T}\,e^{-E_A/RT}$$

By use of statistical mechanics (see Chapter 12 and Equation (1-6)), k_c is found to be proportional to

$$T e^{-E_A/RT}$$

In most cases, the variation of k_c with temperature over limited temperature ranges may be represented by Equation (1-33), because the strong exponential function masks the effect of the weaker functions T and $\sqrt{T}$.

When evaluating the constants A and E_A in Equation (1-33), it is desirable from a statistical point of view [14] to put Equation (1-33) in the following form:

$$\log_e k_c = -\frac{E_A}{R}\left(\frac{1}{T} - \frac{1}{T_0}\right) + \log_e k_c^0 \tag{1-34}$$

where T_0 = mid-temperature for temperature range considered
$k_c^0 = Ae^{-E_A/RT_0}$

Thus E_A/R may be evaluated from the slope of a plot of $\log_e k_c$ versus $(1/T - 1/T_0)$. The intercept $\log_e k_c^0$ is evaluated at $(1/T - 1/T_0) = 0$, and A is calculated from the definition of k_c^0.

EXAMPLE 1-4

In a study of the isomerization of n-propylidene-cyclopropylamine to 5-ethyl-1-pyroline, Cocks and Egger [15] found the rate to be first order. The rate constants were

T (K)	573	600	623.5	635
$k_c \times 10^3$ (min^{-1})	4.15	26.8	114	246

Evaluate the activation energy E_A and the Arrhenius constant A.

SOLUTION

First, the quantity T_0 in Equation (1-34) is evaluated as follows:

$$T_0 = \frac{573 + 600 + 623.5 + 635}{4} \cong 608 \text{ K}$$

For each value of T, the corresponding values of the independent variable $(1/T - 1/T_0)$ are as follows:

T (K)	573	600	623.5	635
$\left(\dfrac{1}{T} - \dfrac{1}{T_0}\right)$	1.005×10^{-4}	2.20×10^{-5}	-4.09×0.0^{-5}	-6.99×10^{-5}

A plot of the values of the dependent variable k_c versus the corresponding values of the independent variable $(1/T - 1/T_0)$ is shown in Figure 1-2. A least-squares fit of

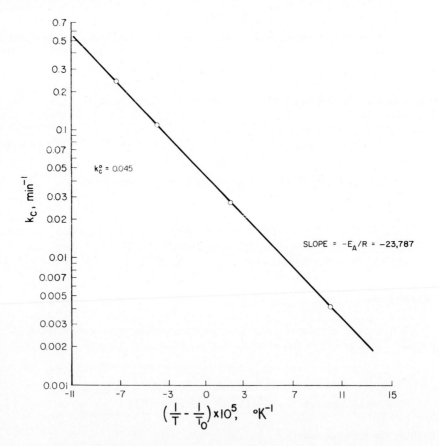

Figure 1-2. Determination of the constants of the Arrhenius equation.

Equation (1-34) yields $k_c^0 = 0.045$ min^{-1}, and $E_A/R = 23,787$ K. Thus

$$k_c \, (\text{min}^{-1}) = 0.045 \exp\left[-23,787\left(\frac{1}{T} - \frac{1}{608} \right) \right]$$

This expression may be restated in the form of Equation (1-33) as follows:

$$k_c \, (\text{min}^{-1}) = 2.1007 \times 10^{15} \exp\left(-\frac{197,778}{RT} \right)$$

or

$$k_c \, (\text{sec}^{-1}) = 7.35 \times 10^{13} \exp\left(-\frac{197,778}{RT} \right)$$

where $R = 8.314$ J/g mol K.

Thus the frequency factor A is given by $A = 7.35 \times 10^{13}$ s^{-1}.

III. BATCH REACTORS

In the analysis that follows, expressions are obtained which give the moles or concentration of each component in the reactor at any time during the reaction period. The use of these expressions in the analysis of data collected with batch reactors is then demonstrated.

A. Material Balances for Batch Reactors

A typical autoclave reactor is shown in Figure 1-3. At the time this picture was taken, this experimental reactor was in the service of hydrogenating lignite. For developmental purposes, the sketch of the batch reactor shown in Figure 1-4 is used. Such a reactor is fixed in space, and the total feed is introduced at the outset, and no withdrawal is made until the reaction has reached the degree of completion desired. The volume V of the reactor is the space filled by the reacting mixture. The mixture in the volume V is commonly mixed throughout the course of the reaction by a stirrer. Thus such reactors are also called *well-stirred tank reactors*. The volume of the reacting mixture may vary during the course of a liquid phase reaction because of either shrinkage or expansion of the liquid phase with reaction. In the case of a gas phase reaction, the volume V may either be held fixed or varied in any prescribed manner. For example, the volume may be varied as required to maintain the pressure constant. The variation of the temperature T of the reaction mixture with time is also permitted.

By the very nature of a batch reactor, it is an *unsteady-state process*, which is defined as a process that has one or more variables which vary with time. On the other hand, a *steady-state process* is one whose variables at all points throughout the process are independent of time.

A material balance on a reactor constitutes an application of the *law of conservation of mass*. For purposes of applications, a convenient statement of this law follows: *Except for the conversion of mass to energy and conversely, mass can neither be created nor destroyed.* Consequently, for a system in which

Figure 1-3. An experimental autoclave reactor in the service of the hydrogenation of lignite.

the conversion of mass to energy and conversely is not involved, it follows that, during the time period from $t = t_n$ to $t = t_n + \Delta t$,

$$
\left\{ \begin{array}{l} \text{Input of mass to the} \\ \text{system during the} \\ \text{time period } \Delta t \end{array} \right\} - \left\{ \begin{array}{l} \text{output of mass from} \\ \text{the system during the} \\ \text{time period } \Delta t \end{array} \right\} = \left\{ \begin{array}{l} \text{accumulation of mass} \\ \text{within the system} \\ \text{during the time period} \\ \Delta t \end{array} \right\} \quad (1\text{-}35)
$$

The accumulation term is defined as follows:

$$
\left\{ \begin{array}{l} \text{Accumulation of mass} \\ \text{within the system} \\ \text{during the time period } \Delta t \end{array} \right\} = \left\{ \begin{array}{l} \text{amount of mass} \\ \text{in the system} \\ \text{at time } t_n + \Delta t \end{array} \right\} - \left\{ \begin{array}{l} \text{amount of mass} \\ \text{in the system} \\ \text{at time } t_n \end{array} \right\} \quad (1\text{-}36)
$$

The particular part of the universe under consideration is called the *system*. A more convenient form of the law of conservation of mass for use in the description of industrial processes follows:

$$
\int_{t_n}^{t_n + \Delta t} \left[\left\{ \begin{array}{l} \text{Input of mass to the} \\ \text{system per unit time} \end{array} \right\} - \left\{ \begin{array}{l} \text{output of mass from the} \\ \text{system per unit time} \end{array} \right\} \right] dt
$$

$$
= \left\{ \begin{array}{l} \text{mass within the system} \\ \text{at time } t_n + \Delta t \end{array} \right\} - \left\{ \begin{array}{l} \text{mass within the} \\ \text{system at time } t_n \end{array} \right\} \quad (1\text{-}37)
$$

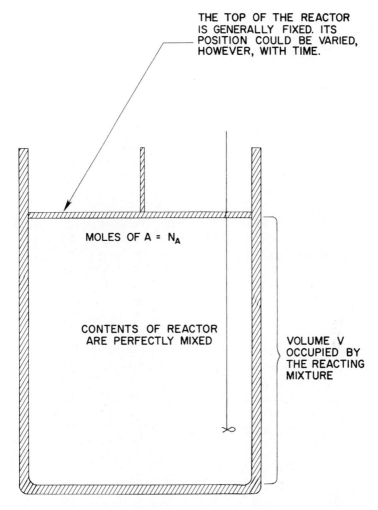

THE TOP OF THE REACTOR IS GENERALLY FIXED. ITS POSITION COULD BE VARIED, HOWEVER, WITH TIME.

MOLES OF A = N_A

CONTENTS OF REACTOR ARE PERFECTLY MIXED

VOLUME V OCCUPIED BY THE REACTING MIXTURE

Figure 1-4. Sketch of a batch reactor.

Suppose that the volume V of the reactor shown in Figure 1-4 is filled with reactant A and other reactants if needed. Let M_A denote the molecular weight of A. Then a mass balance on component A in the system (the volume V) over the time period from t_n to $t_n + \Delta t$ is given by

$$\int_{t_n}^{t_n + \Delta t} \underbrace{[\ \ 0}_{\substack{\text{input of } A \\ \text{per unit time}}} - \underbrace{M_A r_A V\ ]}_{\substack{\text{output of } A \\ \text{by reaction} \\ \text{per unit time}}} dt = \underbrace{M_A N_A|_{t_n + \Delta t}}_{\substack{\text{mass of } A \\ \text{within the system} \\ \text{at time } t_n + \Delta t}} - \underbrace{M_A N_A|_{t_n}}_{\substack{\text{mass of } A \\ \text{within the system} \\ \text{at time } t_n}}$$

$$(1\text{-}38)$$

where N_A is equal to the moles of A within the volume V at any time.

(Throughout this book, the term "mole" is used in the presentation of the developments. It is to be understood that a particular choice such as gram moles, kilogram moles, or pound moles must be made in the solution of numerical problems. Also, it should be emphasized that t_n and Δt are arbitrarily selected fixed values which are nonzero, finite, and positive.) The appearance of the product $M_A r_A V$ in Equation (1-38) implies that the rate of reaction r_A is the same for every element of reactor volume throughout the volume V, regardless of how small the elements of volume are taken to be.

Since the molecular weight of A is independent of time, Equation (1-38) reduces to

$$\int_{t_n}^{t_n + \Delta t} [-r_A V]\, dt = N_A\big|_{t_n + \Delta t} - N_A\big|_{t_n} \tag{1-39}$$

Equation (1-39) may be converted to its corresponding differential equation by use of the mean-value theorems of calculus in the following manner. Consider first the left side of Equation (1-39). This integral may be reduced to algebraic form by use of the *mean-value theorem of integral calculus* (see Appendix at the end of this chapter, Theorem 1), as follows:

$$-\int_{t_n}^{t_n + \Delta t} [r_A V]\, dt = -\left[(r_A V)\big|_{t_n + \alpha \Delta t} \right] \Delta t \tag{1-40}$$

where $0 \le \alpha \le 1$. The quantity $(r_A V)\big|_{t_n + \alpha \Delta t}$ represents the mean value of the integrand; it is that particular value of $r_A V$ for which the rectangle of height $(r_A V)\big|_{t_n + \alpha \Delta t}$ and base Δt has an area equal to that represented by the integral on the left side of Equation (1-40).

Application of the *mean-value theorem of differential calculus* (Appendix, Theorem 2) to the right side of Equation (1-39) gives

$$N_A\big|_{t_n + \Delta t} - N_A\big|_{t_n} = \left(\frac{dN_A}{dt}\bigg|_{t_n + \beta \Delta t} \right) \Delta t \tag{1-41}$$

where $0 < \beta < 1$. The time $t_n + \beta \Delta t$ is that particular value of t in the interval $t_n \le t \le t_n + \Delta t$ at which the slope of the tangent to the curve of N_A versus t is equal to the slope of the secant drawn from t_n to $t_n + \Delta t$.

From Equations (1-38), (1-40), and (1-41) it follows that

$$-(r_A V)\big|_{t_n + \alpha \Delta t} = \frac{dN_A}{dt}\bigg|_{t_n + \beta \Delta t} \tag{1-42}$$

In the limit as Δt goes to zero, Equation (1-42) reduces to

$$-(r_A V)\big|_{t_n} = \frac{dN_A}{dt}\bigg|_{t_n} \tag{1-43}$$

Since t_n was arbitrarily selected as any t in the time domain other than $t = 0$ (the time of the beginning of the reaction), Equation (1-43) holds for all

$t > 0$, or

$$r_A = -\frac{1}{V}\frac{dN_A}{dt} \tag{1-44}$$

This equation represents the limit of a material balance on compound A in the volume V as the time period over which the balance is made is allowed to go to zero. In the interest of simplicity, this equation is referred to hereafter as simply the *material balance on component A*.

It should be remarked that, if the rate of reaction is defined in terms of the appearance of A,

$$r_A = \frac{\text{moles of } A \text{ appearing}}{(\text{unit time})(\text{unit volume})} \tag{1-45}$$

Then, instead of Equation (1-44), the following equation is obtained for the material balance on A:

$$r_A = \frac{1}{V}\frac{dN_A}{dt} \tag{1-46}$$

Note that the sign appearing on the right side of Equations (1-44) and (1-46) does not depend on whether or not A is in fact disappearing or appearing by reaction, but on the definition of the rate of reaction.

B. Variation of the Moles Reacted with Time (Conditions: Isothermal Operation, Perfectly Mixed, and Constant Volume)

By *isothermal operation of a batch reactor* is meant that the temperature of the reacting mixture is maintained constant throughout the course of the reaction. A batch reactor is said to be *perfectly mixed* if the composition of the reacting mixture is uniform throughout the reactor at any time t. That is, the concentration of each element of volume ΔV of the reactor volume V is the same at any time t, regardless of how small the element of volume is taken to be. The mixing is generally accomplished by use of a stirrer. By *constant volume* is meant that the volume of the reacting mixture within the reactor remains constant throughout the course of the reaction. For the volume of the reacting mixture to remain constant, it is necessary for the mass density to remain constant.

To demonstrate the manner in which the expressions relating the moles reacted to the reaction time are obtained, the steps involved in the development of the expressions for three types of unidirectional reactions are presented.

1. First-Order Unidirectional Reactions: $A \rightarrow$ Products. For the first-order reaction $A \rightarrow$ products, the rate expression for r_A (moles of A disappearing per unit time per unit volume) is given by

$$r_A = k_c C_A$$

and the material balance for component A for a batch reactor is given by

$$r_A = -\frac{1}{V}\frac{dN_A}{dt}$$

Elimination of r_A from these two equations yields

$$-\frac{1}{V}\frac{dN_A}{dt} = k_c C_A \qquad (1\text{-}47)$$

Since the concentration C_A is defined by

$$C_A = \frac{N_A}{V} \qquad (1\text{-}48)$$

for a batch reactor (see Figure 1-4), then Equation (1-47) may be stated in terms of N_A and t to give

$$-\frac{1}{V}\frac{dN_A}{dt} = k_c \frac{N_A}{V} \qquad (1\text{-}49)$$

[Since the volume V cancels out, the final expression, Equation (1-52), is independent of whether or not the volume remains constant throughout the course of the first-order reaction.] Separation of variables in Equation (1-49) yields

$$-\int_{N_A^0}^{N_A}\frac{dN_A}{N_A} = \int_0^t k_c\, dt \qquad (1\text{-}50)$$

where N_A^0 = total moles of A in the reactor at time $t = 0$

Since the reaction is carried out isothermally, the rate constant k_c is constant. If it is further supposed that the reaction remains first order over the entire range of concentration, Equation (1-50) may be integrated and rearranged.

$$\log_e N_A = -k_c t + \log_e N_A^0 \qquad (1\text{-}51)$$

For a first-order reaction, the plot of $\log_e N_A$ versus t is a straight line with the intercept of $\log_e N_A^0$ as illustrated in Figure 1-5. Equation (1-51) may be stated in exponential form as follows:

$$N_A = N_A^0 e^{-k_c t} \qquad (1\text{-}52)$$

If it is supposed that the volume V remains constant throughout the course of the reaction, then division of both sides of Equation (1-52) by V yields

$$C_A = C_A^0 e^{-k_c t} \qquad (1\text{-}53)$$

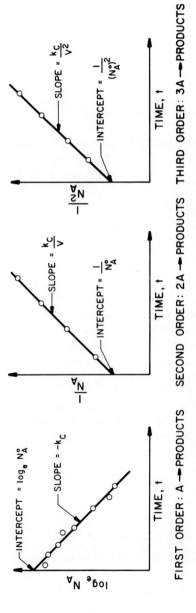

Figure 1-5. Determination of the order and rate constant for a reaction from experimental data taken with a batch reactor.

23

2. Second-Order Unidirectional Reactions: $2A \rightarrow$ Products. Elimination of r_A from the rate expression for a second-order reaction [Equation (1-12)] and the material balance [Equation (1-44)] gives

$$-\frac{1}{V}\frac{dN_A}{dt} = k_c C_A^2 \tag{1-54}$$

If it is now supposed that the volume of the reacting mixture remains constant throughout the course of the reaction, then Equation (1-54) may be restated as follows:

$$-\frac{dC_A}{dt} = k_c C_A^2 \tag{1-55}$$

Separation of variables followed by integration and rearrangement yields

$$\frac{1}{C_A} = k_c t + \frac{1}{C_A^0} \tag{1-56}$$

Since $C_A = N_A/V$, it follows that

$$\frac{1}{N_A} = \frac{k_c t}{V} + \frac{1}{N_A^0} \tag{1-57}$$

As illustrated in Figure 1-5, the plot of $1/N_A$ versus t is a straight line for second-order reactions.

3. Second-Order Unidirectional Reactions: $A + B \rightarrow$ Products. If $N_A^0 \neq N_B^0$, then one equation in the three variables N_A, N_B, and t is obtained by eliminating r_A from Equations (1-13) and (1-44)

$$-\frac{1}{V}\frac{dN_A}{dt} = k_c C_A C_B = \frac{k_c}{V^2} N_A N_B \tag{1-58}$$

If $N_A^0 = N_B^0$, then $N_A = N_B$ for all t and Equation (1-58) reduces to Equation (1-54). Equation (1-58) may be reduced to one equation in two variables by expressing N_A and N_B in terms of a single variable, conversion, which is defined herein as follows:

$$X = \frac{N_A^0 - N_A}{N_A^0} \tag{1-59}$$

Rearrangement gives

$$N_A = N_A^0 - N_A^0 X \tag{1-60}$$

Thus $N_A^0 X$ is seen to be the moles of A that have reacted since the initiation of the reaction. Since one mole of B disappears for each mole of A which disappears, the following stoichiometric relationship is obtained:

$$N_B = N_B^0 - N_A^0 X \tag{1-61}$$

From Equation (1-60), it follows that

$$\frac{dN_A}{dt} = -N_A^0 \frac{dX}{dt} \tag{1-62}$$

Thus Equation (1-58) may be restated in terms of the variables X and t to give

$$N_A^0 \frac{dX}{dt} = \frac{k_c}{V}\left(N_A^0 - N_A^0 X\right)\left(N_B^0 - N_A^0 X\right) \qquad (1\text{-}63)$$

Separation of the variables followed by integration and rearrangement gives

$$t = \left(\frac{V}{k_c\left(N_A^0 - N_B^0\right)}\right)\log_e\left[\frac{N_B^0\left(N_A^0 - N_A^0 X\right)}{N_A^0\left(N_B^0 - N_A^0 X\right)}\right] \qquad (1\text{-}64)$$

where it has been assumed that k_c and V remain constant. By use of the relationships given by Equations (1-60) and (1-61), the result given by Equation (1-64) may be stated in the form

$$t = \frac{V}{k_c\left(N_A^0 - N_B^0\right)}\log_e\left(\frac{N_B^0 N_A}{N_A^0 N_B}\right) = \frac{1}{k_c\left(C_A^0 - C_B^0\right)}\log_e\frac{C_B^0 C_A}{C_A^0 C_B} \qquad (1\text{-}65)$$

If the reaction is second order, a plot of $\log_e N_A/N_B$ versus t gives a straight line with the slope $k_c(N_A^0 - N_B^0)/V = k_c(C_A^0 - C_B^0)$.

4. Third-Order Unidirectional Reactions: $3A \rightarrow$ Products. In the same manner demonstrated for second-order reactions with respect to A, the following equation is obtained for reactions which are third order with respect to A.

$$-\frac{1}{V}\frac{dN_A}{dt} = k_c C_A^3 \qquad (1\text{-}66)$$

Separation of the variables followed by integration and rearrangement yields

$$\frac{1}{C_A^2} = 2k_c t + \frac{1}{\left(C_A^0\right)^2} \qquad (1\text{-}67)$$

where it has been supposed that k_c and V are constant. When this result is expressed in terms of the moles of A, one obtains

$$\frac{1}{N_A^2} = \left(\frac{2k_c}{V^2}\right)t + \frac{1}{\left(N_A^0\right)^2} \qquad (1\text{-}68)$$

As illustrated in Figure 1-5, the plot of $1/N_A^2$ versus t is a straight line if the reaction is third order. Third-order reactions are, however, relatively rare. Additional design equations for batch reactors are presented in Table 1-1.

EXAMPLE 1-5

A saponification reaction which is easily carried out in the laboratory is the reaction of ethyl acetate in an aqueous solution of sodium hydroxide,

$$CH_3COOC_2H_5 + NaOH \rightarrow CH_3COONa + C_2H_5OH$$

In 1906, Walker [16] investigated this reaction at 25°C by commencing with equal concentrations (0.01 g mol/liter) of ethyl acetate and sodium hydroxide and following

TABLE 1-1
DESIGN EQUATIONS FOR CARRYING OUT UNIDIRECTIONAL REACTIONS
IN A BATCH REACTOR
(Conditions: isothermal operation, perfectly mixed, and constant volume)

Reaction	Design Equation
1. $A \rightarrow$ products	$t = \dfrac{1}{k_c} \log_e \dfrac{N_A^0}{N_A}$
2. $2A \rightarrow$ products	$t = \dfrac{V}{k_c} \left[\dfrac{1}{N_A} - \dfrac{1}{N_A^0} \right]$
3. $A + B \rightarrow$ products $\quad N_A^0 \neq N_B^0$	$t = \dfrac{V}{k_c \left(N_A^0 - N_B^0 \right)} \log_e \dfrac{N_B^0 N_A}{N_A^0 N_B}$
4. $3A \rightarrow$ products	$t = \dfrac{V^2}{2k_c} \left[\left(\dfrac{1}{N_A} \right)^2 - \left(\dfrac{1}{N_A^0} \right)^2 \right]$
5. $A + B + C \rightarrow$ products $\quad (N_A^0 \neq N_B^0 \neq N_C^0)$	$t = \dfrac{V^2}{k_c} \left[\dfrac{\log_e N_A^0/N_A}{\left(N_B^0 - N_A^0 \right)\left(N_C^0 - N_A^0 \right)} + \dfrac{\log_e N_B^0/N_B}{\left(N_A^0 - N_B^0 \right)\left(N_C^0 - N_B^0 \right)} \right.$ $\left. + \dfrac{\log_e N_C^0/N_C}{\left(N_A^0 - N_C^0 \right)\left(N_B^0 - N_C^0 \right)} \right]$
6. $2A + B \rightarrow$ products $\quad (N_A^0 \neq 2N_B^0)$	$t = \dfrac{V^2}{k_c} \left[\dfrac{2}{2N_B^0 - N_A^0} \left(\dfrac{1}{N_A} - \dfrac{1}{N_A^0} \right) \right.$ $\left. + \dfrac{2}{\left(2N_B^0 - N_A^0 \right)^2} \log_e \dfrac{N_B^0 N_A}{N_A^0 N_B} \right]$
7. $NA \rightarrow$ products $\quad (N \neq 1)$	$t = \dfrac{1}{k_c(N-1)} \left[C_A^{1-N} - \left(C_A^0 \right)^{1-N} \right]$

the course of the reaction by the electrical conductance method. Some of his results follow:

t (min)	5	9	13	20
g mol of NaOH per liter	0.00755	0.00633	0.00541	0.00434

t (min)	25	33	37
g mol of NaOH per liter	0.00385	0.00320	0.00296

On the basis of these data, determine the order of the reaction and evaluate the rate constant.

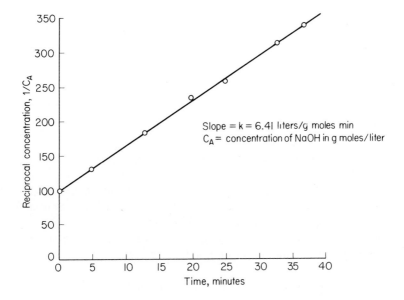

Figure 1-6. Determination of the order and rate constant for the saponification of ethyl acetate [16].

SOLUTION

The saponification reaction may be represented by

$$A + B \rightarrow C + D$$

Since the initial concentrations of ethyl acetate and sodium hydroxide are equal, it follows from the stoichiometry of the saponification reaction that their concentrations remain equal for all time; that is, $C_A = C_B$. Assume the reaction is second order. Then

$$r_A = k_c C_A C_B = k_c C_A^2$$

For a second-order reaction which is carried out isothermally and at constant volume in a batch reactor, Equation (1-56) is applicable.

$$\frac{1}{C_A} = k_c t + \frac{1}{C_A^0}$$

To test the assumption of second order, the original data are plotted as shown in Figure 1-6. Since a straight line is obtained, the rate of reaction may be represented by a second-order rate expression. A least-squares fit of the data illustrated in Figure 1-6 yields a value of $k_c = 6.41$ liter/(min g mol) with a correlation coefficient of 0.99995.

C. Determination of the Order and the Rate Constant for a Unidirectional Reaction

Suppose that a given reaction is carried out isothermally in a batch reactor at constant volume and that the progress of the reaction is followed by analyzing samples of the reacting mixture at selected time intervals throughout the course of the reaction. Furthermore, suppose that these results indicate that the stoichiometry of the reaction is satisfied by an equation of the form

$$A \rightarrow \text{products} \tag{1-69}$$

If the stoichiometry is satisfied by Equation (1-69), it is also satisfied by the equation

$$NA \rightarrow N \text{ products} \tag{1-70}$$

for all $N > 0$. If, however, a plot of $\log_e N_A$ versus t is a straight line, then the reaction is first order ($N = 1$) (see Figure 1-5). If, however, a plot of $1/C_A$ or $1/N_A$ versus t yields a straight line, the reaction is said to be second order ($N = 2$). The reaction is said to be third order ($N = 3$) if a plot of $1/C_A^2$ or $1/N_A^2$ versus t yields a straight line. If none of these plots yields a straight line, the rate of reaction is probably fractional order. For this case, the order may be determined by the following graphical procedure.

1. Graphical Determination of the Order and Rate Constant. Consider the case where the experimental results suggest that the rate of reaction is proportional to the concentration of A raised to some power N; that is,

$$r_A = k_c C_A^N$$

Since the material balance expression for r_A for a batch reactor in which the volume is constant is given by $r_A = -dC_A/dt$, it follows that

$$r_A = -\frac{dC_A}{dt} = k_c C_A^N \tag{1-71}$$

A graphical procedure for determining k_c and N follows. Let

$$C_A|_{t_n} = \text{concentration of } A \text{ at time } t_n$$

$$C_A|_{t_{n+1}} = \text{concentration of } A \text{ at time } t_{n+1}$$

Now let the average values of C_A and r_A be approximated as follows:

$$C_A|_{t_n + \Delta t/2} \cong \frac{C_A|_{t_n} + C_A|_{t_{n+1}}}{2} \tag{1-72}$$

$$r_A|_{t_n + \Delta t/2} \cong -\left[\frac{C_A|_{t_{n+1}} - C_A|_{t_n}}{\Delta t}\right] = \frac{C_A|_{t_n} - C_A|_{t_{n+1}}}{\Delta t} \tag{1-73}$$

Then

$$\log r_A|_{t_n + \Delta t/2} = \log k_c + N \log C_A|_{t_n + \Delta t/2} \tag{1-74}$$

A graph of this equation is shown in Figure 1-7. The slope gives the order N and the intercept gives the logarithm of k. For some reactions a constant value of N cannot be obtained over the concentration range of interest. (The treatment of reactions for which N does not remain constant is considered in

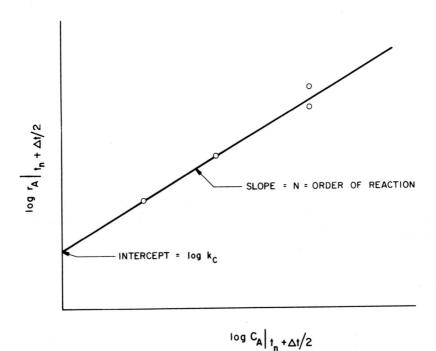

Figure 1-7. Determination of the order N and rate constant for a reaction from data taken with a batch reactor.

Chapter 4.) For additional techniques for the treatment of rate data, the work of Churchill is recommended [17].

2. Use of the Half-Life in the Determination of the Order and Rate Constant. The half-life, denoted by $t_{1/2}$, of a reaction is the time required for one-half of the initial amount of a specified reactant to be consumed by the reaction. Consider the first-order reaction $A \rightarrow$ products. At

$$t = t_{1/2}, \qquad N_A = \frac{N_A^0}{2} \qquad (1\text{-}75)$$

Thus the expression given by Equation (1-51) for a first-order reaction reduces to

$$\log_e \frac{N_A^0}{N_A^0/2} = k_c t_{1/2}$$

or
$$t_{1/2} = \frac{\log_e 2}{k_c} \cong \frac{0.693}{k_c} \qquad (1\text{-}76)$$

Thus, if the rate of reaction is first order, then the half-life time $t_{1/2}$ is independent of the amount of reactant A present initially, and the value for the rate constant is found by solving Equation (1-76) for k_c.

Many reactions involving radioactive decay are first order, and the speeds of such reactions are commonly compared by use of half-life times. From Equation (1-76), it is seen that a short half-life time corresponds to a large rate constant.

For a second-order reaction of the type $2A \rightarrow$ products, Equation (1-56) reduces to

$$t_{1/2} = \frac{1}{k_c C_A^0} \tag{1-77}$$

at $t = t_{1/2}$. Thus, for second-order reactions of the type stated, the half-life time is inversely proportional to the amount of A present initially, and k_c is readily computed by use of Equation (1-77).

The formula for the half-life time for a third-order reaction of the type $3A \rightarrow$ products may be obtained from Equation (1-68):

$$t_{1/2} = \frac{1}{\frac{2}{3}k_c\left(C_A^0\right)^2} \tag{1-78}$$

Thus the half-life time is inversely proportional to the square of the initial concentration of A, and the value of the rate constant k_c is readily computed by use of Equation (1-78).

A second-order reaction of the form $A + B \rightarrow$ products has a half-life both with respect to A and B (see Problem 1-9).

D. Variable-Volume Batch Reactors

Reactors of this type are used to take advantage of the fact that, for certain reactions, the number of moles in the reacting mixture either increases or decreases as the reaction proceeds. For example, suppose the second-order gas phase reaction

$$2A \rightarrow R$$

is carried out at constant pressure and temperature. Thus, as indicated in Figure 1-4, the volume V must be decreased as the reaction proceeds in order to maintain the pressure of the reacting mixture at a specified value. To treat this type of reactor, it is necessary to state the volume V that appears in Equation (1-54) in terms of a variable related to the amount of A present at any time t. Since C_A at any time t is equal to N_A/V, Equation (1-54) may be restated as follows:

$$-\frac{dN_A}{dt} = \frac{k_c}{V}N_A^2 \tag{1-79}$$

This equation, which contains three variables, N_A, t, and V, may be solved by first expressing N_A and V in terms of the conversion X. Consider the case where the pressure–volume behavior of the gas mixture may be approximated by the perfect gas law,

$$PV = N_T RT \qquad (1\text{-}80)$$

This equation makes it possible to state V in terms of conversion, since N_T may be expressed in terms of conversion by use of the stoichiometric relationships. For the reaction $2A \rightarrow R$ and for the case where the feed contains only A and R, the following stoichiometric relationships are obtained:

$$
\begin{aligned}
N_A &= N_A^0 - N_A^0 X \\
N_R &= N_R^0 + \tfrac{1}{2} N_A^0 X \\
\hline
N_T &= N_T^0 - \tfrac{1}{2} N_A^0 X
\end{aligned}
\qquad (1\text{-}81)
$$

where N_T = total number of moles in reacting mixture at any time t

N_T^0 = value of N_T at $t = 0$

Then, by substituting this expression for N_T into Equation (1-80), one obtains

$$V = \left[N_T^0 - \frac{N_A^0 X}{2} \right] \frac{RT}{P} \qquad (1\text{-}82)$$

for the case where it is assumed that the mixture obeys the perfect gas law. At $t = 0$, $V = V^0$, and $V^0 = N_T^0 RT/P$. Thus

$$\frac{V}{V^0} = 1 - \frac{y_A^0 X}{2} \qquad (1\text{-}83)$$

where $y_A^0 = N_A^0 / N_T^0$, the mole fraction of A in the reactor at time $t = 0$. By use of the previous relationships, it is possible to reduce Equation (1-79) to the following form:

$$\frac{dX}{dt} = \left(\frac{k_c N_A^0}{V^0} \right) \frac{(1 - X)^2}{\left[1 - (y_A^0 X/2) \right]} \qquad (1\text{-}84)$$

Separation of variables followed by integration over the time interval from 0 to t yields

$$t = \frac{1}{k_c C_A^0} \left[\left(1 - \frac{y_A^0}{2} \right) \left(\frac{X}{1 - X} \right) + \frac{y_A^0}{2} \log_e \frac{1}{1 - X} \right] \qquad (1\text{-}85)$$

where it has been assumed that k_c remains constant throughout the course of the reaction. Alternately, the time t may be expressed as a function of the reactor volume V (see Problem 1-10).

E. Use of Total Pressure to Follow the Progress of a Reaction

For gas reactions involving a change in the number of moles, say

$$A \rightarrow 2R$$

the progress of the reaction may be followed by measuring the total pressure as a function of time at constant volume and temperature.

Consider the case where the gas mixture obeys the perfect gas law, Equation (1-80), and the initial mixture contains only A and R. The stoichiometric relationships for A and R are

$$N_A = N_A^0 - N_A^0 X$$

$$N_R = N_R^0 + 2N_A^0 X \qquad (1\text{-}86)$$

$$N_T = N_T^0 + N_A^0 X$$

Thus

$$\frac{P}{P^0} = \frac{N_T}{N_T^0} = 1 + y_A^0 X \qquad (1\text{-}87)$$

Since

$$\frac{N_A^0}{N_A} = \frac{1}{1 - X}$$

and since Equation (1-51) holds, it follows that

$$t = \frac{1}{k_c} \log_e \frac{y_A^0}{1 + y_A^0 - P/P^0} \qquad (1\text{-}88)$$

F. Reversible Reactions

Consider the reversible reaction

$$A \underset{k_c'}{\overset{k_c}{\rightleftharpoons}} R \qquad (1\text{-}89)$$

which is first order in both directions. The net rate of disappearance of A is given by

$$r_A = r_{Af} - r_{Ar}$$

or the rate expression for r_A is given by

$$r_A = k_c C_A - k_c' C_R \qquad (1\text{-}90)$$

Thus

$$-\frac{dN_A}{dt} = k_c N_A - k_c' N_R \qquad (1\text{-}91)$$

This equation, which contains the three variables N_A, N_R, and t, may be reduced to one equation in two variables by expressing the moles of A and R at any time t in terms of X through the use of the stoichiometric relationships.

$$N_A = N_A^0 - N_A^0 X$$
$$N_R = N_R^0 + N_A^0 X \tag{1-92}$$

Thus Equation (1-91) may be restated as follows:

$$N_A^0 \frac{dX}{dt} = k_c\left(N_A^0 - N_A^0 X\right) - k_c'\left(N_R^0 + N_A^0 X\right) \tag{1-93}$$

Now suppose that k_c and k_c' remain constant throughout the course of the reaction. Separation of the variables of Equation (1-93) followed by integration and rearrangement gives

$$t = \frac{1}{k_c + k_c'} \log_e \frac{k_c - k_c'\beta}{(k_c - k_c'\beta) - (k_c + k_c')X} \tag{1-94}$$

where

$$\beta = \frac{N_R^0}{N_A^0}$$

Integral expressions for other first- and second-order reactions are given in Table 1-2.

G. Cycle Time

When batch reactors are used commercially, the times required to charge and to empty the reactor become significant operating variables as well as the reaction time. The time required to complete one sequence of events is called the cycle time t_c; that is,

$$t_c = t_{\text{charge}} + t_{\text{discharge}} + t_{\text{reaction}}$$

EXAMPLE 1-6

For the reaction of $A \rightarrow R$, $k_c = 0.02$ min^{-1}. It is desired to produce 4752 g mol of R per 10-h day, and 99% of A entering the reactor is to be converted. To charge the reactor and to heat it to reaction temperature requires 0.26 h. To discharge the reactor and to prepare it for the next run takes 0.9 h. Calculate the volume of the reactor required. Pure A with a molar density of 8 g mol/liter is charged to the reactor.

SOLUTION

From Equations (1-51) and (1-60) the reaction time is calculated:

$$t_{\text{reaction}} = \frac{-1}{0.02} \log_e(1 - 0.99) = \frac{1}{0.02} \log_e 100 = \frac{4.606}{0.02} = 230.2 \text{ min}$$

or

$$= 3.84 \text{ h}$$

TABLE 1-2

DESIGN EQUATIONS FOR CARRYING OUT A REVERSIBLE REACTION IN A BATCH REACTOR
(Conditions: isothermal operation, perfectly mixed, and constant volume)

$$t = \frac{1}{k_c} \int_0^x \frac{dX}{aX^2 + bX + c}$$

Reaction		Constants	
	a	b	c
1. $A \underset{k_c'}{\overset{k_c}{\rightleftharpoons}} 2R$ $N_T^0 = N_A^0 + N_R^0$	$-4\dfrac{C_A^0}{K_c}$	$-\left(1 + 4\beta\dfrac{C_A^0}{K_c}\right)$	$\left(1 - \dfrac{C_A^0\beta^2}{K_c}\right)$
2. $2A \underset{k_c'}{\overset{k_c}{\rightleftharpoons}} R$ $N_T^0 = N_A^0 + N_R^0$	C_A^0	$-\left(2C_A^0 + \dfrac{1}{2K_c}\right)$	$C_A^0 - \dfrac{\beta}{K_c}$
3. $A + B \underset{k_c'}{\overset{k_c}{\rightleftharpoons}} R$ $N_A^0 \neq N_B^0$ $N_T^0 = N_A^0 + N_B^0 + N_R^0$	C_A^0	$-\left(C_A^0(1 + \gamma) + \dfrac{1}{K_c}\right)$	$C_A^0\gamma - \dfrac{\beta}{K_c}$
4. $A + B \underset{k_c'}{\overset{k_c}{\rightleftharpoons}} R + S$ $N_A^0 \neq N_B^0,\ N_R^0 \neq N_S^0$ $N_T^0 = N_A^0 + N_B^0 + N_R^0 + N_S^0$	$C_A^0\left(1 - \dfrac{1}{K_c}\right)$	$-C_A^0\left[1 + \gamma + \dfrac{1}{K_c}(\beta + \alpha)\right]$	$C_A^0\left(\gamma - \dfrac{\beta\alpha}{K_c}\right)$
5. $2A \underset{k_c'}{\overset{k_c}{\rightleftharpoons}} R + S$ $N_R^0 \neq N_S^0,\ N_T^0 = N_A^0 + N_R^0 + N_S^0$	$C_A^0\left(1 - \dfrac{1}{4K_c}\right)$	$-C_A^0\left[2 + \dfrac{1}{2K_c}(\beta + \alpha)\right]$	$C_A^0\left(1 - \dfrac{\beta\alpha}{K_c}\right)$
6. $2A \underset{k_c'}{\overset{k_c}{\rightleftharpoons}} 2R$ $N_T^0 = N_A^0 + N_R^0$	$C_A^0\left(1 - \dfrac{1}{K_c}\right)$	$-2C_A^0\left(1 + \dfrac{\beta}{K_c}\right)$	$C_A^0\left(1 - \dfrac{\beta^2}{K_c}\right)$

where

$$\gamma = \frac{N_B^0}{N_A^0}, \qquad \beta = \frac{N_R^0}{N_A^0}, \qquad \alpha = \frac{N_S^0}{N_A^0}, \qquad K_c = \frac{k_c}{k_c'}$$

The integrated forms of the above equations are:

$$t = \frac{1}{k_c} \int_0^x \frac{dX}{aX^2 + bX + c} = \frac{1}{k_c\sqrt{b^2 - 4ac}} \log_e \frac{\left(2aX + b - \sqrt{b^2 - 4ac}\right)\left(b + \sqrt{b^2 - 4ac}\right)}{\left(2aX + b + \sqrt{b^2 - 4ac}\right)\left(b - \sqrt{b^2 - 4ac}\right)},$$
$$\text{if } b^2 > 4ac$$

$$t = \frac{1}{k_c} \int_0^x \frac{dX}{aX^2 + bX + c} = \frac{2}{k_c\sqrt{4ac - b^2}} \left[\tan^{-1} \frac{2aX + b}{\sqrt{4ac - b^2}} - \tan^{-1} \frac{b}{\sqrt{4ac - b^2}}\right],$$
$$\text{if } b^2 < 4ac$$

$$t = \frac{1}{k_c}\left[\frac{4aX}{(2aX + b)b}\right], \quad \text{if } b^2 = 4ac$$

Thus the total cycle time is

$$t_c = 0.26 + 0.9 + 3.84 = 5.0 \, \text{h}$$

To produce 4752 g mol in a 10-h day, two cycles can be completed, and in each cycle 2376 g mol of R must be produced.

$$N_A^0 = \frac{N_R}{0.99} = \frac{2376}{0.99} = 2400 \, \text{g mol}$$

Since $C_A^0 = N_A^0/V$, the reactor volume V is given by

$$V = \frac{2400 \, \text{g mol}}{8 \, \text{g mol/liter}} = 300 \, \text{liters}$$

or
$$= 79.3 \, \text{gal}$$

Most mixing vessels are available in standard sizes. If it is supposed that the preceding reaction occurs in the liquid phase, then the next largest standard-size reactor should be specified. (A batch reactor should *never be charged completely full of a reacting liquid*.)

PROBLEMS

1-1. Begin with a material balance on reactant A and obtain the result given by Equation (1-46).

1-2. For the reaction

$$A + 2B \rightleftarrows 2C + 3D + F$$

the rate of loss of A is 10 mol per unit time per unit volume.
(a) What is the rate of disappearance of B, $r_B = $?
    Answer: $r_B = 20$.
(b) What is the rate of appearance of C, D, and F, respectively?
    Answer: $r_C = 20$, $r_D = 30$, $r_F = 10$.

1-3. The reactions

$$2A \xrightarrow{k_{c1}} R + 2S$$

$$A + R \xrightarrow{k_{c2}} 2C + D$$

occur simultaneously as written. The rate constants k_{c1} and k_{c2} are defined with respect to component A. That is, the rate of disappearance of A by the first reaction is denoted by r_{A1} and is given by

$$r_{A1} = k_{c1}C_A^2$$

and the rate of disappearance of A by the second reaction is given by

$$r_{A2} = k_{c2}C_A C_R$$

On the basis of these definitions of k_{c1} and k_{c2}, formulate the rate expressions for the following rates:

(a) r_A, where r_A denotes the net rate of disappearance of A
(b) r_A, where r_A denotes the net rate of appearance of A
(c) r_R, where r_R denotes the net rate of appearance of R
(d) r_C, where r_c denotes the net rate of appearance of C
(e) r_S, where r_s denotes the net rate of appearance of S
(f) r_D, where r_D denotes the net rate of appearance of D.

1-4. The hydrogenation of α-(1-methyl-1-nitroethyl) benzyl alcohol over a 10% palladium on charcoal catalyst has been reported as a first-order reaction. The following data were reported by Marquardt [18]:

$10^2 k_c$ (min^{-1})	0.88	1.20	2.16
T (°C)	47	50	58

Calculate the activation energy and frequency factor A for the Arrhenius equation by utilizing the midpoint temperature of 52.5°C. What is the value of k_c at the mid-point temperature? Is the enthalpy of activation the same as E_A in the Arrhenius equation? Calculate the entropy of activation at 52.5°C.
Answer: $E_A = 70{,}641$ J/(mol K), $A = 3.02 \times 10^9$ min^{-1}, and $\Delta S_A^\ddagger = -106.5$ e.u. (entropy units).

1-5. The polymerization of 4,4'-dichlorodiphenylsulfone with the potassium diphenoxide salt of bisphenol-A was found by Schulze and Baron [19] to be a two-step series of second-order reactions. The rate constants for each step were reported as follows:

k_1 liter/(mol)(min)	0.086	0.060	0.091	0.170	0.176	0.350
k_2 liter/(mol)(min)	0.28	0.19	0.31	0.43	0.54	0.75
T (°C)	80.0	79.8	80.0	90.1	89.5	100.0

Evaluate the standard enthalpy and entropy changes at 25°C required to activate these reactions.
Answer: $\Delta H_1^\ddagger = 80{,}040$ and $\Delta H_2^\ddagger = 61{,}000$ J/mol, $\Delta S_1^\ddagger = -74.86$ e.u. and $\Delta S_2^\ddagger = -119.13$ e.u.

1-6. The following data were obtained by Worsfold and Bywater [20] in a study of the polymerization of styrene in the solvent benzene by use of a butyllithium initiator.

k_c (liter/g mol)$^{1/2}$ min^{-1}	T (°C)
0.929	30.3
0.563	25.0
0.387	20.0
0.265	15.0
0.155	10.0

Use these data to evaluate the activation energy E_A and the frequency factor A of the Arrhenius equation.
Answer: $E_A = 14{,}640.2$ cal/g mol, $A = 2.793 \times 10^{10}$ (liters g mol)$^{1/2}$ min^{-1}.

1-7. Hairston and O'Brien [21] studied the reaction

$$(CH_3)_3 - Si - \underset{\underset{CH_3}{|}}{\overset{\overset{CH_3}{|}}{C}} - Cl + SbF_5 \xrightarrow{k_1}$$

$$(CH_3)_3 Si - \underset{\underset{CH_3}{|}}{\overset{\overset{CH_3}{|}}{C^{(+)}}} + SbX_6^{(-)} \xrightarrow[\xrightarrow{k_3}]{\xrightarrow{k_2}} \begin{array}{l} (CH_3)_6 \overset{\overset{F}{|}}{Si} - C(CH_3)_3 + SbX_5 \\[2em] (CH_3)_3 Si - F + SbX_5 \end{array}$$

at 40°C in nitromethane by use of nuclear magnetic resonance. From the following data, determine the order of the first reaction and evaluate the rate constant k_1.

Time (min)	2-Trimethylsilyl-2-chloropropane (g mol/liter)	SbF₅ (g mol/liter)
2	0.1259	0.0900
4.25	0.1225	0.0866
9.25	0.1150	0.0791
16.75	0.1010	0.0696
25.75	0.0968	0.0609
36.25	0.0882	0.0523

Compare your answer with the value of k_1 obtained in Problem 1-23. Which is correct? Are they statistically the same?
Answer: $k_1 = 0.1486$ liter/min g mol

1-8. A second-order reaction may be made to appear first order with respect to A by taking $C_B^0 \gg C_A^0$. (Such reactions are sometimes called "pseudo first order.") Show how the rate constant k_c may be determined, where $r_A = k_c C_A C_B$.

1-9. (a) For the second-order reaction $A + B \rightarrow$ products, show that the half-life time with respect to A is given by

$$(t_{1/2})_A = \frac{1}{k_c(C_A^0 - C_B^0)} \log_e \left[\frac{C_B^0}{2(C_B^0 - \frac{1}{2}C_A^0)} \right]$$

Hint: At time $(t_{1/2})_A$, show that the concentration of B is given by

$$C_B = C_B^0 - \frac{1}{2}C_A^0$$

(b) Show that a necessary condition for the reaction in part (a) to proceed to the point where $\frac{1}{2}$ of A has been reacted in finite time is that

$$C_B^0 > \frac{1}{2}C_A^0$$

1-10. By use of Equation (1-83), state t as given by Equation (1-85) in terms of V.

1-11. Show how a variable-volume batch reactor may be used to follow the first-order gas phase reaction $A \rightarrow 2B$.

1-12. Neumann and Lynn [22] have studied the reaction of hydrogen sulfide and sulfur dioxide in organic solvents and found that it is a second-order reaction. The reaction is given as follows:

$$2H_2S + SO_2 \xrightarrow{k} \frac{3}{x}S_x + 2H_2O, \qquad r_{SO_2} = kC_{H_2S}C_{SO_2}$$

Traditionally, hydrogen sulfide has been removed from gas streams by the Claus process and gas phase reactions over an alumina catalyst at high temperatures. The high-temperature reaction is limited by equilibrium. The low-temperature reaction in organic solvents is irreversible. The second-order rate constant in n,n-dimethylaniline/triglyme/methanol solvent is reported to be 23.4 liters/(s g mol) at 22°C. Calculate $C_{SO_2}^0 t$ required for 99% conversion for a stoichiometric feed of hydrogen sulfide and sulfur dioxide.

Answer: $C_{SO_2}^0 t = 2.12$ s g mol/liter.

1-13. Swain [23] investigated the reaction of triphenyl methyl chloride with methanol in a dry benzene solution in a batch reactor.

$$CH_3OH + (C_6H_5)_3CCl \rightarrow (C_6H_5)_3COCH_3 + HCl$$
$$A \quad + \quad B \quad \rightarrow \quad C \quad + \quad D$$

HCl was removed by reaction with pyridine with subsequent precipitation of pyridine hydrochloride in order to prevent the reverse reaction from occurring. From the data in the following table, show that this reaction is second order with respect to methanol and first order with respect to triphenyl methyl chloride (that is, $r_A = k_c C_A^2 C_B$) and evaluate the rate constant. The initial concentrations of $(C_6H_5)_3COCH_3$ and HCl are zero.

Data: Initial conditions, 0.106 molar chloride and 0.054 molar methanol in dry benzene solution in the presence of pyridine at 25°C.

Time (min)	Triphenyl Methylchloride Reacted (g mol/liter)
168	0.0091
174	0.0110
418	0.0181
426	0.0189
444	0.0207
1150	0.0318
1440	0.0334
1510	0.0345
1660	0.0354
2890	0.0418
2900	0.0414
3120	0.0416
193,000	0.0514

Answer: $k_c = 0.2687$ liter^2/min (g mol)^2

1-14. Kistiakowsky and Lacher [24] investigated the condensation of acrolein and butadiene at 291.2°C. This reaction was found to be second order, and at this reaction temperature (291.2°C), the following rate constant was reported: $k_p = 2.71 \times 10^{-11}$ g mol/s mm^2 liter. Find the value of k_c at these conditions.
Answer: $k_c = 3.356 \times 10^{-5}$ m^3/(s g mol)

1-15. In the rate equations

$$r_{A1} = k_{c1}C_A^2, \qquad k_{c1} = 0.004 \quad \text{at} \quad 700 \text{ K}$$
$$r_{A2} = k_{c2}C_A, \qquad k_{c2} = 0.05 \quad \text{at} \quad 400 \text{ K}$$

where r_A is expressed in lb mol/h ft^3 and C_A in lb mol/ft^3.
(a) Calculate k_{c1} in the units of m^3/(s g mol) and k_{c2} in (s)$^{-1}$.
(b) Calculate the values of k_p for these same reactions when r_A is stated in terms of partial pressures as follows:

$$r_{A1} = k_{p1} p_A^2$$
$$r_{A2} = k_{p2} p_A$$

Again r_A is expressed in m^3/(s g mol). The partial pressure of p_A is in atmospheres.
Answer: $k_{p1} = 2.1 \times 10^{-8}$ g mol/(s m^3 atm^2) and $k_{p2} = 4.237 \times 10^{-4}$ g mol/(s m^3 atm)

1-16. A third-order reaction in a perfect gas system has a reaction velocity constant $k_c = 300$ (ft^3)^2/(h)(lb mol)^2 at 700°R. Calculate the value of k_p and express it in the units of h, atm, ft^3, and lb mol.
Answer: $k_p = 2.2468 \times 10^{-6}$ lb mol/h ft^3 atm^3

1-17. Cain and Nicoll [25] obtained the following data for the decomposition of diazobenzene chloride in aqueous solution at 50°C.

$$C_6H_5N_2Cl \rightarrow C_6H_5Cl + N_2$$

Time (min)	6	9	12	14	18	20	22	24	26	30
Gas liberated (cc)	19.3	26.0	32.6	36.0	41.3	43.3	45.0	46.5	48.5	50.35

After a very long period of time ($t \rightarrow \infty$), a sample with an initial concentration of 10 g/liter liberated 58.3 cm^3 of nitrogen. Determine the reaction velocity constant k_c.
Answer: $k_c = 0.0675$ min^{-1}

1-18. Determine the rate constant for the thermal decomposition of dimethyl ether in the gas phase by use of measurements reported by Hinshelwood and Askey [26]. The reaction is carried out isothermally at 504°C in a constant volume reactor. At time $t = 0$, the reactor is filled to a pressure of 312 mm with pure dimethyl ether.

Time (s)	390	777	1195	3155	∞
Pressure increase (mm Hg)	96	176	250	476	619

$$(CH_3)_2O \rightarrow CH_4 + H_2 + CO$$

Answer: $k_c = 4.13 \times 10^{-4}$ s^{-1}

1-19. Consider the decomposition of di-t-butyl peroxide to acetone and ethane. Use the data obtained by Raley et al. [27] to determine whether the reaction rate is first or second order, and evaluate the rate constant k_c. These data were collected at 154.6°C.

$$(CH_3)_3COOC(CH_3)_3 \rightarrow 2(CH_3)_2CO + C_2H_6$$

Time (min)	Total Pressure (mm Hg)	Time (min)	Total Pressure (mm Hg)
0	173.5	11	239.8
2	187.3	12	244.4
3	193.4	14	254.5
5	205.3	15	259.2
6	211.3	17	268.7
8	222.9	18	273.9
9	228.6	20	282.0
		21	286.8
		∞	491.8

Answer: $k_c = 2.095 \times 10^{-2}$ min^{-1}

1-20. Calculate the value of the activation energy required for the rate to double for the following three cases.
(a) 300 to 310 K
(b) 400 to 410 K
(c) 740 to 750 K
Answer: (a) 53,600 J/g mol; (b) 94,600 J/g mol; (c) 320,000 J/g mol.

1-21. In an investigation of the isomerization of n-propylidene-cyclopropylamine to 5-ethyl-1-pyroline by Cocks and Egger [15], the following data were obtained in a batch reactor which was operated isothermally and at constant volume.

n-propylidene-cyclopropylamine   5-ethyl-pyroline

Temperature (K)	Time (min)	Pressure (torr)	Conversion (%)
572.9	30	15.4	11.13
572.9	100	18.3	33.13
573.0	20	55.2	7.83
573.1	60	18.1	21.04
573.3	210	17.5	57.84

Temperature (K)	Time (min)	Pressure (torr)	Conversion (%)
599.9	20	5.9	39.04
600.1	6	2.5	14.70
600.3	40	6.6	65.79
600.5	60	4.4	81.02
622.9	4	13.6	38.17
623.0	5	15.1	43.37
624.4	6	6.8	50.55
623.5	10	8.0	67.99
635.4	5	9.0	70.32
635.4	6	7.8	77.07
635.5	2	13.8	40.00
635.5	3	14.9	52.89
635.5	4	13.8	63.20

Determine whether the reaction is first or second order. Evaluate the rate constant k_c, the activation energy, and the Arrhenius constant.
Answer: $A = 10.421 \times 10^{14}$, $E_A = 45,581$ cal/g mol

1-22. (a) For a reaction having a rate which is nth order with respect to A,

$$r_A = k_c C_A^n$$

where $n \neq 1$, show that the half-life time is given by

$$t_{1/2} = \frac{(2^{n-1} - 1)}{(n-1)k_c \left(C_A^0\right)^{n-1}}$$

(b) Suppose that two experiments are conducted for the reaction of $A \rightarrow$ products and that the following data are obtained.

$$\text{Experiment 1} \quad t_{1/2} = \tfrac{1}{8} \text{ h}, \quad C_A^0 = 2 \text{ g mol/liter}$$

$$\text{Experiment 2} \quad t_{1/2} = \tfrac{1}{2} \text{ h}, \quad C_A^0 = 1 \text{ g mol/liter}$$

On the basis of these data, predict the order of the reaction.

1-23. Hairston and O'Brien [21, 28] studied the reaction of 2-trimethylsilyl-2 chloroprane with antimony pentafluoride by use of nuclear magnetic resonance. To determine the first step of a proposed mechanism, the following data were obtained at a temperature of 40°C. The initial concentration of $A[(CH_3)_3 SiC(CH_3)_2 Cl]$, was 0.1841 g mol/liter, and the initial concentration of

$B[SbF_5]$ was 0.1846 g mol/liter.

t (min)	C_A (g mol/liter)
8	0.1501
10	0.1429
21	0.112
44	0.0812
76	0.0579
114	0.0433
151	0.0344
153	0.0341

The following mechanism was proposed.

$$(CH_3)_3Si—C(CH_3)_2Cl + SbF_5 \xrightarrow[40°C]{k_1} [(CH_3)_3Si—C(CH_3)_2]^+$$

$$+[SbF_5Cl]^- \underset{k_3}{\overset{k_2}{<}} \begin{array}{l} (CH_3)_2 \overset{\displaystyle F}{\underset{}{Si}} C(CH_3)_3 + SbF_4Cl \\[1em] (CH_3)_3 SiF + \overset{\displaystyle CH_2}{\underset{\displaystyle CH_3}{\overset{\|}{CH}}} + SbF_4Cl \end{array}$$

Determine the order of the first step of the proposed mechanism for this reaction and evaluate the rate constant k_1.

1-24. The gas phase reaction of acrolein with cyclopentadiene to produce endomethylenetetrahydrobenzaldehyde was also studied by Kistiakowsky and Lacher [24]. The reaction may be represented symbolically by

$$A + B \rightleftarrows C$$

with the rate of reaction being given by

$$r_A = k_c C_A C_B - k'_c C_C$$

On the basis of an initial pressure of 1 atm and $C_A^0 = C_B^0$, $C_C^0 = 0$, calculate the time required to obtain a conversion X equal to 90% of the equilibrium conversion at 200°C. The rate constants are given by

$$k_c = 1.5 \times 10^9 \exp\frac{-15,200}{RT}, \text{ cm}^3/\text{g mol s}$$

$$k'_c = 2.2 \times 10^{12} \exp\frac{-33,600}{RT}, \text{ s}^{-1}$$

where T is in K.

Answer: $t = 14.3$ min, $R = 1.9872$ cal/mol K

1-25. A young junior engineer is trying to convince his supervisor that the rate constants, activation energy, and order of a reaction can be determined with at most four data points plus two replicate experiments at two of the three temperatures to be studied. The reaction in question is a liquid phase reaction and is believed to be first or second order. From the following data, determine the order and activation energy. The reaction is represented as $A \rightarrow$ products.

For each temperature, experiments were conducted as follows. At 440 K with an initial concentration of 0.5 g mol/liter, the conversion of A was found to be 20% for a reaction time of 10 min.

(a) With this data point, calculate first- and second-order rate constants.

(b) The next experiment at 440 K is to be conducted with an initial concentration of A of 1.0 g mol/liter, and the reaction time, t, to be used is that one which satisfies the following condition. Let $\phi(t)$ denote the absolute difference in the conversions computed for first and second order rates at any time, t, on the basis of $C_A^0 = 1.0$ g mol/liter and the rate constants found in Part (a). Show that the reaction time which maximizes $\phi(t)$ is $t = 15.7$ min, and compute the corresponding conversions for first and second order reactions at this time.

(c) An experiment was conducted at the conditions given in Part (b) with a reaction time of $t = 15.7$ min, and the experimentally observed conversion was 44%. Determine the order and corresponding rate constant by comparing this experimental observation with the results of Part (b).

(d) A replicate experiment at 15.7 min yielded a conversion of 43%. Calculate the rate constant and determine the expected error.

(e) The next experiments were conducted at 420 K and 460 K with initial concentrations of 0.75 and 0.25 g mol/liter, respectively. For these two experiments conversions of 30% at 50 min and 69% at 60 min were obtained. A replicate run at 460 K for 60 min yielded a conversion of 68%. From these data, calculate the activation energy and an estimated error in the calculated rate constants.

(f) To check the model, a seventh experiment has been proposed at 450 K, initial concentration of A of 1.2, and a reaction time of 20 min. What range of conversions should be expected for this experiment?

Answer: (a) For first order, $k = 0.0223$ min^{-1}, and for second order, $k = 0.05$ liters/(min mol). (c) Expected conversion for second- and first-order reactions is 0.44 and 0.295, respectively. Therefore, the reaction is second order since the experimental conversion is 44%. (d) Expected error in k is 0.0012. (e) $k = 0.01143$ liters/(min mol) at 420 K, and 0.1450 liters/(min mol) at 460 K. Expected error in k is 0.0047 at 460 K. Activation Energy = 24.3 Kcal/mol. (f) Expected range in conversions is 66% to 68%.

1-26. The following liquid phase reaction with arbitrary stoichiometry has been found to be second order; that is,

$$aA + bB \rightarrow cC + dD \quad \text{and} \quad r_A = k_c C_A C_B$$

(a) Show that, when this reaction is carried out at constant volume and temperature in a batch reactor, the following expression relating conversion and time is obtained:

$$\log_e \frac{aC_B^0(1 - X)}{bC_A^0[(aC_B^0/bC_A^0) - X]} = \left(\frac{b}{a}C_A^0 - C_B^0\right)k_c t$$

or

$$\log_e \frac{C_B^0 C_A}{C_A^0 C_B} = \left(\frac{b}{a}C_A^0 - C_B^0\right)k_c t$$

where

$$\frac{aC_B^0}{bC_A^0} > 1$$

(b) Repeat part (a) for the case where

$$aC_B^0 = bC_A^0$$

1-27. Carbinol methyl ether (methyl-trityl ether) is to be produced in a batch process by reacting triphenyl methyl chloride with methanol in a dry benzene and pyridine solution at 25°C and 1 atm. The reaction is

$$CH_3OH + (C_6H_5)_3CCl \rightarrow (C_6H_5)_3COCH_3 + HCl$$

$$A \quad + \quad B \quad \rightarrow \quad C \quad + D$$

Initial conditions are 0.106 g mol/liter for triphenyl methyl chloride and 0.054 g mol/liter for methanol. The rate equation [23] at 25°C is

$$r_A = k_c C_A^2 C_B$$

$$k_c = 4.48 \times 10^{-3} \, (\text{liter/g mol})^2 (\text{s})^{-1} \text{ at } 25°C$$

The addition of pyridine allows the fast reaction of pyridine with HCl to form pyridine hydrochloride, which precipitates. This also allows the reaction to be treated as an irreversible reaction. A 100-gal autoclave is available to conduct the reaction. A contract has been signed to produce 1000 kg of crystalline carbinol methyl ether with a purity of 99$^+$%.

(a) Draw a flow chart to produce the carbinol.
(b) Assume that the time between each batch, which includes charging the reactor and discharging the reactor, is 60 min. A maximum of two 8-h shifts/day will be utilized. The autoclave can only be filled to 75% of capacity for each batch because of safety regulations. Calculate the conversion for each batch and the minimum number of days required to produce the carbinol. How many batches are required?
(c) How much pyridine, methanol, benzene, and triphenyl methyl chloride will be required?
(d) What do you propose to do with the pyridine hydrochloride? Is there a market for it?
(e) Is there a better way (i.e., operating procedure) to produce the carbinol?
For data, see handbooks. Assume ideal solution behavior for the reacting mixture.

1-28. The reaction rate constant of dibromobutenes with potassium iodine is used in an old procedure to evaluate the concentration of butenes in a solution. As part of a broader investigation on the kinetics of the reactions of dibromides with potassium iodine, Dillon [29] reports the following data for the reaction of ethylene dibromide with potassium iodine.

Temperature:	59.72°C						
Initial concentrations:	[KI] = 0.1531 g mol/liter and $[C_2H_4Br_2]$ = 0.02864						
	g mol/liter						
Time (min)	495	675	795	930	1035	1215	1395
Conversion of							
dibromide	0.2863	0.3630	0.4099	0.4572	0.4890	0.5396	0.5795

The stoichiometry of the reaction is as follows:

$$C_2H_4Br_2 + 3KI \rightarrow C_2H_4 + 2KBr + KI_3$$

(a) From these data, evaluate the order of the reaction and the rate constant.
   Answer: $k_c = 0.305$ liter/(h g mol)
(b) The activation energy for the reaction was found to be 22.7 kcal/g mol. Determine the Arrhenius factor, the entropy of activation, and the rate constant at 70°C.

1-29. The kinetics and mechanism of the ketone-bisulfite addition reaction in an aqueous solution was studied by Rao and Salunke [30] by use of a polarographic technique. The specific rate was found to be a function of pH. For an equimolar solution of bisulfite and acetone in a deaerated solution of potassium hydrogen phthalate buffer containing potassium chloride as the supporting electrolyte, the following data were obtained:

Initial concentrations: $[HSO_3^-] = 8.0*10^{-3}$ g mol/liter, pH = 4								
Temp. = 20°C								
Time (s)	0	30	45	60	75	90	105	120	150
10^3 conc. $[HSO_3^-]$ g mol/liter	8.00	7.54	7.35	7.10	6.89	6.70	6.54	6.39	6.14
Time (s)	180	210	240						
10^3 conc. (HSO_3^-) g mol/liter	5.92	5.62	5.37						

Determine the order of reaction and the specific rate constant. The stoichiometry is as follows:

$$\begin{array}{ccc} H & D & O \\ | & | & \| \\ H-C-C-C-H & + & HSO_3^- \rightarrow \text{products} \\ | & | \\ H & H \end{array}$$

Answer: 0.25 liter/(s g mol)

NOTATION

C_A = concentration of component A, moles/unit volume

E_A = activation energy [see Equation (1-33)]

k_c, k_p = rate constants where concentrations and partial pressures are used in the rate expressions, respectively; in the case of multiple reactions, the subscripts c and p are sometimes dropped and the rate constants denoted by $k_1, k_2, \ldots$

k_c', k_p' = rate constants for the reverse reaction as distinguished from the rate constants for the forward reactions

M_A = molecular weight of A

N = order of a reaction

N_A^0, N_A = total number of moles of component A in the reactor at time $t = 0$ and the total number of moles of A in the reactor at any time t

N_T = total number of moles at any time t

p_A = partial pressure of component A, force/unit area

P = total pressure, force/unit area

r_A = net rate of reaction of component A; defined by Equation (1-1)

r_{Af} = forward rate of reaction of component A

r_{Ar} = reverse rate of reaction of component A

R = constant in the perfect gas law; defined by Equation (1-25)

t = time

$t_{1/2}$ = half-life

T = temperature, °R or kelvins

V = reactor volume or more precisely that portion of the reactor volume occupied by the reacting mixture

X = moles of the base component A converted per mole of A in the feed

y_A^0 = mole fraction of component A in the feed

REFERENCES

1. Wenzel, K. F., *Lehre von der Chemischen Affinitat der Korper* (1777).

2. Van't Hoff, J. H., *Etudes de Dynamique Chimique*, F. Muller & Co., Amsterdam (1884).

3. Arrhenius, Svante, *Z. Physikal Chem.*, *4*:226 (1889).

4. Ostwald, W., *Z. Physikal Chem.*, *15*:704 (1894).

5. Lewis, W. K., and E. D. Ries, *Ind. Eng. Chem.*, *19*:830 (1923).

6. Hougen, O. A., *Reaction Kinetics in Chemical Engineering*, Chemical Engineering Progress Monograph Series, *No. 1*, Vol. 47 (1951).

7. Hougen, O. A., and K. M. Watson, *Chemical Process Principles*, Part III, *Kinetics and Catalysis*, John Wiley & Sons, Inc., New York (1947).

8. Busby, G. W., "Soap," *Encyclopedia of Chemical Technology*, *12*:573, Wiley/Interscience, New York (1954).

9. Godfrey, T. M., *Glycerol*, edited by C. S. Miner and N. N. Dalton, Van Nostrand Reinhold, New York (1953).

10. Hinshelwood, C. N., and C. R. Prichard, *Proc. Roy. Soc. A108*:211 (1925).

11. Groggins, P. H., editor-in-chief, *Unit Processes in Organic Synthesis*, 5th ed., McGraw-Hill Book Company, New York (1958).

12. Seestrom, H. E., J. P. Hughes, and W. H. Shearon, Jr., *Modern Chemical Processes*, Vol. II, W. J. Murphy, editor, *Ind. Eng. Chem.*, Van Nostrand Reinhold, New York (1954).

13. Gumz, Wilhelm, *Gas Producers and Blast Furnaces*, Theory and Methods of Calculation, John Wiley & Sons, Inc., New York (1950).

14. Himmelblau, D. M., *Process Analysis by Statistical Methods*, John Wiley & Sons, Inc., New York (1970).

15. Cocks, A. T., and K. W. Egger, *Int. J. of Chem. Kin.*, 4:169 (1972).

16. Walker, J., *Proc. Roy. Soc.*, A78:157 (1906).

17. Churchill, S. W., *The Determination and Use of Rate Data: The Rate Concept*, McGraw-Hill Book Company, New York (1974).

18. Marquardt, Fritz-Hans, "Fourth Conference on Catalytic Hydrogenation and Analogous Pressure Reactions," edited by Paul N. Rylander, *Ann. N.Y. Acad. Sci.*, 214:110 (1973).

19. Schulze, S. R., and A. L. Baron, *Addition and Condensation Polymerization Processes*, edited by Norbert A. J. Platzer, American Chemical Society, *Advances in Chemistry Series*, 91:692 (1969).

20. Worsfold, S., and S. Bywater, *Canadian J. of Chem.*, 38:1891 (1960).

21. Hairston, T. J., and D. H. O'Brien, *J. Organometallic Chem.*, 29:79 (1971).

22. Neumann, D. W., and S. Lynn, *Ind. Eng. Chem. Process Des. Dev.*, 25:248 (1986).

23. Swain, C. G., *J. Am. Chem. Soc.*, 70:1119 (1948).

24. Kistiakowsky, G. B., and J. R. Lacher, *J. Am. Chem. Soc.*, 58:123 (1936).

25. Cain, J. C., and F. Nicoll, *Proc. Chem. Soc.*, 24:282 (1909).

26. Hinshelwood, C. H., and P. S. Askey, *Proc. Roy. Soc.*, A115:215 (1927).

27. Raley, J. H., R. R. Rust, and W. E. Vaughan, *J. Am. Chem. Soc.*, 70:98 (1948).

28. O'Brien, D. H., *Personal communication*, Texas A & M University (1972).

29. Dillon, R. T., *J. Am. Chem. Soc.*, 70:952 (1932).

30. Rao, T. S., and S. B. Salunke, *Reaction Kinetics and Catalysis Letters*, 26(3–4):273 (1984).

APPENDIX

Theorem 1. *Mean-Value Theorem of Integral Calculus*

If the function $f(x)$ is continuous in the interval $a \le x \le b$, then

$$\int_a^b f(x)\,dx = f(\varepsilon)(b - a)$$

where $\varepsilon = a + \alpha(b - a)$ and $0 \le \alpha \le 1$.

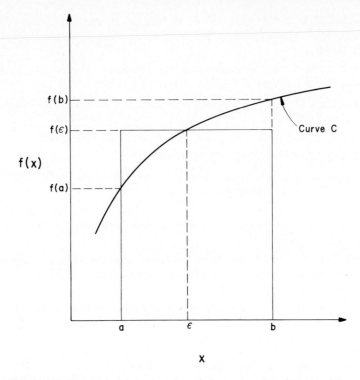

Figure A-1. Geometrical interpretation of the mean-value theorem of integral calculus.

Figure A-1 is useful in the geometrical interpretation of this theorem. The theorem asserts that there exists a number ε lying in the closed interval $a \leq \varepsilon \leq b$ such that the area of the rectangle of height $f(\varepsilon)$ and base $(b - a)$ is equal to the area bounded by curve C and the x-axis from $x = a$ to $x = b$.

Theorem 2. *Mean-Value Theorem of Differential Calculus*

If the function $f(x)$ is continuous in the interval $a \leq x \leq b$, and differentiable at every point in the interval $a < x < b$, then there exists at least one value of β such that

$$\left. \frac{df(x)}{dx} \right|_{\varepsilon} = \frac{f(b) - f(a)}{b - a}$$

where $\varepsilon = a + \beta(b - a)$ and $0 < \beta < 1$.

Figure A-2 is useful in the geometrical interpretation of this theorem. This theorem asserts that there exists a number ε lying in the open interval

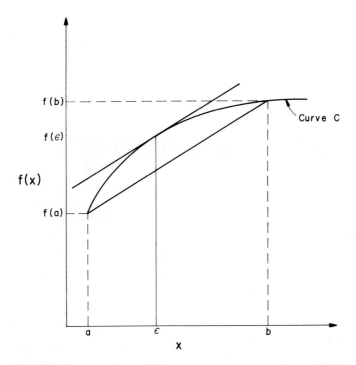

Figure A-2. Geometrical interpretation of the mean-value theorem of differential calculus.

$a < \varepsilon < b$ at which the slope of the tangent line to curve C is equal to the slope of the secant line drawn through the points $(a, f(a))$ and $(b, f(b))$. The slope of the secant is equal to $[f(b) - f(a)]/[b - a]$. At the point $(\varepsilon, f(\varepsilon))$ the slope of the tangent line is equal to $df(x)/dx$ evaluated at ε.

Flow Reactors at Isothermal Operation

2

The modern chemical industry is characterized by the almost exclusive use of continuous processes. This type of operation calls for continuous, or flow, reactors. The chemical reactions occur while the reacting stream is flowing through the reactor. Provisions are commonly made to heat the reactants to the reacting temperature just before they enter the reactor and to terminate the reaction by cooling the mixture as it leaves the reactor. Unless it is desired to operate the reactor *adiabatically*, provisions are made to maintain some control over the temperature of the reacting mixture by adding or removing heat from the reacting mixture as it passes through the reactor. The reacting mixture is customarily passed through the reactor at an appreciable velocity to insure good mixing and heat transfer.

The conversion per pass of the reacting stream through the reactor is proportional to the length of time the reacting mixture resides in the reactor, which is in turn proportional to the reactor volume. Thus, to achieve a significant conversion per pass, the reactor must be operated at conditions such that the rate constant for the desired reaction is appreciable. Many commercial reactors are operated, however, at relatively low conversions per pass, and yet relatively high overall conversions are achieved by recycling a portion of the effluent stream. In many reactors, the selectivity of one particular reaction over other possible reactions becomes a more important consideration than does the conversion per pass. In fact, many industrial processes are based on producing a valuable intermediate product rather than the final products which would be obtained at thermodynamic equilibrium. For example, in the dehydrogenation of ethane, the desirable product is the intermediate ethylene rather than carbon, hydrogen, and methane, which are more thermodynamically favorable.

The analysis and design of flow reactors are introduced in this chapter by the consideration of isothermally operated reactors in which single reactions occur. The resulting simplification of the chemical kinetics makes it relatively easy to predict the behavior which can be expected for various types of flow patterns and mixing mechanisms. The methods developed in this chapter in the treatment of the relatively simple systems are useful in the preliminary analysis of commercial reactors in which any number of reactions may occur at nonisothermal conditions.

The steady-state operation of flow reactors is considered in Part I, and unsteady-state operation is considered in Part II.

I. STEADY-STATE OPERATION OF FLOW REACTORS

Steady-state operation means that the variables at each point within the system do not vary with time. Therefore, the total amount of component A within the system at time $t_n + \Delta t$ is equal to the amount at time t_n. Since the rates of input and output of mass to and from the system, respectively, are independent of time, Equation (1-37) reduces to

$$\left\{ \begin{array}{c} \text{Input of mass to the} \\ \text{system per unit time} \end{array} \right\} - \left\{ \begin{array}{c} \text{output of mass from the} \\ \text{system per unit time} \end{array} \right\} = 0 \qquad (2\text{-}1)$$

or simply

$$\text{Input} - \text{output} = 0 \qquad (2\text{-}2)$$

Equation (2-1) [or (2-2)] is used to make the material balances for the various types of reactors considered in Part I.

The application of Equation (2-1) in the making of a material balance on a given reactor depends on the particular mixing model assumed for the fluid within the reactor. The development that follows is subdivided according to the classic mixing models: (1) plug flow, (2) perfectly mixed, (3) partial axial mixing, (4) partial axial and radial mixing, and (5) laminar-flow reactors.

A. Plug-Flow Reactors

The first model used to describe the mixing in tubular reactors is the plug-flow model (see Figure 2-1). The reacting stream flowing through a reactor is said to be in plug flow if the contents of the reactor are perfectly mixed in the radial direction, while no mixing occurs in the axial direction (the

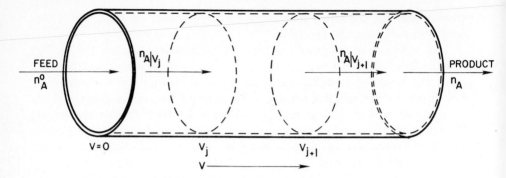

Figure 2-1. Sketch of a tubular reactor in which the mixing is described by the plug-flow model (perfect mixing in the radial direction, and no mixing in the horizontal direction).

direction of flow). These mixing assumptions are sometimes called the *plug-flow assumption* [1, 2].

Near perfect mixing in the radial direction may be achieved by passing the fluid through the reactor at relatively high flow rates where the Reynolds numbers are large. The mixing in the axial direction is generally small, provided that the length of the reactor is large relative to its diameter. Denbigh [3] has suggested that the mixing assumptions for plug-flow reactors are valid, provided that the length-to-diameter ratio is greater than 50 and that the Reynolds number of the flowing stream is greater than 10^4.

1. Material Balances for Plug-Flow Reactors. Let n_A denote the flow rate of A in moles per unit time. Let V_j and $V_j + \Delta V$ be selected arbitrarily $(0 < V_j < V_j + \Delta V < V_T$, where V_T is the total volume of the reactor). Since r_A generally depends upon the temperature and concentration of the reactants, it will take on a different value at each V in the interval $V_j < V < V_j + \Delta V$. Then, by Equation (2-1), it follows that a material balance for component A over the element of volume ΔV is given by

$$\underbrace{M_A n_A|_{V_j}}_{\substack{\text{input of } A \text{ per unit time by flow} \\ \text{to the element of volume } \Delta V}} - \underbrace{\left[M_A n_A|_{V_j + \Delta V} + \int_{V_j}^{V_j + \Delta V} M_A r_A \, dV \right]}_{\substack{\text{output of } A \text{ per unit time from the element of} \\ \text{volume by flow and by reaction}}} = 0 \qquad (2\text{-}3)$$

Since the molecular weight M_A of a pure component does not vary, Equation (2-3) reduces to

$$\left[n_A|_{V_j} - n_A|_{V_j + \Delta V} \right] - \int_{V_j}^{V_j + \Delta V} r_A \, dV = 0 \qquad (2\text{-}4)$$

The difference in the flow rate of A at V_j and $V_j + \Delta V$ may be replaced by its equivalent, as given by the *mean-value theorem of differential calculus*. (See

Theorem 2 in the Appendix of Chapter 1.)

$$\left[n_A|_{V_j} - n_A|_{V_j + \Delta V} \right] = -\Delta V \frac{dn_A}{dV}\bigg|_{V_j + \alpha \Delta V} \tag{2-5}$$

where $0 < \alpha < 1$. (That is, at the point $V_j + \alpha \Delta V$, the slope of the tangent line to the curve of n_A versus V is equal to the slope of the straight line connecting $n_A|_{V_j}$ and $n_A|_{V_j + \Delta V}$.) Next the *mean-value theorem of integral calculus* may be used to replace the integral of r_A over the element of volume from V_j to $V_j + \Delta V$ by a rectangle having the same area.

$$\int_{V_j}^{V_j + \Delta V} (-r_A)\, dV = \left(-r_A|_{V_j + \beta \Delta V} \right) \Delta V \tag{2-6}$$

where $0 \le \beta \le 1$. Substitution of the results given by Equations (2-5) and (2-6) into Equation (2-4) yields

$$-\frac{dn_A}{dV}\bigg|_{V_j} - r_A|_{V_j} = 0 \tag{2-7}$$

Since V_j was arbitrarily selected in the open interval $(0 < V < V_T)$, it follows that the material-balance expression for r_A is given by

$$r_A = -\frac{dn_A}{dV} \tag{2-8}$$

For the case where the rate of reaction is defined in terms of the appearance of A [see Equation (1-45)], the expression corresponding to Equation (2-8) is

$$r_A = \frac{dn_A}{dV} \tag{2-9}$$

2. Variation of Conversion with Reactor Volume for Liquid Phase Reactions. (Other conditions: isothermal and steady-state operation, constant mass density.) Consider first the case where the first-order reaction

$$A \rightarrow \text{products} \quad r_A = k_c C_A$$

is carried out in the liquid phase in a plug-flow reactor. If it is further supposed that the mass density of the reacting stream is constant throughout the reactor, then the volumetric flow rate v_T (volume/time) of the reacting mixture throughout the reactor is also constant. (*Note:* The volumetric flow rate is equal to the mass-flow rate divided by the mass density, and the mass-flow rate is constant at steady-state operation.)

Elimination of r_A from the material balance [Equation (2-8)] and the first-order rate expression given above yields

$$-\frac{dn_A}{dV} = k_c C_A \tag{2-10}$$

Since the mixing is assumed to be perfect in the radial direction, the concentration C_A may be expressed in terms of the flow rates,

$$C_A = \frac{n_A}{v_T} \qquad (2\text{-}11)$$

and this expression may be used to restate Equation (2-10) in terms of n_A

$$-\frac{dn_A}{dV} = k_c \frac{n_A}{v_T} \qquad (2\text{-}12)$$

For a first-order reaction which is carried out isothermally, the rate constant k_c is independent of n_A and V. Since v_T is also constant throughout the reactor, Equation (2-12) may be solved by separation of the variables. Thus

$$-\int_{n_A^0}^{n_A} \frac{dn_A}{n_A} = \frac{k_c}{v_T} \int_0^V dV$$

and, upon carrying out the integrations indicated, it is found that

$$V = \frac{v_T}{k_c} \log_e \frac{n_A^0}{n_A} \qquad (2\text{-}13)$$

or

$$n_A = n_A^0 e^{-N_{Da}} \qquad (2\text{-}14)$$

where $\quad N_{Da} = \dfrac{k_c V}{v_T}, \quad$ the Damköhler number for a first-order, liquid phase reaction

Since the volumetric flow rate is constant, it follows that $n_A/v_T = C_A$ (the concentration of A at a given volume V) and $n_A^0/v_T = C_A^0$ (the concentration of A at the inlet of the reactor). Thus, at any given V or N_{Da}, Equation (2-14) may be stated in terms of concentrations as follows:

$$C_A = C_A^0 e^{-N_{Da}} \qquad (2\text{-}15)$$

For the case where v_T remains constant throughout the reactor, C_A may be expressed in terms of the *true residence time*, which is equal to the actual time (V/v_T) that a unit of mass remains in the volume V. (The quantity V/v_T is also equal to the time required to sweep out the volume V one time.) Let the *true residence time* be denoted by θ; that is,

$$\theta = \frac{V}{v_T} \qquad (2\text{-}16)$$

Thus, $N_{Da} = k_c \theta$, and for the case under consideration C_A at any V may be expressed in terms of the true residence time θ:

$$C_A = C_A^0 e^{-k_c \theta} \qquad (2\text{-}17)$$

This equation is identically the same as Equation (1-53), the material-balance

TABLE 2-1
DESIGN EQUATIONS FOR CARRYING OUT UNIDIRECTIONAL REACTIONS
IN A PLUG-FLOW REACTOR
(Conditions: isothermal and steady-state operation, plug flow,
and constant mass density)

Reaction	Design Equation
1. $A \rightarrow$ products	$V = \dfrac{v_T}{k_c} \log_e \dfrac{n_A^0}{n_A}$
2. $A + B \rightarrow$ products	$V = \dfrac{v_T^2}{k_c} \left[\dfrac{1}{n_A^0 - n_B^0} \log_e \dfrac{n_B^0 n_A}{n_A^0 n_B} \right]$
3. $2A \rightarrow$ products	$V = \dfrac{v_T^2}{k_c} \left[\dfrac{1}{n_A} - \dfrac{1}{n_A^0} \right]$
4. $3A \rightarrow$ products	$V = \dfrac{v_T^3}{2k_c} \left[\left(\dfrac{1}{n_A} \right)^2 - \left(\dfrac{1}{n_A^0} \right)^2 \right]$
5. $NA \rightarrow$ products	$V = \dfrac{v_T^N}{(N-1)k_c} \left[\left(\dfrac{1}{n_A} \right)^{N-1} - \left(\dfrac{1}{n_A^0} \right)^{N-1} \right]$

By comparing Tables 1-1 and 2-1, one notes that if the concentration $C_i = N_i/V$ of component i in Table 1-1 is replaced by n_i/v_T, and t is replaced by V/v_T, then the two tables are equivalent. These substitutions are not permissible unless the mass density remains constant.

equation for a batch reactor. This identity leads immediately to the interpretation that the amount of A reacted in each element of fluid that resides in the plug-flow reactor an amount of time θ is exactly equal to the amount of A which would have reacted in an isothermal batch reactor in the same amount of time. This one-to-one correspondence between batch and plug-flow reactors holds for all reactions which are carried out isothermally and at constant mass density. Expressions for the variation of the reactor volume with conversion (or molar flow rate of A) for first-, second-, and third-order reactions are presented in Table 2-1.

EXAMPLE 2-1

When bromine cyanide is reacted with methylamine,

$$CNBr + CH_3NH_2 \rightarrow CNNH_2 + CH_3Br$$

in the liquid phase with water as a solvent, the rate of disappearance of bromide cyanide was found to be second order by Griffith et al. [4] and Reference [5].

$$r_{CNBr} = k_c C_{CNBr} C_{CH_3NH_2}$$

At 10°C, Griffith et al. found that

$$k_c = 2.22 \text{ liters}/(s)(g \text{ mol})$$

For a residence time of 4 s and the initial concentration $C_{CNBr}^0 = C_{CH_3NH_2}^0 = 0.1$ g mol/liter, calculate the concentration of bromine cyanide in the effluent stream from the reactor, as well as the corresponding percentage conversion of bromine cyanide which can be expected when the reaction is carried out in a plug-flow reactor operated isothermally at 10°C.

SOLUTION

For convenience, let the compounds participating in the preceding reaction be denoted by A, B, C, and D to give

$$A + B \rightarrow C + D$$

Since the initial concentrations of the reactants are equal, their concentrations remain equal throughout the reactor, and thus

$$r_A = k_c C_A^2 = k_c \frac{n_A^2}{v_T^2}$$

When the second-order reaction is carried out in a plug-flow reactor, the concentration of A in the outlet stream may be found by rearranging Equation (3) of Table 2-1 as follows:

$$\theta = \frac{V}{v_T} = \frac{1}{k_c} \left[\frac{1}{n_A/v_T} - \frac{1}{n_A^0/v_T} \right] = \frac{1}{k_c} \left[\frac{1}{C_A} - \frac{1}{C_A^0} \right]$$

$$C_A = \frac{C_A^0}{1 + k_c C_A^0 \theta} = \frac{0.1}{1 + (2.22)(0.1)(4)} = 0.05297$$

$$\text{Percentage conversion} = \frac{C_A^0 - C_A}{C_A^0} \times 100 = \left(\frac{0.1 - 0.05297}{0.1} \right) 100 = 47.03$$

3. Volume versus Conversion for Gas Phase Reactions. (Other conditions: isothermal and steady-state operation, negligible pressure drop across the reactor, and perfect gas behavior.) For reactions in which there is no change in the total number of moles, such as

$$A \rightarrow R, \qquad 2A \rightarrow 2R, \qquad A + B \rightarrow C + D$$

the total molar flow rate, n_T, is independent of conversion, and consequently it remains constant throughout the reactor. Thus, for isothermal and isobaric operation of a plug-flow reactor, it follows from the perfect gas law that the volumetric flow rate v_T is constant throughout the reactor:

$$v_T = n_T \frac{RT}{P}$$

For reactions which do not produce a change in the total number of moles, the expressions for volume versus conversion (or flow rate of A) are the same as those for liquid phase reactions in Table 2-1.

EXAMPLE 2-2

Fisher and Smith [6] studied the thermal reaction of methane with sulfur,

$$CH_4 + 2S_2 \rightarrow CS_2 + 2H_2S$$

in the gas phase and found that at 600°C the rate of disappearance of sulfur is given by

$$r_{S_2} = k_c C_{CH_4} C_{S_2}$$

where
$$k_c = 11.98 \times 10^6 \text{ ml}/(\text{h})(\text{g mol})$$

Suppose that this reaction is to be carried out in a plug-flow reactor which is to be operated isothermally at 600°C and at a pressure of 1 atm. The molar flow rate of sulfur in the feed is to be twice that of methane, whose flow rate is 23.8 g mol/h. Calculate the residence time required to convert 18% of the methane.

SOLUTION

For convenience, let the chemical compounds in the above reaction be represented by the letters A = methane, B = sulfur, C = carbon disulfide, and D = hydrogen sulfide,

$$A + 2B \rightarrow C + 2D$$

Since the reaction does not produce a change in the number of moles, the volumetric flow rate v_T remains constant throughout the reactor. Furthermore, since 2 moles of B react for each mole of A that reacts and the amount of B in the feed is twice that of A, it follows that $C_B = 2C_A$ throughout the reactor. Also, the stoichiometry of the reaction requires that r_A (the rate of disappearance of A) be equal to one-half of r_B (the rate of disappearance of B). Thus, since k_c is defined with respect to S_2 in the statement of the problem,

$$r_A = \frac{1}{2} r_B = \frac{k_c}{2} C_A C_B = \frac{k_c}{2} C_A(2C_A) = k_c C_A^2$$

The following relationship between residence time and the outlet flow rate of A is obtained by rearranging Equation (3) of Table 2-1 as follows:

$$\theta = \frac{V}{v_T} = \frac{v_T}{k_c}\left(\frac{1}{n_A} - \frac{1}{n_A^0}\right)$$

Let x denote the fraction of A converted,

$$x = \frac{n_A^0 - n_A}{n_A^0} \quad \text{or} \quad n_A = n_A^0 - n_A^0 x$$

Use of this expression to eliminate n_A from the above formula for θ gives

$$\theta = \frac{v_T}{k_c n_A^0}\left(\frac{x}{1-x}\right)$$

The volumetric flow rate v_T is computed by use of the perfect gas law as follows:

$$v_T = \frac{n_T RT}{P} = \frac{(3)(23.8)(82.05)(873.15)}{1} = 5.115 \times 10^6 \text{ ml/h}$$

Then

$$\theta = \frac{5.115 \times 10^6}{(11.98 \times 10^6)(23.8)} \left(\frac{0.18}{1 - 0.18} \right) = 0.003938 \text{ h} = 14.2s$$

If the reaction produces a change in the total number of moles, then the volumetric flow rate v_T varies with the conversion x. For example, consider the first-order unidirectional reaction

$$A \rightarrow 2R$$

The rate expression is given by

$$r_A = k_c C_A = \frac{k_c n_A}{v_T}$$

and the material balance on A is given by

$$r_A = -\frac{dn_A}{dV}$$

Thus

$$-\frac{dn_A}{dV} = \frac{k_c n_A}{v_T} \qquad (2\text{-}18)$$

which constitutes one equation in the three unknowns, V, n_A, and v_T. The conversion of A in a flow system is defined in a manner analogous to that for batch systems,

$$x = \frac{n_A^0 - n_A}{n_A^0} \qquad (2\text{-}19)$$

The flow rates n_A and v_T are expressed in terms of conversion by use of the following stoichiometric relationships:

$$n_A = n_A^0 - n_A^0 x$$
$$n_R^0 = n_R^0 + 2n_A^0 x \qquad (2\text{-}20)$$
$$n_T = n_T^0 + n_A^0 x$$

Since the gaseous mixture is assumed to obey the perfect gas law, it follows that

$$Pv_T = n_T RT \quad \text{and} \quad v_T = n_T \frac{RT}{P} = \left(n_T^0 + n_A^0 x \right) \frac{RT}{P}$$

Then Equation (2-18) may be expressed in terms of x and V as follows:

$$n_A^0 \frac{dx}{dV} = \frac{k_c}{RT} \left[\frac{\left(n_A^0 - n_A^0 x \right) P}{n_T^0 + n_A^0 x} \right] = k_p \left[\frac{\left(n_A^0 - n_A^0 x \right) P}{n_T^0 + n_A^0 x} \right] \qquad (2\text{-}21)$$

where k_c/RT has been replaced by its equivalent k_p as given by Equation (1-26). Separation of variables followed by integration and rearrangement yields

$$V = \left(\frac{n_T^0}{k_p P}\right)\left[\left\{(1 + y_A^0)\log_e\frac{1}{1 - x}\right\} - y_A^0 x\right] \qquad (2\text{-}22)$$

where it has been assumed that k_p is constant. This result suggests that one could start with the rate expressed in terms of k_p and partial pressures rather than k_c and concentrations. In fact, this alternate approach is the one most commonly employed in the analysis of gas phase reactions.

To demonstrate this approach, the previous example is solved again. Since $r_A = k_c C_A = k_p p_A$, it follows that

$$-\frac{dn_A}{dV} = k_p p_A \qquad (2\text{-}23)$$

TABLE 2-2
DESIGN EQUATIONS FOR CARRYING OUT UNIDIRECTIONAL REACTIONS
IN THE GAS PHASE IN PLUG-FLOW REACTORS
(Conditions: isothermal and steady-state operation, negligible pressure drop, and perfect gas behavior)

Reactions	Design Equation
1. $A \rightarrow R$ $n_T^0 = n_A^0 + n_R^0 + n_I^0$	$V = \dfrac{n_T^0}{k_p P}\log_e\dfrac{1}{1 - x}$
2. $2A \rightarrow 2R$ $n_T^0 = n_A^0 + n_R^0 + n_I^0$	$V = \dfrac{n_T^0}{k_p P^2 y_A^0}\left[\dfrac{x}{1 - x}\right]$
3. $3A \rightarrow 3R$ $n_T^0 = n_A^0 + n_R^0 + n_I^0$	$V = \dfrac{n_T^0}{2k_p P^3\left(y_A^0\right)^2}\left[\dfrac{2x - x^2}{(1 - x)^2}\right]$
4. $A + B \rightarrow R$ $n_A^0 \neq n_B^0$ $n_T^0 = n_A^0 + n_B^0 + n_I^0 + n_R^0$ $\alpha = \dfrac{n_B^0}{n_A^0}$	$V = \dfrac{n_A^0}{k_p\left(y_A^0 P\right)^2}\left[\left(y_A^0\right)^2 x + \dfrac{\left(1 - \alpha y_A^0\right)^2}{\alpha - 1}\log_e\left(\dfrac{\alpha - x}{\alpha}\right)\right.$ $\left. -\dfrac{\left(1 - y_A^0\right)^2}{\alpha - 1}\log_e(1 - x)\right]$
5. $A + B \rightarrow R + S$ $n_A^0 \neq n_B^0$ $n_T^0 = n_A^0 + n_B^0 + n_R^0 + n_S^0 + n_I^0$ $\alpha = \dfrac{n_B^0}{n_A^0}$	$V = \dfrac{n_A^0}{k_p(\alpha - 1)\left(y_A^0 P\right)^2}\log_e\dfrac{\alpha - x}{(1 - x)\alpha}$

For perfect gas mixtures,

$$p_A = \frac{n_A}{n_T} P = \left(\frac{n_A^0 - n_A^0 x}{n_T^0 + n_A^0 x} \right) P \tag{2-24}$$

Thus Equation (2-23) reduces to

$$n_A^0 \frac{dx}{dV} = k_p \left(\frac{n_A^0 - n_A^0 x}{n_T^0 + n_A^0 x} \right) P \tag{2-25}$$

which upon integration yields the final result given by Equation (2-22).

Expressions for reactor volume as a function of conversion for other gas phase reactions are given in Table 2-2.

EXAMPLE 2-3

Marek and McCluer [7] investigated the thermal decomposition of ethane and found that the stoichiometry of the decomposition could be represented by

$$C_2H_6 \rightarrow C_2H_4 + H_2$$

The rate of decomposition of ethane was found to be first order with respect to ethane, and the following expression for the rate constant was given:

$$\log_{10} k_c = 15.12 - \frac{15{,}970}{T}$$

where T is in kelvins and k_c has the units of reciprocal seconds. Calculate the conversion of ethane which can be achieved in a plug-flow reactor which is operated isothermally at 750°C and 1 atm pressure with $V/v_T^0 = 4$ s, where V is the total volume of the reactor and v_T^0 is the volumetric flow rate of the feed at the reactor conditions, 750°C and 1 atm. The feed consists of pure ethane.

SOLUTION

Let the preceding reaction be represented by

$$A \rightarrow B + C$$

Then

$$n_A = n_A^0 - n_A^0 x$$

$$n_B = n_A^0 x$$

$$n_C = n_A^0 x$$

$$n_T = n_A^0 (1 + x)$$

$$r_A = n_A^0 \frac{dx}{dV} = k_c C_A = k_c \frac{n_A}{v_T}$$

For a perfect gas mixture,

$$\frac{v_T}{v_T^0} = \frac{n_T}{n_T^0} = \frac{n_A^0(1 + x)}{n_A^0} = 1 + x$$

or

$$v_T = v_T^0(1 + x)$$

Thus

$$n_A^0 \frac{dx}{dV} = \frac{k_c n_A^0(1 - x)}{v_T^0(1 + x)}$$

Separation of variables followed by integration over the volume V gives

$$-2\log_e(1 - x) - x = k_c \frac{V}{v_T^0} \tag{A}$$

At 750°C,

$$\log_{10} k_c = 15.12 - \frac{15{,}970}{1023.15}$$

$$k_c = 0.3229 \text{ s}^{-1}$$

Since V/v_T^0 is given, Equation (A) becomes

$$-2\log_e(1 - x) - x = (0.3229)(4) = 1.292$$

Since this equation cannot be solved explicitly for x, it must be solved by a trial and error process. Newton's method [8] will be used. In the application of Newton's method, it is convenient to restate the problem. Let

$$f(x) = 1.292 + x + 2\log_e(1 - x)$$

Then it is desired to find that $x(0 < x < 1)$ such that $f(x) = 0$. Newton's method consists of the repeated use of the equation

$$x_{n+1} = x_n - \frac{f(x_n)}{f'(x_n)}$$

where x_n is the assumed value of x, and

$$f'(x) = 1 - \frac{2}{1 - x}$$

A good approximation of the solution value of x may be found by neglecting the increase in the number of moles (or the change in V_T). Instead of Equation (A), one obtains

$$-\log_e(1 - x) = k_c \frac{V}{v_T^0}$$

Thus

$$x = 1 - e^{-k_c \theta} = 1 - e^{-(0.3229)(4)}$$
$$= 0.725$$

When this starting value of x is used, the following results are obtained by application of Newton's method.

x_n	$f(x_n)$	$f'(x_n)$	$f(x_n)/f'(x_n)$	x_{n+1}
0.725	-0.565	-6.273	0.0900	0.6349
0.635	-0.0884	-4.478	0.0197	0.6152
0.6152	-0.00282	-4.197	0.0006	0.6145
0.6145	-0.000003	-4.1887	0.7×10^{-7}	0.6145

Thus the solution is $x = 0.614$, which is substantially different from the value obtained when the change in the total molar flow rate n_T with conversion is neglected. It should be pointed out that the pyrolysis of ethane has received considerable attention in recent years, and models of the pyrolysis which are more complete than the one used for this example have been proposed, as noted in References [9] and [10] and Chapter 4.

4. Volume versus Conversion for Reversible Reactions Carried out in the Gas Phase. (Other conditions: isothermal and steady-state operation, negligible pressure drop, and perfect gas behavior.) As an example, consider the following reaction, which has a first-order rate in each direction.

$$A \rightleftarrows R$$

For illustrative purposes, suppose that the feed contains components A, R, and chemically inert components in the gas phase. The flow rate (moles/unit time) of these inert components is denoted by n_I. Also, suppose that the gaseous mixture obeys the perfect gas law. The stoichiometric relationships are as follows:

$$n_A = n_A^0 - n_A^0 x$$
$$n_R = n_R^0 + n_A^0 x$$
$$\underline{n_I = n_I}$$
$$n_T = n_T^0 \tag{2-26}$$

Thus

$$n_A^0 \frac{dx}{dV} = \frac{k_p \left(n_A^0 - n_A^0 x \right) P}{n_T^0} - \frac{k_p' \left(n_{Ra}^0 + n_A^0 x \right) P}{n_T^0} \tag{2-27}$$

or

$$\frac{dx}{dV} = \frac{k_p' P}{n_T^0} \left[K_p (1 - x) - (\beta + x) \right] \tag{2-28}$$

where

$$\beta = \frac{n_R^0}{n_A^0}$$

Separation of variables followed by integration and rearrangement yields

$$V = \left[\frac{n_T^0}{Pk_p'(K_p + 1)} \right] \log_e \frac{K_p - \beta}{(K_p - \beta) - (K_p + 1)x} \tag{2-29}$$

An alternate form of this expression may be obtained by introducing the conversion at thermodynamic equilibrium (see Problem 2-5). Expressions for reactor volume as a function of conversion for other reactions are given in Table 2-3.

5. Determination of the Order and Rate Constant for a Reaction. The use of plug-flow reactors for the determination of the order and rate constant for a unidirectional reaction is similar to that described for batch reactors. For example, suppose the stoichiometry is satisfied by

$$NA \rightarrow NR$$

for any $N > 0$. The problem is to find the N that satisfies the rate expression

$$r_A = k_p p_A^N$$

For $N = 1$, 2, and 3, the corresponding integral expressions are given in Table 2-2. The plots of these expressions are presented in Figure 2-2. Note, however, for any one sequence of runs, P and y_A^0 must be held fixed; otherwise, a family of curves may be obtained since both P and y_A^0 are parameters. Furthermore, the ratio V/n_T^0, which is plotted on the abscissa in Figure 2-2, may be varied experimentally by varying either the reactor volume V or the feed rate n_T^0 to the reactor.

The reaction rate is said to be first order if the experimental data fall on a straight line when plotted as indicated for a first-order reaction in Figure 2-2. If, instead of being first order, the reaction rate is second or third order, then a straight line is obtained when the data are plotted as indicated for second- and third-order reactions in Figure 2-2. Fractional order and complex reactions are considered in a subsequent chapter.

EXAMPLE 2-4

Fisher and Smith [6] investigated the formation of carbon disulfide from methane and sulfur by the reaction

$$CH_4 + 2S_2 \rightarrow CS_2 + 2H_2S$$

in a flow reactor that consisted of a stainless steel pipe 1 in. in diameter and 6 in. long. The reactor was packed with varying amounts of rock salt to promote heat transfer and

TABLE 2-3
DESIGN EQUATIONS FOR CARRYING OUT REVERSIBLE REACTIONS IN THE GAS PHASE IN PLUG-FLOW REACTORS
(Conditions: isothermal and steady-state operation, negligible pressure drop, and perfect gas behavior)

Reaction	Design Equation
1. $A \underset{k'_p}{\overset{k_p}{\rightleftarrows}} 2R$	$$\frac{k_p PV}{K_p n_T^0} = \frac{(y_A^0)^2 x}{a} + \left[2y_A^0 - \frac{(y_A^0)^2 b}{a}\right]\int_0^x \frac{x\,dx}{ax^2+bx+c} + \left[1 - \frac{(y_A^0)^2 c}{a}\right]\int_0^x \frac{dx}{ax^2+bx+c}$$ where $a = -y_A^0(4P+K_p)$, $\;\; b = K_p(y_A^0-1) - 4y_A^0\beta P$ $$c = K_p - y_A^0 P\beta^2, \quad y_A^0 = \frac{n_A^0}{n_T^0}, \quad \beta = \frac{n_R^0}{n_A^0}$$
2. $2A \underset{k'_p}{\overset{k_p}{\rightleftarrows}} 2R$	$$\frac{k_p(y_A^0 P)^2 V}{n_A^0} = \int_0^x \frac{dx}{ax^2+bx+c}$$ where $a = 1 - \dfrac{1}{K_p}, \;\; b = -\left(\dfrac{2\beta}{K_p}+2\right), \;\; c = 1 - \dfrac{\beta^2}{K_p}, \;\; \beta = \dfrac{n_R^0}{n_A^0}$
3. $A + B \underset{k'_p}{\overset{k_p}{\rightleftarrows}} 2R$ $n_A^0 \neq n_B^0$	$$\frac{k_p(y_A^0 P)^2 V}{n_A^0} = \int_0^x \frac{dx}{ax^2+bx+c}$$ where $a = 1 - \dfrac{4}{K_p}, \;\; b = -(1+\gamma) - \dfrac{4\beta}{K_p}, \;\; c = \gamma - \dfrac{\beta^2}{K_p}, \;\; \gamma = \dfrac{n_B^0}{n_A^0}, \;\; \beta = \dfrac{n_R^0}{n_A^0}$

TABLE 2-3 (*Continued*)

4. $A + B \xrightleftharpoons{} R$

$$\frac{k_p y_A^0 PV}{K_p n_A^0} = \int_0^x \frac{(1 - y_A^0 x)^2}{ax^2 + bx + c} \, dx$$

$n_B^0 \neq n_A^0$
$n_T^0 = n_A^0 + n_B^0 + n_R^0 + n_I^0$

where $a = y_A^0(PK_p + 1)$, $b = -PK_p y_A^0(\gamma + 1) - (1 - \beta y_A^0)$, $c = y_A^0 K_p P\gamma - \beta$

$$\beta = \frac{n_B^0}{n_A^0}, \quad \gamma = \frac{n_R^0}{n_A^0}, \quad K_p = \frac{k_p}{k_p'}$$

The integrated forms of the above integrals are:

$$\int_0^x \frac{x \, dx}{ax^2 + bx + c} = \frac{1}{2c} \log_e\left(\frac{ax^2 + bx + c}{c}\right) - \frac{b}{2a} \int_0^x \frac{dx}{ax^2 + bx + c}$$

$$\int_0^x \frac{x^2 \, dx}{ax^2 + bx + c} = \frac{x}{a} - \frac{b}{2a^2} \log_e\left(\frac{ax^2 + bx + c}{c}\right) + \frac{b^2 - 4ac}{2a^2} \int_0^x \frac{dx}{ax^2 + bx + c}$$

where the integrated forms of the integral

$$\int_0^x \frac{dx}{ax^2 + bx + c}$$

are given in Table 1-2.

65

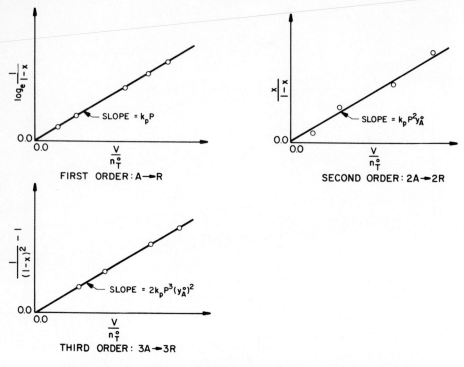

Figure 2-2. Determination of the order and rate constant for gas phase reactions from data taken with a plug-flow reactor.

to ensure homogeneity of the reacting mixture. The following data were obtained at 600°C and 1 atm pressure.

Void Volume of Reactor (ml)	Methane Fed (g mol/h)	Conversion of Methane
67	0.238	0.105
67	0.417	0.075
67	0.119	0.180
35.2	0.0595	0.133
35.2	0.02975	0.269
35.2	0.119	0.058
35.2	0.0298	0.268
35.2	0.238	0.025
35.2	0.119	0.066
35.2	0.0595	0.144
35.2	0.119	0.060

In all runs, the feed consisted of methane and sulfur alone in the stoichiometric ratio of one to two. Evaluate the rate constant in the following rate expression for the disappearance of sulfur:

$$r_{S_2} = k_c C_{CH_4} C_{S_2}$$

Assume that the reactor is described by the plug-flow model.

SOLUTION

For convenience, let the chemical species in the reaction be denoted by A, B, C, and D to give

$$A + 2B \rightarrow C + 2D$$

Then

$$r_B = k_c C_A C_B = 2r_A$$

and the stoichiometric relationships are

$$n_A = n_A^0 - n_A^0 x$$

$$n_B = n_B^0 - 2n_A^0 x$$

$$n_C = \quad n_A^0 x$$

$$n_D = \quad 2n_A^0 x$$

$$n_T = n_T^0$$

Since $n_B^0 = 2n_A^0$ for all runs, the flow rate of B is twice that of A throughout the reactor, and

$$r_A = \frac{1}{2} r_B = \frac{1}{2} k_c C_A (2C_A) = k_c C_A^2 = k_c \frac{n_A^2}{v_T^2}$$

The expression for the outlet flow rate of A versus the reactor volume V given by item 2 of Table 2-2 also applies for the preceding reaction, since they are both second order and no change in the number of moles occurs with reaction. Thus

$$\frac{V}{n_T^0} = \left(\frac{1}{k_p P^2 y_A^0} \right) \left(\frac{x}{1 - x} \right)$$

Since $n_T^0 = 3n_A^0$, this expression may be rearranged to form

$$\frac{x}{1 - x} = \left(\frac{k_p P^2 y_A^0}{3} \right) \frac{V}{n_A^0}$$

A plot of $x/(1 - x)$ versus V/n_A^0 is shown in Figure 2-3. The straight line shown was found by the method of least squares (see Reference [11]). The slope of this line is equal to 2.96×10^{-4}. Thus

$$\frac{k_p P^2 y_A^0}{3} = 2.96 \times 10^{-4}$$

Since $P = 1$ atm, $y_A^0 = \frac{1}{3}$, it follows that

$$k_p = 26.64 \times 10^{-4} \text{ g mol}/(\text{ml h atm}^2)$$

and, from Equation (1-26),

$$k_c = k_p (RT)^2 = (26.64 \times 10^{-4})[(82.05)(873.15)]^2 = 13.67 \times 10^6 \left(\frac{\text{ml}}{\text{g mol h}} \right)$$

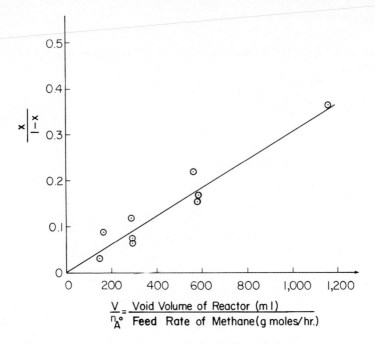

$$\frac{V}{n_A^0} = \frac{\text{Void Volume of Reactor (m l)}}{\text{Feed Rate of Methane (g moles/hr.)}}$$

Figure 2-3. Evaluation of the rate constant for Example 2-4.

Instead of using the ratio V/n_T^0 (or V/n_A^0 as was done in the preceding example) as an independent variable, it has become increasingly popular to use the variable called space velocity.

The *gas space velocity* S_V is defined as follows:

$$S_V = \frac{\text{volumetric flow rate of feed as a gas at } P_s \text{ and } T_s}{\text{reactor volume}} \qquad (2\text{-}30)$$

where P_s, T_s = standard conditions, which may be selected arbitrarily. Suppose that P_s and T_s are selected such that the feed mixture obeys the perfect gas law. Then

$$S_V = \frac{n_T^0 (RT_s/P_s)}{V} \qquad (2\text{-}31)$$

and

$$\frac{V}{n_T^0} = \left(\frac{RT_s}{P_s} \right) \left(\frac{1}{S_V} \right) \qquad (2\text{-}32)$$

Thus the integral expressions given in Table 2-2 may be stated in terms of $1/S_V$, which is frequently referred to as the *space time*. For example, for the first-order reaction $A \rightarrow R$, the integral expression is given by

$$\frac{1}{S_V} = \frac{P_s/RT_s}{k_p P} \log_e \frac{1}{1 - x} \qquad (2\text{-}33)$$

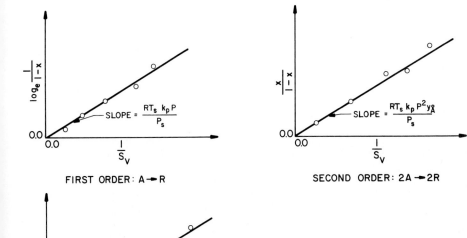

Figure 2-4. Use of the gas space velocity and data taken with a plug-flow reactor in the determination of the order and rate constant for a gas phase reaction.

When gas space velocities are used, the plots for first-, second-, and third-order reactions are as shown in Figure 2-4.

Alternately, one may choose P_s and T_s such that the feed is all liquid even though the reaction occurs in the gas phase. The *liquid space velocity* S_L is defined in the following manner.

$$S_L = \frac{\text{volumetric flow rate of the feed in the liquid state at } P_s \text{ and } T_s}{\text{reactor volume}} \quad (2\text{-}34)$$

Then

$$S_L = \frac{n_T^0}{CV} \quad (2\text{-}35)$$

where C = molar density of feed in liquid state at P_s and T_s, moles per unit volume

When the liquid space velocity is employed, the integral expression for the first-order gas phase reaction $A \rightarrow R$ becomes

$$\frac{1}{S_L} = \left(\frac{C}{k_p P} \right) \log_e \frac{1}{1 - x} \tag{2-36}$$

EXAMPLE 2-5

Kunzru et al. [12] investigated the thermal cracking of 2-pentene over the temperature range of 670 to 725°C. At 670°C, the following data were obtained:

	Set I	Set II
Temperature, °C	670	670
$1/S_V$, s	0.26	0.87
Conversion, %	8.8	22.0

Moles of product per 100 moles of reactant consumed		
H_2	20.6	27.1
CH_4	61.2	69.1
C_2H_4	15.6	20.6
C_2H_6	6.1	6.5
C_3H_6	12.4	13.2
$2\text{-}C_4H_8$	30.5	32.9
$1,3\text{-}C_4H_6$	10.8	15.9
C_5^+	14.0	9.1
Total	171.2	194.5

The data were obtained by use of a plug-flow reactor, and $1/S_V$ was based on a P_s and T_s equal to those used in the experiments. Determine the order of reaction and the corresponding value of the rate constant.

SOLUTION

Assume that the reaction rate is first order with respect to A:

$$A \rightarrow bB$$

where B represents the products and is referred to as a *lumped* product. Then, for plug flow, the following equation (for the case where the feed is pure A) is obtained in a manner analogous to that demonstrated for Equation (2-22).

$$-b \log_e (1 - x) - (b - 1)x = k_c \left(\frac{1}{S_V} \right)$$

The lumped stoichiometric coefficient b is equal to the moles of products produced per mole of A reacted. Thus, for the first set of data,

$$b = \frac{171.2}{100} = 1.712$$

and, for the second set,

$$b = \frac{194.5}{100} = 1.945$$

Then, for the first set of data, the left side of the preceding equation has the value

$$-1.712 \log_e(1 - 0.088) - (0.712)(0.088) = 0.0950$$

and, for the second set, the value

$$-1.945 \log_e(1 - 0.22) - (0.945)(0.22) = 0.2754$$

These data points are plotted in Figure 2-5. The fact that the two experimental points and the origin all fall on the same straight line suggests that the rate of reaction is first order with respect to A. Although Kunzru et al. present a minimum of four data points for each isotherm studied, only two sets of data were reported in tabular form. The additional points were included, however, on a graph. Since several reactions occurred in the system investigated by Kunzru et al., the lumped stoichiometry coefficient b varied from run to run. In particular, for their sets of experiments, b varied between 1.71 to 2.21.

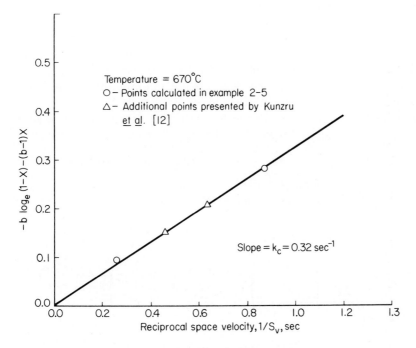

Figure 2-5. Use of the experimental results from a tubular reactor and the plug-flow model in the determination of the order and rate constant for the reactor in Example 2-5.

B. Perfectly Mixed Flow Reactors and Differential Reactors

A flow reactor of volume V is said to be perfectly mixed if the composition, temperature, and pressure are the same throughout each element of the volume V regardless of how small the element is taken to be.

Because of end effects which may produce appreciable backmixing, many tubular reactors closely approach the limiting condition of perfect mixing. Backmixing is quite pronounced where the length and diameter of a reactor are of the same order of magnitude. Also, if the diameters of the inlet and outlet are considerably smaller than the diameter of the reactor, considerable backmixing can be expected. A sketch of a perfectly mixed tubular reactor is shown in Figure 2-6. Tubular reactors which are partially backmixed are treated in a subsequent section.

Reactors having the geometry illustrated in Figure 2-7 are called *continuous stirred tank reactors*. If the mixing is perfect in all directions, then the reactor is called a *perfectly mixed flow reactor*. Again, the definition of a perfectly mixed reactor is the same as the one given previously. Continuous stirred tank reactors are used primarily for the purpose of controlling the reaction temperature. A continuous stirred tank reactor used for the polymerization of styrene is shown in Figure 2-8.

1. Material Balances for Perfectly Mixed Flow Reactors. For a flow reactor whose mixing effects are described by the perfectly mixed model, the composition, temperature, and pressure are uniform throughout the total volume V. Thus the rate of reaction r_A is also the same at every point throughout the volume V. The material balance on component A over the volume V (see Figures 2-6 and 2-7) at steady-state operation is given by

$$\underbrace{n_A^0}_{\substack{\text{input of } A \text{ to the} \\ \text{reactor per unit time}}} - \underbrace{[n_A + r_A V]}_{\substack{\text{output of } A \text{ by} \\ \text{flow and by reaction}}} = 0 \qquad (2\text{-}37)$$

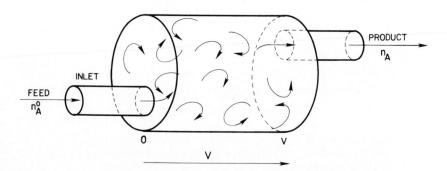

Figure 2-6. Perfectly mixed tubular reactor.

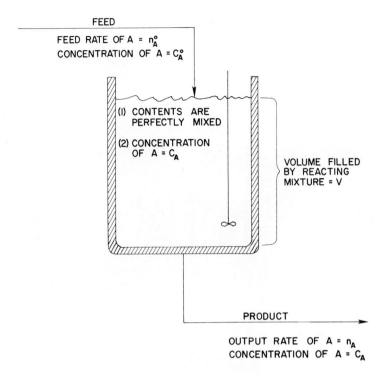

FEED

FEED RATE OF A = n_A^0
CONCENTRATION OF A = C_A^0

(1) CONTENTS ARE PERFECTLY MIXED

(2) CONCENTRATION OF A = C_A

VOLUME FILLED BY REACTING MIXTURE = V

PRODUCT

OUTPUT RATE OF A = n_A
CONCENTRATION OF A = C_A

Figure 2-7. Perfectly mixed tank reactor.

where n_A^0 = flow rate of A into the reactor, moles/time. This material balance on A may be restated in the following form, which is more convenient for subsequent use:

$$r_A = \frac{n_A^0 - n_A}{V} \tag{2-38}$$

2. Material Balances for Differential Reactors. Although differential reactors are used primarily in the study of reaction behavior as shown in Chapter 4, they are introduced here because the material-balance expressions for these reactors are of the same general form as those for perfectly mixed reactors. A differential reactor is defined as one in which the conversion per pass through the reactor is small as a consequence of a relatively small reactor volume to feed ratio, V/n_T^0. Integration of the material-balance expression [Equation (2-8)] followed by the application of the *mean-value theorem of integral calculus* gives

$$V = -\int_{n_A^0}^{n_A} \frac{dn_A}{r_A} = \left(\frac{1}{r_A}\right)_m \left(n_A^0 - n_A\right)$$

or

$$\left(r_A\right)_m = \frac{n_A^0 - n_A}{V} \tag{2-39}$$

Figure 2-8. An experimental glass reactor used in the study of the kinetics of the polymerization of styrene (operated as batch or flow reactor).

which is seen to be of the same general form as Equation (2-38). The subscript m is used to denote the mean value of $1/r_A$. For small conversions per pass, the mean value of $1/r_A$ may be approximated with good accuracy by use of the arithmetic average of the values of $1/r_A$ at n_A and n_A^0, respectively; see Equation (4-106).

3. Volume versus Conversion for Perfectly Mixed Flow Reactors. For reactors of this type, which are perfectly mixed in all directions, the determination of the reactor volume as a function of conversion is simpled by the fact that at steady state the component material balance [Equation (2-38)] for reactant A is an algebraic equation.

Consider again the first-order reaction

$$A \rightarrow \text{products}$$

which is carried out in the liquid phase. Furthermore, suppose that the mass density is constant. Then the rate of reaction at every point within the reactor is given by

$$r_A = k_c C_A = k_c \frac{n_A}{v_T} \tag{2-40}$$

Elimination of r_A from Equations (2-38) and (2-40), followed by rearrangement, yields

$$V = \frac{v_T}{k_c}\left(\frac{n_A^0 - n_A}{n_A}\right) \tag{2-41}$$

or

$$n_A = \frac{n_A^0}{1 + N_{Da}} \quad \text{and} \quad x = \frac{N_{Da}}{1 + N_{Da}} \tag{2-42}$$

where, again, $N_{Da} = k_c V / v_T$. Expressions for the volumes required to achieve a given conversion are presented in Table 2-4. These relationships are valid for reactions which are carried out in either the vapor or liquid phase.

For gas phase reactions, expressions relating reactor volume and conversion are developed as demonstrated by the following example. Consider the unidirectional gas phase reaction

$$A \rightarrow 2R$$

TABLE 2-4

DESIGN EQUATIONS FOR CARRYING OUT UNIDIRECTIONAL AND REVERSIBLE REACTIONS
IN THE LIQUID OR VAPOR PHASE IN A PERFECTLY MIXED FLOW REACTOR
AT STEADY-STATE OPERATION

Reactions	Design Equation
1. $2A \rightarrow R$ $n_T^0 = n_A^0 + n_{Da}^0 + n_I^0$	$\dfrac{V}{v_T} = \left[\dfrac{v_T}{n_A^0 k_c}\right]\left[\dfrac{x}{(1-x)^2}\right]$
2. $A + B \rightarrow R$ $n_T^0 = n_A^0 + n_B^0 + n_{Da}^0 + n_I^0$ $n_A^0 \neq n_B^0$	$\dfrac{V}{v_T} = \left[\dfrac{v_T}{n_A^0 k_c}\right]\left[\dfrac{x}{(1-x)(\beta-x)}\right], \quad \beta = \dfrac{n_B^0}{n_A^0}$
3. $3A \rightarrow R + S$ $n_T^0 = n_A^0 + n_{Da}^0 + n_S^0$	$\dfrac{V}{v_T} = \dfrac{v_T^2}{\left(n_A^0\right)^2 k_c}\left[\dfrac{x}{(1-x)^3}\right]$
4. $A \overset{k_c}{\underset{k_c'}{\rightleftharpoons}} R$ $n_T^0 = n_A^0 + n_{Da}^0 + n_I^0$	$\dfrac{V}{v_T} = \dfrac{x}{k_c\{[1 - \beta/K_c] - [1 + 1/K_c]x\}}, \quad K_c = \dfrac{k_c}{k_c'}$
5. $A \rightarrow 3R$	$\dfrac{V}{v_T} = \dfrac{1}{k_c}\left[\dfrac{x}{1-x}\right]$

where v_T is the flow rate at the outlet of the reactor

whose rate is given by

$$r_A = k_p p_A$$

Again the component material balance on A is given by

$$r_A = \frac{n_A^0 - n_A}{V} = \frac{n_A^0 x}{V}$$

Elimination of r_A from these two expressions, followed by rearrangement, gives

$$\frac{V}{n_T^0} = \frac{y_A^0 x}{k_p p_A} \qquad (2\text{-}43)$$

By use of the stoichiometric relationship

$$n_A = n_A^0 - n_A^0 x$$

$$n_R = n_R^0 + 2n_A^0 x$$

$$n_T = n_T^0 + n_A^0 x$$

the partial pressure p_A may be stated in terms of conversion,

$$p_A = \frac{n_A}{n_T} P = \frac{y_A^0 (1 - x) P}{(1 + y_A^0 x)} \qquad (2\text{-}44)$$

for a perfect gas mixture. Elimination of p_A from Equations (2-43) and (2-44), followed by rearrangement, gives

$$\frac{V k_p P}{n_T^0} = \frac{x(1 + y_A^0 x)}{1 - x} \qquad (2\text{-}45)$$

The expressions for V as a function of conversion for other gas phase reactions are presented in Table 2-5.

EXAMPLE 2-6

Hinshelwood and Askey [13] studied the thermal decomposition of dimethyl ether in a batch reactor. For pressures above 400 mm Hg, the decomposition was first order and proceeded as follows:

$$(CH_3)_2O \rightarrow [CH_4 + CH_2O] \rightarrow CH_4 + CO + H_2$$

The presence of formaldehyde was transitory in that it decomposed to carbon monoxide and hydrogen as soon as it was formed. The expression found for the rate constant for the first-order decomposition of dimethyl ether was

$$\log_e k_c = 30.36 - \frac{58,500}{RT}$$

where k_c has the units of s^{-1}, T is in K, and $R = 1.9872$ cal/(g mol K).

TABLE 2-5

DESIGN EQUATIONS FOR CARRYING OUT UNIDIRECTIONAL AND REVERSIBLE REACTIONS IN THE GAS PHASE IN A PERFECTLY MIXED FLOW REACTOR AT STEADY-STATE OPERATION

Reaction	Design Equation
1. $2A \rightarrow R$ $n_T^0 = n_A^0 + n_R^0 + n_I^0$	$\dfrac{k_p \left(y_A^0 P \right)^2 V}{n_A^0} = \dfrac{x \left[1 - (1/2) y_A^0 x \right]^2}{(1-x)^2}, \quad y_A^0 = \dfrac{n_A^0}{n_T^0}$
2. $A + B \rightarrow R$	$\dfrac{k_p \left(y_A^0 P \right)^2 V}{n_A^0} = \dfrac{x \left(1 - y_A^0 x \right)^2}{(1-x)(\beta - x)}, \quad \beta = \dfrac{n_B^0}{n_A^0}$
3. $3A \rightarrow R + S$	$\dfrac{k_p \left(y_A^0 P \right)^3 V}{n_A^0} = \dfrac{x \left(1 - (1/3) y_A^0 x \right)^3}{(1-x)^3}$
4. $A \underset{k_p'}{\overset{k_p}{\rightleftharpoons}} R$	$\dfrac{k_p \left(y_A^0 P \right) V}{n_A^0} = \dfrac{x}{\left[\left(1 - \beta/K_p \right) - \left(1 + 1/K_p \right) x \right]}, \quad \gamma = \dfrac{n_R^0}{n_A^0}$
5. $2A \underset{k_p'}{\overset{k_p}{\rightleftharpoons}} R$	$\dfrac{V \left(y_A^0 P \right)^2 k_p}{n_A^0} = \dfrac{x \left(1 - (1/2) y_A^0 x \right)^2}{x^2 \left[1 + (1/4 K_p P) \right] - x \left[2 - (1/2 P K_p) \left(\gamma - (1/2 y_A^0) \right) \right] + \left(1 - \dfrac{\gamma}{y_A^0 P K_p} \right)}$ $K_p = \dfrac{k_p}{k_p'}, \quad \gamma = \dfrac{n_R^0}{n_A^0}$

It is proposed to carry this same reaction out in a perfectly mixed flow reactor operated at 550°C and 1 atm pressure. Calculate the respective space velocities (evaluated at the temperature and pressure of the reactor) which are required to achieve conversions of 20%, 50%, and 80% of dimethyl ether in the perfectly mixed flow reactor. Assume that the mixture behaves as a perfect gas.

SOLUTION

Let the reactant dimethyl ether be denoted by A and the products methane, carbon monoxide, and hydrogen by B, C, and D, respectively, to give

$$A \rightarrow B + C + D$$

The stoichiometric relationships are

$$n_A = n_A^0 - n_A^0 x$$

$$n_B = n_A^0 x$$

$$n_C = n_A^0 x$$

$$n_D = n_A^0 x$$

$$\overline{n_T = n_A^0 + 2n_A^0 x}$$

For a perfectly mixed flow reactor in which the preceding first-order reaction is carried out,

$$r_A = \frac{n_A^0 x}{V} = k_c C_A = k_c \frac{n_A}{v_T} = \frac{k_c n_A^0 (1 - x)}{v_T}$$

For a perfect gas mixture,

$$\frac{n_T}{n_T^0} = \frac{v_T}{v_T^0}, \quad v_T = v_T^0 \frac{n_T}{n_T^0} = v_T^0 (1 + 2x)$$

Thus

$$\frac{v_T^0}{V} = S_V = \frac{k_c(1 - x)}{x(1 + 2x)}$$

At 550°C,

$$\log_e k_c = 30.36 - \frac{58{,}500}{(1.9872)(823.15)}$$

$$k_c = 4.50 \times 10^{-3} \text{ s}^{-1}$$

Thus, for $x = 0.2$, 0.5, and 0.8, the space velocities are computed as follows:

x	$\dfrac{1 - x}{x(1 + 2x)}$	$S_V = \dfrac{k_c(1 - x)}{x(1 + 2x)}$ (s^{-1})	$\dfrac{1}{S_V}$ (s)
0.2	2.857	12.857×10^{-3}	77.78
0.5	0.5	2.250×10^{-3}	444.4
0.8	0.09615	0.433×10^{-3}	2309

A perfectly mixed flow reactor may be used to determine the order and the rate constant for a reaction. For example, if the reaction $A \rightarrow 2R$ (considered previously) is first order, the mixing is perfect, and the perfect gas law is obeyed, a plot of V/n_T^0 versus the right side of Equation (2-45) should give a straight line with the slope $1/k_p P$. The experimental points may be determined by varying either the reactor volume or the total flow rate of the feed, n_T^0, with the composition of the feed held fixed for all runs.

C. Comparison of the Plug-Flow and the Perfectly Mixed Flow Reactor Models

The difference between the plug-flow and the perfectly mixed models is perhaps best illustrated by use of Figure 2-9, which shows the concentration profiles predicted by each of these models for a first-order reaction in a tubular reactor. More precisely, the curves were computed for the case where the first-order reaction

$$A \rightarrow \text{products}$$

is carried out isothermally in the liquid phase at constant mass density with the parameter k_c/v_T fixed such that at $V = V_T$ (the total volume of the reactor), the Damköhler number $N_{Da} = 2$. (It should be noted that in the interest of convenience the symbol V is used throughout this book to represent both the variable and the total volume of the reactor. However, when both of these quantities appear in the same set of equations, the variable is denoted by V and the total volume by V_T.)

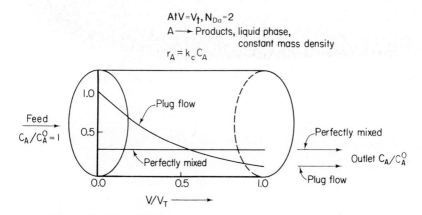

Figure 2-9. Concentration profiles predicted by the plug-flow and perfectly mixed models for a first-order liquid phase reaction carried out in a tubular reactor of volume V_T and at a reaction number, $N_{Da} = k_c V_T/v_T = 2$.

For this case, Equation (2-17) for a plug-flow reactor becomes

$$C_A = C_A^0 \exp(-k_c\theta) = C_A^0 \exp\left(-k_c \frac{V}{v_T}\right) = C_A^0 \exp\left(-k_c \frac{V_T}{v_T} \cdot \frac{V}{V_T}\right)$$

$$= C_A^0 \exp\left(-N_{Da} \frac{V}{V_T}\right) = C_A^0 \exp\left(-2\frac{V}{V_T}\right)$$

For a Damköhler number equal to 2, the concentration profile is shown in Figure 2-9. Observe that at the outlet of the plug-flow reactor $V = V_T$, and

$$C_A = C_A^0 e^{-N_{Da}} = C_A^0 e^{-2} = 0.1353 C_A^0$$

For the perfectly mixed reactor, the concentration of A at each point within the reactor is equal to the concentration of A in the stream leaving the reactor. The exit concentration is found by first restating Equation (2-41) in the form

$$C_A = \frac{C_A^0}{1 + N_{Da}}$$

and then, for $N_{Da} = 2$,

$$C_A = \frac{C_A^0}{3}$$

Now suppose that the reaction considered is carried out in two perfectly mixed reactors in series with reactor volumes equal to one-half the volume of the tubular reactor. The reaction number for each of the perfectly mixed reactors is then equal to one-half of the N_{Da} at $V = V_T$ for the tubular reactor. Then the terminal composition, C_{A1}, of the first perfectly mixed reactor is given by

$$C_{A1} = \frac{C_A^0}{1 + N_{Da1}} = \frac{C_A^0}{1 + N_{Da}/2} = \frac{C_A^0}{2}$$

For the second perfectly mixed reactor, the outlet concentration, C_{A2}, is given by

$$C_{A2} = \frac{C_{A1}}{1 + N_{Da2}} = \frac{C_{A1}}{1 + N_{Da}/2} = \frac{C_A^0}{(1 + N_{Da}/2)^2} = \frac{C_A^0}{4}$$

A comparison of the concentration profiles given by the two perfectly mixed reactors with the concentration profile for the plug-flow reactor is depicted in Figure 2-10. Comparison of Figures 2-9 and 2-10 suggests that, as the number of perfectly mixed reactors is increased, the terminal concentration from the last reactor of the series tends to approach the outlet concentration of a plug-flow reactor having a total volume equal to the sum of the volumes of the perfectly mixed reactors.

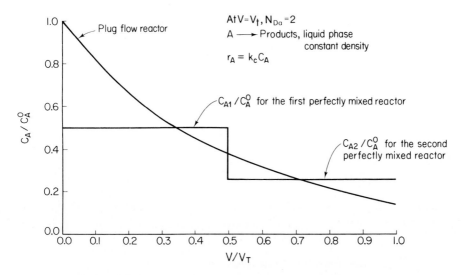

Figure 2-10. Comparison of the concentration profile for two perfectly mixed reactors with the concentration profile for a single plug-flow reactor.

An expression for the concentration of A in the effluent stream from the last reactor in a series of N reactors is developed as follows. By use of the previous procedure to compute C_{A1} and C_{A2}, it follows that, for N perfectly mixed reactors connected in series in which the same first-order reaction considered previously occurs,

$$C_{A1} = \frac{C_A^0}{1 + N_{Da}/N}$$

$$C_{A2} = \frac{C_{A1}}{1 + N_{Da}/N} = \frac{C_A^0}{(1 + N_{Da}/N)^2}$$

$$C_{A, N-1} = \frac{C_{A, N-2}}{1 + N_{Da}/N} = \frac{C_{A, N-3}}{(1 + N_{Da}/N)^2} = \cdots = \frac{C_A^0}{(1 + N_{Da}/N)^{N-1}} \quad (2\text{-}46)$$

$$C_{A, N} = \frac{C_{A, N-1}}{1 + N_{Da}/N} = \frac{C_{A, N-2}}{(1 + N_{Da}/N)^2} = \cdots = \frac{C_A^0}{(1 + N_{Da}/N)^N}$$

where N_{Da} is based on the total volume of the N reactors.

To find the limit of $C_{A, N}$ as the number of reactors N is increased without bound while the total volume of the reactor is held fixed, it is convenient to restate Equation (2-46) in the following form:

$$\log_e \frac{C_{A,N}}{C_A^0} = \frac{-\log_e(1 + N_{Da}/N)}{1/N}$$

As N is increased without bound, the term on the right side becomes indeterminate. Application of l'Hôpital's rule gives

$$\lim_{N \to \infty} \log_e \frac{C_{A,N}}{C_A^0} = \lim_{N \to \infty} \left[\frac{-\log_e(1 + N_{Da}/N)}{1/N} \right] = -N_{Da} \qquad (2\text{-}47)$$

Upon comparison of Equations (2-15) and (2-47), it is evident that the concentration $C_{A,N}$ of A in the stream leaving the Nth perfectly mixed reactor is equal to the concentration of A in the terminal stream leaving a plug-flow reactor as the number of perfectly mixed reactors is increased without bound.

D. Plug-Flow Reactors with Recycle

To maintain adequate control of the reaction temperature, it may be necessary to operate at a small conversion per pass of the reacting stream through the reactor. Yet from an economic point of view a relatively high conversion of the feed may be desirable. By recycling of a portion of the product stream back through the reactor, both of these objectives (a small conversion per pass with a relatively large conversion of the feed) may be achieved.

Commercial reactors with recycle are also used to increase reactivity and shorten reaction time by efficiently removing the heat of reaction and improving mass transfer when more than one phase is involved, as described by Leuteritz et al. [14]. Also, the yield of the desired products may be improved in certain cases where complex reactions are involved, as demonstrated by Carberry [15].

A sketch of a tubular reactor with a recycle stream is shown in Figure 2-11. Most of the notation to be used in the analysis of reactors with recycle streams is also shown in Figure 2-11. The fraction of the product stream which is recycled to the inlet of the reactor is denoted by α, and the fraction of the product stream which leaves the system is denoted by $1 - \alpha$. The symbols are perhaps best described by following component A through the system. The molar flow rates n_A^0, n_{Ai}, and n_{Af} denote the rates at which component A enters the system, enters the reactor, and leaves the reactor, respectively. The molar flow rate of A at any point in the reactor is denoted by n_A.

1. Conversion per Pass. The equations developed for conversion versus volume for plug-flow reactors (see Tables 2-1, 2-2, and 2-3) are applicable to reactors with recycle provided that the superscript zero on all symbols is replaced by the subscript i and the subscript i is added to the symbol for the volumetric flow rate to give v_{Ti}. Also, in the interest of clarity, for reactors

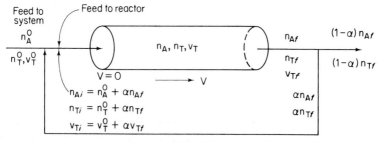

Figure 2-11. Sketch of a tubular reactor with recycle.

with recycle all values of the variable at the outlet of the reactor are identified by the symbol f.

To illustrate the treatment of plug-flow reactors in recycle systems, the equations are developed for two relatively simple cases. Consider again the first-order reaction

$$A \rightarrow \text{products}$$

which is carried out in the liquid phase at constant mass density. Again the limit of a material balance on A over the element of volume ΔV in the reactor shown in Figure 2-11 takes the form

$$r_A = -\frac{dn_A}{dV}$$

as the element over which the balance is made is allowed to go to zero. Elimination of r_A from this material-balance equation by use of the rate expression $r_A = k_c C_A$ gives

$$-\frac{dn_A}{dV} = k_c C_A = k_c \frac{n_A}{v_T}$$

Since the mass density is constant, $v_{Ti} = v_{Tf} = v_T$, the variables may be separated, and the resulting equation integrated.

$$-\int_{n_A = n_{Ai}}^{n_A = n_{Af}} \frac{dn_A}{n_A} = \frac{k_c}{V_{ti}} \int_{V=0}^{V=V_T} dV$$

The subscript T has been added to the volume to emphasize that the upper limit is equal to the total volume of the reactor. After the indicated integration has been performed, the result so obtained may be rearranged to the form given by Equation (2-14) or the first expression of Table 2-1:

$$n_{Af} = n_{Ai} \exp\left(\frac{-k_c V_T}{v_{Ti}}\right) \tag{2-48}$$

or

$$V_T = \frac{v_{Ti}}{k_c} \log_e \frac{n_{Ai}}{n_{Af}} \tag{2-49}$$

Consider next the case where the reaction

$$A \to R$$

is carried out in the gas phase in a plug-flow reactor with recycle as shown in Figure 2-11. Let the conversion at any V of the reactor be defined in the usual way as the moles of A reacted per mole of A fed to the reactor; that is,

$$x = \frac{n_{Ai} - n_A}{n_{Ai}} \qquad (2\text{-}50)$$

Then the stoichiometric relationships follow:

$$n_A = n_{Ai} - n_{Ai}x$$
$$n_R = n_{Ri} + n_{Ai}x \qquad (2\text{-}51)$$
$$n_T = n_{Ti}$$

By use of Equation (2-50), the material balance on A may be stated in terms of x,

$$n_{Ai}\frac{dx}{dV} = r_A \qquad (2\text{-}52)$$

Elimination of r_A by use of the rate expression $r_A = k_p p_A$ gives

$$n_{Ai}\frac{dx}{dV} = k_p p_A = k_p\frac{(n_{Ai} - n_{Ai}x)P}{n_{Ti}} \qquad (2\text{-}53)$$

where it is assumed that the mixture obeys the perfect gas law. Then

$$\int_{x=0}^{x=x_f} \frac{dx}{(1-x)} = \frac{k_p P}{n_{Ti}}\int_0^{V_T} dV \qquad (2\text{-}54)$$

and integration, followed by rearrangement, gives

$$V_T = \frac{n_{Ti}}{k_p P}\log_e\frac{1}{1 - x_f} \qquad (2\text{-}55)$$

which is seen to be the same form as the first expression of Table 2-2.

In conclusion then, the equations given in Tables 2-1, 2-2, and 2-3 for reactors without recycle also apply to reactors with recycle provided that the appropriate changes are made in the subscripts and superscripts.

2. Total Conversion for the System. Let the total conversion χ for the system be defined as the moles of the base component A reacted in the system per mole of A that enters the system. Component A enters the system at the rate n_A^0 and leaves at the rate $(1 - \alpha)n_{Af}$. Then

$$\chi = \frac{n_A^0 - (1 - \alpha)n_{Af}}{n_A^0} \qquad (2\text{-}56)$$

Since $n_A^0 \chi$ is equal to the moles of A reacted in the system per unit time, the following stoichiometric relationships for the system are readily deduced:

$$(1 - \alpha)n_{Af} = n_A^0 - n_A^0 \chi$$

$$(1 - \alpha)n_{Rf} = n_R^0 + n_A^0 \chi \qquad (2\text{-}57)$$

$$(1 - \alpha)n_{Tf} = n_T^0$$

The conversion per pass and the total conversion for the system may be related by commencing with Equations (2-50) and (2-56), and the fact that $n_{Ai} = n_A^0 + \alpha n_{Af}$:

$$x_f = \frac{(1 - \alpha)\chi}{1 - \alpha\chi} \qquad (2\text{-}58)$$

or

$$\chi = \frac{x_f}{1 - \alpha(1 - x_f)} \qquad (2\text{-}59)$$

3. Variation of the Conversion per Pass and the Total Conversion with the Fraction of the Product Stream Recycled. Equation (2-55) may be restated in terms of the total conversion and n_T^0 by use of Equation (2-58) and the fact that

$$n_{Ti} = n_T^0 + \alpha n_{Tf} = n_T^0 + \left(\frac{\alpha}{1 - \alpha}\right)n_T^0 = \frac{n_T^0}{1 - \alpha} \qquad (2\text{-}60)$$

The result so obtained may be solved for the total conversion to give

$$\chi = \frac{1 - \exp[-N_{Da}(1 - \alpha)]}{1 - \alpha\exp[-N_{Da}(1 - \alpha)]} \qquad (2\text{-}61)$$

In this case

$$N_{Da} = \frac{V_T k_p P}{n_T^0} \qquad (2\text{-}62)$$

At $\alpha = 0$, Equation (2-61) reduces to

$$\chi = 1 - e^{-N_{Da}} \qquad (2\text{-}63)$$

which is the equation for a plug-flow reactor without recycle.

As $\alpha \to 1$, the right side of Equation (2-61) becomes indeterminate. This indeterminate form may be evaluated by use of l'Hôpital's rule:

$$\lim_{\alpha \to 1} \chi = \lim_{\alpha \to 1} \frac{1 - \exp[-N_{Da}(1 - \alpha)]}{1 - \alpha\exp[-N_{Da}(1 - \alpha)]} = \frac{N_{Da}}{1 + N_{Da}} \qquad (2\text{-}64)$$

Comparison of this result with Equation (2-42) shows that as $\alpha \to 1$ the system conversion χ approaches that which could be obtained by operating a perfectly mixed reactor with the same total volume V_T and total feed rate n_T^0 as those used for the plug-flow reactor of the system.

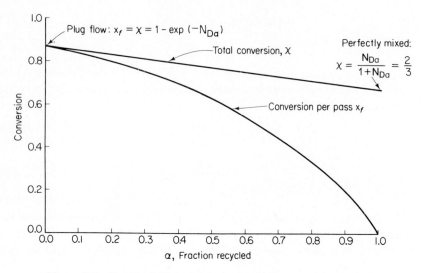

Figure 2-12. Variation of the total conversion of the system and the conversion per pass with the fraction of the product stream recycled.

The results obtained are summarized by the graphs of χ and x_f versus α, which are shown in Figure 2-12.

E. Laminar-Flow Reactors

When the fluid passes through the reactor in laminar flow, there is no mixing in either the radial or the horizontal directions. As indicated in Figure 2-13, laminar flow may be visualized as consisting of a set of concentric cylinder walls of the incremental thickness Δr which pass through the reactor at successively greater velocities as the radii of the cylinders decrease. In the limit as Δr goes to zero, the linear velocity u at which a shell of fluid of radius r passes through the reactor is given by the following relationship, which may be found in standard texts on fluid dynamics:

$$ u = u_{max}\left[1 - \left(\frac{r}{R}\right)^2\right] \tag{2-65} $$

where r = radius of the reactor; measured from center, as in Figure 2-13,
         $0 \leq r \leq R$

u_{max} = maximum velocity

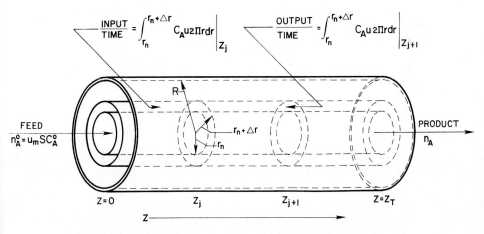

Figure 2-13. Laminar flow of the reacting stream through a tubular reactor.

The maximum velocity occurs at $r = 0$ and is given by the following form of the Poiseuille equation:

$$u_{max} = \left(\frac{P_1 - P_2}{4\mu z_T}\right) R^2 \tag{2-66}$$

where P_1 is the pressure at the inlet of the reactor, P_2 is the pressure at the outlet of the reactor, and μ is the viscosity, where all quantities are expressed in appropriate units. For laminar flow through circular tubes, the average velocity u_m is related to the volumetric flow rate v_T and the maximum velocity u_{max} as follows

$$u_m = \frac{v_T}{S} = \frac{u_{max}}{2} \tag{2-67}$$

1. Material Balances for Laminar-Flow Reactors. Consider the element of volume $2\pi r_n \Delta r \Delta z$ shown in Figure 2-13, where z_j, Δz, r_n, and Δr are arbitrarily selected in the domains $[0 < z_j < (z_j + \Delta z) < z_T; 0 < r_n < (r_n + \Delta r) < R]$ of interest. A material balance on component A over this element of volume at steady-state operation is given by

$$\underbrace{\int_{r_n}^{r_n+\Delta r} C_A u 2\pi r\, dr\Big|_{z_j}}_{\text{input by flow}} - \underbrace{\int_{r_n}^{r_n+\Delta r} C_A u 2\pi r\, dr\Big|_{z_j+\Delta z}}_{\text{output by flow}} - \underbrace{\int_{z_j}^{z_j+\Delta z}\int_{r_n}^{r_n+\Delta r} r_A 2\pi r\, dr\, dz}_{\text{output by reaction}} = 0 \tag{2-68}$$

The first two integrals may be written as a single integral and the *mean-value theorem of differential calculus* applied to give

$$2\pi \int_{r_n}^{r_n+\Delta r}\left[(ruC_A)\big|_{z_j, r} - (ruC_A)\big|_{z_j+\Delta z, r}\right] dr$$

$$= -2\pi\Delta z \int_{r_n}^{r_n+\Delta r} \frac{\partial(ruC_A)}{\partial z}\bigg|_{z_j+\alpha(r)\cdot\Delta z, r} dr \tag{2-69}$$

where $\alpha(r)$ depends on r and $0 < \alpha(r) < 1$. Application of the *mean-value theorem of integral calculus* to the integral on the right side of Equation (2-69) follows:

$$-2\pi\Delta z \int_{r_n}^{r_n+\Delta r} \left.\frac{\partial(ruC_A)}{\partial z}\right|_{z_j+\alpha(r)\cdot\Delta z,\, r} dr = -2\pi\Delta z\,\Delta r \left.\frac{\partial(ruC_A)}{\partial z}\right|_{z_j+\alpha(r_p)\cdot\Delta z,\, r_p}$$

(2-70)

where

$$r_p = r_n + \gamma_1\Delta r, \qquad 0 \le \gamma_1 \le 1$$

Application of the *mean-value theorem of integral calculus* first with respect to r and then with respect to z to the double integral appearing in Equation (2-68) gives

$$+\Delta r \left| \int_{z_j}^{z_j+\Delta z}\int_{r_n}^{r_n+\Delta r} r_A 2\pi r\, dr\, dz = 2\pi\Delta r\Delta z(r_A r) \right|_{r_n+\beta(z_p)\Delta r,\, z_p}$$

(2-71)

where $0 \le \beta(z_p) \le 1$, $z_p = z_j + \gamma_2\Delta z$, and $0 \le \gamma_2 \le 1$. When the results given by Equations (2-70) and (2-71) are substituted into Equation (2-68), the following result is obtained:

$$-\left[\left.\frac{\partial(ruC_A)}{\partial z}\right]\right|_{z_j+\alpha(r_p)\cdot\Delta z,\, r_p} - (r_A r)|_{r_n+\beta(z_p)\cdot\Delta r,\, z_p} = 0$$

(2-72)

If Δz and Δr are allowed to go to zero in any order whatsoever, one obtains

$$\left[\left.\frac{\partial(ruC_A)}{\partial z}\right]\right|_{z_j,\, r_n} + (r_A r)|_{r_n,\, z_j} = 0$$

(2-73)

Since z_j and r_n were arbitrarily selected in the domains of interest, Equation (2-73) holds for all such choices of z_j and r_n in these domains. Thus the subscripts z_j and r_n may be removed and Equation (2-73) restated to give the following form of the material-balance expression for r_A:

$$r_A = -\frac{1}{r}\frac{\partial(ruC_A)}{\partial z} = -\frac{\partial(uC_A)}{\partial z} \qquad \begin{matrix}(0 < r < R)\\(0 < z < z_T)\end{matrix}$$

(2-74)

2. Variation of Conversion with Reactor Volume for First-Order Liquid Phase Reactions in Laminar Flow. (Other conditions: isothermal and steady-state operation, constant mass density, and constant viscosity.) At any given radius r, Equation (2-74) reduces to

$$r_A = -u\frac{dC_A}{dz}$$

(2-75)

where u at this fixed value of r is given by Equation (2-65) for all z. For the first-order reaction $r_A = k_c C_A$, Equation (2-75) may be solved to give

$$C_A = C_A^0\exp\left(-\frac{k_c z}{u}\right)$$

(2-76)

The rate at which A enters the reactor (see Figure 2-13) is given by

$$n_A^0 = u_m \pi R^2 C_A^0 \tag{2-77}$$

where $\qquad u_m = \dfrac{v_T}{S} = \dfrac{u_{max}}{2} \qquad$ [see Equation (2-67)]

The molar flow rate n_A at which A leaves the reactor shown in Figure 2-13 may be computed as follows:

$$n_A = \int_0^R C_A u(2\pi r\, dr) = 2\pi C_A^0 \int_0^R \left[\exp\left(\frac{-k_c z_T}{u} \right) \right] ur\, dr \tag{2-78}$$

where u as a function of r is given by Equation (2-65). The integral appearing in Equation (2-78) may be restated in the following form (see Problem 2-8):

$$n_A = n_A^0 \left[\left(1 - \frac{N_{Da}}{2} \right) e^{-N_{Da}/2} + \left(\frac{N_{Da}}{2} \right)^2 E\left(\frac{N_{Da}}{2} \right) \right] \tag{2-79}$$

where $N_{Da} = \dfrac{2k_c z_T}{u_{max}} = \dfrac{k_c z_T}{u_m} = \dfrac{k_c V}{v_T} = k_c \theta,$ the Damköhler number for a first-order reaction

The function $E(y)$ is defined by

$$E(y) = \int_y^\infty \frac{e^{-\phi}\, d\phi}{\phi} \tag{2-80}$$

The function $E(y)$ is commonly tabulated in standard sets of tables as $- Ei(-x)$ (see, for example, Reference [16]).

This general approach used to relate conversion and reaction number for the particular reaction considered may be used for other types of reactors, but it may be necessary to evaluate the integral in Équation (2-78) numerically.

As shown in a subsequent section, the conversion which can be obtained per pass by the laminar-flow model is less than that which can be obtained by use of a plug-flow model and more than that which can be obtained by use of the perfectly mixed model.

EXAMPLE 2-7

Styrene in the solvent dioxane is to be polymerized in a tubular reactor at 25°C by use of sodium naphthalene as the initiator. Laminar flow through the reactor may be assumed. The concentration of styrene in the feed is 1.0 g mol/liter. The polymerization reaction of styrene (denoted by A) in the presence of the initiator (denoted by I) may be represented by

$$A \xrightarrow{(I)} \text{polymer}$$

In the presence of 0.0005 g mol/liter of sodium naphthalene, the rate of reaction of styrene [17] is given by

$$r_A = kC_A$$

where

$$k = 1.7 \times 10^{-3} \text{ s}^{-1}$$

The mass density may be assumed to remain constant throughout the reactor. For a residence time of 20 min, calculate the conversion of styrene which can be expected in a laminar-flow reactor at the conditions outlined.

SOLUTION

First calculate the Damköhler number N_{Da} as follows:

$$N_{Da} = k\theta = (1.7 \times 10^{-3})(20)(60) = 2.04$$

Next the conversion is calculated by use of Equation (2-79) as follows:

$$\frac{n_A}{n_A^0} = 1 - x = \left[\left(1 - \frac{2.04}{2} \right) \exp\left(\frac{-2.04}{2} \right) + \left(\frac{2.04}{2} \right)^2 E\left(\frac{2.04}{2} \right) \right]$$

$$x = 1 - [-0.007212 + (1.0404) E(1.02)] = 0.7859$$

F. Partial Axial Mixing Model

A stream in turbulent flow is characterized by point velocities which deviate about a mean velocity, just as the velocities of the molecules of a body of fluid deviate about a mean velocity. Experience has shown that the molecules of, say, type A of such a system tend to have a net movement in the direction in which molecules of type A are least abundant. *Fick's first law* is a quantification of this experience for molecular diffusion. To account for the mass transfer resulting from molecular motion and turbulent motion, the following form of Fick's first law is commonly used:

$$J_A = -D \frac{dC_A}{dz} \qquad (2\text{-}81)$$

where C_A = concentration of A, moles per unit volume

   D = dispersion coefficient

   J_A = moles of component transferred in the positive direction of z per unit time per unit of area perpendicular to the direction of transfer

1. Material Balances. A sketch of a tubular reactor in which axial diffusion is considered is shown in Figure 2-14. The component material balance for component A over the element of volume $S \Delta z$ is given by

$$uSC_A|_{z_j} + SJ_A|_{z_j} - \left[uSC_A|_{z_{j+1}} + SJ_A|_{z_{j+1}} + \int_{z_j}^{z_{j+1}} r_A S \, dz \right] = 0 \qquad (2\text{-}82)$$

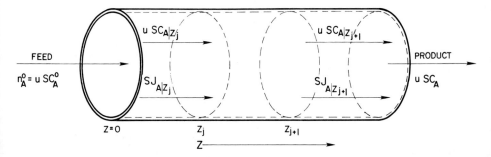

Figure 2-14. Partial axial mixing model: perfect mixing in the radial direction and partial mixing in the axial direction are assumed.

where $u = v_T/S$, the average linear velocity. Application of the mean-value theorems to Equation (2-82) followed by the limiting process, wherein Δz is allowed to go to zero, yields

$$\frac{d(uC_A)}{dz} + \frac{dJ_A}{dz} + r_A = 0 \tag{2-83}$$

Evaluation of dJ_A/dz by use of Equation (2-81), followed by rearrangement, gives the material balance for component A.

$$\frac{d(uC_A)}{dz} - D\frac{d^2C_A}{dz^2} + r_A = 0 \tag{2-84}$$

2. Variation of Conversion with Reactor Volume for a First-Order Reaction. (Other conditions: isothermal operation, constant mass density, and diffusivity.) To compare the behavior of the model for partial axial mixing with that exhibited by other mixing models, an expression for conversion versus reactor volume is developed for the case where the first-order reaction $A \rightarrow R$ is carried out in the liquid phase at constant mass density.

Suppose that the volumetric flow rate v_T is constant over all z. Since $v_T = uS$ and S is assumed to be the same for all z, it follows that u is a constant. Then, upon elimination of r_A from Equation (2-84) by use of the rate expression $r_A = k_cC_A$, the following result is obtained:

$$\frac{d^2C_A}{dz^2} - \left(\frac{u}{D}\right)\frac{dC_A}{dz} - \left(\frac{k_c}{D}\right)C_A = 0 \tag{2-85}$$

This equation is recognized as a second-order, linear differential equation with constant coefficients. The general solution of equations of this type is obtained in the following manner. Let

$$C_A = Ce^{mz} \tag{2-86}$$

where C and m are finite and nonzero constants. After the first and second derivatives of C_A with respect to z have been found by use of Equation (2-86) and substituted into Equation (2-85), the following result is obtained upon rearrangement:

$$\left[m^2 - \left(\frac{u}{D} \right) m - \left(\frac{k_c}{D} \right) \right] = 0 \qquad (2-87)$$

The two values of m that make the left side equal to zero are the roots of the quadratic equation

$$m_1 = \frac{u}{2D} (1 - \sqrt{1 + 4\alpha})$$

$$m_2 = \frac{u}{2D} (1 + \sqrt{1 + 4\alpha})$$

where

$$\alpha = \frac{k_c D}{u^2} = \frac{k_c z_T}{u} \frac{D}{u z_T} = \frac{N_{Da}}{N_{Pe}}$$

$$u = \frac{v_T}{S}$$

$$N_{Pe} = \frac{u z_T}{D}$$

Since $C_1 e^{m_1 z}$ and $C_2 e^{m_2 z}$ are both linearly independent solutions of Equation (2-85), their sum is also a solution, a fact which is readily shown by direct substitution. Thus the general solution of Equation (2-85) is given by

$$C_A = C_1 e^{m_1 z} + C_2 e^{m_2 z} \qquad (2-88)$$

The values of C_1 and C_2 are to be selected such that the boundary conditions are satisfied. Reactant A enters the reactor at the boundary $z = 0$ at the rate $n_A^0 = SuC_A^0$ and leaves this boundary at the rate $[SuC_A(0 +) + SJ_A(0 +)]$, where $C_A(0 +)$ and $J_A(0 +)$ are the values of C_A and J_A, respectively, at $z = 0 +$. Thus the boundary condition at the inlet consists of the material balance on A at the inlet:

$$SuC_A^0 - [SuC_A(0 +) + SJ_A(0 +)] = 0 \qquad (2-89)$$

A second boundary condition may be deduced by commencing with a material balance on A at $z = z_T$.

$$SuC_A(z_T -) + SJ_A(z_T -) - SuC_A(z_T +) = 0 \qquad (2-90)$$

This balance is based on the supposition that the mechanism of axial diffusion is not involved for all $z > z_T$. This condition is satisfied if, for example, the reactor empties into a large tank. If it is further supposed that the concentration of A at $z = z_T -$ is approximately equal to the concentration at $z = z_T +$, then, by Equation (2-90),

$$J_A(z_T -) \cong 0$$

or
$$\frac{dC_A}{dz} = 0 \quad \text{at} \quad z = z_T - \qquad (2\text{-}91)$$

This boundary condition is commonly referred to as the *closed-system* or *closed-end* boundary condition because axial diffusion across the boundary at $z = z_T$ has been ruled out. Thus the end of the reactor is closed with respect to axial diffusion.

When the arbitrary constants C_1 and C_2 are evaluated by use of these two boundary conditions [Equations (2-89) and (2-91)], the following solution to the differential equation [Equation (2-85)] is obtained [18]:

$$C_A = \frac{4\beta C_A^0}{(1 + \beta)^2 \exp[(-N_{Pe}/2)(1 - \beta)] - (1 - \beta)^2 \exp[(-N_{Pe}/2)(1 + \beta)]}$$

$$(2\text{-}92)$$

where $\qquad N_{Pe} = \dfrac{u z_T}{D},\quad$ the Peclet number; $\quad \beta = \sqrt{1 + 4\alpha}$

To obtain expressions for the outlet concentration versus the reactor volume for other types of reactions, it will be necessary, generally, to solve Equation (2-84) by use of an appropriate numerical method. A comparison of the results predicted by the partial axial mixing model with other models is presented next.

G. Comparison of the Plug-Flow, Perfectly Mixed, Laminar-Flow, and Partial Axial Mixing Models

The effect of mixing on the behavior of a tubular reactor is perhaps best visualized by use of a numerical example in which it is supposed that the first-order reaction

$$A \rightarrow R$$

is carried out in the liquid phase in each of four reactors. Further suppose that the mixing in each of the reactors may be represented by:

1. Plug flow:

$$x = 1 - e^{-N_{Da}}$$

$$N_{Da} = k_c\theta = \frac{k_c V}{v_T} = \frac{k_c z_T}{u}, \qquad \begin{array}{l}\text{the Damköhler number for} \\ \text{a first-order reaction.}\end{array} \quad (2\text{-}93)$$

2. Partially mixed:

$$x = 1 - \frac{4\beta}{(1 + \beta)^2\exp[(-N_{Pe}/2)(1 - \beta)] - (1 - \beta)^2\exp[(-N_{Pe}/2)(1 + \beta)]} \tag{2-94}$$

3. Laminar flow:

$$x = 1 - \left[(1 - N_{Da}/2)\exp(-N_{Da}/2) + (N_{Da}/2)^2 E(N_{Da}/2)\right] \tag{2-95}$$

4. Perfectly mixed:

$$x = \frac{N_{Da}}{1 + N_{Da}} \tag{2-96}$$

Graphs of these equations for the plug-flow, laminar-flow, and the perfectly mixed models for tubular reactors are presented in Figure 2-15. These graphs show that for a given value of N_{Da} or θ, the plug-flow model gives the largest conversion, followed by the laminar-flow model and then the perfectly

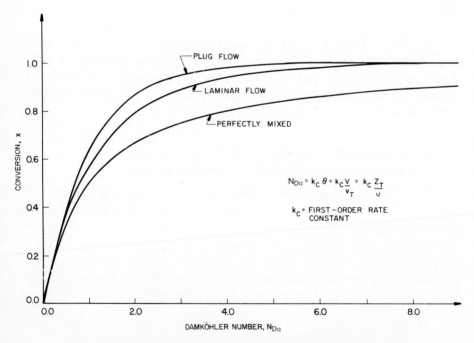

Figure 2-15. Comparison of the exit conversions for plug-flow, laminar-flow, and perfectly mixed models for tubular reactors.

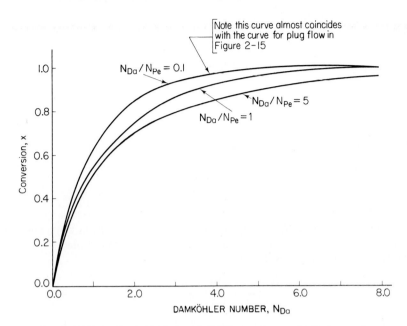

Figure 2-16. As the ratio of the Damköhler number to the Peclet number decreases, the exit conversion given by the partial axial mixing model approaches that given by a plug-flow model.

mixed model. The behavior of the partially mixed model is depicted in Figure 2-16. As the ratio of N_{Da} to N_{Pe} is decreased, the conversion at a given N_{Da} is increased. For example, the curve for $N_{Da}/N_{Pe} = 0.1$ in Figure 2-16 closely approximates the plug-flow model shown in Figure 2-15.

Also, at relatively small conversions all the mixing models are seen to predict very nearly the same conversions. In fact, in this region of conversion, the differences in conversion predicted by the respective models are, in many cases, the same order of magnitude as the inaccuracies of the experimental data, and one mixing model will be found to fit the data about as well as another, as illustrated by the following example.

EXAMPLE 2-8

On the basis of the two data points given in Example 2-5 and the two interior data points given in Figure 2-5, repeat Example 2-5 for the case where it is assumed that the behavior of the tubular reactor may be represented by the perfectly mixed model. For the two interior points shown in Figure 2-5, the values of b and x were not given by Kunzru [12]. On the basis of the assumption of a linear variation of b between the two values found in Example 2-5, the values of b and the conversions for each of the two interior points shown in Figure 2-5 were estimated. A summary of the data given in

Example 2-5 plus the estimated values of b and x follows.

$\dfrac{1}{S_V}$	b	x
0.26	1.712	0.088
0.47	1.787	0.133
0.64	1.86	0.177
0.87	1.945	0.22

SOLUTION

Assume the reaction $A \rightarrow bB$ is first order. On the basis of this assumption, the following expression may be obtained in a manner similar to that demonstrated in the development of Equation (2-45).

$$\frac{VPk_p}{n_T^0} = \frac{x[1 + (b - 1)x]}{1 - x}$$

When the space velocity is based on the temperature and pressure of the system, the left side of this expression reduces to k_c/S_V. Thus, if the reaction is first order and the reactor is perfectly mixed, a straight line should be obtained when the data are plotted, as shown in Figure 2-17.

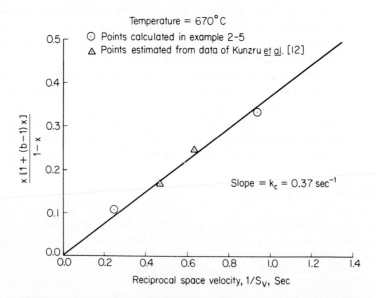

Figure 2-17. Use of the experimental results from a tubular reactor and the perfectly mixed model in the determination of the order and rate constant for the reaction in Example 2-5.

A comparison of Figures 2-5 and 2-17 shows that the data are fit by each model about equally well. However, the rate constants differ by more than 13%. That equally good fits should be obtained at values of $x \leq 0.22$ is suggested by the curves shown in Figure 2-15.

H. The Partial Axial and Partial Radial Mixing Models and Other Mixing Models

For the sake of completeness, the material-balance equation is developed for the case where a reactor is partially mixed in the axial and radial directions.

1. The Partial Axial and Partial Radial Mixing Model. In this generalized mixing model, depicted in Figure 2-18, mixing in both the radial and axial directions is assumed to occur. It is further supposed that this mixing can be described by the forms of Fick's first law for the diffusion of A by mixing in the positive directions of r and z. For the rate of diffusion of A by mixing in the radial direction, the following form of Fick's first law is used.

$$J_r \left[\frac{\text{moles of } A \text{ transferred}}{(\text{unit time})(\text{unit area})} \right] = -D_r \frac{\partial C_A}{\partial r} \tag{2-97}$$

The rate of diffusion of A by mixing in the positive direction of z is given by

$$J_z \left[\frac{\text{moles of } A \text{ transferred}}{(\text{unit time})(\text{unit area})} \right] = -D_z \frac{\partial C_A}{\partial z} \tag{2-98}$$

The symbols D_r and D_z are used to emphasize that different effective diffusivities are permitted in the r and z directions. Let $u(r, z)$ denote the velocity of the fluid at any point (r, z) inside the boundaries $(0 < r < R; 0 < z < z_T)$ of the reactor. At steady-state operation, a material balance on A must take into account the amount of component A entering and leaving the element of volume $2\pi r \Delta r \Delta z$ in Figure 2-18 by flow and by diffusion, as well as the

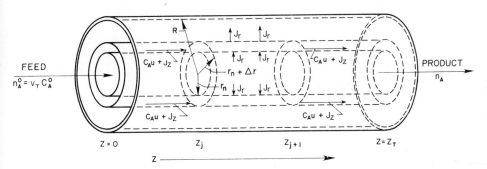

Figure 2-18. Partial mixing in the radial direction and partial mixing in the axial direction of a tubular reactor.

amount of A reacted within the element of volume as follows:

$$\int_{r_n}^{r_n+\Delta r}\left[(C_A u 2\pi r + J_z 2\pi r)|_{z_j,r} - (C_A u 2\pi r + J_z 2\pi r)|_{z_j+\Delta z,r}\right]\,dr$$

$$\underbrace{\qquad\qquad\qquad\qquad\qquad\qquad\qquad\qquad\qquad\qquad\qquad\qquad\qquad\qquad\qquad}$$

input by flow plus diffusion at z_j − output by flow plus diffusion at $z_j + \Delta z$

$$+ \int_{z_j}^{z_j+\Delta z}\left[J_r 2\pi r|_{r_n,z} - J_r 2\pi r|_{r_n+\Delta r,z}\right]\,dz$$

$$\underbrace{\qquad\qquad\qquad\qquad\qquad\qquad\qquad\qquad\qquad\qquad\qquad\qquad\qquad}$$

input by diffusion at r_n − output by diffusion at $r_n + \Delta r$

$$- \int_{r_n}^{r_n+\Delta r}\int_{z_j}^{z_j+\Delta z} r_A 2\pi r\,dr\,dz = 0 \qquad (2\text{-}99)$$

$$\underbrace{\qquad\qquad\qquad\qquad\qquad\qquad\qquad\qquad\qquad\qquad}$$

output by reaction

To reduce Equation (2-99) to its corresponding partial differential equation, the *mean-value theorem of differential calculus* is first applied to each difference appearing in the integrals. Then the *mean-value theorem of integral calculus* is applied once to the single integrals and twice to the double integral to give

$$+2\pi\Delta r\Delta z\left[\frac{\partial(C_A u r + J_z r)}{\partial z}\bigg|_{①} + \frac{\partial(J_r r)}{\partial r}\bigg|_{②} + (r_A r)|_{③}\right] = 0 \quad (2\text{-}100)$$

where

$$① = \left[(z_j + \alpha_1(r_p)\cdot\Delta z), r_p\right],\quad 0 < \alpha_1(r_p) < 1,$$

$$r_p = r_n + \gamma_1\Delta r,\quad 0 \le \gamma_1 \le 1$$

$$② = \left[(r_n + \alpha_2(z_p)\cdot\Delta r), z_p\right],\quad 0 < \alpha_2(z_p) < 1,$$

$$z_p = z_j + \gamma_2\Delta z,\quad 0 \le \gamma_2 \le 1$$

$$③ = \left[(r_n + \alpha_3(z_k)\cdot\Delta r), z_k\right],\quad 0 < \alpha_3(z_k) < 1,$$

$$z_k = z_j + \gamma_3\Delta z,\quad 0 \le \gamma_3 \le 1$$

Division of each member of Equation (2-100) by $(2\pi\Delta r\Delta z)$, followed by the limiting process wherein Δz and Δr are allowed to go to zero, gives the following result upon recognition of the fact that r_n and z_j were arbitrarily selected within the domains $(0 < r < R; 0 < z < z_T)$ of interest.

$$\frac{\partial(C_A u r + J_z r)}{\partial z} + \frac{\partial(J_r r)}{\partial r} + r_A r = 0 \qquad (2\text{-}101)$$

After J_z and J_r have been replaced by their equivalents as given by Equations (2-97) and (2-98), respectively, and the partial differentiation implied by Equation (2-101) hs been performed, the following result is obtained upon rearrangement:

$$D_z \frac{\partial^2 C_A}{\partial z^2} + D_r \frac{\partial^2 C_A}{\partial r^2} + \frac{D_r}{r} \frac{\partial C_A}{\partial r} - \frac{\partial(C_A u)}{\partial z} - r_A = 0 \qquad (2\text{-}102)$$

A comparison of Equations (2-8), (2-74), and (2-84) with Equation (2-102) shows that the plug-flow model, the laminar-flow model, and the partial axial mixed model are all special cases of the generalized model represented by Equation (2-102).*

2. Other Mixing Models. Beginning with the work of Danckwerts [19] and Zwietering [20], the degree of mixing in chemical reactors has received considerable attention. Summaries of recent work in this area are presented by Himmelblau and Bishoff [21] and by Wen and Fan [22].

II. UNSTEADY-STATE OPERATION

The behavior of continuous reactors at unsteady-state operation becomes of importance in the continuous control and in the start-up of reactors. The purpose of the treatment presented is to introduce this important subject by developing the material-balance equations for the plug-flow and perfectly mixed reactor models at unsteady-state operation and to demonstrate the use of these equations in the prediction of the dynamic behavior of these respective reactor models.

A. Plug-Flow Reactors

Suppose that the tubular reactor shown in Figure 2-1 is adequately represented by the plug-flow model. Further suppose that an upset in the operation of the reactor has just occurred, such as a change in the concentration of reactant A in the feed. The development of the equations needed to describe the behavior of this reactor is initiated by making a material balance on A in a given element of reactor volume over a given time period.

---

*The boundary conditions are deduced in the following manner. Since A does not pass through the wall, the radial flux is zero for all z at the wall; that is,

$$J_r(R) = 0, \qquad 0 + \leq z \leq z_T -$$

The condition of symmetry requires

$$\frac{\partial C_A}{\partial r}\bigg|_{r=0}, \qquad 0 + \leq z \leq z_T -$$

The boundary conditions with respect to z are given by Equations (2-89) and (2-90). For a first-order reaction, an analytical solution can be obtained, but development of this is beyond the scope of this book.

To make a material balance on A over the time period from t_n to t_{n+1}, Equation (1-37) is applied. Let the volume from V_j to $V_j + \Delta V$ (see Figure 2-1) be selected as the system. Then the material balance on component A over the element of volume from V_j to V_{j+1} and over the time period from t_n to t_{n+1} is given by

$$\int_{t_n}^{t_{n+1}} \left[n_A|_{V_j,t} - n_A|_{V_{j+1},t} - \int_{V_j}^{V_{j+1}} r_A \, dV \right] dt = \int_{V_j}^{V_{j+1}} \left[C_A|_{t_{n+1},V} - C_A|_{t_n,V} \right] dV$$

$$(2\text{-}103)$$

where
$$t_{n+1} = t_n + \Delta t$$
$$V_{j+1} = V_j + \Delta V$$

By application of the mean-value theorems followed by the limiting process wherein ΔV and Δt are allowed to go to zero, the following form of the material-balance equation is obtained:

$$-\frac{\partial n_A}{\partial V} - r_A = \frac{\partial C_A}{\partial t}, \qquad 0 < V < V_T \qquad (2\text{-}104)$$

The material-balance equation is used to find the variation of C_A or n_A with time and position in the reactor. The concentration of A in the reactor at a given position at any time t depends on the initial conditions and the nature of the chemical reaction. For purposes of illustration, consider the relatively simple situation wherein the first-order reaction

$$A \to R, \qquad r_A = kC_A$$

is being carried out in the liquid phase at constant mass density. Suppose that at time $t = 0$ the concentration of A at each point in the reactor and in the feed is C_A^0. (This condition implies that the reaction is not occurring at time $t = 0$.) Now suppose that at time $t = 0 +$ the reactor is heated instantaneously to the reaction temperature and held constant at this temperature for all $t > 0 +$.

For convenience, let the following changes of variables be made:

$$n_A = C_A v_T, \qquad x = \frac{V}{v_T}, \qquad Y(x,t) = \frac{C_A - C_A^0}{C_A^0}$$

to give

$$\frac{\partial Y}{\partial x} + \frac{\partial Y}{\partial t} = -k(1 + Y), \qquad \begin{array}{l} t > 0 + \\ 0 < x < x_f \end{array} \qquad (2\text{-}105)$$

where
$$x_f = \frac{V_T}{v_T}$$

The initial and boundary conditions are

$$Y(x,0) = 0, \qquad Y(0,t) = 0 \qquad (2\text{-}106)$$

The first of these follows immediately from the preceding definition of $Y(x, t)$. The boundary condition is obtained by first observing that, as V approaches zero, the concentration C_A approaches the inlet value C_A^0.

Let the Laplace transform of $Y(x, t)$ with respect to t be denoted by $y(x, s)$; that is,

$$L\{Y(x, t)\} = \int_0^\infty e^{-st}Y(x, t)\, dt = y(x, s) \tag{2-107}$$

After the Laplace transform of each term of Equation (2-105) has been taken by use of the procedures described in standard texts on Laplace transforms (see, for example, [23]), the following ordinary differential equation with constant coefficients is obtained.

$$\frac{dy}{dx} + sy = -k\left(\frac{1}{s} + y\right) \tag{2-108}$$

The conditions given by Equation (2-106) may be transformed to give

$$y(x, 0) = 0, \qquad y(0, s) = 0 \tag{2-109}$$

Equation (2-108) may be solved by separation of the variables followed by integration:

$$\int_0^y \frac{dy}{k/s + (k + s)y} = -\int_0^x dx \tag{2-110}$$

After the integrals have been evaluated, the result so obtained may be rearranged to give

$$y(x, s) = -\frac{1}{s} + \frac{e^{-xs}}{s}e^{-kx} + \frac{1}{s+k} - \frac{e^{-(s+k)x}}{s+k} \tag{2-111}$$

To obtain $Y(x, t)$, the inverse Laplace transform of each term of Equation (2-111) is taken as described by Churchill [23].

$$Y(x, t) = -1 + (e^{-kx} - e^{-kt})S_x(t) + e^{-kt} \tag{2-112}$$

where $S_x(t)$ is the step function defined by

$$S_x(t) = \begin{cases} 0, & \text{when } 0 < t < x \\ 1, & \text{when } t > x \end{cases}$$

Restatement of Equation (2-112) in terms of concentration gives

$$C_A = C_A^0[e^{-kx} - e^{-kt}]S_x(t) + C_A^0 e^{-kt} \tag{2-113}$$

The equation may be restated as two cases.

1. For $0 < t < x$, $S_x(t) = 0$, and Equation (2-113) reduces to

$$C_A = C_A^0 e^{-kt}, \qquad 0 < t < x \tag{2-114}$$

2. For $t > x$, $S_x(t) = 1$, and Equation (2-113) reduces to

$$C_A = C_A^0 \exp\left(-k\frac{V}{v_T}\right) = C_A^0 e^{-k\theta}, \qquad t > x \qquad (2\text{-}115)$$

where $\theta = V/v_T$, the true residence time.

Comparison of Equations (2-115) and (2-15) leads to the conclusion that steady-state operation is achieved after the reactor has been swept out one time. This characteristic results from the fact that, in the case of plug flow at constant mass density and isothermal operation, each element of fluid passes through the volume V independently of the elements of fluid ahead and behind it. As a consequence of this fact, it is possible to deduce from intuitive reasoning the solutions given by Equations (2-114) and (2-115), as well as the solutions for other sets of boundary conditions and other reactions.

B. Perfectly Mixed Reactors

Suppose that the tank reactor in Figure 2-6 may be represented by the perfectly mixed model. Further suppose that an upset in the operation of the reactor has just occurred.

The component material balance on A in the volume V over the time period Δt is given by

$$\int_{t_n}^{t_n + \Delta t} \left[n_A^0 - n_A - r_A V \right] dt = N_A|_{t_n + \Delta t} - N_A|_{t_n} \qquad (2\text{-}116)$$

In this expression, n_A^0 and n_A represent the inlet and outlet flow rates, respectively, to the reactor at any time t. The total moles of A in the reactor at any time t is denoted by N_A, and the volume occupied by the reacting mixture at any time t is V. Application of the mean-value theorems, followed by the limiting process wherein Δt is allowed to go to zero, gives the following form of the material-balance equation:

$$n_A^0 - n_A - r_A V = \frac{dN_A}{dt} \qquad (2\text{-}117)$$

1. Variation of the Outlet Concentration with Time. To illustrate the use of the material-balance equation in predicting the variation of the outlet flow rate of A with respect to time, one simple example is solved. Suppose that the first-order reaction $A \rightarrow R$ is being carried out in a perfectly mixed tank reactor. Further suppose that at time $t = 0$ a step change in the concentration of A in the feed occurs, and, for $t = 0 +$, the flow rate at which A enters the reactor is n_A^0 while the total volumetric flow rate remains constant. Let the moles of A in the reactor at time $t = 0$ be denoted by N_A^0. Further suppose that the volume V of the reactor which is filled by the reacting mixture remains constant.

Elimination of r_A from Equation (2-117) by use of the rate expression yields

$$\frac{n_A^0 - n_A}{V} - \frac{1}{V}\frac{dN_A}{dt} = k_c C_A \tag{2-118}$$

Since the reactor is perfectly mixed,

$$C_A = \frac{n_A}{v_T} = \frac{N_A}{V} \tag{2-119}$$

When C_A and n_A are expressed in terms of N_A by use of this relationship, Equation (2-118) reduces to

$$\frac{n_A^0 - (v_T/V)N_A}{V} - \frac{1}{V}\frac{dN_A}{dt} = k_c \frac{N_A}{V} \tag{2-120}$$

Separation of variables followed by integration and rearrangement yields

$$N_A = \frac{n_A^0}{\beta}\left[1 - \left(1 - \frac{N_A^0 \beta}{n_A^0}\right)e^{-\beta t}\right] \tag{2-121}$$

where N_A^0 = moles of A in the reactor at time $t = 0$

$$\beta = \frac{v_T}{V} + k_c$$

Equation (2-121) may be rearranged to the following form:

$$n_A = \frac{n_A^0}{\theta\beta}\left[1 - \left\{\left(1 - \frac{N_A^0 \beta}{n_A^0}\right)\exp(-k_c t)\right\}e^{-t/\theta}\right] \tag{2-122}$$

where $\theta = V/v_T$, the residence time. The correct steady-state value of n_A is given by taking the limit of both sides of Equation (2-122) as the process time t is allowed to increase without bound.

$$\lim_{t \to \infty} n_A = \frac{n_A^0}{\theta\beta} = \frac{n_A^0}{1 + k_c\theta} \tag{2-123}$$

This steady-state value of n_A is closely approached after the reactor has been in operation for a period of time equal to three to five times the residence time. That is, for $t = 3\theta$,

$$e^{-t/\theta} = e^{-3\theta/\theta} = 0.05$$

and for $t = 5\theta$,

$$e^{-t/\theta} = e^{-5\theta/\theta} = 0.007$$

For $t = 3\theta$, Equation (2-122) gives a value of n_A which is slightly greater than 95% of the steady-state value, and for $t = 5\theta$, a value of n_A slightly greater than 99.3% of the steady-state value is given by Equation (2-122). In this particular case, the term $\exp(-k_c t)$ improves the steady-state approximation since k_c is always positive.

Aris [1] has shown that after a process time t equal to 3θ to 5θ has elapsed, the steady state is closely approximated for reactions of all orders.

PROBLEMS

2-1. Let r_A be defined in terms of the appearance of A; that is, r_A is equal to the moles of A appearing per unit time per unit volume. On the basis of this definition of r_A, make a material balance on A over the element ΔV of a plug-flow reactor. Apply the mean-value theorems and take the limit of the result so obtained as $\Delta V \to 0$, and show that the resulting expression for the component material balance is given by Equation (2-9).

2-2. For the case where the following liquid phase reactions
   (a) $2A \to$ products
   (b) $A + B \to$ products
   (c) $3A \to$ products
   are carried out isothermally at constant mass density in a plug-flow reactor at steady-state operation, show that the flow rate n_A at which A leaves the reactor is related to the reactor volume V as follows:

   (a) $\dfrac{k_c V}{v_T} = k_c \theta = \dfrac{1}{C_A} - \dfrac{1}{C_A^0}$

   (b) $k_c \theta = \dfrac{1}{C_A^0 - C_B^0} \log \dfrac{C_A C_B^0}{C_B C_A^0}$

   (c) $2k_c \theta = \dfrac{1}{C_A^2} - \dfrac{1}{\left(C_A^0\right)^2}$

2-3. Suppose that the following data are obtained for the gas phase reaction $2A \to R$ in a perfectly mixed batch reactor which is operated isothermally at $100°C$ and at constant volume. Initially, the reactor is filled with pure A with a concentration of 3 g mol/liter. The following data are obtained:

Time (min)	Conversion X
10	0.310
30	0.674
60	0.894
80	0.950

   (a) Determine whether the reaction is first or second order, and evaluate the rate constant.
   (b) Calculate the space velocity required to obtain a 95% conversion of A in a plug-flow reactor. The reactor is to be operated at 50 atm pressure and $100°C$, and pure A is fed to the reactor. In the evaluation of the space velocities, take $T_s = 273.15$ K and $P_s = 1$ atm.

(c) If the viscosity of the gas is 130 μP, the reactor length ≤ 3 m, and the pressure drop is to be less than 1 atm, find the tube diameter required for plug flow. Is plug flow feasible at the reaction conditions? Use an arithmetic average of the inlet and exit velocities in this calculation.
Answer: (b) $1/S_V = 1.4412$ min

2-4. Over most metal catalysts, the rate of hydrogenation of ethylene is

$$C_2H_4 + H_2 \rightarrow C_2H_6$$

zero order with respect to ethylene and first order with respect to hydrogen [24, 25]; that is, the rate of disappearance of ethylene is given by

$$r_{C_2H_4} = k_p p_{H_2}$$

If $k_p = 0.17$ g mol/(s liter atm) at 100°C and if a reactor pressure of 10 atm and a feed rate of 100 g mol/s are used, calculate the reactor volume required to obtain 90% conversion at a reaction temperature of 100°C. The feed is equimolar, and plug flow prevails throughout the reactor.
Answer: $V = 94.2$ liters

2-5. For the case where the reversible reaction $A \rightleftarrows R$ is carried out in the gas phase in a plug-flow reactor, the reactor volume V required to obtain a conversion x may be expressed in terms of x and the equilibrium conversion x^*. At equilibrium,

$$K_p = \frac{p_R^*}{p_A^*} = \frac{n_R^*}{n_A^*} = \frac{n_R^0 + n_A^0 x^*}{n_A^0 - n_A^0 x^*}$$

for a perfect gas mixture.

(a) Show that this same expression for K_p may be obtained by imposing the condition on Equation (2-28) that at equilibrium

$$\frac{dx}{dV} = 0$$

What does this condition imply about the reactor volume?
(b) For the preceding reversible reaction show that

$$\frac{dx}{dV} = \bar{k}(x^* - x)$$

where

$$\bar{k} = \left(\frac{k_p' P}{n_T^0}\right)(K_p + 1)$$

What other rate process is sometimes described by an equation of this type?
(c) Show that when the preceding reaction is carried out as described

$$V = \frac{1}{\bar{k}} \log_e \frac{x^*}{x^* - x}$$

2-6. For the second-order reaction, $2A \rightarrow$ products, which occurs in the liquid phase in a tubular reactor, show that for the case of plug flow and constant mass density

$$\theta = \frac{V}{v_T} = \left(\frac{1}{k_c C_A^0}\right)\left(\frac{x}{1-x}\right)$$

For an $x = 0.6$, what will be the ratio of the volume for the plug-flow reactor to that of a perfectly mixed flow reactor? The feed rate and concentration to each reactor are the same.

Answer: 0.4

2-7. Suppose that the following half-life data are available from an investigation of the isothermal decomposition of reactant A at constant mass density in the liquid phase in a plug-flow reactor.

Half-life (s)	C_A^0 (g mol/liter)
1.8	5
2.8	4
11.3	2

(a) Determine the order of the reaction and evaluate the rate constant.
(b) Calculate the residence time θ required to reduce the concentration of A ($C_A^0 = 3.0$ g mol/liter) to one-half of its initial value in a perfectly mixed flow reactor and in a plug-flow tubular reactor.

Answer: (b) perfectly mixed: 13.3 s; plug flow: 5 s.

2-8. Begin with Equation (2-78) and obtain the result given by Equation (2-79).
Hints: (a) By use of Equation (2-65), show that

$$\left(\frac{R^2}{-2u_{\max}}\right) u\, du = ur\, dr$$

(b) First make the change in variable,

$$\psi = \frac{1}{u/u_{\max}}$$

and then make the change in variable,

$$\phi = y\psi, \quad \text{where} \quad y = \frac{k_c z_T}{u_{\max}}$$

and integrate two times by parts.

2-9. Suppose that it is desired to convert 90% of reactant A to R by the reaction $A \rightarrow R$ in the liquid phase. The reaction is first order, $r_A = k_c C_A$.
(a) Show that if this reaction is carried out isothermally at constant mass density in a perfectly mixed flow reactor,

$$\frac{V}{v_T} = \frac{9}{k_c}$$

(b) Show that if the same reaction is carried out isothermally in a plug-flow reactor,

$$\frac{V}{v_T} = \frac{2.303}{k_c}$$

2-10. Begin with Equation (2-103) and apply the mean-value theorems and a limiting process in a manner analogous to that demonstrated in the text for partial mixing in the axial and radial mixing and show that the material-balance expression given by Equation (2-104) is obtained.

2-11. Begin with Equation (2-116) and show that the material-balance expression given by Equation (2-117) may be obtained by following the procedure described following Equation (2-116).

2-12. Show that when Equation (2-84) is restated in dimensionless form the following result is obtained:

$$-\frac{1}{N_{Pe}}\frac{d_0}{L}\frac{d^2\overline{C}_A}{d\overline{z}^2} + \frac{d\left[\overline{u}\,\overline{C}_A\right]}{d\overline{z}} + \frac{Lr_{A0}}{u_0 C_{A0}}\overline{r}_A = 0$$

where $\quad \overline{C}_A = \dfrac{C_A}{C_{A0}}, \qquad \overline{u} = \dfrac{u}{u_0} \qquad \overline{r}_A = \dfrac{r_A}{r_{A0}}$

$\qquad \overline{z} = \dfrac{z}{L}, \qquad N_{Pe} = \dfrac{u_0 d_0}{D}$

The subscript zero denotes the reference values.

For Reynolds number, N_{Re}, greater than 10^4 (based on tube or reactor diameter), the Peclet number approaches a constant which is equal to approximately 1.4 to 10 for both gas and liquid phase reactions [26]. Hence, for $N_{Re} > 10^4$ and $L/d_0 \geq 100$ (where d_0 is the diameter of the reactor), the coefficient $d_0/(N_{Pe}L)$ in the above equation becomes so small that the first term can be neglected.

2-13. Show that Equation (2-102) reduces to the following dimensionless form when the dimensionless variables are defined as in Problem 2-12 and as given here.

$$\frac{1}{N_{Pe,z}}\frac{d_0}{L}\frac{\partial^2\overline{C}_A}{\partial\overline{z}^2} + \frac{4}{N_{Pe,r}}\frac{L}{d_0}\frac{\partial^2\overline{C}_A}{\partial\overline{r}^2} + \frac{4}{N_{Pe,r}}\frac{L}{d_0}\frac{1}{\overline{r}}\frac{\partial\overline{C}_A}{\partial\overline{r}} - \frac{\partial\left(\overline{C}_A\overline{u}\right)}{\partial\overline{z}} - \frac{r_{A0}L}{u_0 C_{A0}}\overline{r}_A = 0$$

where $\quad N_{Pe,z} = \dfrac{u_0 d_0}{D_z}, \qquad N_{Pe,r} = \dfrac{u_0 d_0}{D_r}, \qquad \overline{r} = \dfrac{r}{r_0}, \qquad d_0 = 2r_0$

2-14. Cyclopropane is to be isomerized to form propylene in a tubular reactor by the reaction

Davis and Scott [27] carried this reaction out in a borosilicate glass reactor and found the reaction to be first order at 1 atm with the following rate constant:

$$\log_e k_c = 36.1923 - \frac{66,950}{RT}$$

where k_c has the units of reciprocal seconds, T is in Kelvin, and $R = 1.9872$ (cal/g mol K). Calculate the space velocity required to achieve a conversion of 60% at a reaction temperature of 570°C and a reaction pressure of 1 atm (absolute). Evaluate the space velocity on the basis of $T_s = 273.15$ K and $P_s = 1$ atm. Assume that plug-flow conditions prevail and that none of the propylene pyrolyzes to other products at this temperature.

Answer: $S_V = 29.466$ h^{-1}

2-15. The decomposition of cyclohexane over a platinum–alumina–mordenite catalyst was studied by Allan and Voorhies [28]. The principal reactions which were found to occur were

$$C_6H_{12} \rightleftarrows C_6H_6 + 3H_2$$
$$\text{Benzene}$$

$$C_6H_{12} \rightleftarrows \quad C_6H_9$$
$$\text{Methyl cyclopentane}$$

At the reaction conditions investigated, isomerization was one-fifth the rate of dehydrogenation. Let the first reaction be denoted by

$$A \rightleftarrows B + 3C$$

For this reaction, Allan and Voohies give the following expression for the rate of appearance of benzene:

$$\frac{1}{\rho}\frac{dn_B}{dV} = \left(\frac{k_1}{\rho}\right)p_A - \left(\frac{k_1'}{\rho}\right)p_B p_C^3$$

where ρ is equal to the bulk density (grams catalyst per cubic centimeter of catalyst bed) of the catalyst bed through which the reacting stream passes. For a reaction temperature of 775°R, a reactor pressure of 85 psia, and a feed composition of 20 mol of hydrogen per mole of cyclohexane, calculate the value of the parameter $\rho V/n_T^0$ required to achieve an 80% conversion of cyclohexane per pass. In this calculation, neglect the isomerization reaction and the variation of the partial pressure of hydrogen at these conditions

$$K_p = \frac{k_1}{k_1'} = \frac{k_1/\rho}{k_1'/\rho} = 8301 \text{ atm}^3$$

$$\frac{k_1}{\rho} = 0.045 \text{ g mol/(min-g catalyst-atm)}$$

Answer: $\rho V/n_T^0 = 6.385$ g/(g mol/min)

2-16. Anthracene was oxidized over a cobalt molybdate catalyst at 300°C and 760 mm Hg by Subramanian and Murthy [29],

$$C_{14}H_{10} + 1.50_2 \rightarrow C_{14}H_8O_2 + H_2O$$
$$\text{Anthracene}$$

For a feed mixture which was 10% (by volume) anthracene and 90% O_2, the reaction was found to be first order with respect to oxygen and zero order with respect to anthracene; that is, the preceding reaction is represented by

$$A + 1.5B \rightarrow C + D$$

$$-\frac{1}{\rho}\frac{dn_A}{dV} = \left(\frac{k_1}{\rho}\right)p_B$$

where ρ is the bulk density of the catalyst bed (grams of catalyst per liter of catalyst bed).

Calculate the value of the parameter $\rho V/n_T^0$ required to achieve a conversion of 90% of the anthracene in a plug-flow reactor at the operating conditions enumerated above. Subramanian and Murthy [29] give the following value for the rate constant:

$$\frac{k_1}{\rho} = 2.61 \times 10^{-7} \text{ g mol}/(\text{min-g catalyst-mm Hg})$$

Answer: $\rho V/n_T^0 = 533.6$ g/(g mol min)

2-17. Styrene monomer is to be polymerized by use of butyllithium as an initiator. At 20°C and a butyllithium concentration of 0.01 g mol/liter, the rate of consumption of the styrene monomer is first order:

$$r_S = kC_s$$

where $k = 0.0387$ min^{-1}. If the polymerization is carried out in each of the following reactor systems at an initial concentration of styrene $C_S^0 = 1.3$ g mol/liter, and it is desired to obtain a conversion of 95%, find:
(a) Reaction time required for a batch reactor.
(b) Residence time $\theta = V/v_T$ for a tubular reactor with plug flow.
(c) Residence time $\theta = V/v_T$ for a tubular reactor with laminar flow.
(d) Residence time (s) for one perfectly mixed flow reactor, for two perfectly mixed flow reactors in parallel, for two perfectly mixed flow reactors in series, and for five perfectly mixed reactors in series.
(e) The value of θ required for the Peclet numbers of 0.01, 10, and 20 in a tubular reactor with a flat velocity profile.
Answer: (a) $t = 1.29$ h, (b) $\theta = 1.29$ h.
(d) $\theta = 8.18$ h for one reactor, $\theta = 8.18$ h for two in parallel, $\theta_1 = \theta_2 = 1.5$ h for two in series, $\theta_1 = \theta_2 = \cdots \theta_5 = 0.353$ h for five in series.
(e) For $N_{Pe} = 0.01$, $\theta = 7.94$ h; $N_{Pe} = 10$, $\theta = 1.64$ h; $N_{Pe} = 20$, $\theta = 1.47$ h.

2-18. The following half-life data were obtained when the reaction

$$A \rightarrow B + C$$

was carried out in the liquid phase in a batch reactor at 25°C.

C_A^0 (g mol/liter)	1	3
$t_{1/2}$ (min)	5	10

(a) Determine the order of the reaction and the rate constant.

(b) If the rate constant is found to double with an increase of the reaction temperature of 10°C, find the energy of activation.

(c) Find the residence time required in a perfectly mixed flow reactor to achieve a conversion of 95% of A when the initial concentration is $C_A^0 = 3$ g mol/liter.

2-19. A continuous stirred tank reactor in which perfect mixing is assumed and a tubular reactor in which plug flow is assumed are being considered for the production of the product R by the gas phase reaction

$$A + B \rightarrow R, \qquad r_A = k_p (p_A p_B)^{1/2}$$

Operating information, design conditions, and physical properties data follow:

- An equal molar feed of A and B is to be used.
- $k_p = 3$ g mol/(min liter atm).
- Reactor pressure = 2 atm
- Inside reactor tube diameter = 0.0775 ft
- $M_A = 84$ (g/g mol), $M_B = 44$ g/(g mol), $M_R = 128$ g/(g mol)
  Fluid viscosity = 0.048 lb/(ft h)
  Reaction temperature = 745 K

Calculate:

(a) The space velocity required at the reaction conditions to achieve a conversion of 99.5% in each type of reactor.

(b) If each tube of the tubular reactor is 6 m long, how many tubes are required? Is the assumption of plug flow satisfied?

(c) Which reactor do you recommend and why?

2-20. Wan et al. [30] studied the dehydration of isopropanol to propylene over a zirconium phosphate catalyst pillared with diphenyl groups. The surface area of the catalyst was 342 m²/g, and the interlayer spacing was 1.39 nm. At 200°C, the dehydration of isopropanol was found to be first order with a rate constant of 1.8 min^{-1} and an energy of activation of 14.9 Kcal/g mol.

(a) Calculate the space velocities for a plug flow and a perfectly mixed flow reactor which are required to achieve conversions of 50% and 99% when pure isopropanol is fed to the reactors which are operated at 525 K and 2 atm.

(b) For equal throughputs for each type of reactor, calculate the percentage difference in the volumes required for each type of reactor at 50% and 99% conversions.

2-21. Riley and Anthony [31] investigated the hydrocracking of heptane over a nickel exchanged ZSM-5 catalyst and found it to be first order with respect to heptane at 650 K with a rate constant of 0.20 ± 0.02 s^{-1} and an energy of activation of 32 ± 3 Kcal/g mol. Within the experimental error, the gas phase reaction could be represented as $A \rightarrow 2B$, where B represents hydrocarbon components ranging from C_3 to C_8. For a feed consisting of 70% hydrogen and 30% heptane and a reaction temperature and pressure of 673 K and 10 atm, respectively, calculate the space velocity at the reaction conditions which is required to achieve a conversion of 80% of the heptane in a plug-flow reactor and in a perfectly mixed reactor.

2-22. The molecular reaction

$$2A = B + C \quad \text{(gas phase)}$$

was studied in a perfectly mixed flow reactor. Equilibrium conversion has been evaluated as 90%. The following data were obtained for a feed of pure A at a reaction temperature and pressure of 600 K and 40 atm, respectively.

V/n_A^0 (liters/mol/s)	Conversion (%)
0.017	5
0.065	15
0.139	25
0.260	35
0.633	50
1.21	60
2.61	70
7.80	80
19.62	85

(a) Evaluate the equilibrium constant K_p.
(b) Evaluate the rate constant k_p.
(c) Calculate the value of V/n_A^0 for a plug-flow reactor for a conversion of 85%.

2-23. The gas phase reaction

$$A \rightarrow 3B$$

is to be carried out in the gas phase in a plug-flow reactor. The feed to the reactor is pure A and the reactor is to be operated at 10 atm. If the reaction is first order with respect to A,

$$r_A = k_p p_A$$

where r_A is in g mol/(s liter), p_A is in atmospheres, and

$$k_p = 2\left(\frac{\text{g mol}}{\text{s liter atm}}\right)$$

at the reaction temperature of 400 K, calculate:
(a) The value of V/n_T^0 required to achieve a conversion x of 95%.
(b) For a reactor containing 1000 tubes each of which is 2.5 cm in diameter and 6 m long, calculate the moles of B produced per mole of A reacted for a conversion x of 95%.

2-24. Nitric oxide is formed in the combustion of fuels from the reaction of nitrogen and oxygen:

$$N_2 + O_2 = 2NO$$

The equilibrium constant is given by

$$K_p = \frac{p_{NO}^2}{p_{N_2} p_{O_2}} = 21.9\, e^{-43,400/RT}$$

where K_p is dimensionless, T is in Kelvin, and $R = 1.9872$ cal/(g mol K). By utilizing the mechanism proposed by Zeldovich [32], Glick et al. [33] and Camac and Feinberg [34] developed the following rate expression:

$$r_{NO} = kC_{N_2}C_{O_2}^{1/2} - k'C_{NO^2}C_{O_2}^{-1/2}$$

$$k = 9 \times 10^{14}e^{-135,000/RT}$$

$$k' = 4.1 \times 10^{13}e^{-91,600/RT}$$

where r_{NO} has the units of g mol/s cm^3, T is in Kelvin, $R = 1.9872$ cal/(g mol K), and concentrations are in g mol/cm^3. Bartok et al. [35] used this rate expression and the equations for the combustion reactions of methane to predict the effect of excess oxygen on the production of NO in a furnace. For the purpose of this problem, only the reaction of nitrogen with oxygen will be considered. Calculate the concentration of NO as a function of reciprocal space velocity evaluated at the reactor conditions for a plug-flow reactor for the following cases. Vary the reciprocal space velocity from zero to the value corresponding to 95% of the equilibrium concentration at each condition stated in parts (a), (b), and (c). The reactor pressure is 1 atm for all parts.

(a) The reactor temperature is constant at 2400 K and the feed composition is 3 mole% oxygen, 79% nitrogen, and 18% inerts.

(b) The feed composition is 10 mole% oxygen, 82% nitrogen, and 8% inerts, and the reactor temperature is 2400 K. Parts (a) and (b) illustrate the effect of feed concentration on the production of NO.

(c) Repeat parts (a) and (b) for a reactor temperature of 1800 K. A comparison of the results of part (c) with the results of parts (a) and (b) illustrates the effect of temperature on the production of NO.

(d) Assume the temperature profile in the furnace (reactor) is given by the following equation:

$$T = 2400 - 149.16S_V^{-1}$$

where T is in K and S_V is in s^{-1}. This part illustrates the effect of a temperature profile on the production of NO.

(e) Devise a graph or a set of graphs that illustrates the various effects illustrated in parts (a), (b), (c), and (d). If possible, use less than four graphs.

2-25. Show that the solution given by Equation (2-92) is obtained for the differential equation given by Equation (2-85) when the arbitrary constants C_1 and C_2 of Equation (2-88) are evaluated by use of the boundary conditions given by Equations (2-89) and (2-91).

NOTATION

A = area perpendicular to the direction of mass transfer, ft^2

C_A = concentration of component A, moles/unit volume

D = eddy diffusivity due to both molecular motion and turbulent motion, (length)^2/unit time; defined by Equation (2-81)

J_A = rate of dispersion of component A as a consequence of molecular diffusion and turbulent motion; see Equation (2-81)

n_A^0, n_A = molar flow rate of component A at the inlet to the reactor and at any point in the reactor, respectively, moles/unit time

n_T^0 = molar flow rate of the feed to the reactor

n_T = total molar flow rate at any point in the reactor

n_A^0 = for reactors with recycle, the molar flow rate of A in the feed to the system (see Figure 2-11)

n_{Ai} = for reactors with recycle, the molar flow rate of A at the inlet of the reactor

n_{Af} = for reactors with recycle, the molar flow rate of A at the outlet of the reactor, $V = V_T$

n_T^0 = for reactors with recycle, the total molar flow rate of A in the feed to the system (see Figure 2-11)

n_{Ti} = for reactors with recycle, the total molar flow rate at the inlet to the reactor

n_T = for reactors with recycle, the total molar flow rate at any point in the reactor

n_{Tf} = for reactors with recycle, the total molar flow rate at the outlet of the reactor, $V = V_T$

N = order of a reaction

N_A^0, N_A = total moles of component A in the reactor at time $t = 0$ and the total moles of A in the reactor at any time t, moles

N_{Da} = Damköhler number; for a first-order reaction, $N_{Da} = k_c\theta = k_c V/v_T = k_c z_T/u$

N_{Pe} = Peclet number, $u z_T/D$

p_A = partial pressure of component A, force/unit area

P = total pressure, force/unit area

r = radius of a tubular reactor (see Figure 2-13)

r_A = net rate of reaction of component A; except where noted, r_A is defined in terms of disappearance; that is, r_A is equal to the moles of A disappearing by reaction per unit time per unit volume

R = constant in the perfect gas law, defined by Equation (2-18); also used to denote the radius of a tubular reactor (see Figure 2-13)

S = cross-sectional area of a reactor

S_V = gaseous space velocity; defined by Equation (2-30)

S_L = liquid space velocity; defined by Equation (2-34)

t = time

u = linear velocity at any given radius r of a tubular reactor in which the contents are in laminar flow; also used to denote the average velocity at any cross section of a tubular reactor, $u = v_T/S$

u_{max} = velocity at the center of a laminar flow reactor (at $r = 0$)

v_T = total volumetric flow rate

V = reactor volume, or more precisely that portion of the reactor volume occupied by the reacting mixture, $0 \le V \le V_T$, where V_T is the total volume

x = moles of the base converted per mole of the base component in the feed

χ = moles of the base component converted in a system consisting of a reactor plus a recycle stream per mole of the base component in the feed to the system

y_A^0 = mole fraction of component A in the feed

z = reactor length; z_T = total length of the reactor

REFERENCES

1. Aris, R., *Elementary Chemical Reactor Analysis*, Prentice-Hall, Inc., Englewood Cliffs, N.J. (1969).

2. Levenspiel, O., *Chemical Reaction Engineering*, John Wiley & Sons, Inc., New York (1972).

3. Denbigh, K. G., *Chemical Reactor Theory*, Cambridge Press, London (1965).

4. Griffith, R. O., R. S. Jobin, and A. McKeown, *Trans. Faraday Soc.*, *34*:316 (1938).

5. NBS Circular 510, *Tables of Chemical Kinetics-Homogeneous Reactions*, United States Department of Commerce, National Bureau of Standards, Table 352, 475 (Sept. 28, 1951).

6. Fisher, R. A., and J. M. Smith, *Ind. Eng. Chem.*, *42*:704 (1950).

7. Marek, L. F., and W. B. McCluer, *Ind. Eng. Chem.*, *23*:878 (1931).

8. Carnahan, B., H. A. Luther, and J. O. Wilkes, *Applied Numerical Methods*, John Wiley & Sons, Inc., New York (1969).

9. Snow, R. H., *J. Phys. Chem.*, *70*:2780 (1966).

10. Snow, R. H., R. E. Peck, and C. G. Von Fredersdorff, *AIChE. J.* *5*:304 (1959).

11. Himmelblau, D. M., *Process Analysis by Statistical Methods*, John Wiley & Sons, Inc., New York (1970).

12. Kunzru, D., Y. T. Shah, and E. B. Stuart, *I & EC Proc. Design and Develop.*, *2*, No. 3, 339 (1923).

13. Hinshelwood, C. H., and P. S. Askey, *Proc. Roy. Soc.*, *A115*:215 (1927).

14. Leuteritz, G. M., P. Reimann, and P. Vergeris, *Hydrocarbon Processing*, *55*:99 (1976).

15. Carberry, J. J., *Applied Kinetics and Chemical Engineering*, American Chemical Society Publications, Washington, D.C. (1976).

16. *Handbook of Mathematical Tables*, Supplement to Handbook of Chemistry and Physics, The Chemical Rubber Company, Cleveland, Ohio (1964).

17. Bhattacharyya, D. N., J. Smid, and M. Szwarc, *J. Phys. Chem.* *69*:624 (1965).

18. Burghardt, A., and T. Zaleski, *Chem. Eng. Sci.* *23*:575 (1968).

19. Danckwerts, P. V., *Chemical Reaction Engineering, 12th Meeting, Eur. Fed. Chem. Eng.*, Amsterdam (1957).

20. Zwietering, Th. N., *Chem. Eng. Sci.*, *11*:1 (1959).

21. Himmelblau, D. M., and K. B. Bishoff, *Process Analysis and Simulation: Deterministic Systems*, John Wiley & Sons, Inc., New York (1968).

22. Wen, C. Y., and C. T. Fan, *Models for Flow Systems and Chemical Reactors*, Marcel Dekker, Inc., New York (1975).

23. Churchill, R. W., *Operational Mathematics*, 2d ed., McGraw-Hill Book Company, New York (1958).

24. Wynkoop, R., and R. Wilhelm, *CEP*, *46*:300 (1950).

25. Horiutu, J., and K. Miyahara, *Hydrogenation of Ethylene on Metallic Catalysts*, National Standard Reference Data Series, N.B.S., *13*: *Category 6, Chem. Kin.*

26. Levenspiel, O. *Ind. Eng. Chem.*, *50*:343 (1948).

27. Davis, B. R., and D. S. Scott, *I & EC Fund.*, *3*:20 (1964).

28. Allan, D. E., and A. Voorhies, Jr., *I & EC Prod. Research and Develop.*, *11*:159 (1972).

29. Subramanian, P., and M. S. Murthy, *Chem. Eng. Sci.*, *29*:25 (1974).

30. Wan, B.-Z., and others, *J. Catalysis*, *101*:19 (1986).

31. Riley, M. G., and R. G. Anthony, *J. Catalysis*, *100*:322 (1986).

32. Zeldovich, J., *Acta Physiochim. U.S.S.R.*, No. 4, *21*:577 (1946).

33. Glick, H. S., J. J. Klein, and W. Squire, *J. Chem. Phys.*, *27*:850 (1957).

34. Camac, M., and R. M. Feinberg, "Formation of NO in Shock-Heated Air." *11th International Symposium on Combustion*, 137–145, Combustion Institute, Pittsburgh (1967).

35. Bartok, W., A. R. Crawford, and A. Skopp, *Chem. Eng. Prog.*, *67*:64–72 (1971).

Simultaneous and Consecutive Reactions

3

In the majority of industrial reactors, more than one chemical reaction occurs. In this chapter, expressions relating reactor volume or reactor time to the conversion of a base component or the extent to which the group of reactions has progressed are treated. Simultaneous reactions are treated in Part I and consecutive reactions are treated in Part II.

I. SIMULTANEOUS REACTIONS

Simultaneous reactions, as the name implies, includes those cases where two or more reactions involving some base component A occur simultaneously. These reactions are also referred to as *parallel reactions*. Expressions relating conversion and the appropriate independent variable, time or reactor volume, are developed for batch reactors, plug-flow reactors, and perfectly mixed flow reactors.

The system of notation used in the analysis of simultaneous reactions occurring in batch reactors is presented in Table 3-1. For flow reactors, the corresponding system of notation is given by replacing N_i, the moles of A reacted by the ith reaction, with the corresponding flow rate n_i, moles per unit time.

Expressions relating conversion or progress of the reactions versus time for batch reactors and versus reactor volume for flow reactors follow. The case where all reactions occur in the gas phase and the case where all reactions occur in the liquid phase are treated.

TABLE 3-1
DEFINITION OF THE MOLES OF COMPOUND A
REACTED BY EACH REACTION OF A SET OF
SIMULTANEOUS REACTIONS

Reactions	Moles of A Reacted	Moles of Product
(1) $A \rightarrow R$	N_1	$N_R = N_R^0 + N_1$
(2) $A \rightarrow 2D$	N_2	$N_D = N_D^0 + 2N_2$
(3) $2A \rightarrow S$	N_3	$N_S = N_S^0 + \frac{1}{2}N_3$
(4) $2A \rightarrow 2M$	N_4	$N_M = N_M^0 + N_4$
(5) $3A \rightarrow U$	N_5	$N_U = N_U^0 + \dfrac{N_5}{3}$
(6) $3A \rightarrow 2V$	N_6	$N_V = N_V^0 + \dfrac{2N_6}{3}$
(7) $3A \rightarrow 3W$	N_7	$N_W = N_W^0 + N_7$

A. Simultaneous Reactions Carried out Isothermally in Batch Reactors

The three types of simultaneous reactions are considered next. Throughout the treatment that follows for batch reactors, it is supposed that the volume of the reactor filled by the reacting mixture remains constant throughout the course of the reaction. Perfect mixing of the contents of the reactor during the reaction period is also assumed. Under these conditions, the equations for gas and liquid phase reactions are identical.

1. First-Order Reactions. Consider first the case where the two simultaneous reactions of Table 3-2,

$$(1) \quad A \xrightarrow{k_1} R$$

$$(2) \quad A \xrightarrow{k_2} 2S$$

occur in a batch reactor. (Throughout this chapter, the subscript c on the rate constant k_c has been dropped in the interest of simplicity.) The rate constant for each reaction is defined with respect to component A. Note that in general $k_1 \neq k_2$. For if k_1 did equal k_2, then 1 mole of R would be produced for each

TABLE 3-2
TWO SIMULTANEOUS REACTIONS AND THE MATERIAL BALANCES

Reaction and Definitions of Rate Constants		Moles of A Reacted	Moles of Product Formed
(1) $A \rightarrow R$,	$r_{A1} = k_1 C_A$	N_1	$N_R = N_R^0 + N_1$
(2) $A \rightarrow 2S$,	$r_{A2} = k_2 C_A$	N_2	$N_S = N_S^0 + 2N_2$

2 moles of S produced, and it would appear that only the single reaction

$$2A \rightarrow R + 2S$$

had occurred.

The net rate of disappearance of A is equal to the sum of the rates at which A disappears by reactions (1) and (2); that is,

$$r_A = r_{A1} + r_{A2} \tag{3-1}$$

where

$$r_{A1} = k_1 C_A$$

$$r_{A2} = k_2 C_A$$

For a batch reactor, the material balance on component A is given by

$$r_A = -\frac{1}{V}\frac{dN_A}{dt} \tag{3-2}$$

Thus

$$-\frac{1}{V}\frac{dN_A}{dt} = k_1 C_A + k_2 C_A = (k_1 + k_2)\frac{N_A}{V} \tag{3-3}$$

Separation of variables followed by integration and rearrangement yields

$$N_A = N_A^0 \exp\left[-(k_1 + k_2)t\right] \tag{3-4}$$

To compute the moles of each of the products formed, the following procedure may be used. Let the rates of reaction of the products be stated in terms of appearance and the rate of reaction of the reactant A in terms of disappearance. Then, for the two simultaneous reactions (1) $A \rightarrow R$ and (2) $A \rightarrow 2S$, component material balances give

$$r_A = -\frac{1}{V}\frac{dN_A}{dt}, \qquad r_R = \frac{1}{V}\frac{dN_R}{dt}, \qquad r_S = \frac{1}{V}\frac{dN_S}{dt} \tag{3-5}$$

The rates of disappearance of A by reactions (1) and (2) (r_{A1} and r_{A2}) are related to the rates of formation of R and S as follows:

$$r_{A1} = r_R, \qquad r_{A2} = \tfrac{1}{2}r_S \tag{3-6}$$

where the expressions for r_{A1} and r_{A2} are given following equation (3-1). Thus, Equations (3-5) and (3-6) may be combined to give

$$\frac{1}{V}\frac{dN_R}{dt} = k_1 C_A \tag{3-7}$$

$$\frac{1}{2V}\frac{dN_S}{dt} = k_2 C_A \tag{3-8}$$

If N_R and N_S in these expressions are expressed in terms of N_1 and N_2, as defined in Table 3-2, the following result is obtained:

$$\frac{dN_1}{dt} = k_1 N_A \tag{3-9}$$

$$\frac{dN_2}{dt} = k_2 N_A \tag{3-10}$$

Division of the members of Equation (3-9) by the corresponding members of Equation (3-10) yields

$$\frac{dN_1}{dN_2} = \frac{k_1}{k_2} \tag{3-11}$$

Then, by integration of this expression, it is found that

$$\frac{N_1}{N_2} = \frac{k_1}{k_2} \tag{3-12}$$

since $N_1 = N_2 = 0$ at $t = 0$.

To obtain N_1 as a function of time, first note that

$$N_A = N_A^0 - N_1 - N_2 \tag{3-13}$$

Then, by use of Equation (3-12), it follows that

$$N_A = N_A^0 - N_1\left(1 + \frac{k_2}{k_1}\right)$$

or

$$N_1 = \frac{k_1\left(N_A^0 - N_A\right)}{k_1 + k_2} \tag{3-14}$$

where N_A is given by Equation (3-4). From Equations (3-12) and (3-14), one obtains

$$N_2 = \frac{k_2\left(N_A^0 - N_A\right)}{k_1 + k_2} \tag{3-15}$$

where N_A is again given by Equation (3-4). Since $N_R = N_R^0 + N_1$ and $N_S = N_S^0 + 2N_2$, Equations (3-4), (3-14), and (3-15) may be used to express N_R and N_S as functions of time.

The equations for the case where n simultaneous, irreversible first-order reactions are carried out isothermally in a perfectly mixed batch reactor at constant volume are given in Table 3-3. (The expressions given in this table for plug-flow and perfectly mixed reactors are discussed in subsequent sections.)

2. Second-Order Reactions. Consider the case of two second-order simultaneous reactions,

$$2A \rightarrow R, \qquad r_{A1} = k_1 C_A^2$$
$$2A \rightarrow 2S, \qquad r_{A2} = k_2 C_A^2 \tag{3-16}$$

The moles of A, R, and S may be stated in terms of N_1 and N_2 (defined in Table 3-1) as follows:

$$N_A = N_A^0 - N_1 - N_2$$
$$N_R = N_R^0 + \tfrac{1}{2}N_1$$
$$N_S = N_S^0 + N_2 \tag{3-17}$$

TABLE 3-3

SIMULTANEOUS AND IRREVERSIBLE FIRST-ORDER REACTIONS

$$A \rightarrow \alpha R_1, \qquad r_{A1} = k_1 C_A$$

$$A \rightarrow \alpha_2 R_2, \qquad r_{A2} = k_2 C_A$$

$$\vdots \qquad \vdots \qquad \vdots \qquad \vdots$$

$$A \rightarrow \alpha_n R_n, \qquad r_{An} = k_n C_A$$

Batch Reactors

(Conditions: perfectly mixed, isothermal, and constant volume)

$$N_A = N_A^0 - N_1 - N_2 - \cdots - N_n \qquad\qquad N_A = N_A^0 e^{-at}$$

$$N_{R1} = N_{R1}^0 + \alpha_1 N_1 \qquad\qquad\qquad\qquad a = \sum_{i=1}^{n} k_i$$

$$\vdots \qquad \vdots \qquad \vdots$$

$$N_{Rn} = N_{Rn}^0 + \alpha_n N_n \qquad\qquad\qquad N_j = \frac{k_j}{a}(N_A^0 - N_A), \quad j = 1, 2, \ldots, n$$

$$-\frac{1}{V}\frac{dN_A}{dt} = r_A = \sum_{i=1}^{n} r_{Ai}$$

$$r_{Ai} = \frac{1}{\alpha i} r_{Ri}, \quad i = 1, 2, \ldots, n$$

$$\frac{N_i}{N_1} = \frac{k_i}{k_1}, \quad i = 1, 2, \ldots, n$$

Plug-Flow Reactors

(Conditions: steady state, plug flow, isothermal operation, liquid phase, and constant mass density)

$$n_A = n_A^0 - n_1 - n_2 - \cdots - n_n \qquad\qquad n_A = n_A^0 e^{-a\theta}$$

$$n_{R1} = n_{R1}^0 + \alpha_1 n_1 \qquad\qquad\qquad\qquad a = \sum_{i=1}^{n} k_i$$

$$\vdots \qquad \vdots \qquad \vdots$$

$$n_{Rn} = n_{Rn}^0 + \alpha_n n_n \qquad\qquad\qquad \theta = \frac{V}{v_T}$$

$$-\frac{dn_A}{dV} = r_A = \sum_{i=1}^{n} r_{Ai} \qquad\qquad n_j = \frac{k_j}{a}(n_A^0 - n_A), \quad j = 1, 2, \ldots, n$$

$$r_{Ai} = \frac{1}{\alpha_i} r_{Ri}, \quad i = 1, 2, \ldots, n$$

$$\frac{n_i}{n_1} = \frac{k_i}{k_1}, \quad i = 1, 2, \ldots, n$$

Perfectly Mixed Flow Reactors

(Conditions: steady state)

The expressions for perfectly mixed flow reactors are obtained by replacing

$$-\frac{dn_A}{dV} = r_A = \sum_{i=1}^{n} r_{Ai} \quad \text{and} \quad n_A = n_A^0 e^{-a\theta}$$

in the expressions for plug-flow reactors by

$$\frac{n_A^0 - n_A}{V} = r_A = \sum_{i=1}^{n} r_{Ai} \quad \text{and} \quad n_A = \frac{n_A^0}{1 + a\theta}$$

Since r_A is equal to the net rate of disappearance of A, it is again related to r_{A1} and r_{A2} as indicated by Equation (3-1). The material balance on component A is given by

$$r_A = -\frac{1}{V}\frac{dN_A}{dt} \qquad (3\text{-}18)$$

Thus

$$-\frac{1}{V}\frac{dN_A}{dt} = k_1 C_A^2 + k_2 C_A^2 = (k_1 + k_2)\left(\frac{N_A}{V}\right)^2 \qquad (3\text{-}19)$$

For reactions which occur at constant volume, Equation (3-19) may be rearranged and integrated to give

$$\frac{1}{N_A} - \frac{1}{N_A^0} = (k_1 + k_2)\frac{t}{V}$$

or

$$N_A = \frac{1}{1/N_A^0 + (k_1 + k_2)(t/V)} \qquad (3\text{-}20)$$

The product distribution may be obtained by use of relationships which are developed as follows. Let r_R and r_S be defined as the net rates of appearance of R and S, respectively. Then

$$r_R = \tfrac{1}{2}r_{A1}, \qquad r_S = r_{A2} \qquad (3\text{-}21)$$

The component material balances on R and S are given by

$$r_R = \frac{1}{V}\frac{dN_R}{dt}, \qquad r_S = \frac{1}{V}\frac{dN_S}{dt} \qquad (3\text{-}22)$$

From the stoichiometric relationships given by Equation (3-17), it is evident that

$$r_R = \frac{1}{V}\frac{dN_R}{dt} = \frac{1}{2V}\frac{dN_1}{dt} \qquad (3\text{-}23)$$

$$r_S = \frac{1}{V}\frac{dN_S}{dt} = \frac{1}{V}\frac{dN_2}{dt} \qquad (3\text{-}24)$$

These relationships may be used to show that

$$\frac{1}{2V}\frac{dN_1}{dt} = \frac{1}{2}k_1 C_A^2 \qquad (3\text{-}25)$$

$$\frac{1}{V}\frac{dN_2}{dt} = k_2 C_A^2 \qquad (3\text{-}26)$$

When the members of Equation (3-25) are divided by the corresponding members of Equation (3-26), one obtains Equation (3-11), which may be integrated to give Equation (3-12). Since $N_A = N_A^0 - N_1 - N_2$, it follows by

TABLE 3-4

SIMULTANEOUS AND IRREVERSIBLE SECOND-ORDER REACTIONS

$$2A \rightarrow 2\alpha_1 R_1, \qquad r_{A1} = k_1 C_A^2$$

$$2A \rightarrow 2\alpha_2 R_2, \qquad r_{A2} = k_2 C_A^2$$

$$\vdots \qquad \vdots \qquad \qquad \vdots \qquad \vdots$$

$$2A \rightarrow 2\alpha_n R_n, \qquad r_{An} = k_n C_A^2$$

Batch Reactors
(Conditions: perfectly mixed, isothermal, and constant volume)

The expressions for second-order irreversible reactions in batch reactors are obtained from those shown for batch reactors in Table 3-3 by replacing

$$N_A = N_A^0 e^{-at}$$

by

$$N_A = \frac{1}{1/N_A^0 + at/V}$$

Plug-flow Reactors
(Conditions: steady state, plug flow, isothermal operation, liquid phase, and constant mass density)

The expressions for second-order irreversible reactions in a plug-flow reactor are obtained from those shown for plug-flow reactors in Table 3-3 by replacing

$$n_A = n_A^0 e^{-a\theta}$$

by

$$n_A = \frac{1}{1/n_A^0 + (a/v_T)\theta}$$

Perfectly Mixed Flow Reactors
(Conditions: steady state)

The expressions for second-order irreversible reactions in perfectly mixed reactors are obtained from those shown for plug-flow reactors in Table 3-3 by replacing

$$-\frac{dn_A}{dV} = r_A = \sum_{i=1}^{n} r_{Ai} \quad \text{and} \quad n_A = n_A^0 e^{-a\theta}$$

by

$$\frac{n_A^0 - n_A}{V} = r_A = \sum_{i=1}^{n} r_{Ai} \quad \text{and} \quad \theta = \frac{v_T}{a}\left(\frac{n_A^0 - n_A}{n_A^2}\right)$$

use of Equation (3-12) that

$$N_1 = \frac{k_1(N_A^0 - N_A)}{k_1 + k_2} \tag{3-27}$$

where N_A is given by Equation (3-20). After a number value for N_1 has been found, the corresponding values of N_2, N_S, and N_R may now be calculated by use of Equations (3-12) and (3-17). The equations for the general case of n simultaneous, irreversible second-order reactions are presented in Table 3-4. (The equations given in this table for plug-flow and perfectly mixed reactors are discussed in subsequent sections.)

TABLE 3-5

SIMULTANEOUS AND IRREVERSIBLE THIRD-ORDER REACTIONS

$$3A \rightarrow 3\alpha_1 R_1, \qquad\qquad r_{A1} = k_1 C_A^3$$
$$3A \rightarrow 3\alpha_2 R_2, \qquad\qquad r_{A2} = k_2 C_A^3$$
$$\vdots$$
$$3A \rightarrow 3\alpha_n R_n, \qquad\qquad r_{A3} = k_3 C_A^3$$

Batch Reactors

(Conditions: perfectly mixed, isothermal, and constant volume)

The expressions for third-order irreversible reactions in batch reactors are obtained from those shown for batch reactors in Table 3-3 by replacing

$$N_A = N_A^0 e^{-at}$$

by

$$N_A = \left[\frac{1}{\left(N_A^0\right)^2} + \frac{2At}{V^2} \right]^{-1/2}$$

Plug-flow Reactors

(Conditions: steady state, plug flow, isothermal operation, liquid phase, and constant mass density).

The expressions for third-order irreversible reactions in plug-flow reactors are obtained from those shown for plug-flow reactors in Table 3-3 by replacing

$$n_A = n_A^0 e^{-a\theta}$$

by

$$n_A = \left[\frac{1}{\left(N_A^0\right)^2} + \frac{2A\theta}{v_T^2} \right]$$

Perfectly Mixed Flow Reactors

(Conditions: steady state)

The expressions for third-order irreversible reactions in perfectly mixed flow reactors are obtained from those shown for plug-flow reactors in Table 3-3 by replacing

$$-\frac{dn_A}{dV} = r_A = \sum_{i=1}^{e} r_{Ai} \quad \text{and} \quad n_A = n_A^0 e^{-a\theta}$$

by

$$\frac{n_A^0 - n_A}{V} = r_A = \sum_{i=1}^{n} r_{Ai} \quad \text{and} \quad \theta = \frac{v_T^2}{a}\left(\frac{n_A^0 - n_A}{n_A^3} \right)$$

3. Third-Order Reactions. The equations for carrying out third-order reactions in a batch reactor are developed in a manner analogous to that demonstrated for second-order reactions. For the general case of n simultaneous, irreversible third-order reactions, the equations are given in Table 3-5.

EXAMPLE 3-1

Let it be supposed that the two first-order reactions given in Table 3-2 are carried out isothermally in a batch reactor. After a reaction time of 50 min, 90% of reactant A has decomposed and the product is found to contain 9.1 mol of R per mole of S.

Neither R nor S is present initially in the reaction vessel. Evaluate the rate constants k_1 and k_2.

SOLUTION

Equation (3-4) may be written in terms of conversion to give

$$-\log_e(1 - X) = (k_1 + k_2)t$$

$$k_1 + k_2 = \frac{-\log_e(1 - 0.90)}{50} = \frac{-\log_e 0.1}{50} = 0.04605 \text{ min}^{-1}$$

Now $N_R = 9.1 N_S$ and, from Table 3-2, $N_R = N_1$. Thus $9.1 N_S = N_1$ and, from Table 3-2, $N_S = 2 N_2$. Therefore, $N_1/N_2 = 18.2$. From Equation (3-12), this ratio is also equal to k_1/k_2. Thus

$$k_1 + k_2 = 18.2 k_2 + k_2 = 0.04606$$

and

$$k_2 = \frac{0.04602}{19.2} = 2.4 \times 10^{-3} \text{ min}^{-1}$$

$$k_1 = 0.0437 \text{ min}^{-1}$$

It should be emphasized that the ratio of the moles of R produced to the moles of S produced is constant throughout the course of the reaction cycle. Generally, this ratio may be changed by altering the temperature of the reaction, since the rate constants are usually different functions of temperature.

B. Simultaneous Reactions Carried out Isothermally in Plug-Flow Reactors at Steady-State Operation

To illustrate the treatment of reactors of this type, consider the case where the following first-order reactions occur in the liquid phase:

$$A \rightarrow R$$
$$A \rightarrow 2S$$

For flow reactors, the stoichiometric relationships are stated in terms of the flow rates of each component, rather than the moles of the component in the reactor. Thus the stoichiometric relationships are as follows:

$$n_A = n_A^0 - n_1 - n_2$$
$$n_R = n_R^0 + n_1 \qquad\qquad (3\text{-}28)$$
$$n_S = n_S^0 + 2 n_2$$

where n_1 is the flow rate of A that has reacted by the first reaction and n_2 is the flow rate of A that has reacted by the second reaction. Since r_{A1} and r_{A2} are the rates at which A reacts by the first and second reactions, respectively, it follows that the net rate of reaction r_A is given by

$$r_A = r_{A1} + r_{A2}$$

A component material balance on A gives

$$r_A = -\frac{dn_A}{dV}$$

Thus

$$-\frac{dn_A}{dV} = k_1 C_A + k_2 C_A = (k_1 + k_2)\frac{n_A}{v_T}$$

Separation of the variables followed by integration yields

$$n_A = n_A^0 \exp\left[-(k_1 + k_2)\theta\right] \tag{3-29}$$

where $n_A = n_A^0$ at $\theta = 0$, and $\theta = V/v_T$, the residence time. Again, as in the case of the batch reactor, the rates of appearance of R and S are related to r_{A1} and r_{A2}, respectively, as follows:

$$r_R = r_{A1}, \qquad r_S = 2r_{A2} \tag{3-30}$$

and from the preceding stoichiometric relationships,

$$\frac{dn_R}{dV} = \frac{dn_1}{dV}, \qquad \frac{dn_S}{dV} = \frac{2dn_2}{dV} \tag{3-31}$$

Thus

$$\frac{dn_1}{dV} = k_1 C_A \tag{3-32}$$

$$2\frac{dn_2}{dV} = 2k_2 C_A \tag{3-33}$$

Division of the members of Equation (3-32) by the corresponding members of Equation (3-33) yields

$$\frac{dn_1}{dn_2} = \frac{k_1}{k_2} \tag{3-34}$$

Separation of the variables followed by integration yields

$$\frac{n_1}{n_2} = \frac{k_1}{k_2} \tag{3-35}$$

where $n_1 = n_2 = 0$ at $V = 0$.

To find the moles of R and S produced in a given reactor of volume V and flow rate v_T, expressions for n_1 and n_2 as a function of V/v_T are first obtained. To obtain such an expression for n_1, consider first the stoichiometric relationship

$$n_A = n_A^0 - n_1 - n_2 = n_A^0 - n_1(1 + n_2/n_1) \tag{3-36}$$

Elimination of n_2/n_1 from this expression by use of Equation (3-35) yields the following result upon rearrangement:

$$n_1 = \frac{k_1\left(n_A^0 - n_A\right)}{k_1 + k_2} \tag{3-37}$$

where n_A is given by Equation (3-29). After n_1 has been determined, n_2 may be calculated by use of Equation (3-35), and then n_R and n_S may be evaluated since $n_R = n_R^0 + n_1$, and $n_S = n_S^0 + 2n_2$.

The equations for the case where n simultaneous, irreversible first-order reactions are carried out isothermally in a tubular reactor are given in Table 3-3. These equations are based on the additional assumptions of plug flow and constant-mass density. The corresponding equations for second- and third-order reactions are shown in Tables 3-4 and 3-5. The equations obtained when gas phase reactions are carried out isothermally in plug-flow reactors are developed in a manner similar to that shown previously (see also Problems 3-2 and 3-3).

C. Simultaneous Reactions Carried out Isothermally in Perfectly Mixed Flow Reactors at Steady-State Operation

A sketch of this type of reactor is shown in Figure 2-7. To demonstrate the development of the material-balance equations, consider the first-order reactions given in Table 3-2. Suppose that these reactions occur in the liquid phase at constant-mass density. The stoichiometric relationships given by Equation (3-28) are also valid for the perfectly mixed flow reactor. Again the net rate of disappearance of A is given by Equation (3-1). The material balance on A for a perfectly mixed flow reactor is given by

$$r_A = \frac{n_A^0 - n_A}{V} \tag{3-38}$$

Thus

$$\frac{n_A^0 - n_A}{V} = \left(k_1 + k_2\right)\frac{n_A}{v_T} = a\frac{n_A}{v_T} \tag{3-39}$$

where $a_1 = k_1 + k_2$. The volumetric flow rate v_T is evaluated at the conditions within (and at the outlet of) the reactor.

Equation (3-39) may be solved for n_A to give

$$n_A = \frac{n_A^0}{1 + a\theta} \tag{3-40}$$

Again let r_A and r_S be defined as the net rates of appearance of R and S, respectively. Then these rates are related to r_{A1} and r_{A2}, as given by Equation

(3-6). Thus

$$\frac{n_R - n_R^0}{V} = k_1 \frac{n_A}{v_T} \tag{3-41}$$

$$\frac{n_S - n_S^0}{V} = 2k_2 \frac{n_A}{v_T} \tag{3-42}$$

After n_R and n_S in Equations (3-41) and (3-42) have been expressed in terms of n_1 and n_2 by use of the stoichiometric relationships given by Equation (3-28), and after the members of the first expression have been divided by the corresponding members of the second expression, the result given by Equation (3-35) is obtained.

To find the moles of R and S produced in a given reactor of volume V and flow rate v_T, n_A may be eliminated from Equations (3-41) and (3-42) by use of Equation (3-40) to give

$$n_R = n_R^0 + n_A^0 \left(\frac{k_1 \theta}{1 + a\theta} \right) \tag{3-43}$$

$$n_S = n_S^0 + n_A^0 \left(\frac{2k_2 \theta}{1 + a\theta} \right) \tag{3-44}$$

For the general case of n simultaneous first-, second-, and third-order reactions, the equations are presented in Tables 3-3, 3-4, and 3-5 (also see Problem 3-4). Gas phase reactions are treated in a manner similar to that shown here for liquid phase reactions.

D. Simultaneous Reactions of Mixed Orders

The special techniques used to solve the differential equations for simultaneous reactions of the same order are no longer applicable when the orders differ. For example, consider the case in which one reaction is first order and the other reaction is second order:

$$A \rightarrow R, \qquad r_{A1} = k_1 C_A \tag{3-45}$$

$$2A \rightarrow 2S, \qquad r_{A2} = k_2 C_A^2 \tag{3-46}$$

Consider first the case where these reactions are carried out isothermally in a batch reactor. The volume V of the reactor filled by the reacting mixture remains constant, and the density of the reacting mixture also remains fixed throughout the course of the reactions.

The net rate of disappearance of A is again given by Equation (3-1). A material balance on A is given by

$$r_A = -\frac{dC_A}{dt} \tag{3-47}$$

where the variation of the volume of the reacting mixture is negligible. Thus

$$-\frac{dC_A}{dt} = k_1 C_A + k_2 C_A^2 \tag{3-48}$$

where $C_A = C_A^0$ at $t = 0$. Separation of variables followed by integration yields

$$t = \frac{-1}{k_1} \log_e \frac{C_A(k_1 + k_2 C_A^0)}{C_A^0(k_1 + k_2 C_A)}$$

or

$$C_A = \frac{C}{f \exp(k_1 t) - 1} \tag{3-49}$$

where

$$C = \frac{k_1}{k_2}$$

$$f = \frac{k_1/k_2}{C_A^0} + 1$$

The variation of C_R with time is obtained in the following manner. First note that

$$r_{A1} = r_R = \frac{dC_R}{dt} = k_1 C_A \tag{3-50}$$

and that $C_R = C_R^0$ at $t = 0$. After C_A has been replaced by its equivalent as given by Equation (3-49) and the equation so obtained is integrated, the following expression is obtained for C_R:

$$C_R = C_R^0 + C\left\{-k_1 t - \log_e \left[\frac{f - 1}{f \exp(k_1 t) - 1}\right]\right\} \tag{3-51}$$

The corresponding integral expression for C_S is

$$C_S = C_S^0 + k_2 C^2 \int_0^t \frac{dt}{[f \exp(k_1 t) - 1]^2} \tag{3-52}$$

The concentration of S may also be obtained from the stoichiometry $C_S = C_A^0 - C_R - C_A$, where C_R and C_A are given by Equations (3-49) and (3-51). The equations describing the use of plug-flow reactors and perfectly mixed flow reactors in carrying out mixed-order reactions are presented in Problems 3-5 and 3-6.

E. General Observations on Simultaneous Reactions

For simultaneous reactions of equal order, it is to be observed that the ratio of the moles of A reacted by any reaction to the moles of A reacted by the reference reaction is equal to the ratio of rate constants. Since the rate constants are functions only of temperature, the rate of one reaction relative to another can be increased by altering the temperature. For example, suppose for the set of second-order reactions given by Equation (3-16) it is desired to increase the moles of R produced relative to the moles of S produced. To accomplish this, the rate of the first reaction must be increased. Examination of Equation (3-12) shows that the ratio of the moles of A reacted by each reaction is a function only of temperature. By combining Equation (3-35) with Equation (1-33), one obtains

$$\frac{n_1}{n_2} = \frac{k_1}{k_2} = \frac{A_1 \exp\left(-E_{A1}/RT\right)}{A_2 \exp\left(-E_{A2}/RT\right)}$$

or

$$\frac{n_1}{n_2} = \frac{A_1}{A_2} \exp\left(\frac{E_{A2} - E_{A1}}{RT}\right)$$

If $E_{A1} > E_{A2}$, an increase in temperature will increase the rate of the first reaction over the second. However, if $E_{A2} > E_{A1}$, then the temperature should be decreased in order to increase the production of R relative to S. This is a general conclusion which is independent of the type of reactor.

For simultaneous reactions which are of mixed order, the yield of one product over another can be increased by altering the temperature and the concentration of the reactants. Consider the reactions given by Equations (3-45) and (3-46). One can show that

$$\frac{n_2}{n_1} = \frac{k_2 C_A}{k_1}$$

Maintaining a high concentration of A in the reactor will favor the production of S over that of R. The effect of temperature on the ratio of the rate constants is the same as that described previously.

II. CONSECUTIVE REACTIONS

Reactions that occur consecutively (one after another) are called *consecutive reactions*. As an example of reactions of this type, the following set of reactions is considered:

$$\begin{aligned} A &\rightarrow R, & r_{R1} = r_A = k_1 C_A \\ R &\rightarrow S, & r_{R2} = r_S = k_2 C_R \end{aligned} \tag{3-53}$$

Each reaction rate is first order, and the stoichiometry is as indicated.

The development of the equations for computing the moles of A, R, and S at any time in a batch reactor follows.

A. Batch Reactors

Suppose the consecutive reactions given by Equation (3-53) are carried out isothermally in either the gas or liquid phase in a perfectly mixed batch reactor. Then, for component A,

$$r_A = -\frac{1}{V}\frac{dN_A}{dt} = k_1 C_A = \frac{k_1 N_A}{V}$$

which may be integrated and rearranged to give

$$N_A = N_A^0 e^{-k_1 t} \tag{1-52}$$

The net rate of disappearance of R is given by

$$r_R = r_{R2} - r_{R1} = k_2 C_R - k_1 C_A \tag{3-54}$$

and a material balance on R gives

$$r_R = -\frac{1}{V}\frac{dN_R}{dt} \tag{3-55}$$

Elimination of r_R and V from Equations (3-54) and (3-55) gives

$$\frac{dN_R}{dt} + k_2 N_R = k_1 N_A$$

When N_A is replaced by its equivalent as given by Equation (1-52), the following differential equation is obtained:

$$\frac{dN_R}{dt} + k_2 N_R = k_1 N_A^0 e^{-k_1 t} \tag{3-56}$$

Equation (3-56) is a special case of the general class of linear, first-order differential equations with variable coefficients which are of the form

$$\frac{dy}{dx} + P(x)y = Q(x) \tag{3-57}$$

where $N_R = y$, $t = x$, $k_2 = P(x)$, and $k_1 N_A^0 e^{-k_1 t} = Q(x)$. A general solution of the differential equation given by Equation (3-57) is

$$y = e^{-\int P(x)\,dx}\left[\int e^{\int P(x)\,dx} Q(x)\,dx + K\right] \tag{3-58}$$

where K is the constant of integration [1]. After the integrations indicated by Equation (3-58) have been carried out for the case where the functions and variables have the meanings given by Equation (3-57), the following result is obtained:

$$N_R = \left(\frac{k_1 N_A^0}{k_2 - k_1}\right) e^{-k_1 t} + K e^{-k_2 t} \tag{3-59}$$

Suppose that at $t = 0$, $N_R = N_S = 0$, and $N_A = N_A^0$. The value of K that satisfies the initial condition is

$$K = -\frac{k_1 N_A^0}{k_2 - k_1}$$

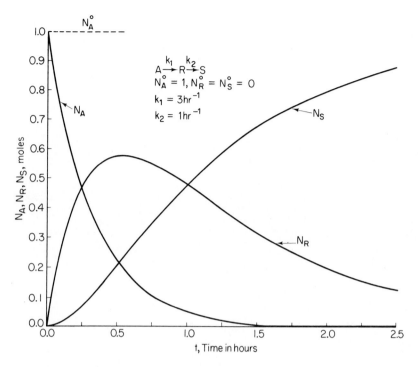

Figure 3-1. Variation of N_A, N_R, and N_S with time in a batch reactor at isothermal operation.

After this value of K has been substituted into Equation (3-59), the following expression for N_R is obtained:

$$N_R = \frac{k_1 N_A^0}{k_2 - k_1} \left[e^{-k_1 t} - e^{-k_2 t} \right] \qquad (3\text{-}60)$$

The expression for N_S as a function of time may be obtained by noting that at any time t the stoichiometry of the reactions requires that

$$N_A^0 = N_A + N_R + N_S \qquad (3\text{-}61)$$

When N_A and N_R are eliminated from Equation (3-61) by use of Equations (1-52) and (3-60), respectively, the following expression for N_S is obtained upon rearrangement:

$$N_S = N_A^0 \left[1 - \left(\frac{k_2}{k_2 - k_1} \right) e^{-k_1 t} + \left(\frac{k_1}{k_2 - k_1} \right) e^{-k_2 t} \right] \qquad (3\text{-}62)$$

Sketches of N_A, N_R, and N_S versus time are shown in Figures 3-1 and 3-2. Formulas for the time at which N_R is a maximum, the time at which N_S passes through its point of injection, and the time at which N_R passes through its point of inflection are given in Problem 3-7.

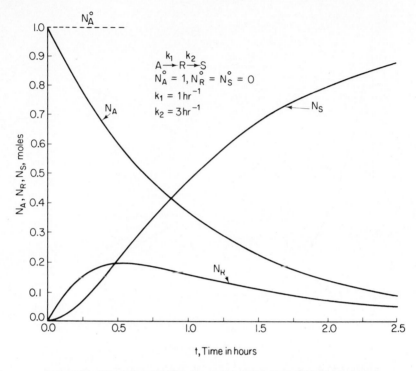

Figure 3-2. Comparison of the above values of k_1 and k_2 with those used for Figure 3-1 shows that the moles of R in the system at any time decreases as the ratio k_2/k_1 increases.

B. Plug-Flow Reactors

Suppose that the reactions given by Equation (3-53) are carried out isothermally in the liquid phase in a plug-flow reactor at steady-state operation. Plug flow of the reacting mixture throughout the reactor is assumed, and it is also supposed that the density of the reacting mixture is constant throughout the reactor and that $n_R^0 = n_S^0 = 0$. In this case,

$$n_A = n_A^0 e^{-k_1\theta} \tag{3-63}$$

and

$$n_R = \frac{k_1 n_A^0}{k_2 - k_1}\left(e^{-k_1\theta} - e^{-k_2\theta}\right) \tag{3-64}$$

$$n_S = n_A^0\left[1 - \left(\frac{k_2}{k_2 - k_1}\right)e^{-k_1\theta} + \left(\frac{k_1}{k_2 - k_1}\right)e^{-k_2\theta}\right] \tag{3-65}$$

These relationships also hold for gas phase reactions if the gaseous mixture behaves as a perfect gas and if the pressure drop across the reactor is negligible.

For a perfectly mixed flow reactor at steady-state operation and isothermal conditions, the corresponding expressions for n_A, n_R, and n_S are as follows:

$$n_A = \frac{n_A^0}{1 + k_1\theta} \tag{3-66}$$

$$n_R = n_A^0 \left[\frac{k_1\theta}{(1 + k_1\theta)(1 + k_2\theta)} \right] \tag{3-67}$$

$$n_S = n_A^0 \left[1 - \left\{ \frac{1 + (k_1 + k_2)\theta}{(1 + k_1\theta)(1 + k_2\theta)} \right\} \right] \tag{3-68}$$

There exists a residence time (called the optimum residence time) for which the production of R is maximized in a given reactor (see Problem 3-10).

EXAMPLE 3-2

The reactions involved in the cracking of gas oils to jet fuels and the subsequent cracking of the jet fuel over an Isomax catalyst were approximated by Strangeland and Kettrell [2] by use of the following simplified kinetic model for operating pressures in the range of 2000 psi:

$$A \xrightarrow{k_1} \nu_1 R \xrightarrow{k_2} \nu_2 S$$

where A represents the gas oil, R represents the jet fuel, and S represents secondary products.

$$r_A = k_1 C_A$$

$$r_R = \nu_1 r_A - k_2 C_R$$

$$k_1 = 1.81 \times 10^{16} \exp\left(\frac{-50{,}000}{RT} \right) \quad \text{in } h^{-1}$$

$$R = 1.9872 \quad \text{and} \quad T \text{ is in Kelvin}$$

$$k_2 = 1.93 \times 10^{16} \exp\left(\frac{-50{,}000}{RT} \right) \quad \text{in } h^{-1}$$

$$R = 1.9872 \quad \text{and} \quad T \text{ is in Kelvin}$$

$$\nu_1 = 0.90$$

(a) Calculate the conversion of A and the moles of R produced per mole of A fed for a reciprocal space velocity of 2 h and a reaction temperature and pressure of 750°F and 2000 psi. The change in the total number of moles may be neglected, and the reciprocal space velocity $1/S_V$ may be taken equal to $V/v_T = \theta$.

(b) Determine the value of the reciprocal space velocity θ at which the moles of R produced per mole of A fed are at their maximum value. At this value of θ, evaluate n_R/n_A^0 and n_A/n_A^0.

SOLUTION

(a) At 750°F,

$$k_1 = 1.005 \text{ h}^{-1}$$

$$k_2 = 0.02538 \text{ h}^{-1}$$

For this case,

$$n_A = n_A^0 e^{-k_1 \theta} \tag{A}$$

and since $x = (n_A^0 - n_A)/n_A^0$, it follows that

$$x = 1 - e^{-k_1 \theta}$$

$$x = 1 - e^{-1.005 \times 2} = 0.866$$

(b) The net rate of appearance of R as given by material balance is

$$r_R = \frac{dn_R}{dV}$$

Since $r_R = v_1 r_A - k_2 C_R$, it follows that

$$\frac{dn_R}{dV} = v_1 k_1 \frac{n_A}{v_T} - \frac{k_2 n_R}{v_T}$$

Then

$$\frac{dn_R}{d\theta} + k_2 n_R = v_1 k_1 n_A$$

After n_A has been expressed in terms of θ by use of Equation (A), the resulting differential equation may be solved in a manner analogous to that demonstrated for Equation (3-56) to give

$$\frac{n_R}{n_A^0} = \frac{v_1 k_1}{k_2 - k_1} (e^{-k_1 \theta} - e^{-k_2 \theta}) \tag{B}$$

where it has been supposed that $n_R^0 = 0$. Thus

$$\frac{n_R}{n_A^0} = \frac{(0.9)(1.005)}{(0.02538 - 1.005)} [\exp(-1.005 \times 2) - \exp(-0.02538 \times 2)]$$

$$= 0.7539$$

(c) At the maximum value of n_R/n_A^0,

$$\frac{d(n_R/n_A^0)}{d\theta} = 0$$

By use of this condition and Equation (B), it is found that

$$\theta = \frac{\log_e k_2/k_1}{k_2 - k_1}$$

$$= \frac{\log_e(0.02538/1.005)}{0.02538 - 1.005} = 3.755$$

At this value of θ,

$$x = 1 - \exp(-1.005 \times 3.755) = 0.977$$

and

$$\frac{n_R}{n_A^0} = \frac{(0.09)(1.005)}{0.02538 - 1.005}[\exp(-1.005 \times 3.755) - \exp(-0.2581 \times 3.755)]$$

$$= 0.817 \text{ mol of jet fuel produced per mole of gas oil fed}$$

C. Solution of Kinetic Problems by Numerical Methods

For most systems involving either simultaneous or consecutive reactions or combination of these types of reactions, it is difficult if not impossible to obtain a solution of closed form. Such problems may be solved, however, by use of numerical methods. A variety of numerical methods which are suitable for solving problems of this type are to be found in books on numerical analysis. For the convenience of the reader, one of the Runge and Kutta methods, called Kutta's fourth-order method, is presented and demonstrated.

Consider first the differential equation

$$\frac{dy}{dt} = f(t, y) \tag{3-69}$$

for which a solution [sensed pairs (t, y) that satisfy both the initial condition $y = y_0$ when $t = t_0$ and the differential equation] is sought. Kutta's fourth-order method consists of the alternate use of Equation (3-69) and the following predictor:

$$y_{n+1} = y_n + \frac{k_1 + 2k_2 + 2k_3 + k_4}{6} \tag{3-70}$$

where

$$k_1 = hf(t_n, y_n)$$

$$k_2 = hf\left(t_n + \frac{h}{2}, y_n + \frac{k_1}{2}\right)$$

$$k_3 = hf\left(t_n + \frac{h}{2}, y_n + \frac{k_2}{2}\right)$$

$$k_4 = hf(t_n + h, y_n + k_3)$$

where h is equal to the time step used in the numerical solution of the differential equation.

The derivations of formulas for the fourth- and higher-order predictors are readily available [3, 4]. A modified form of Kutta's fourth-order method that reduces the storage requirement has been presented by Gill [5] and is to be found in most recent books on numerical analysis.

To demonstrate the application of Kutta's fourth-order method, the following example is used.

EXAMPLE 3-3

The first-order reaction

$$A \rightarrow R$$

is carried out isothermally in a batch reactor. At $t = 0$, $N_A = N_A^0 = 1$ g mol. The rate constant $k = 3 \text{ h}^{-1}$.

(a) Use a time step $h = 0.2$ h to compute the moles of A present at the end of the first time step, $t = 0.2$ h and at $t = 0.4$ h.

(b) Compare the result found in part (a) with that given by the exact solution of the differential equation.

SOLUTION

(a) The differential equation is

$$\frac{-dN_A}{dt} = 3N_A$$

When this equation is restated in terms of the notation of the Runge–Kutta method, it becomes

$$y' = -3y$$

Thus

$$f(t, y) = -3y$$

At $t = t_0 = 0$, $y = y_0 = 1$, and Equation (3-70) is applied as follows:

$$k_1 = 0.2[(-3)(1)] = -0.6$$

$$k_2 = 0.2\left[(-3)\left(1 - \frac{0.6}{2}\right)\right] = -0.42$$

$$k_3 = 0.2\left[(-3)\left(1 - \frac{0.42}{2}\right)\right] = -0.474$$

$$k_4 = 0.2[(-3)(1 - 0.474)] = -0.3156$$

Thus

$$y_1 = 1 + \frac{(-0.6) + 2(-0.42) + 2(-0.474) + (-0.3156)}{6}$$

$$y_1 = 0.5494 \quad \text{at} \quad t_1 = 0.2 \text{ h}$$

At $t_1 = 0.2$ h, $y_1 = 0.5494$, and y_2 at $t_2 = 0.4$ h is computed by use of Equation (3-70) as follows:

$$k_1 = 0.2[(-3)(0.5494)] = -0.32964$$

$$k_2 = 0.2\left[(-3)\left(0.5494 - \frac{0.32964}{2}\right)\right] = -0.23075$$

$$k_3 = 0.2\left[(-3)\left(0.5494 - \frac{0.23075}{2}\right)\right] = -0.26042$$

$$k_4 = 0.2[(-3)(0.5494 - 0.26042)] = -0.17339$$

Thus

$$y_2 = 0.5494 + \frac{(-0.32964) + 2(-0.23075) + 2(-0.264042) + (-0.17339)}{6}$$

$$= 0.3018 \quad \text{at} \quad t_2 = 0.4 \text{ h}$$

(b) At $t = 0.2$ h,

$$N_A = N_A^0 e^{-kt} = (1)e^{-(3)(0.2)} = 0.5488$$

which is to be compared with the approximate value of 0.5494 found by the Runge–Kutta method. At $t = 0.4$ h,

$$N_A = N_A^0 e^{-kt} = (1)e^{-(3)(0.4)} = 0.3012$$

The value of 0.3018 found previously compares well with the correct value of 0.3012.

SOLUTION OF SYSTEMS OF DIFFERENTIAL EQUATIONS BY THE RUNGE–KUTTA METHOD

A system of differential equations may be solved numerically by use of a generalization of the Runge–Kutta method; see, for example, Conte and de Boor [6]. For the case of two differential equations,

$$y' = f(t, y, z)$$
$$z' = g(t, y, z)$$

$$(3\text{-}71)$$

the predictor formulas corresponding to Equation (3-70) are

$$y_{n+1} = y_n + \frac{k_1 + 2k_2 + 2k_3 + k_4}{6}$$

$$z_{n+1} = z_n + \frac{l_1 + 2l_2 + 2l_3 + l_4}{6}$$

$$(3\text{-}72)$$

where

$$k_1 = hf(t_n, y_n, z_n)$$

$$l_1 = hg(t_n, y_n, z_n)$$

$$k_2 = hf\left(t_n + \frac{h}{2}, y_n + \frac{k_1}{2}, z_n + \frac{l_1}{2}\right)$$

$$l_2 = hg\left(t_n + \frac{h}{2}, y_n + \frac{k_1}{2}, z_n + \frac{l_1}{2}\right)$$

$$k_3 = hf\left(t_n + \frac{h}{2}, y_n + \frac{k_2}{2}, z_n + \frac{l_2}{2}\right)$$

$$l_3 = hg\left(t_n + \frac{h}{2}, y_n + \frac{k_2}{2}, z_n + \frac{l_2}{2}\right)$$

$$k_4 = hf(t_n + h, y_n + k_3, z_n + l_3)$$

$$l_4 = hg(t_n + h, y_n + k_3, z_n + l_3)$$

The ks and ls with the lower subscripts must be computed before proceeding to those having the next higher subscript. Extension of this method to a system of three or more differential equations is carried out in a manner analogous to that of going from one to two differential equations. The use of this predictor method is demonstrated by applying it to the reaction system shown in Figure 3-1.

EXAMPLE 3-4

The first-order consecutive reaction

$$A \xrightarrow{k_1} R \xrightarrow{k_2} S$$

is carried out isothermally in a batch reactor. At $t = 0$, $N_A = N_A^0 = 1$ g mol, and $N_R = N_R^0 = 0$. The rate constants are $k_1 = 3$ h^{-1} and $k_2 = 1$ h^{-1}.

(a) Use a time step $h = 0.2$ h to compute the moles of A and R present at the end of the first time step, $t = 0.2$ h.

(b) Compare the results found in part (a) with those given by the exact solution of the differential equations for this system.

SOLUTION

(a) The differential equations are

$$\frac{-dN_A}{dt} = 3N_A$$

$$\frac{dN_R}{dt} = 3N_A - N_R$$

Restatement of these equations in terms of the notation of the Runge–Kutta method gives

$$\frac{dy}{dt} = -3y$$

$$\frac{dz}{dt} = 3y - z$$

Thus

$$f(t, y, z) = -3y$$
$$g(t, y, z) = 3y - z$$

At $t = t_0 = 0$, $y = y_0 = 1$, and $z = z_0 = 0$. Equation (3-72) is applied as follows:

$$k_1 = 0.2[(-3)(1)] = -0.6$$

$$l_1 = 0.2[(3)(1) - 0] = 0.6$$

$$k_2 = 0.2\left[(-3)\left(1 - \frac{0.6}{2}\right)\right] = -0.42$$

$$l_2 = 0.2\left[(3)\left(1 - \frac{0.6}{2}\right) - \left(0 + \frac{0.6}{2}\right)\right] = 0.36$$

$$k_3 = 0.2\left[(-3)\left(1 - \frac{0.42}{2}\right)\right] = -0.474$$

$$l_3 = 0.2\left[(3)\left(1 - \frac{0.42}{2}\right) - \left(0 + \frac{0.36}{2}\right)\right] = 0.438$$

$$k_4 = 0.2[(-3)(1 - 0.474)] = -0.3156$$

$$l_4 = 0.2[(3)(1 - 0.474) - (0 + 0.438)] = 0.228$$

Thus

$$y_1 = 1 + \frac{(-0.6) + 2(-0.42) + 2(-0.474) + (-0.3156)}{6}$$

$$y_1 = 0.5494$$

$$z_1 = 0 + \frac{(0.6) + 2(0.36) + 2(0.438) + 0.228}{6}$$

$$z_1 = 0.404$$

(b) Again as in Example 3-3,

$$N_A = N_A^0 e^{-k_1 t} = (1) e^{-(3)(0.2)} = 0.5488$$

and N_R is found by use of Equation (3-60):

$$N_R = \frac{(3)(1)}{1 - 3} [e^{-(3)(0.2)} - e^{-(1)(0.2)}] = 0.4049$$

When the rate constant k_1 differs appreciably from k_2 (say $k_1/k_2 > 50$), the numerical techniques described generally become unstable, and the corresponding equations are called "stiff or moderately stiff differential equations." Such equations may be solved by use of the simple one-point implicit method, the two-point trapezoidal rule, or multipoint methods such as those described by Gear [7] and Michelson [8].

PROBLEMS

3-1. Develop the equations presented in (a) Table 3-3, (b) Table 3-4, and (c) Table 3-5.

3-2. (a) Suppose that the following reactions are carried out isothermally in the gas phase in a plug-flow reactor:

$$A \rightarrow 2R, \qquad r_{A1} = k_1 p_A$$
$$A \rightarrow 3S, \qquad r_{A2} = k_2 p_A$$

(The subscript p on the rate constant k_p has been omitted in the interest of simplicity.) Suppose that the pressure drop across the reactor is negligible and that the reacting mixture obeys the perfect gas law. Show that

$$V = \frac{b}{aP}\left[\left(\frac{n_A^0}{a} + \frac{n_T^0}{b}\right)\log_e \frac{n_A^0}{n_A} - \left(\frac{n_A^0 - n_A}{a}\right)\right]$$

where
$$a = k_1 + k_2$$
$$b = k_1 + 2k_2$$

(b) For the general case where n simultaneous first-order reactions,

$$A = \alpha_1 R_1, \qquad r_{A1} = k_1 p_A$$
$$A = \alpha_2 R_2, \qquad r_{A2} = k_2 p_A$$
$$\vdots$$
$$A = \alpha_n R_n, \qquad r_{An} = k_n p_A$$

are carried out in the gas phase in a plug-flow reactor under the same conditions stated in part (a), show that the preceding expression is again obtained for the reactor volume V, provided that a and b are given the following meanings:

$$a = \sum_{i=1}^{n} k_i$$

$$b = \sum_{i=1}^{n} (\alpha_i - 1)k_i$$

3-3. The following set of reactions is to be carried out in the gas phase in a plug-flow reactor.

$$2A \rightarrow 2\alpha_1 R_1, \qquad r_{A1} = k_1 p_A^2$$
$$2A \rightarrow 2\alpha_2 R_2, \qquad r_{A2} = k_2 p_A^2$$
$$\vdots \qquad \vdots \qquad \qquad \vdots \qquad \vdots$$
$$2A \rightarrow 2\alpha_n R_n, \qquad r_{An} = k_n p_A^2$$

If the reactor is operated isothermally, the gas behaves as perfect gas, and the pressure drop across the reactor is negligible. Show that

$$V = \frac{1}{aP^2} \left[\frac{2bc}{a} \log_e \frac{n_A}{n_A^0} + \left\{ \left(\frac{b}{a} \right)^2 (n_A^0 - n_A) \right\} + c^2 \left(\frac{1}{n_A} - \frac{1}{n_A^0} \right) \right]$$

where $c = n_T^0 + (bn_A^0/a)$ and a and b are defined in Problems 3-2(b).

3-4. (a) For the case where the n simultaneous first-order reactions listed in Table 3-3 are carried out in the gas phase in a perfectly mixed flow reactor at steady-state operation, show that

$$V = \frac{n_1 (n_T^0 + bn_1/k_1)}{k_1 (n_A^0 - an_1/k_1) P}$$

where $a = \sum\limits_{i=1}^{n} k_i$ (in this case the k_is represent the k_{pi}s)

$$b = \sum_{i=1}^{n} (\alpha_i - 1) k_i$$

(b) Repeat part (a) for the case of the n simultaneous second-order reactions listed in Table 3-4.

$$V = \frac{n_1 (n_T^0 + bn_1/k_1)^2}{k_1 (n_A^0 - an_1/k_1)^2 P^2}$$

(c) Repeat part (b) for the case of the n simultaneous third-order reactions listed in Table 3-5.

$$V = \frac{n_1 (n_T^0 + bn_1/k_1)^3}{k_1 (n_A^0 - an_1/k_1)^3 P^3}$$

3-5. The mixed-order reactions given by Equations (3-45) and (3-46) are to be carried out isothermally in the liquid phase in a plug-flow reactor at steady-state operation. Also suppose that the total volumetric flow rate v_T is constant throughout the reactor. Show that

$$\theta = \frac{1}{k_1} \log_e \frac{n_A^0 [1 + k_2 n_A/(k_1 v_T)]}{n_A [1 + k_2 n_A^0/(k_1 v_T)]}$$

and that $n_R = n_R^0 + C \left[-k_1\theta - \log_e \frac{f - 1}{f \exp(k_1\theta) - 1} \right]$

where $C = \dfrac{k_1 v_T}{k_2}$

$$f = \frac{k_1 v_T/k_2}{n_A^0} + 1$$

3-6. (a) If the mixed-order reactions given by Equations (3-45) and (3-46) are carried out in the liquid phase in a perfectly mixed flow reactor at steady-state operation, show that

$$\theta = \frac{n_A^0 - n_A}{k_1 n_A + (k_2/v_T) n_A^2}$$

(b) For the reactions and reactor of part (a), show that

$$n_R = n_R^0 + k_1 \theta n_A$$

where

$$n_A = \frac{-(k_1\theta + 1) + \sqrt{(k_1\theta + 1)^2 + 4(k_2\theta/v_T) n_A^0}}{(2k_2\theta/v_T)}$$

3-7. (a) When the reaction

$$A \xrightarrow{k_1} R \xrightarrow{k_2} S$$

is carried out isothermally in a perfectly mixed batch reactor, show that the maximum values of N_R occur at $t = t_m$ when $N_R^0 = N_S^0 = 0$.

$$t_m = \frac{\log_e k_1/k_2}{k_1 - k_2}$$

(b) Show that, at $t = t_m$, N_S has a point of inflection.
(c) Show that N_R has a point of inflection at $t = 2t_m$.
(d) Obtain an expression for N_R/N_A^0 for $t = t_m$.

3-8. Develop the expressions given by Equations (3-64) and (3-65).

3-9. Develop the expressions given by Equations (3-66), (3-67), and (3-68).

3-10. When the reaction given by Equation (3-53) is carried out in a perfectly mixed flow reactor, show that n_R takes on its maximum value at

$$\theta_m = \frac{1}{\sqrt{k_1 k_2}}$$

where the initial conditions are $n_R^0 = n_S^0 = 0$, and $n_A^0 = n_T^0$

3-11. The following liquid phase reactions are carried out isothermally at plug-flow conditions in a tubular reactor. The mass density of the mixture remains constant throughout the reactor.

$$A \xrightarrow{k_1} 2R, \qquad r_{A1} = k_1 C_A$$

$$A \xrightarrow{k_2} 3S, \qquad r_{A2} = k_2 C_A$$

For a conversion of A of 70% at $\theta = V/v_T = 10$ s, the moles of R produced per mole of S produced are 4 to 1. The feed to the reactor consists of pure A. Calculate the value of k_1 and k_2.
Answer: $k_1 = 0.103 \text{ s}^{-1}$, $k_2 = 0.0172 \text{ s}^{-1}$

3-12. The intermediate R is to be produced in a flow reactor by the following reaction

$$A \xrightarrow{k_1} R \xrightarrow{k_2} S$$

For $k_1/k_2 = 4$ and $k_1 = 2$ s^{-1}, calculate the value of θ for a perfectly mixed flow and a plug-flow reactor for maximum production of R per mole of A fed. On the basis of the optimum value of θ for each reactor, calculate the conversion of A and the values of n_R/n_A^0 and n_S/n_A^0 for each reactor.

Answer: Perfectly mixed: $x = 0.667$, $n_R/n_A^0 = 0.444$, $n_S/n_A^0 = 0.223$
Plug flow: $x = 0.8424$, $n_R/n_A^0 = 0.63$, $n_S/n_A^0 = 0.2124$

3-13. For the reactions

$$A \xrightarrow{\ k_1\ } R$$
$$A \xrightarrow{\ k_2\ } S$$

$k_1 = 3$ s^{-1} and $k_2 = 0.5$ s^{-1}

(a) Calculate the moles of R and the moles of S produced per mole of A fed for a plug flow and a perfectly mixed flow reactor. The conversion of A is 95% for both reactors.

(b) Calculate the residence time θ required to achieve 99% conversion in each type of reactor.

Answer: (a) $n_R/n_A^0 = 0.8486$, $n_S/n_A^0 = 0.2828$
(b) Plug flow, $\theta = 1.316$ s; perfectly mixed, $\theta = 28.28$

3-14. For the reactions,

$$2A \xrightarrow{\ k_1\ } B$$
$$2A \xrightarrow{\ k_2\ } 2C$$

$$k_1 = 10^{14} \exp\left(\frac{-15,000}{RT}\right)\left(\frac{\text{liters}}{\text{g mol s}}\right)$$

$$k_2 = 10^{15} \exp\left(\frac{-10,000}{RT}\right)\left(\frac{\text{liters}}{\text{g mol s}}\right)$$

where T is in Kelvin and $R = 1.9872$ cal/g mol K. The reactions occur in the temperature range of 200° to 500°C. At what temperature should the reactions be carried out in order to maximize the moles of B per mole of C produced for a plug flow and a perfectly mixed flow reactor?

3-15. The reactions

$$A \xrightarrow{\ k_1\ } R$$
$$A \xrightarrow{\ k_2\ } 2S$$

are to be carried out isothermally in a plug-flow reactor. Because of market considerations, it is desired to make two moles of R per mole of S. If the rate constants for these two first-order reactions are

$$k_1 = 6 \times 10^{14} \exp\left(-\frac{40,000}{RT}\right)$$

$$k_2 = 3.7 \times 10^{13} \exp\left(-\frac{37,000}{RT}\right)$$

find the temperature at which the reactor should be operated.

Answer: $T = 1078.5$ K

3-16. For the reactions $A \xrightarrow{k_1} R \xrightarrow{k_2} S$:

(a) Show that, for the case where the feed contains A, R, and S, the expressions for outlet flow rates for plug flow and perfectly mixed flow reactors are as follows:

(1) Plug-flow reactors:

$$n_A = n_A^0 e^{-k_1\theta}, \qquad \theta = V/v_T$$

$$\frac{n_R}{n_A^0} = \left(\frac{n_R^0}{n_A^0} \right) e^{-k_2\theta} + \left(\frac{1}{k_2/k_1 - 1} \right)(e^{-k_1\theta} - e^{-k_2\theta})$$

$$\frac{n_S}{n_A^0} = \frac{n_S^0}{n_A^0} + (1 - e^{-k_1\theta}) + \left(\frac{n_R^0}{n_A^0} \right)(1 - e^{-k_2\theta})$$

$$- \left(\frac{1}{k_2/k_1 - 1} \right)(e^{-k_1\theta} - e^{-k_2\theta})$$

(2) Perfectly mixed flow reactors:

$$\frac{n_A}{n_A^0} = \frac{1}{1 + k_1\theta}$$

$$\frac{n_R}{n_A^0} = \frac{n_R^0/n_A^0}{1 + k_2\theta} + \frac{k_1\theta}{(1 + k_1\theta)(1 + k_2\theta)}$$

$$\frac{n_S}{n_A^0} = \frac{n_S^0}{n_A^0} + \frac{(n_R^0/n_A^0)k_2\theta}{1 + k_2\theta} + \frac{k_1 k_2 \theta^2}{(1 + k_1\theta)(1 + k_2\theta)}$$

(b) Show that, if a maximum in n_R/n_A^0 occurs at any V/v_T, it occurs at

$$\theta_m = \frac{1}{k_1 - k_2} \log_e \frac{1}{\alpha(1 + \beta) - \beta\alpha^2}$$

for a plug-flow reactor, and at

$$\theta_m = -\frac{\beta}{k_1(1 + \beta)} + \sqrt{\frac{1}{k_1 k_2(1 + \beta)} - \frac{\beta}{(1 + \beta)^2}}$$

for a perfectly mixed flow reactor, where $\beta = n_R^0/n_A^0$ and $\alpha = k_2/k_1$.

(c) Examine the effect of the parameter β on θ_m.

3-17. Determine the effect of mixing on the selectivity and yield of R and the conversion of A for the consecutive reactions $A \xrightarrow{k_1} R \xrightarrow{k_2} S$. Selectivity is defined as moles of R produced per mole of A consumed by reaction, and yield is defined as moles of R produced per mole of A fed. The feed to the reactor consists of pure A, and the rate constants have the values, $k_1 = 2 \ s^{-1}$ and $k_2 = 1 \ s^{-1}$.

(a) To evaluate the effect of mixing compare the results obtained for the plug-flow reactor with the results for the perfectly mixed flow reactor when each is operated at a V/v_T to produce the maximum yield of R, respectively.

(b) To evaluate the effect of partial axial mixing, compare the results for a partially mixed reactor with the results obtained for the plug-flow and perfectly mixed flow reactors with the conversion of A, selectivity and yield of R for Peclet numbers of 0.1, 1.0, and 20, and for each case the residence time, $V/v_T = 0.7$ seconds. (*Hint*: Use numerical methods to obtain the solutions where necessary.)

Answer: (a) $(\theta\,\text{max})_{\text{plug flow}} = 0.693$ s, $x_A = 0.75$, Yield $= 0.5$, Selectivity $= 0.667$. $(\theta\,\text{max})_{\text{perfectly mixed flow}} = 0.707s$, $x_A = 0.585$, Yield $= 0.343$, Selectivity $= 0.586$.

(b)

N_{Pe}	x_A	Yield	Selectivity
0.1	0.59	0.35	0.59
1.0	0.62	0.38	0.61
20.0	0.73	0.48	0.66

3-18. The degradation of diethanolamine (DEA) in the presence of CO_2 concentrations greater than 0.2 g CO_2/g DEA was represented by Kennard and Meisen [9] by the following set of reactions:

$$\text{DEA} \xrightarrow{k_1} \text{HEOD}$$
$$\text{DEA} \xrightarrow{k_2} \text{THEED} \xrightarrow{k_3} \text{BHEP}$$

where HEOD = 3-(hydroxyethyl)-2-oxazoliden, THEED = $n,n,n,n,$-tetra(hydroxyethyl)-ethylenediamine, and BHEP = $n,n,$-*bis*(hydroryethyl) piperazine.

The reaction chemistry as discussed by Kennard and Meisen is far more complex than this kinetic sequence. The rate constants k_1 and k_2 were found to be functions of the initial DEA concentrations. The temperature and pressure ranged from 90° to 150°C and from 1.5 to 6.9 MPa. Calculate the concentrations of each component as a function of time over the time period from $t = 0$ to 20 h when the reactions are carried out in a batch reactor at constant volume. Data: $k_1 = 0.0015\,\text{h}^{-1}$, $k_2 = 0.0075\,\text{h}^{-1}$, $k_3 = 0.0007\,\text{h}^{-1}$, and $C_{\text{DEA}}^0 = 2.5$ g mol/liter.

NOTATION
(see also Chapters 1 and 2)

N_i = moles of the reactant A reacted by reaction i $(i = 1, 2, \ldots, n)$

n_i = moles per unit time of the reactant A reacted by reaction i $(i = 1, 2, \ldots, n)$

n = total number of reactions

$$a = \sum_{i=1}^{n} k_i$$

$$b = \sum_{i=1}^{c} (\alpha_i - 1)k_i$$

α_i = stoichiometric numbers in the reactions $A \to \alpha_i R_i$, $2A \to 2\alpha_i R_i$, and $3A = 3\alpha_i R_i$

$\theta = V/v_T$, residence time

REFERENCES

1. Rainville, E. D., *Elementary Differential Equations*, 2d ed., Macmillan, Inc., New York (1958).

2. Strangeland, B. E., and J. R. Kettrell, *I & EC Proc. Design Develop.*, *11*:15 (1972).

3. Carnahan, B., H. A. Luther, and J. O. Wilkes, *Applied Numerical Methods*, John Wiley & Sons, Inc., New York (1964).

4. Stanton, R. G., *Numerical Methods for Science and Engineering*, Prentice-Hall, Inc., Englewood Cliffs, N.J. (1961).

5. Gill, S., *Proc. Cambridge Phil. Soc.*, *47*:96 (1951).

6. Conte, S. D., and C. de Boor, *Elementary Numerical Analysis*, McGraw-Hill Book Company, New York (1965).

7. Gear, C. W., *Numerical Initial Value Problems in Ordinary Differential Equations*, Prentice-Hall Inc., Englewood Cliffs, N.J., 1971.

8. Michelsen, M. L., *AIChE J.*, *22*, 594 (1976).

9. Kennard, M. L., and A. Meisen, *Ind. Eng. Chem. Fund*, *24*:129 (1985).

Complex Reactions and Interpretation of Experimental Results

4

Reactions which cannot be classified as a unidirectional reaction, a reversible reaction, a set of simultaneous reactions, or a set of consecutive reactions are called complex reactions. Many reactions which are carried out commercially fall under the classification of complex reactions. In Part I of this chapter, reaction mechanisms involving a combination of simultaneous and consecutive reactions are presented. Simplifying assumptions commonly used in the treatment of complex reactions are presented and demonstrated in Part II. In Part III, some methods and techniques which are useful in the interpretation of experimental results are presented.

I. COMBINATIONS OF SIMULTANEOUS AND CONSECUTIVE REACTIONS

One of the simplest combinations of simultaneous and consecutive reactions is the set

$$A \underset{k_1'}{\overset{k_1}{\rightleftharpoons}} R \underset{k_2'}{\overset{k_2}{\rightleftharpoons}} S \tag{4-1}$$

where the rate constants are defined as indicated by Equation (4-3). The reactions $A \rightarrow R \rightarrow S$ constitute consecutive reactions, and the reactions $R \rightarrow A$ and $R \rightarrow S$ constitute simultaneous reactions. Equation (4-1) is representative of isomerization reactions. Again, throughout this chapter, the symbol k is used to represent all rate constants regardless of whether the rate expression is stated in terms of concentrations or pressures. It is supposed in each case, however, that the rate constants possess the appropriate dimensions and corresponding numerical values required for the rate expressions to be numerically and dimensionally consistent.

First, suppose that the preceding set of reactions is carried out isothermally in a batch reactor. Furthermore, suppose that initially the reactor is filled with pure A. Since the set of reactions given by Equation (4-1) does not result in any change in the total moles present at any time, it follows that

$$N_A^0 = N_A + N_R + N_S \tag{4-2}$$

Let r_A be the net rate of disappearance of A, and r_R and r_S be the net rates of appearance of R and S, respectively. Then

$$
\begin{aligned}
r_A &= k_1 C_A - k_1' C_R \\
r_R &= k_1 C_A + k_2' C_S - k_2 C_R - k_1' C_R \\
r_S &= k_2 C_r - k_2' C_S
\end{aligned}
\tag{4-3}
$$

Component material balances on A, R, and S give

$$r_A = -\frac{1}{V}\frac{dN_A}{dt}, \qquad r_R = \frac{1}{V}\frac{dN_R}{dt}, \qquad r_S = \frac{1}{V}\frac{dN_S}{dt} \tag{4-4}$$

These two sets of equations may be combined to give

$$-\frac{dN_A}{dt} = k_1 N_A - k_1' N_R \tag{4-5}$$

$$\frac{dN_R}{dt} = k_1 N_A + k_2' N_S - k_1' N_R - k_2 N_R \tag{4-6}$$

$$\frac{dN_S}{dt} = k_2 N_R - k_2' N_S \tag{4-7}$$

To obtain N_A as a function of time, the following procedure may be used. First, N_S is eliminated from Equation (4-6) by use of Equation (4-2):

$$\frac{dN_R}{dt} = (k_1 - k_2') N_A - (k_1' + k_2' + k_2) N_R + k_2' N_A^0 \tag{4-8}$$

An expression for dN_R/dt as a function of the derivatives of N_A is obtained by differentiating each member of Equation (4-5) with respect to t and rearranging to give

$$\frac{dN_R}{dt} = \frac{1}{k_1'}\frac{d^2 N_A}{dt^2} + \frac{k_1}{k_1'}\frac{dN_A}{dt} \tag{4-9}$$

Use of this expression and Equation (4-5) to eliminate dN_R/dt and N_R, respectively, from Equation (4-8) yields, upon rearrangement,

$$\frac{d^2 N_A}{dt^2} + a\frac{dN_A}{dt} + bN_A - c = 0 \tag{4-10}$$

where
$$a = k_1 + k_1' + k_2 + k_2'$$
$$b = k_1 k_2 + k_1 k_2' + k_1' k_2'$$
$$c = k_1' k_2' N_A^0$$

Equation (4-10) may be transformed to a homogeneous differential equation by making the following change of variable. Let

$$y = b N_A - c \tag{4-11}$$

This change of variable reduces Equation (4-10) to

$$\frac{d^2 y}{dt^2} + a \frac{dy}{dt} + by = 0 \tag{4-12}$$

This is an ordinary, linear, homogeneous differential equation with constant coefficients. Such equations may be solved in the following manner. Assume a solution of the form

$$y = C e^{mt} \tag{4-13}$$

where m and C are nonzero constants. Since

$$\frac{dy}{dt} = C m e^{mt} \quad \text{and} \quad \frac{d^2 y}{dt^2} = C m^2 e^{mt}$$

Equation (4-12) reduces to

$$m^2 + am + b = 0 \tag{4-14}$$

The values of m that satisfy this equation are

$$m_1 = \frac{-a + \sqrt{a^2 - 4b}}{2}$$
$$m_2 = \frac{-a - \sqrt{a^2 - 4b}}{2} \tag{4-15}$$

For the case where m_1 and m_2 are real and unequal, the general solution of Equation (4-12) is as follows:

$$y = \bar{C}_1 e^{m_1 t} + \bar{C}_2 e^{m_2 t} \tag{4-16}$$

where $\bar{C}_1$ and $\bar{C}_2$ are arbitrary constants. In view of Equation (4-11), the result given by Equation (4-16) may be restated in the form

$$N_A = C_1 e^{m_1 t} + C_2 e^{m_2 t} + \frac{c}{b} \tag{4-17}$$

where C_1 and C_2 are arbitrary constants, which are determined by the following initial conditions:

(1) At $t = 0$, $N_A = N_A^0$, $N_R = N_S = 0$

(2) At $t = 0$, $-\dfrac{dN_A}{dt} = k_1 N_A^0$

The second initial condition follows from Equation (4-5), since $N_R = 0$ at $t = 0$. On the basis of these initial conditions, the following values are obtained for C_1 and C_2:

$$C_1 = \frac{-k_1 N_A^0 - \left(N_A^0 - c/b\right)m_2}{m_1 - m_2} \tag{4-18}$$

$$C_2 = \frac{k_1 N_A^0 + \left(N_A^0 - c/b\right)m_1}{m_1 - m_2} \tag{4-19}$$

The formula for N_R as a function of time is found by use of Equations (4-5) and (4-17):

$$N_R = \frac{1}{k_1'}\left[(m_1 + k_1)C_1 e^{m_1 t} + (m_2 + k_1)C_2 e^{m_2 t} + \frac{k_1 c}{b}\right] \tag{4-20}$$

The formula for N_S as a function of time is readily found by eliminating N_A and N_R from the material balance [Equation (4-2)] by use of Equations (4-17) and (4-20).

If the complex reactions given by Equation (4-1) are carried out isothermally in the liquid phase in a plug-flow reactor at steady-state operation and the total volumetric flow rate is constant, then the corresponding relationships for n_A and n_R are obtained by replacing t by V/v_T, N_A by n_A, N_A^0 by n_A^0, and N_R by n_R in Equations (4-17) through (4-20). These relationships also hold for the case where the complex reaction given by Equation (4-1) are carried out in the gas phase, provided, of course, that the pressure drop is negligible and the gaseous mixture obeys the perfect gas law. [Note that under these conditions the total volumetric flow rate is independent of conversion, since no change in the number of moles is involved in the reactions given by Equation (4-1).]

When the complex reactions given by Equation (4-1) are carried out isothermally in a perfectly mixed flow reactor at steady-state operation, the expression for n_A as a function of V is obtained by the same general approach as demonstrated previously for these reactors (see Problem 4-3).

Another example of a complex reaction is the combination of simultaneous and consecutive reactions of the form

$$B + C \xrightarrow{k_1} M + H$$
$$M + C \xrightarrow{k_2} D + H \tag{4-21}$$
$$D + C \xrightarrow{k_3} T + H$$

These reactions are seen to be simultaneous with respect to C and consecutive with respect to B, M, D, and T.

An industrial example of reactions of the type given by Equation (4-21) is the chlorination of benzene,

$$C_6H_6 + Cl_2 \rightarrow C_6H_5Cl + HCl$$

$$C_6H_5Cl + Cl_2 \rightarrow C_6H_4Cl_2 + HCl$$

$$C_6H_4Cl_2 + Cl_2 \rightarrow C_6H_3Cl_3 + HCl$$

where ferric chloride ($FeCl_3$) is used as a catalyst [1].

Suppose that the reactions under consideration are carried out isothermally in a batch reactor in which the reactions go as written; that is, they are unidirectional second-order reactions. Then

$$r_B = -\frac{1}{V}\frac{dN_B}{dt} = \frac{k_1}{V^2}N_B N_C \tag{4-22}$$

$$r_M = \frac{1}{V}\frac{dN_M}{dt} = \frac{k_1}{V^2}N_B N_C - \frac{k_2}{V^2}N_M N_C \tag{4-23}$$

$$r_D = \frac{1}{V}\frac{dN_D}{dt} = \frac{k_2}{V^2}N_M N_C - \frac{k_3}{V^2}N_D N_C \tag{4-24}$$

$$r_T = \frac{1}{V}\frac{dN_T}{dt} = \frac{k_3}{V^2}N_D N_C \tag{4-25}$$

Note that r_B is defined in terms of the disappearance of B, while r_M, r_D, and r_T are defined in terms of the appearance of M, D, and T, respectively. The relationships between N_B, N_C, N_M, N_D, and N_T at any time t are developed in the following manner. Suppose that the reactants consist of N_B^0 moles of benzene in the presence of chlorine. Since $N_M^0 = N_D^0 = N_T^0 = 0$,

$$N_B^0 = N_B + N_M + N_D + N_T \tag{4-26}$$

$$\left.\begin{array}{c}\text{Moles of } C \\ \text{consumed}\end{array}\right\} = N_M + 2N_D + 3N_T \tag{4-27}$$

The method presented next for solving Equations (4-22) through (4-25) is similar to the one shown by Smith [2]. This problem was originally solved by MacMullin [1] in a somewhat different manner.

Division of the members of Equation (4-23) by the corresponding members of Equation (4-22) yields, upon rearrangement,

$$\frac{dN_M}{dN_B} - \left(\frac{\alpha}{N_B}\right)N_M = -1 \tag{4-28}$$

where

$$\alpha = \frac{k_2}{k_1}$$

The general solution to this linear differential equation with the variable coefficient (α/N_B) is given by Equation (3-58). The following result is obtained on the basis of the initial condition that when $N_B = N_B^0$, $N_M = N_M^0 = 0$.

$$\frac{N_M}{N_B} = \left(\frac{1}{\alpha - 1}\right)\left[1 - \left(\frac{N_B}{N_B^0}\right)^{\alpha - 1}\right] \tag{4-29}$$

Next, division of the members of Equation (4-24) by the corresponding members of Equation (4-22) yields, upon rearrangement,

$$\frac{dN_D}{dN_B} - \left(\frac{\beta}{N_B}\right)N_D = -\alpha\frac{N_M}{N_B} \tag{4-30}$$

where

$$\beta = \frac{k_3}{k_1}$$

When N_M/N_D is eliminated from Equation (4-30) by use of Equation (4-29), the resulting expression is again seen to be a member of the general class of differential equations given by Equation (3-57). After the integrations indicated by Equation (3-58) have been carried out and the constant of integration has been selected such that the initial condition $N_D = 0$ when $N_B = N_B^0$ is satisfied, the resulting expression found for N_D is

$$N_D = \left(\frac{\alpha N_B^0}{1 - \alpha}\right)\left[\frac{N_B/N_B^0}{1 - \beta} - \frac{(N_B/N_B^0)^\alpha}{\alpha - \beta}\right] + \frac{\alpha N_B^0 (N_B/N_B^0)^\beta}{(1 - \beta)(\alpha - \beta)} \tag{4-31}$$

Next N_T may be stated in terms of N_B by use of Equations (4-26), (4-29), and (4-31).

For a given amount of B consumed, the corresponding amount of C consumed is given by use of Equations (4-26), (4-27), (4-29), and (4-31). If the concentration of C is maintained constant throughout the course of the reaction by supplying the batch reactor with C at the same rate at which it is consumed, then the expression given by Equation (4-22) may be solved to give

$$N_B = N_B^0 \exp\left[-\left(\frac{k_1 N_C}{V}\right)t\right] \tag{4-32}$$

By substitution of this result into Equations (4-29) and (4-31), expressions for N_M and N_D as functions of time are obtained. By use of the expressions so obtained, the remaining quantities (N_T and the moles of C consumed) are readily obtained as functions of time.

EXAMPLE 4-1

Suppose that the reactions shown for the chlorination of benzene go as written in the liquid phase; that is, benzene, monochlorobenzene, and dichlorobenzene react with absorbed chlorine in the liquid phase. The chlorination is to be carried out isothermally at 55°C in a perfectly mixed semibatch reactor. This semibatch reactor consists of a batch reactor (see Figure 1-4) into which chlorine gas is introduced continuously. Chlorine gas is to be introduced continuously in a manner such that its concentration in the liquid phase remains essentially constant throughout the chlorination. Also, it may be supposed that the volume of the reacting mixture does not vary during the chlorination.

MacMullin [1] has reported the following values for the ratio of the rate constants at 55°C:

$$\frac{k_1}{k_2} = 8, \qquad \frac{k_2}{k_3} = 30$$

For an initial feed of N_B^0 moles of pure benzene, find the moles of chlorine required to maximize the production of monochlorobenzene. Also, find the corresponding product distribution.

SOLUTION

Since monochlorobenzene is produced by the first reaction and consumed by the second reaction of Equation (4-21), there exists the possibility for a maximum concentration of monochlorobenzene to occur. In particular, since $N_M^0 = 0$ and $k_1 > k_2$, N_M increases initially as the moles of benzene reacted ($N_B^0 - N_B$) increase. This process continues until the concentration of monochlorobenzene builds up to a value such that $r_M = 0$ [Equation (4-23)]. At the point where $r_M = 0$, it can be shown by use of Equations (4-23) and (4-28) that

$$\frac{dN_M}{dN_B} = 0$$

When the expression given by Equation (4-29) for N_M is differentiated with respect to N_B and set equal to zero, the following formula is obtained. This formula gives the moles of benzene remaining when the number of moles of monochlorobenzene is maximized:

$$N_B = N_B^0 \left(\frac{1}{\alpha}\right)^{1/(\alpha-1)}$$

Since $\alpha = \frac{1}{8}$, it follows that

$$N_B = N_B^0 (8)^{-8/7} = 0.09287 N_B^0$$

This value of N_B may be used in Equation (4-29) to compute the maximum value of N_M:

$$(N_M)_{max} = \frac{0.09287 N_B^0}{-7/8} \left[1 - (0.09287)^{-7/8}\right] = 0.743 N_B^0$$

The fraction of benzene converted when $N_M = (N_M)_{max}$ is given by

$$\frac{N_B^0 - N_B}{N_B^0} = 1 - 0.09287 = 0.90713$$

The moles of dichlorobenzene formed when $N_M = (N_M)_{max}$ may be found by use of Equation (4-31). First note that

$$\beta = \frac{k_3}{k_1} = \frac{k_2}{k_1} \cdot \frac{k_3}{k_2} = \left(\frac{1}{8}\right)\left(\frac{1}{30}\right) = \frac{1}{240}$$

Thus

$$N_D = \frac{N_B^0(1/8)}{7/8}\left[\frac{0.09287}{1 - 1/240} - \frac{(0.09287)^{1/8}}{1/8 - 1/240}\right] + \frac{N_B^0(1/8)(0.09287)^{1/240}}{(1 - 1/240)(1/8 - 1/240)}$$

$$= 0.1635 N_B^0$$

The corresponding moles of trichlorobenzene may be found by use of the total material balance [Equation (4-26)].

$$N_T = N_B^0\left(1 - \frac{N_B}{N_B^0} - \frac{N_M}{N_B^0} - \frac{N_D}{N_B^0}\right) = N_B^0(1 - 0.09287 - 0.7430 - 0.1635)$$

$$= 0.0006 N_B^0$$

Then the moles of chlorine reacted when $N_M = (N_M)_{max}$ is given by

$$N_C = N_M + 2N_D + 3N_T = N_B^0[0.743 + 2(0.1635) + 3(0.0006)] = 1.0718 N_B^0$$

Other examples of the combination of simultaneous and consecutive reactions are polymerization reactions. One type of polymerization reaction may be represented by the following set of reactions:

$$R_1 + M \xrightarrow{k_1} R_2$$

$$R_2 + M \xrightarrow{k_2} R_3$$

$$\vdots \qquad\quad \vdots \qquad\qquad\qquad (4\text{-}33)$$

$$R_j + M \xrightarrow{k_3} R_{j+1}$$

where R_1 represents the monomer which has been activated instantaneously by an initiator, and R_j denotes the polymer molecule which is composed of j monomer units. The symbol M denotes the monomer involved, such as styrene, propylene, or ethylene. (Because of the commercial importance of polymerization reactions, Chapter 10 is devoted to this topic.)

Let $R_1, R_2, \ldots, R_{j+1}$ represent concentrations of the polymer molecules $R_1, R_2, \ldots, R_{j+1}$, and let M denote the concentration of the monomer. If the reactions given by Equation (4-33) are assumed to go as written, and the propagation rate constants, k_j, are independent of chain length, then k_j is equal to a constant k for all j.

$$\frac{dR_1}{dt} = -kMR_1$$

$$\frac{dR_2}{dt} = kM(R_1 - R_2)$$

$$\vdots \qquad \vdots \tag{4-34}$$

$$\frac{dR_j}{dt} = kM(R_{j-1} - R_j)$$

$$\vdots \qquad \vdots$$

Let the new variable τ be defined by

$$\tau = \int_0^t M \, d\xi \tag{4-35}$$

where ξ is the dummy variable of integration. This definition implies that when $t = 0$, $\tau = 0$, and that

$$\frac{d\tau}{dt} = M \tag{4-36}$$

With this change of variable, the expressions given by Equation (4-34) reduce to the following set:

$$\frac{dR_1}{d\tau} + kR_1 = 0$$

$$\frac{dR_j}{d\tau} + kR_j = kR_{j-1} \tag{4-37}$$

$$\vdots \qquad \vdots \qquad \vdots$$

where $j \geq 2$. At $t = 0$,

$$R_1 = R_1^0 \quad \text{and} \quad R_j = 0, \qquad \text{for all } j > 1$$

The solution that satisfies both the set of differential equations and the initial conditions is

$$R_1 = R_1^0 e^{-k\tau}$$

$$R_2 = R_1^0 (k\tau) e^{-k\tau}$$

$$R_3 = R_1^0 \frac{(k\tau)^2}{2} e^{-k\tau}$$

$$\vdots \qquad \vdots \tag{4-38}$$

$$R_j = R_1^0 \frac{(k\tau)^{j-1}}{(j-1)!} e^{-k\tau}$$

$$\vdots \qquad \vdots$$

The expressions given by Equation (4-38) are recognized as the Poisson distribution. A solution has been given by Aris [3] for the case where k depends on the reaction number j.

An expression for the monomer concentration as a function of time is developed as follows. A material balance on the monomer gives

$$r_M = -\frac{dM}{dt}$$

The rate of disappearance r_M of the monomer M is given by

$$r_M = kMR_1 + kMR_2 + kMR_3 + \cdots$$

Thus

$$-\frac{dM}{dt} = kM \sum_{j=1}^{\infty} R_j \tag{4-39}$$

Since the polymers $R_2, R_3, \ldots, R_j \ldots$ present at any time t were formed from an activated monomer R_1, it follows by material balance that

$$R_1^0 = R_1 + R_2 + R_3 + \cdots + R_j + \cdots \tag{4-40}$$

The sum of the infinite series appearing in Equation (4-39) may be replaced by its equivalent as given by Equation (4-40). Consequently, Equation (4-39) reduces to

$$-\frac{dM}{dt} = kR_1^0 M \tag{4-41}$$

Separation of variables followed by integration and rearrangement gives

$$M = M^0 e^{-kR_1^0 t} \tag{4-42}$$

where M^0 is the concentration of the monomer at $t = 0$.

By first replacing M in Equation (4-35) by its equivalent as given by Equation (4-42) and then integrating, the following expression for τ as a function of t is obtained:

$$\tau = \frac{M^0}{kR_1^0}\left(1 - e^{-kR_1^0 t}\right) \tag{4-43}$$

It is of interest to note that the variable τ takes on a maximum value in the limit as t increases without bound. In particular,

$$\tau_{\max} = \frac{M^0}{kR_1^0}$$

A numerical example which illustrates the use of the preceding expressions follows.

EXAMPLE 4-2

Styrene in the solvent dioxane is to be polymerized in a batch reactor at 25°C by use of sodium naphthalene as the initiator. The initial concentration of sodium naphthalene is 0.0005 g mol/liter. It may be assumed that the initiation reaction of styrene and sodium naphthalene is exceedingly fast and goes to completion; that is, it may be assumed that

$$R_1^0 = 0.0005 \text{ g mol/liter}$$

The following value of the rate constant has been reported [4, 5]:

$$k = 3.4 \text{ liters}/(\text{g mol})(\text{s})$$

(a) Find the time required to convert 99.9% of the monomer styrene. (b) For the time t found in part (a), compute the corresponding value of τ. (c) Construct the distribution curve of R_j/R_1^0 versus j, where j ranges from $j = 0$ to $j = 10^6$.

SOLUTION

(a) The time required for 99.9% of the monomer to react may be found by use of Equation (4-42), which may be restated as follows:

$$t = \frac{1}{kR_1^0} \log_e \frac{M^0}{M}$$

Since $M = (1 - 0.999)M^0$,

$$t = \frac{1}{(3.4)(0.0005)} \log_e \frac{1}{1 - 0.999} = 4063.39 \text{ s} = 67.72 \text{ min}$$

(b) By use of this value of t and the known values for k, R_1^0 and M^0, the value of τ may be found by use of Equation (4-43).

$$\tau = \frac{1}{(3.4)(0.0005)} \{1 - \exp[-(3.4)(0.0005)(4063.39)]\}$$

$$= 587.65 \text{ (g mol s/liter)}$$

(c) On the basis of the value of τ found in part (b), the values of R_j/R_1^0 used in the construction of the distribution curve in Figure 4-1 were found by use of Equation (4-38). For large values of j, the factorial in Equation (4-38) may be evaluated with good accuracy by use of Sterling's formula,

$$(j - 1)! \cong \sqrt{2\pi(j - 1)}\,(j - 1)^{j-1} e^{-(j-1)}$$

From the results presented in Figure 4-1, it is evident that most of the polymer molecules formed have chain lengths ranging from 1850 to 2150 monomer units; that is, each polymer molecule is composed of from 1850 to 2150 styrene molecules.

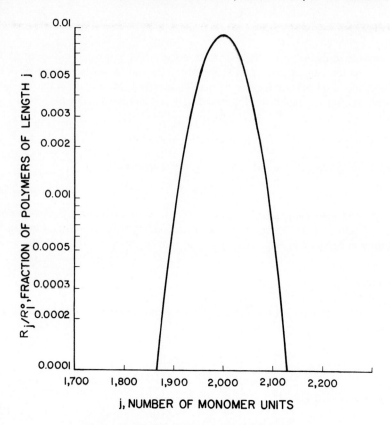

Figure 4-1. Distribution of polymer for Example 4-2.

II. SIMPLIFICATIONS USED IN THE ANALYSIS OF COMPLEX REACTIONS

Many rate expressions for complex reactions may be formulated on the basis of one or more simplifying assumptions. Two of the most widely used assumptions are presented next and their use is demonstrated. The first of these is known by two names, the *pseudo-steady-state* assumption and the *stationary-state* assumption. The second approximation is known as the *rate-controlling-step* assumption.

A. The Pseudo-Steady-State Assumption

Consider the two consecutive reactions given by Equation (3-53):

$$A \rightarrow R, \quad r_A = k_1 C_A$$
$$R \rightarrow S, \quad r_S = k_2 C_R$$

$$(3\text{-}53)$$

where r_A is the rate of disappearance of A and r_S is the rate of formation of S. If $k_1 \ll k_2$, then the expressions for N_R and N_S given by Equations (3-60) and (3-62), respectively, reduce to the following expressions in the limit as k_1/k_2 approaches zero:

$$\lim_{k_1/k_2 \to 0} N_R = 0 \tag{4-44}$$

$$\lim_{k_1/k_2 \to 0} N_S = N_A^0(1 - e^{-k_1 t}) \tag{4-45}$$

This same expression for N_S may be obtained by use of the pseudo-steady-state assumption in which it is supposed that

$$k_1 \ll k_2 \tag{4-46}$$

and that R is consumed by reaction at approximately the same rate at which it is formed:

$$\frac{dN_R}{dt} \cong 0$$

Now consider the case where A reacts first to form R, which then reacts to form S as indicated by Equation (3-53). For the case where the amount of A in the form of R at any time is negligible, the stoichiometry which can be expected experimentally is given by

$$A \to S \tag{4-47}$$

Suppose that the rate of reaction of A and the rate of formation of S are found by experiment to be of the form

$$r_S = r_A = kC_A \tag{4-48}$$

Now it will be shown that the stoichiometric relationship given by Equation (4-47) and the first-order rate expression given by Equation (4-48) may be explained on the basis of the mechanism given by Equation (3-53), the assumption $k_1 \ll k_2$, and the assumption $dN_R/dt \cong 0$.

Since $r_A = -dC_A/dt$ by material balance, and since $r_A = k_1 C_A$ by the first step of the mechanism [Equation (3-53)], the following well-known result is readily obtained, as shown in Chapter 1:

$$N_A = N_A^0 e^{-k_1 t} \tag{1-52}$$

Thus the proposed mechanism [Equation (3-53)] satisfies the observed first-order rate, Equation (4-48).

Next an approximate expression for the moles of R present at any time t is obtained. Since the net rate of appearance of R is assumed to be negligible, it follows that

$$\frac{dN_R}{dt} = k_1 N_A - k_2 N_R \cong 0 \tag{4-49}$$

or

$$N_R = \frac{k_1}{k_2} N_A = \frac{k_1}{k_2} N_A^0 e^{-k_1 t} \tag{4-50}$$

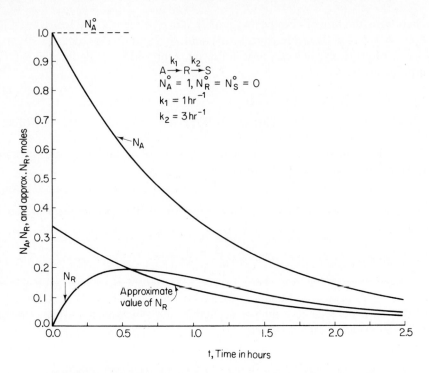

Figure 4-2. Comparison of the correct and approximate values of N_R ($k_2 = 3k_1 = 3$ h^{-1}).

A comparison of the pseudo-steady-state value of N_R and the correct value of N_R [given by Equation (3-60)] is presented in Figures 4-2 and 4-3. The maximum error in the pseudo-steady-state value of N_R is seen to occur at $t = 0$. Also, Figures 4-2 and 4-3 show that the pseudo-steady-state value of N_R tends to approach the correct value of N_R for $t > 0$ as k_2/k_1 increases.

Next it will be shown that the expression given by Equation (4-45) is obtained for N_S by use of the pseudo-steady-state assumption. Since

$$\frac{dN_S}{dt} = k_2 N_R \tag{4-51}$$

it follows from Equation (4-50) that

$$\frac{dN_S}{dt} = k_2\left(\frac{k_1}{k_2}N_A^0 e^{-k_1 t}\right) = k_1 N_A^0 e^{-k_1 t} \tag{4-52}$$

Again, as in the development of Equation (3-62), it will be supposed that only A is present initially. Then, after the variables in Equation (4-52) have been

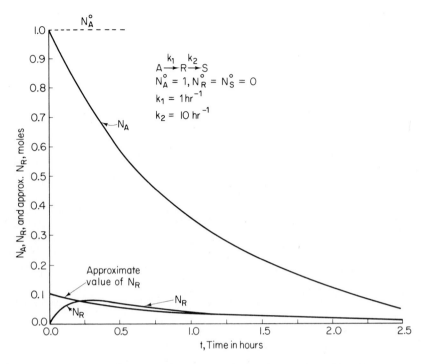

Figure 4-3. Comparison of the correct and approximate values of N_R ($k_2 = 10k_1 = 10 \text{ h}^{-1}$).

separated, the following expression is obtained for N_S upon integration and rearrangement.

$$N_S = N_A^0(1 - e^{-k_1 t}) \tag{4-53}$$

$$= N_A^0 - N_A \tag{4-54}$$

Thus the approximate value of N_S is seen to be equal to the limiting value given by Equation (4-45). The relationship given by Equation (4-54) implies that A is either in the form of A or S; that is, $N_R \cong 0$, which is precisely the implication of Equation (4-47).

The pseudo-steady-state concept may be used in the analysis of reactions having widely different stoichiometric relationships and mechanisms. For example, consider the case of a reaction whose stoichiometry is given by

$$A + B \rightarrow D \tag{4-55}$$

and whose rate is given by the first-order expression

$$r_A = kC_A \tag{4-56}$$

To explain both the stoichiometry and first-order rate expression, the following mechanism is postulated:

$$A \xrightarrow{k_1} C \tag{4-57}$$

$$C + B \xrightarrow{k_2} D \tag{4-58}$$

where $k_1 \ll k_2$ and $dN_C/dt \cong 0$. The first step of the mechanism satisfies the observed rate expression, Equation (4-56). Thus the first step leads to

$$N_A = N_A^0 e^{-k_1 t} \tag{1-52}$$

To show that the proposed mechanism satisfies the stoichiometry [Equation 4-55)], the pseudo-steady-state postulate may be used to obtain

$$-\frac{dN_C}{dt} = \frac{k_2}{V} N_C N_B - k_1 N_A \cong 0 \tag{4-59}$$

Thus

$$N_C = \frac{k_1 V}{k_2} \frac{N_A}{N_B} = \frac{k_1 V N_A^0}{k_2 N_B} e^{-k_1 t} \tag{4-60}$$

As a consequence of the second reaction of the proposed mechanism, it follows that

$$\frac{dN_D}{dt} = \frac{k_2}{V} N_C N_B \tag{4-61}$$

Elimination of N_C by use of the relationship given by Equation (4-60) yields

$$\frac{dN_D}{dt} = k_1 N_A^0 e^{-k_1 t} \tag{4-62}$$

Since $N_D = 0$ at $t = 0$, integration of Equation (4-62) yields

$$N_D = N_A^0 (1 - e^{-k_1 t}) \tag{4-63}$$

which may be stated in the form

$$N_D = N_A^0 - N_A \tag{4-64}$$

Since Equation (4-64) implies that $N_C = 0$, it follows that all of A is either in the form of A or D, which is precisely the implication of Equation (4-55). Consequently, the proposed mechanism satisfies both the stoichiometry and first-order rate expression.

To account for the fact that certain reactions are second order at relatively low pressures and first order at high pressures, Lindermann [6] and Hinshelwood [7] proposed the following example and mechanism. Suppose that the stoichiometry of a reaction is given by

$$A \rightarrow B + C \tag{4-65}$$

At low pressures,

$$r_A = k C_A^2 \tag{4-66}$$

At high pressures,

$$r_A = \bar{k}C_A \tag{4-67}$$

The following mechanism was proposed:

$$A + A \underset{k_1'}{\overset{k_1}{\rightleftarrows}} A^* + A \tag{4-68}$$

$$A^* \overset{k_2}{\longrightarrow} B + C \tag{4-69}$$

where the amount of A in the activated form of A^* at any time is negligible. (The forward reaction $A + A \overset{k_1}{\to} A^* + A$ represents the formation of an activated molecule A^* by means of a bimolecular collision. The reverse reaction $A^* + A \overset{k_1'}{\to} A + A$ represents the deactivation of A^* by means of a bimolecular collision.) In this case the pseudo-steady-state assumptions are taken to be

$$k_1 \ll k_2 \quad \text{and} \quad \frac{dC_A^*}{dt} \cong 0$$

The first assumption complements the second assumption and also satisfies the condition that the amount of A in the form of A^* at any time is negligible. Note that it is not necessary to place any restriction on the size of k_1' relative to k_1 and k_2. Any value of k_1' supports the observation that A^* is negligible at any time by providing another path by which A^* may disappear.

On the basis of this mechanism, the stoichiometry and the rate expressions may be explained as shown by the following analysis. First

$$\frac{dC_A^*}{dt} = k_1 C_A^2 - k_1' C_A^* C_A - k_2 C_A^* \cong 0$$

Thus

$$C_A^* = \frac{k_1 C_A^2}{k_1' C_A + k_2} \tag{4-70}$$

The net rate of disappearance of A is given by

$$r_A = -\frac{dC_A}{dt} = k_1 C_A^2 - k_1' C_A^* C_A \tag{4-71}$$

Elimination of C_A^* from this expression by use of Equation (4-70) gives

$$r_A = -\frac{dC_A}{dt} = \frac{k_1 k_2 C_A^2}{k_1' C_A + k_2} \tag{4-72}$$

At high pressures $k_1' C_A \gg k_2$, and Equation (4-72) reduces to

$$r_A = -\frac{dC_A}{dt} = \frac{k_1 k_2}{k_1'} C_A$$

which is of the same form as Equation (4-67). At low pressures $k_1'C_A \ll k_2$, and Equation (4-72) reduces to

$$r_A = -\frac{dC_A}{dt} = k_1 C_A^2$$

which is of the same form as Equation (4-66). Since $k_1 \ll k_2$, essentially all of A^* must be in the form of either A or B and C at any time. Thus the stoichiometry [Equation (4-65)] is satisfied.

B. Concept of the Rate-Controlling Step

Although the concepts of the *rate-controlling step* and the *pseudo steady state* are fundamentally different, they do have some similarities and they frequently lead to the same results. In the application of the concept of the rate-controlling step, it is supposed that one reaction of the mechanism is the rate-controlling step and that the remaining reactions of the mechanism are in dynamic equilibrium.

To illustrate the concept of the rate-controlling step, the reaction with the stoichiometry given by Equation (4-65) and rate expression given by Equation (4-67) are used. In particular, it is supposed that the second reaction [Equation (4-69)] of the mechanism is the rate-contolling step, while the first reaction [Equation (4-68)] is in dynamic equilibrium, and again the amount of A in the form of A^* is negligible. By dynamic equilibrium, it is meant that even though the ratios of the concentrations remain fixed as required by the respective equilibrium constants, the individual concentrations are continually changing as required by the rate-controlling step. Such a dynamic equilibrium is approached as k_1 and k_1' become very large relative to k_2. Then suppose $k_1 \gg k_2$ and $k_1' \gg k_2$. The fact that the amount of A in the form of A^* at any time is negligible implies that $K_1 \ll 1$. Since the first reaction is assumed to be in a state of dynamic equilibrium,

$$K_1 = \frac{C_A^* C_A}{C_A^2} = \frac{C_A^*}{C_A} \tag{4-73}$$

Since the second reaction is assumed to be the rate-controlling step, it follows that

$$r_B = r_C = k_2 C_A^* \tag{4-74}$$

Elimination of C_A^* from Equations (4-73) and (4-74) gives

$$r_B = r_C = K_1 k_2 C_A \tag{4-75}$$

For the mechanism to satisfy the stoichiometry $(A \rightarrow B + C)$, it is evident that A must be in the form of either A, B, or C, but not A^*. This result has been assured by taking $K_1 \ll 1$. Consequently, the stoichiometry is satisfied, and the rate of reaction of A is equal to the rate of formation of B or

C; that is, $r_A = r_B = r_C$. The concept of the rate-controlling step has been found to be most helpful in the analysis of reaction mechanisms composed of large numbers of reactions.

EXAMPLE 4-3

An experimental investigation of the liquid phase hydrochlorination of octyl and dodecyl alcohols in a batch reactor by Date et al. [8] revealed the following stoichiometry:

$$R_1OH + HCl \rightarrow R_1Cl + H_2O \qquad (1)$$

$$R_2OH + HCl \rightarrow R_2Cl + H_2O \qquad (2)$$

where the symbols R_1OH and R_1Cl are used to represent octyl alcohol and octyl chloride, respectively, and R_2OH and R_2Cl are used to represent dodecyl alcohol and dodecyl chloride, respectively. It was further found that the rates of reaction could be represented at 100.5°C as follows:

$$r_{R_1OH} = k_1[R_1OH][HCl] \qquad (A)$$

$$r_{R_2OH} = k_2[R_2OH][HCl] \qquad (B)$$

where
$$k_1 = 1.6 \times 10^{-3} \text{ liter/g mol min}$$
$$k_2 = 1.92 \times 10^{-3} \text{ liter/g mol min}$$
$$[A] = C_A = \text{concentration of compound } A$$

(a) Propose an ionic mechanism that satisfies both the stoichiometry and the rate expressions.

(b) On the basis of the initial concentrations in gram moles per liter,

$$[HCl]^0 = 1.3, \qquad [R_1OH]^0 = [R_2OH]^0 = 2.21$$

calculate the gram moles of dodecyl chloride and octyl chloride produced per liter of solution at the time 20% of the octyl alcohol has been consumed by reaction. Reactions (1) and (2) are carried out at 100.5°C in the liquid phase in a batch reactor, and it may be assumed that the density of the liquid phase remains essentially constant throughout the course of the reaction. Also, calculate the percentage of the HCl reacted when the 20% of the octyl alcohol has reacted.

SOLUTION

(a) The following mechanism was proposed by Date et al. [8] for the hydrochlorination of each alcohol:

$$ROH + HCl \underset{k_3'}{\overset{k_3}{\rightleftharpoons}} R\overset{\overset{\displaystyle H}{|}}{\underset{\underset{\displaystyle +}{}}{O}}-H + Cl^- \qquad (3)$$

$$Cl^- + R\overset{\overset{\displaystyle H}{|}}{\underset{\underset{\displaystyle +}{}}{O}}-H \overset{k_4}{\longrightarrow} RCl + H_2O \qquad (4)$$

where R is used to represent either the octyl or the dodecyl group. It was further supposed that reaction (3) was in equilibrium and that $K_3 \ll 1$. Reaction (4) was assumed to be the rate-controlling step. Then the rate of appearance of RCl (R_1Cl or R_2Cl) is given by

$$r_{RCl} = k_4[Cl^-][ROH_2^+] \tag{C}$$

Since reaction (3) is assumed to be at equilibrium,

$$[ROH_2^+][Cl^-] = K_3[ROH][HCl] \tag{D}$$

Equations (C) and (D) may be combined to give

$$r_{RCl} = k_4 K_3[ROH][HCl] \tag{E}$$

Since $K_3 \ll 1$, it follows that essentially all the ROH is in the form of either ROH or RCl, but not ROH_2^+. Consequently, the stoichiometry is satisfied, with the result that the rate of formation of RCl must be equal to the rate of disappearance of ROH. Thus

$$r_{ROH} = r_{RCl} = k_4 K_3[ROH][HCl] = k[ROH][HCl] \tag{F}$$

Equation (F) is seen to be in agreement with Equations (A) and (B); the constant $k_4 K_3 = k$ in Equation (F) is of course different for each alcohol.

(b) Since the mass density is constant, the material balances on R_1OH and R_2OH in a batch reactor may be stated in the form

$$r_{R_1OH} = -\frac{d[R_1OH]}{dt} \tag{G}$$

$$r_{R_2OH} = -\frac{d[R_2OH]}{dt} \tag{H}$$

Elimination of r_{R_1OH} and r_{R_2OH} from Equations (A), (B), (G), and (H) yields

$$-\frac{d[R_1OH]}{dt} = k_1[R_1OH][HCl] \tag{I}$$

$$-\frac{d[R_2OH]}{dt} = k_2[R_2OH][HCl] \tag{J}$$

By division of the members of Equation (I) by the corresponding members of Equation (J) and by integration of the result so obtained, it is found upon rearrangement that

$$\frac{[R_2OH]}{[R_2OH]^0} = \left(\frac{[R_1OH]}{[R_1OH]^0}\right)^{k_2/k_1} \tag{K}$$

For 20% conversion of octyl alcohol, $[R_1OH] = 0.8[R_1OH]^0$. For this condition, Equation (K) gives

$$\frac{[R_2OH]}{[R_2OH]^0} = 0.8^{(1.92\times10^{-3})/(1.6\times10^{-3})} = 0.8^{1.2} = 0.765$$

Thus

$$[R_2OH] = 0.765[R_2OH]^0 = (0.765)(2.21) = 1.69$$

Then, on the basis of 1 liter of the reacting mixture, it follows that the gram moles of dodecyl chloride formed $= 2.21 - 1.69 = 0.52$.

The gram moles of octyl chloride formed = (2.21)(0.2) = 0.442.

The gram moles of dodecyl chloride formed per gram mole of octyl chloride formed = 0.52/0.442 = 1.18.

The gram moles of HCl reacted are equal to the gram moles of R_1Cl plus the gram moles of R_2Cl formed = 0.52 + 0.442 = 0.962.

The gram moles of HCl reacted per gram mole of HCl in the feed = 0.962/1.30 = 0.74.

In the interest of simplicity, relatively simple reaction mechanisms were used in the demonstration of the applications of the pseduo-steady-state assumption and the rate-controlling-step assumption. The student may demonstrate the utility of these assumptions in the analysis of relatively complex reaction mechanisms by solving the problems at the end of this chapter.

III. INTERPRETATION OF EXPERIMENTAL RESULTS

When only one reaction is involved and the reaction proceeds at a relatively slow rate, the reaction rate constant may be determined in a batch reactor operated at isothermal conditions. The mechanism may be established and the rate constant or constants determined by use of the integral expressions and plots presented in Chapter 1.

If the reaction proceeds too fast for an investigation of the reaction in a batch reactor to be feasible, a flow-type reactor may be employed. In the following analysis, batch reactors and plug-flow reactors are used to demonstrate some possible techniques which may be employed in the analysis of the experimental results for complex reactions.

If a single reaction is involved in an isothermally operated flow reactor in which a negligible drop in pressure occurs, then the integral expressions presented in Chapter 2 may be used for the determination of the mechanism and rate constant (or constants where reversible reactions are involved). Also, because of the simplicity of the rate expressions, perfectly mixed flow reactors (see Figure 2-7) could prove useful in the study of relatively fast reactions.

A. Determination of the Stoichiometry

Actually, there is no unique technique for establishing the stoichiometry of any arbitrary mechanism. For example, in the case of some relatively complex mechanisms, it may be necessary to determine the mechanism and stoichiometry simultaneously. Although the following techniques are not applicable for all types of reactions, they should prove helpful in the analysis of experimental results.

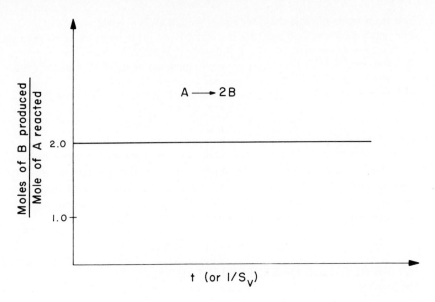

Figure 4-4. Determination of the stoichiometric numbers for a single reaction.

Suppose that data have been collected from a batch reactor or a plug-flow reactor in the form of product compositions as a function of t for batch reactors and $1/S_V$ for plug-flow reactors. Furthermore, suppose that only t (or $1/S_V$) has been varied and that all remaining variables, such as the composition of the feed, the reaction temperature, and the total pressure, have been held fixed for all experiments. First, suppose that the single reaction

$$A \xrightarrow{k_1} 2B \tag{4-76}$$

occurs. If this is the only reaction occurring in the system, a plot of the moles of product B formed per mole of A reacted versus t (or $1/S_V$) would give a horizontal line with an ordinate of 2, as shown in Figure 4-4.

Instead of a single reaction, suppose that A also reacts to give $2B$ and C as follows:

$$\begin{aligned} A &\xrightarrow{k_1} 2B \\ A &\xrightarrow{k_2} C \end{aligned} \tag{4-77}$$

For this case, plots of B and C formed per mole of A reacted would yield horizontal lines with ordinates having the values shown in Figure 4-5.

Next, consider the case where B further reacts to a secondary product D; that is,

$$\begin{aligned} A &\xrightarrow{k_1} 2B \\ A &\xrightarrow{k_2} C \\ B &\xrightarrow{k_3} D \end{aligned} \tag{4-78}$$

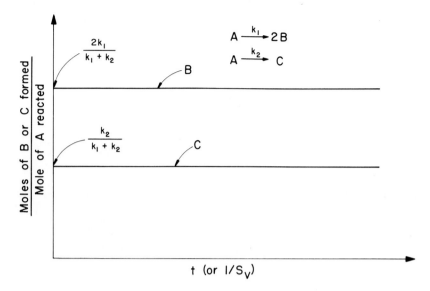

Figure 4-5. Determination of the stoichiometric numbers for the two simultaneous reactions.

Then the moles of B formed per mole of A reacted decrease as t or $1/S_V$ increases. Also, the plot of the moles of D formed per mole of A reacted begins at the origin and increases as t or $1/S_V$ increases, as shown in Figure 4-6.

The curves for B and C are typical of those for compounds produced by *primary* reactions, and the curve for D is typical of those for compounds produced by *secondary* reactions. Graphs of this type are helpful in establishing the primary and secondary reactions involved in a particular reaction mechanism.

B. Interpretation of Integral Reactor Data

In the following analysis the parameters of temperature, pressure, and feed composition are held fixed for all runs. Suppose that a series of runs has been made in a plug-flow reactor and that product distributions have been determined for each run and plotted as indicated in Figure 4-7. Before further consideration of the particular reaction mechanism needed to satisfy the results presented in Figure 4-7, the relationships needed in the interpretation of *integral reactor* data are developed. At each value of V for a given run, the component material balance on A is given by

$$r_A = -\frac{dn_A}{dV} \tag{4-79}$$

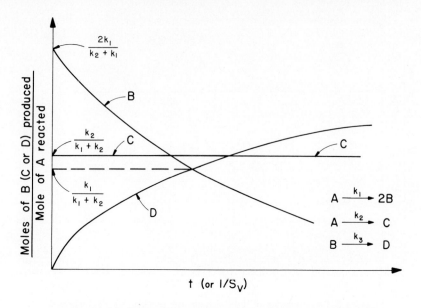

Figure 4-6. Determination of primary and secondary reactions.

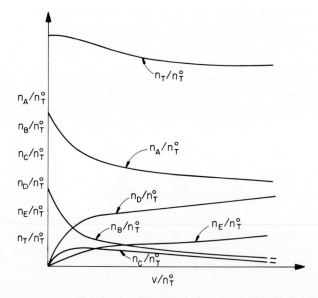

Figure 4-7. Results obtained in an integral plug-flow reactor at the conditions (1) isothermal operation at the same temperature for all runs, (2) same pressure for all runs, and (3) same feed composition for all runs.

Then, for this run,

$$V = -\int_{n_A^0}^{n_A} \frac{dn_A}{r_A} \tag{4-80}$$

If the run under consideration is made at the feed rate n_T^0, then both sides of Equation (4-80) can be divided by the constant n_T^0 (for the run under consideration) to give

$$\frac{V}{n_T^0} = -\int_{n_A^0}^{n_A} \frac{dn_A}{n_T^0 r_A} = -\int_{y_A^0}^{n_A/n_T^0} \frac{d\left(n_A/n_T^0\right)}{r_A} \tag{4-81}$$

where $y_A^0 = n_A^0/n_T^0$. Now, Equation (4-81) may be regarded as representing the result achieved by a sequence of runs in which V/n_T^0 is varied. Note either or both V and n_T^0 may be varied from run to run, but that n_A/n_T^0 depends on only the ratio of V to n_T^0. The sequence of runs may be represented as follows:

$$\left(\frac{V}{n_T^0}\right)_1 = -\int_{y_A^0}^{(n_A/n_T^0)_1} \frac{d\left(n_A/n_T^0\right)}{r_A}$$

$$\left(\frac{V}{n_T^0}\right)_2 = -\int_{y_A^0}^{(n_A/n_T^0)_2} \frac{d\left(n_A/n_T^0\right)}{r_A} \tag{4-82}$$

$$\vdots \qquad\qquad \vdots \qquad\qquad \vdots$$

$$\left(\frac{V}{n_T^0}\right)_n = -\int_{y_A^0}^{(n_A/n_T^0)_n} \frac{d\left(n_A/n_T^0\right)}{r_A}$$

If V/n_T^0 is now regarded as a dependent variable and n_A/n_T^0 as an independent variable, Equation (4-81) may be differentiated to give

$$\frac{d\left(V/n_T^0\right)}{d\left(n_A/n_T^0\right)} = -\frac{1}{r_A}$$

or

$$r_A = -\frac{d\left(n_A/n_T^0\right)}{d\left(V/n_T^0\right)} \tag{4-83}$$

[Note that the result given by Equation (4-83) is based on the assumption that the lower limits in the integrals in Equation (4-82) remain constant for all runs.] Then the slope of the curve n_A/n_T^0 versus V/n_T^0 in Figure 4-7 gives the numerical value of r_A in a reactor in which the partial pressures of A, B, C, and D are given by expressions of the form

$$p_A = \frac{n_A}{n_T} P = \frac{\left(n_A/n_T^0\right)P}{n_T/n_T^0} \tag{4-84}$$

n_A/n_T^0 and n_T/n_T^0 are read from the graph at the V/n_T^0 having the slope r_A [see Equation (4-83)].

Alternately, r_A may be stated in terms of the derivative of n_A/n_T^0 with respect to $1/S_V$. Since

$$S_V = \frac{n_T^0}{V} \frac{RT_s}{P_s}$$

it follows that

$$r_A = -\left(\frac{P_s}{RT_s}\right) \frac{d\left(n_A/n_T^0\right)}{d(1/S_V)} \tag{4-85}$$

Since r_A may be expressed in terms of the rate constants and partial pressures of the proposed mechanism, it is evident that the results of integral reactor tests may be used to determine the rate constants corresponding to a proposed mechanism. For example, suppose that the results presented in Figure 4-7 have been obtained by making a series of runs and that it is desired to determine the mechanism and rate constants. The following approach may be used. First, the data shown in Figure 4-7 are plotted in the same form demonstrated in Figure 4-6. For example, for component C,

$$\frac{\left(n_C/n_T^0\right) - y_C^0}{y_A^0 - \left(n_A/n_T^0\right)} \quad \text{is plotted versus} \quad \frac{V}{n_T^0}$$

From such a plot, it may be established that C and D are primary products and E is a secondary product. The fact that the curve for C in Figure 4-7 passes through a maximum and has a shape similar to compound B in Figure 4-6 further suggests that C is involved in the reaction that produces E. Furthermore, suppose that the reactions $A + B \to C + D$ and $2C \to E$ constitute balanced stoichiometric equations. Then a mechanism which is in agreement with stoichiometric experimental results is as follows:

$$A + B \underset{k_1'}{\overset{k_1}{\rightleftarrows}} C + D \tag{4-86}$$

$$2C \underset{k_2'}{\overset{k_2}{\rightleftarrows}} E \tag{4-87}$$

If a single set of four rate constants can be found which fits the experimental results over the range of interest, then it can be concluded that the mechanism given by Equations (4-86) and (4-87) is a satisfactory one. If sufficient thermodynamic data are available, the equilibrium constants K_1 and K_2 may be computed, and the number of rate constants to be determined experimentally may be reduced to two, say k_1 and k_2, since

$$k_1' = \frac{k_1}{K_1} \tag{4-88}$$

$$k_2' = \frac{k_2}{K_2} \tag{4-89}$$

The following differential equations may be stated for components A, B, C, D, and E:

$$-\frac{d(n_A/n_T^0)}{d(V/n_T^0)} = r_1 - r_1' \tag{4-90}$$

$$-\frac{d(n_B/n_T^0)}{d(V/n_T^0)} = r_1 - r_1' \tag{4-91}$$

$$\frac{d(n_C/n_T^0)}{d(V/n_T^0)} = (r_1 - r_1') - (2r_2 - 2r_2') \tag{4-92}$$

$$\frac{d(n_D/n_T^0)}{d(V/n_T^0)} = r_1 - r_1' \tag{4-93}$$

$$\frac{d(n_E/n_T^0)}{d(V/n_T^0)} = r_2 - r_2' \tag{4-94}$$

For a gas phase reaction,

$$\begin{aligned} r_1 &= k_1 p_A p_B, & r_1' &= k_1' p_C p_D \\ r_2 &= k_2 p_C^2, & r_2' &= k_2' p_E \end{aligned} \tag{4-95}$$

If the proposed mechanism is in agreement with the experimental results shown in Figure 4-7, then at each V/n_T^0, it follows from Equations (4-90), (4-91), and (4-93) that the following equalities must be satisfied:

$$-\frac{d(n_A/n_T^0)}{d(V/n_T^0)} = -\frac{d(n_B/n_T^0)}{d(V/n_T^0)} = \frac{d(n_D/n_T^0)}{d(V/n_T^0)} \tag{4-96}$$

Suppose that the experimental results are in agreement with Equation (4-96). Then, of the three expressions given by Equations (4-90), (4-91), and (4-93), only one of them is independent. Let Equation (4-90) be selected as the independent equation of the set. However, of the three remaining equations [Equations (4-90), (4-92), and (4-94)], only two of these at most are independent because Equation (4-92) may be obtained by multiplying both sides of Equation (4-94) by -2 and adding the result to Equation (4-90); that is,

$$-2\frac{d(n_E/n_T^0)}{d(V/n_T^0)} - \frac{d(n_A/n_T^0)}{d(V/n_T^0)} = r_1 - r_1' - 2(r_2 - r_2') \tag{4-97}$$

The right side of this expression is seen to be identical to the right side of Equation (4-92). The left side may be shown to be identical by means of material balances. Let z denote the moles of A that have reacted at any point in a reactor. Then

$$n_A = n_A^0 - z \tag{4-98}$$

and the moles of C at any point in a reactor are given by

$$n_C = n_C^0 + z - 2(n_E - n_E^0) \qquad (4\text{-}99)$$

Addition of Equations (4-98) and (4-99) followed by rearrangement yields

$$n_C = (n_A^0 + n_C^0 + 2n_E^0) - 2n_E - n_A \qquad (4\text{-}100)$$

Division of each member of Equation (4-100) by n_T^0 followed by differentiation with respect to V/n_T^0 yields

$$\frac{d(n_C/n_T^0)}{d(V/n_T^0)} = -\frac{2d(n_E/n_T^0)}{d(V/n_T^0)} - \frac{d(n_A/n_T^0)}{d(V/n_T^0)} \qquad (4\text{-}101)$$

Therefore, the left sides of Equations (4-97) and (4-92) are identical, and consequently only two, at most, of the three expressions [Equations (4-90), (4-92), and (4-94)] are independent. (For the general case of n rate equations, the theory of linear algebra may be used to determine the number of linearly independent equations.) Let Equations (4-90) and (4-94) be selected for consideration as the independent set. Obviously, neither of these expressions can be produced from the other. Consequently, they are independent. Since Equations (4-90) and (4-94) constitute two equations in four unknowns (k_1, k_1', k_2, and k_2'), each equation must be stated at each of two different values of V/n_T^0. The equations so obtained may be represented by the following matrix equations:

$$\begin{bmatrix} a_1 & -b_1 \\ a_2 & -b_2 \end{bmatrix} \begin{bmatrix} k_1 \\ k_1' \end{bmatrix} = \begin{bmatrix} m_1 \\ m_2 \end{bmatrix} \qquad (4\text{-}102)$$

and

$$\begin{bmatrix} c_1 & -d_1 \\ c_2 & -d_2 \end{bmatrix} \begin{bmatrix} k_2 \\ k_2' \end{bmatrix} = \begin{bmatrix} m_3 \\ m_4 \end{bmatrix} \qquad (4\text{-}103)$$

where $a_1 = (p_A p_B)_1$. [The subscript 1 means that the product enclosed by parentheses is to be evaluated at the particular value of V/n_T^0 denoted by $(V/n_T^0)_1$.]

$$a_1 = (p_A p_B)_1, \qquad a_2 = (p_A p_B)_2$$
$$b_1 = (p_C p_D)_1, \qquad b_2 = (p_C p_D)_2$$
$$c_1 = (p_C^2)_1, \qquad c_2 = (p_C^2)_2$$
$$d_1 = (p_E)_1, \qquad d_2 = (p_E)_2$$
$$m_1 = \left[-\frac{d(n_A/n_T^0)}{d(V/n_T^0)} \right]_1, \qquad m_2 = \left[-\frac{d(n_A/n_T^0)}{d(V/n_T^0)} \right]_2$$
$$m_3 = \left[\frac{d(n_E/n_T^0)}{d(V/n_T^0)} \right]_1, \qquad m_4 = \left[\frac{d(n_E/n_T^0)}{d(V/n_T^0)} \right]_2$$

Matrix Equation (4-102) is readily solved for k_1 and k'_1, and k_2, and k'_2 may be determined by solving matrix Equation (4-103). If after the procedure described has been repeated for other choices of V/n^0_T within the range of interest and the values of the rate constants so obtained are in reasonable agreement, a suitable mechanism has been found. (Note that, if equilibrium constants are available for each of the two reactions, the number of equations and unknowns is reduced to two, since k'_1 and k'_2 may be expressed in terms of k_1 and k_2, respectively.)

C. Interpretation of Data from Other Types of Reactors

At small values of V/n^0_T, where the conversion per pass through the reactor is small, the reactor is commonly called a *differential reactor*. The small conversion per pass in a differential reactor simplifies the evaluation of the integral appearing in Equation (4-80). In particular, Equation (4-80) may be restated as follows by use of the *mean-value theorem of integral calculus*:

$$V = -\int_{n^0_A}^{n_A} \frac{dn_A}{r_A} = -\left(\frac{1}{r_A}\right)_m (n_A - n^0_A) \qquad (4\text{-}104)$$

where the subscript m is used to denote the mean value of $1/r_A$. For small conversions per pass, the mean value of $1/r_A$ may be approximated with good accuracy by use of the arithmetic average of $1/r_A$ at n_A and n^0_A. That is,

$$\left(\frac{1}{r_A}\right)_m \cong \frac{1}{2}\left(\frac{1}{r_A}\bigg|_{n_A} + \frac{1}{r_A}\bigg|_{n^0_A}\right) \qquad (4\text{-}105)$$

In the interest of simplicity, the subscript m is dropped in the following analysis of the case where the reactions given by Equations (4-86) and (4-87) are carried out in a differential reactor. For a differential reactor, the equations corresponding to Equations (4-90) and (4-94) are as follows:

$$\frac{n^0_A - n_A}{V} = r_1 - r'_1 \quad \text{or} \quad \frac{n^0_A/n^0_T - n_A/n^0_T}{V/n^0_T} = r_1 - r'_1 \qquad (4\text{-}106)$$

$$\frac{n_E - n^0_E}{V} = r_2 - r'_2 \quad \text{or} \quad \frac{n_E/n^0_T - n^0_E/n^0_T}{V/n^0_T} = r_2 - r'_2 \qquad (4\text{-}107)$$

Since the feed composition is held fixed for all runs, it follows that n^0_A/n^0_T and n^0_E/n^0_T are constant for all choices of V/n^0_T. Equations (4-106) and (4-107) can be applied within their range of applicability to determine the rate constants in the same manner as described for Equations (4-90) and (4-94).

Data collected in batch reactors may be analyzed in a manner analogous to that described. In this case, Figure 4-7 would consist of a plot of the ratios

TABLE 4-1

SELECTED EXPERIMENTAL RESULTS REPORTED BY BILLINGSLEY AND HOLLAND [9] FOR THE PARTIAL OXIDATION OF PROPYLENE BY AIR OVER A COPPER OXIDE CATALYST

W/n_T^0	Partial Pressure (in. of Hg)		Feed Rates (g mol/h)					Effluent Rates (g mol/h)				
	C_3H_6	O_2	O_2	N_2	CO_2	C_3H_8	C_3H_6	O_2	CO_2	C_3H_8	C_3H_6	C_3H_4O
4.04	10.90	11.07	0.3819	1.5279	0.0006	0.0132	0.3663	0.3641	0.0109	0.0142	0.3606	0.0022
2.84	10.91	11.23	0.5440	2.1761	0.0007	0.0189	0.5217	0.5242	0.0120	0.0218	0.5151	0.0029
2.07	10.94	10.79	0.7404	2.9617	0.0010	0.0269	0.7437	0.7197	0.0130	0.0261	0.7368	0.0029
1.69	12.03	11.67	0.9023	3.6096	0.0013	0.0334	0.9213	0.8797	0.0146	0.0261	0.9142	0.0027
1.29	10.83	11.66	1.2032	4.8128	0.0016	0.0402	1.1099	1.1769	0.0169	0.0401	1.1015	0.0034
4.09	19.19	10.01	0.3240	1.2961	0.0006	0.0221	0.6105	0.3101	0.0073	0.0199	0.6045	0.0038
2.95	17.75	10.26	0.4629	1.8515	0.0008	0.0285	0.7881	0.4483	0.0079	0.0257	0.7819	0.0038
2.22	18.76	10.05	0.6017	2.4070	0.0010	0.0402	1.1100	0.5826	0.0105	0.0386	1.1020	0.0049
1.55	19.13	10.01	0.8563	3.4250	0.0015	0.0590	1.6316	0.8280	0.0160	0.0552	1.6204	0.0065
4.31	25.90	8.54	0.2660	1.0643	0.0004	0.0285	0.7881	0.2527	0.0060	0.0280	0.7823	0.0036
2.76	24.82	8.65	0.4210	1.6843	0.0008	0.0430	1.1876	0.4050	0.0080	0.0471	1.1800	0.0052
1.93	25.36	8.72	0.6015	2.4062	0.0010	0.0623	1.7204	0.5818	0.0102	0.0679	1.7114	0.0061
1.19	25.38	8.64	0.9717	3.8871	0.0017	0.1020	2.8194	0.9428	0.0154	0.1073	2.8064	0.0083
3.92	28.33	7.49	0.2660	1.0643	0.0005	0.0358	0.9879	0.2539	0.0056	0.0154	0.9818	0.0044
2.66	30.76	7.65	0.3817	1.5272	0.0007	0.0542	1.4985	0.3618	0.0090	0.0558	1.4883	0.0074
1.90	29.50	7.64	0.5437	2.1750	0.0010	0.0747	2.0646	0.5223	0.0097	0.0804	2.0534	0.0083
1.20	29.90	7.65	0.8560	3.4244	0.0015	0.1193	3.2967	0.8255	0.0138	0.1186	3.2805	0.0120
4.04	11.06	11.36	0.3816	1.5261	0.0007	0.0132	0.3663	0.3638	0.0110	0.0134	0.3606	0.0023
4.04	11.03	11.30	0.3813	1.5255	0.0007	0.0132	0.3663	0.3630	0.0112	0.0139	0.3601	0.0025
1.55	18.91	9.82	0.8552	3.4209	0.0016	0.0591	1.6316	0.8329	0.0124	0.0584	1.6220	0.0060
1.21	18.40	9.64	1.0863	4.3455	0.0020	0.0743	2.0535	1.0585	0.0160	0.0740	2.0420	0.0068
4.04	7.98	8.20	0.3812	1.5247	0.0007	0.0132	0.3663	0.3664	0.0092	0.0129	0.3614	0.0021
4.04	8.03	8.33	0.3812	1.5247	0.0007	0.0132	0.3663	0.3671	0.0087	0.0131	0.3615	0.0021

of N_A, N_B, N_C, N_D, and N_E with respect to N_T^0 versus time. Data for perfectly mixed flow reactors are analyzed in a manner analogous to that shown for differential reactors except that the procedure is valid over all values of V/n_T^0.

To determine the variation of the rate constants with temperature, it is necessary to repeat the experiments required to plot Figure 4-7 for each of two or three other temperatures. The values of each k so obtained can generally be expressed as an exponential function of temperature of the form of Equation (1-33).

EXAMPLE 4-4

Propylene was oxidized with air by Billingsley and Holland [9] over a copper oxide catalyst in a fixed-bed, flow-type reactor which is described in some detail in Chapter 9. Selected data obtained by Billingsley are presented in Table 4-1. These data are to be used to establish the stoichiometry of the reactions involved in the catalytic oxidation of propylene by use of the techniques illustrated by Figures 4-4, 4-5, and 4-6. In this case, however, the variable W/n_T^0 [grams of catalyst (gram mole of feed/hour)] may be used as the independent variable rather than t or $1/S_V$.

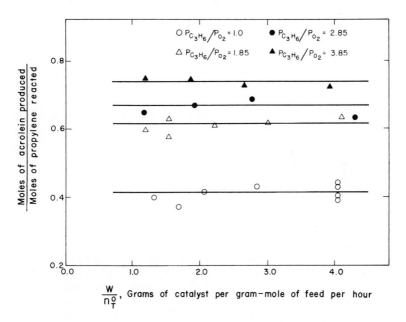

Figure 4-8. The horizontal lines suggest that acrolein is a primary product which does not undergo either further oxidation or polymerization to any significant extent. [Taken from D. S. Billingsley and C. D. Holland, *I & EC Fundamentals*, 2:252 (1963). Used with permission of the American Chemical Society.]

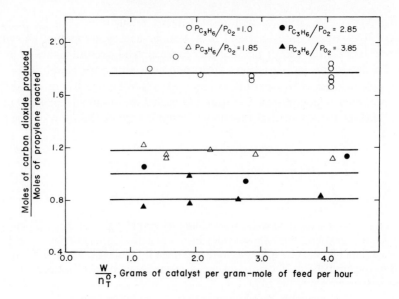

Figure 4-9. The horizontal lines suggest that carbon dioxide is produced by a primary reaction and that any carbon dioxide produced by secondary reactions is insignificant. [Taken from D. S. Billingsley and C. D. Holland, *I & EC Fundamentals*, 2:252 (1963). Used with permission of the American Chemical Society.]

(a) By use of the data presented in Table 4-1, construct the following graphs:

(1) Moles of acrolein produced per mole of propylene reacted versus W/n_T^0.

(2) Moles of carbon dioxide produced per mole of propylene reacted versus W/n_T^0.

(b) From the graphs obtained in part (a), propose a stoichiometry for the oxidation of propylene, and give supporting arguments for the proposed stoichiometry.

SOLUTION

(a) From the data presented in Table 4-1, the graphs presented in Figures 4-8 and 4-9 were constructed.

(b) Comparison of Figures 4-8 and 4-9 with Figures 4-5 and 4-6 suggest that acrolein and carbon dioxide are produced by primary reactions. If a secondary reaction, such as the further oxidation of acrolein to carbon dioxide, were significant, then the curves in Figure 4-8 would have exhibited negative slopes and those in Figure 4-9 would have exhibited positive slopes. If the polymerization of acrolein to the dimer were significant, then the curves in Figure 4-8 would have exhibited negative slopes. The horizontal lines shown in Figures 4-8 and 4-9, plus the fact that other possible products did not appear in the effluent in significant amounts, suggest that the catalytic oxidation of propylene over a copper oxide catalyst is adequately described by the following stoichiometric equations:

$$C_3H_6 + O_2 \rightleftarrows CH_2CHCHO + H_2O$$
$$C_3H_6 + 4.5O_2 \rightleftarrows 3CO_2 + 3H_2O$$

PROBLEMS

4-1. (a) Begin with Equation (4-17) and the initial conditions stated below it and develop the expressions given by Equations (4-18) and (4-19).

(b) Begin with Equations (4-5) and (4-17) and develop Equation (4-20).

(c) Begin with Equations (4-28), (3-58), and the initial condition that at $N_B = N_B^0$, $N_M = 0$, and obtain the result given by Equation (4-29).

(d) Beginning with Equations (4-30), (3-58), (4-29), and the initial condition that at $N_B = N_B^0$, $N_D = 0$, obtain the result given by Equation (4-31).

(e) Show that Equation (4-22) may be integrated to give Equation (4-32), provided that the moles of C are held fixed by feeding C to the reactor at the same rate at which it is consumed by the reaction.

4-2. For the case where the reaction given by Equation (4-1) is carried out isothermally in the gas phase in a plug-flow reactor at steady-state operation, obtain n_A, n_R, and n_S as functions of V/v_T. Assume that the pressure drop is negligible and that the gaseous mixture obeys the perfect gas law.

4-3. If the complex reaction given by Equation (4-1) is carried out isothermally in the liquid phase in a perfectly mixed flow reactor at steady-state operation, show that when the feed contains A alone ($n_R^0 = n_S^0 = 0$),

$$n_A = \frac{n_A^0[1/\theta + (k_1' k_2'/a)]}{1/\theta + k_1 - k_1'/a(k_1 - k_2')}$$

where $\quad a = \dfrac{1}{\theta} + k_2' + k_2 + k_1'$, and the rate constants are based on concentration units

$$\theta = \frac{V}{v_T}$$

4-4. The chloride–chlorate reaction for the production of chlorine dioxide was investigated by Hong et al. [10], who found that the stoichiometry of the reactions involved could be represented by the following equation:

$$2H^+ + ClO_3^- + Cl^- \rightarrow ClO_2 + \tfrac{1}{2}Cl_2 + H_2O \qquad (A)$$

In a strong acid solution at 25°C and at high ratios of chlorate (ClO_3^-) to chloride (Cl^-). Hong et al. proposed the following mechanism

$$3H^+ + 2Cl^- + ClO_3^- \rightleftarrows Cl_2 + HClO_2 + H_2O \qquad (1)$$

$$HClO_2 + H^+ + ClO_3^- \rightarrow 2ClO_2 + H_2O \qquad (2)$$

$$HClO_2 + H^+ + Cl^- \rightarrow 2HOCl \qquad (3)$$

$$HOCl + H^+ + Cl^- \rightarrow Cl_2 + H_2O \qquad (4)$$

and assumptions:

1. The accumulation of $HClO_2$ and $HOCl$ with time is negligible.
2. $k_2[H^+][ClO_3^-] \gg k_3[H^+][Cl^-]$.
3. $k_2[H^+][ClO_3^-] \gg k_1'[Cl_2]$.

On the basis of the proposed mechanism and assumptions, show that

$$\frac{d[\text{ClO}_2]}{dt} = 2k_1[\text{H}^+]^3[\text{Cl}^-]^2[\text{ClO}_3^-]$$

and

$$\frac{d[\text{Cl}_2]}{dt} = k_1[\text{H}^+]^3[\text{Cl}^-]^2[\text{ClO}_3^-]$$

4-5. Acrylonitrile is produced by reaction of propylene, ammonia, and oxygen over a bismuth phosphomolybdate catalyst [11, 12]. (This catalyst was developed by SOHIO for producing acrylonitrile.) The reaction rates were found to be independent of ammonia or oxygen concentrations and could be represented as

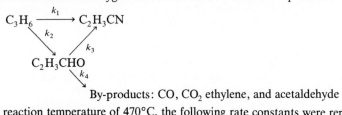

By-products: CO, CO_2 ethylene, and acetaldehyde

For a reaction temperature of 470°C, the following rate constants were reported:

$k_1 = 0.195 \text{ s}^{-1}$

$k_2 = 0.005 \text{ s}^{-1}$

$k_3 = 0.4 \text{ s}^{-1}$

k_4 = a small bimolecular rate constant $(r_4 = k_4[\text{O}_2][\text{C}_2\text{H}_3\text{CHO}])$

The rates of the first and second reactions are first order with respect to propylene, and the third is first order with respect to acrolein.

    Acrylonitrile is to be produced by use of the SOHIO process in a plug-flow reactor. The reactor is to be operated isothermally at 470°C, and a reciprocal space velocity $V/v_T^0 = 9.1$ s. (Actually, Farkas [11] and Callahan et al. [12] state that they used a residence time of 9.1 s. Their residence time was taken to mean space time or reciprocal space velocity.) The feed to the reactor contains 6.6% propylene, 86.1% air, and 7.3% ammonia. Since the amount of propylene in the feed is relatively small, the change in the total number of moles resulting from the reaction may be neglected, and the volumetric flow rate v_T may be assumed constant throughout the reactor volume V. If the by-product reaction is neglected, calculate:

(a) Percent of the propylene converted

(b) Moles of acrolein produced per mole of propylene reacted

(c) Moles of acrylonitrile produced per mole of propylene reacted

Answer: (a) 83.8% (b) 0.00405 (c) 0.996

4-6. Gulyaev and Polak [13] proposed the following mechanism for the thermal decomposition of methane:

$$CH_4 \xrightarrow{k_1} \ddot{C}H_2 + H_2$$

$$\ddot{C}H_2 + CH_4 \xrightarrow{\overline{k_1}} C_2H_6 \tag{1}$$

$$C_2H_6 \xrightarrow{k_2} C_2H_4 + H_2 \tag{2}$$

$$C_2H_4 \xrightarrow{k_3} C_2H_2 + H_2 \tag{3}$$

$$C_2H_2 \xrightarrow{k_4} 2C + H_2 \tag{4}$$

where
$$k_1 = 2.25 \times 10^{13} \exp\left(\frac{-91{,}000}{RT}\right) s^{-1}$$

$$k_2 = 9 \times 10^{13} \exp\left(\frac{-69{,}000}{RT}\right) s^{-1}$$

$$k_3 = 2.57 \times 10^8 \exp\left(\frac{-40{,}000}{RT}\right) s^{-1}$$

$$k_4 = 1.7 \times 10^6 \exp\left(\frac{-30{,}000}{RT}\right) s^{-1}$$

where T is in Kelvin and $R = 1.9872$.

To approximate the residence times required to maximize the conversions of methane to ethylene and to acetylene, respectively, in a plasma jet, the equations for isothermal operation of a constant-volume batch reactor were used [13]. Carry out the analysis outlined below in (a) through (g) on the basis of a reaction temperature of 3000 K and a feed consisting of pure methane.

(a) On the basis of the assumption that the accumulation of $\ddot{C}H_2$ is negligible, show that $[\ddot{C}H_2] = k_1/\overline{k}_1$, where $[\ddot{C}H_2]$ denotes the concentration of $\ddot{C}H_2$ in moles per unit volume.

(b) Show that
$$[CH_4] = [CH_4]^0 e^{-2k_1 t}$$

(c) If the accumulation of ethane is negligible, show that the concentration of ethane at any time t may be approximated as follows:
$$[C_2H_6] = \frac{k_1}{k_2}[CH_4]$$

(d) Show that
$$[C_2H_4] = \frac{k_1[CH_4]^0}{(2k_1 - k_3)}\left(e^{-k_3 t} - e^{-2k_1 t}\right)$$

(e) Show that
$$[C_2H_2] = \frac{k_1 k_3[CH_4]^0}{(2k_1 - k_3)(2k_1 - k_4)(k_3 - k_4)}$$
$$\times\left[(k_3 - k_4)\, e^{-2k_1 t} - (2k_1 - k_4)\, e^{-k_3 t} + (2k_1 - k_3)\, e^{-k_4 t}\right]$$

(f) Show that the maximum concentration of ethylene occurs at time
$$t = \frac{\log_e(2k_1/k_3)}{2k_1 - k_3}$$

(g) Calculate the moles of ethylene, ethane, acetylene, and carbon produced per mole of methane charged at 3000 K at the reaction time corresponding to the maximum production of ethylene.
Answer: $[C_2H_4]/[CH_4]^0 = 0.449$, $[C_2H_6]/[CH_4]^0 = 0.000166$, $[C_2H_2]/[CH_4]^0 = 0.0365$, $[C]/[CH_4]^0 = 0.00196$

4-7. Bollard [14] proposed the following mechanism for the liquid phase oxidation of hydrocarbons:

$$\text{Initiation:} \quad x \xrightarrow{k_1} R \cdot$$

$$\text{Propagation:} \quad R \cdot + O_2 \xrightarrow{k_2} RO_2 \cdot$$

$$RO_2 \cdot + RH \xrightarrow{k_3} ROOH + R \cdot$$

$$\text{Termination:} \quad 2R \cdot \xrightarrow{k_4} R_2$$

$$R \cdot + RO_2 \cdot \xrightarrow{k_5} \text{stable products}$$

$$2RO_2 \cdot \xrightarrow{k_6} \text{stable products}$$

where x = an initiator such as benzoyl peroxide

R = any saturated hydrocarbon

$R \cdot$ = free radical of saturated hydrocarbon

On the assumptions that there is no appreciable accumulation of $[R \cdot]$ and $[RO_2 \cdot]$ and that $k_5 = \sqrt{k_4 k_6}$, show that the rate of reaction of oxygen is given by

$$r_{O_2} = \frac{k_2[O_2]\{ r_i \sqrt{k_6} + k_3[RH]\sqrt{r_i} \}}{k_2[O_2]\sqrt{k_6} + k_3[RH]\sqrt{k_4} + \sqrt{k_4 k_6 r_i}}$$

where r_i is the rate of initiation (the rate of formation of $R \cdot$).

4-8. The following mechanism has been reported by Pacey and Purnell [15] for the pyrolysis of ethane.

$$\text{Initiation:} \quad C_2H_6 \rightarrow 2CH_3 \cdot \tag{1}$$

$$\text{Propagation:} \quad CH_3 \cdot + C_2H_6 \rightarrow CH_4 + C_2H_5 \cdot \tag{2}$$

$$C_2H_5 \cdot \rightarrow C_2H_4 + H \cdot \tag{3}$$

$$H \cdot + C_2H_6 \rightarrow H_2 + C_2H_5 \cdot \tag{4}$$

$$\text{Termination:} \quad C_2H_5 \cdot + C_2H_5 \cdot \rightarrow n\text{-}C_4H_{10} \tag{5}$$

$$C_2H_5 \cdot + C_2H_5 \cdot \rightarrow C_2H_6 + C_2H_4 \tag{6}$$

$$\text{Propylene formation:} \quad C_2H_5 \cdot + C_2H_4 \rightarrow C_3H_6 + CH_3 \cdot \tag{7}$$

$$\text{Inhibition:} \quad H \cdot + C_2H_4 \rightarrow C_2H_5 \cdot \tag{8}$$

No significant accumulation of free radicals was observed experimentally, and thus the pseudo-steady-state assumption may be used in the analysis. Also, it was found experimentally that $k_5 = 7k_6$. Show that the rate of disappearance of ethane and the rate of appearance of ethylene are given by

$$r_{C_2H_6} = k_1[C_2H_6] + k_2[CH_3 \cdot][C_2H_6] + k_4[H \cdot][C_2H_6] - \frac{k_6}{2}[C_2H_5 \cdot]^2$$

$$r_{C_2H_4} = k_3[C_2H_5 \cdot] - k_7[C_2H_5 \cdot][C_2H_4] - k_8[H \cdot][C_2H_4] + \frac{k_6}{2}[C_2H_5 \cdot]^2$$

and show that

$$[CH_3^{\cdot}] = \frac{2k_1}{k_2} + \frac{k_7}{k_2}\frac{[C_2H_4][C_2H_5^{\cdot}]}{[C_2H_6]}$$

$$[H^{\cdot}] = \frac{k_3[C_2H_5^{\cdot}]}{k_4[C_2H_6] + k_8[C_2H_4]}$$

$$[C_2H_5^{\cdot}] = \frac{1}{2}\left(\frac{k_1[C_2H_6]}{k_6}\right)^{1/2}$$

4-9. The following free radical mechanism for the oxidation of hydrocarbons was proposed by Alagy et al. [16].

$$\text{Initiation:}\quad \text{Initiator} \rightarrow R^{\cdot} \qquad\qquad (1)$$

$$R^{\cdot} + O_2 \rightarrow RO_2^{\cdot} \qquad\qquad (2)$$

$$\text{Propagation:}\quad RO_2^{\cdot} + RH \begin{array}{l} \nearrow ROOH + R^{\cdot} \qquad (3) \\ \searrow ROH + RO^{\cdot} \qquad (4) \end{array}$$

$$RO^{\cdot} + RH \rightarrow ROH + R^{\cdot} \qquad\qquad (5)$$

$$\text{Termination:}\quad RO_2^{\cdot} + RO_2^{\cdot} \rightarrow \text{stable products} \qquad (6)$$

$$RO_2^{\cdot} + R^{\cdot} \rightarrow \text{stable products} \qquad (7)$$

$$R^{\cdot} + R^{\cdot} \rightarrow \text{stable products} \qquad (8)$$

On the basis of the assumption that there is no appreciable accumulation of the free radicals, and the assumption that $k_7 = \sqrt{k_6 k_8}$, show that the rate of disappearance of RH and the rates of appearance of ROH and $ROOH$ are given by the following expressions:

$$r_{RH} = (k_3 + k_4)[RO_2^{\cdot}][RH] + k_5[RO^{\cdot}][RH]$$

$$r_{ROH} = k_4[RO_2^{\cdot}][RH] + k_5[RO^{\cdot}][RH]$$

$$r_{ROOH} = k_3[RO_2^{\cdot}][RH]$$

and show that

$$[RO_2^{\cdot}] = \sqrt{r_i/k_6} - \sqrt{k_8/k_6}[R^{\cdot}]$$

$$[R^{\cdot}] = \frac{r_i + (k_3 + k_4)(\sqrt{r_i/k_6})[RH]}{k_2[O_2] + (k_3 + k_4)(\sqrt{k_8/k_6})[RH] + \sqrt{k_8 r_i}}$$

$$[RO^{\cdot}] = \frac{k_4}{k_5}[RO_2^{\cdot}]$$

4-10. For the reaction

$$A + B \rightarrow 2D$$

it has been observed that at high concentrations of B relative to A the rate of reaction is given by

$$r_A = kC_A^2$$

while at high concentrations of A relative to B the rate of reaction is given by

$$r_A = \bar{k}C_A C_B$$

Propose a mechanism that will satisfy both the stoichiometry and the rate expressions.

4-11. Suppose that the following reactions were investigated in a perfectly mixed flow reactor:

$$A \xrightarrow{k_1} R \xrightarrow{k_2} S$$
$$A \xrightarrow{k_3} P$$

For a feed rate of 100 g mol/min of pure A and a residence time $\theta = V/v_T = 20$ min, 80% of A was converted. The product compositions in mole percent were as follows:

Component	%
A	20
R	50
S	20
P	10

(a) Find k_1, k_2, and k_3.
   Answer: $k_1 = 0.175$ min^{-1}; $k_2 = 0.002$ min^{-1}; $k_3 = 0.025$ min^{-1}
(b) For the set of rate constants obtained in part (a), determine the value of θ which should be used to maximize the moles of R produced per mole of A fed. Calculate the composition of the product at this optimum value of θ.
   Answer: $\theta = 15.8$ min, and the mole fractions of A, R, S, and P were 0.240, 0.505, 0.160, and 0.0950, respectively.

4-12. Habelt and Selker [19] utilized the Zeldovich mechanism [20] and the rate constants reported by Baulch et al. [21] and Jones [22] to predict the NO emissions from tangentially fired oil and gas furnaces. The Zeldovich mechanism is as follows:

$$O_2 + M \rightleftarrows 2O + M \tag{1}$$
$$N_2 + O \rightleftarrows NO + N \tag{2}$$
$$O_2 + N \rightleftarrows NO + O \tag{3}$$

where M is any molecule in the system. The rate constants are:

$$k_1 = 5.5 \times 10^{17} \, T^{-1/2} \exp(-59{,}400/T), \text{cm}^3/\text{mol s}$$
$$k_1' = 2.2 \times 10^{16} \, T^{-1/2}, (\text{cm}^3/\text{mol})^2/\text{s}$$
$$k_2 = 1.3 \times 10^{14} \exp(-37{,}943/T), \text{cm}^3/\text{mol s}$$
$$k_2' = 3.10 \times 10^{13} \exp(-166/T), \text{cm}^3/\text{mol s}$$
$$k_3 = 6.43 \times 10^9 \, T \exp(-3145/T), \text{cm}^3/\text{mol s}$$
$$k_3' = 1.55 \times 10^9 \, T \exp(-19{,}445/T), \text{cm}^3/\text{mol s}$$

where T is in Kelvin.

(a) On the basis of the pseudo-steady-state assumption for the atomic species of N and the further assumption that reaction (1) is in dynamic equilibrium, show that the following rate expression is obtained for the formation of NO:

$$r_{NO} = \frac{2k_2[O][N_2]}{1 + k_2'[NO]/k_3[O_2]}\left(1 - \frac{k_2'k_3'[NO]^2}{k_2k_3[N_2][O_2]}\right)$$

(b) Calculate the equilibrium concentration of NO on the basis of a feed consisting of air, a furnace temperature of 2000 K, and a pressure of 1 atm. Give your answer in parts per million (ppm) based on volume.

   (*Hint*: Set $r_{NO} = 0$ or ignore concentrations of atomic nitrogen and atomic oxygen.)

(c) Since the equilibrium conversion is small, the composition of nitrogen and oxygen can be regarded as constant with good accuracy. Show that on the basis of this assumption the preceding expression for the rate of formation of NO reduces to

$$r_{NO} = C_1[NO]*\frac{1 - x^2}{1 + C_2 x}$$

where $[NO]*$ = equilibrium concentration of NO

$$C_1 = 2k_2[N_2][O]/[NO]*$$
$$C_2 = k_2'[NO]*/k_3[O_2]$$
$$x = [NO]/[NO]*$$

(d) On the basis of the rate expression found in part (c), calculate the residence time required to reach 95% of the equilibrium concentration of NO for plug-flow and perfectly mixed flow reactor models for the case where the feed to the furnace is air and the furnace is to operate at 2000 K and 1 atm.

   Answers: (b) 7819 ppm, (c) For perfectly mixed flow reactor, $\theta = 33.61s$, and for plug-flow reactor, $\theta = 5.1s$.

4-13. Show that the ordinates for B and C have the values shown in Figure 4-5.

4-14. (a) Show that the intercepts for B, C, and D have the values shown in Figure 4-6.

(b) Show that at the point of intersection the ordinates of B and C have the value of $k_2/(k_1 + k_2)$. Show that the ordinates of B and D at the point of intersection have the value shown in Figure 4-6.

(c) Show that in the limit as t (or $1/S_V$) approaches infinity the y intercept of the curve for D in Figure 4-6 approaches $2k_1/(k_1 + k_2)$.

4-15. The decomposition of cyclohexane over a platinum–aluminum–mordenite catalyst was investigated by Allan and Voorhies [17]. The principal reactions were the dehydrogenation and the isomerization of cyclohexane. The results could be explained by the following mechanism.

$$C_6H_{12} \rightleftarrows C_6H_6 + 3H_2$$
Benzene

$$C_6H_{12} \rightleftarrows C_6H_{12}$$
Methyl cyclopentane

At 413°C, $k_1/\rho = 0.045$ g mol/(min-g catalyst-atm) and $k_2/\rho = 0.0078$ g mol/(min-g catalyst-atm). The equilibrium constants at 413°C are as follows:

$$K_{p1} = 8301 \text{ atm}^3$$

$$K_{p2} = 8.58$$

These reactions are to be carried out in a plug-flow reactor at 413°C, and 6 atm. The feed to the reactor consists of 15 moles of hydrogen per mole of cyclohexane.

(a) On the basis of these conditions, compute the value of $\rho V/n_T^0$ required to achieve a conversion of 70% of the cyclohexane.

(b) At these conditions, compute the moles of methyl cyclopentane produced per mole of cyclohexane in the feed. The change in the total number of moles as well as the change in the moles of hydrogen may be neglected.

(c) Repeat part (b) where the change in the total number of moles and the change in the moles of hydrogen in the reactor are not neglected.

Answer: (a) $\rho V/n_T^0 = 3.93$ g min/mole.

4-16. Adams et al. [18] used the following network of first-order reactions for the oxidation of propylene over a bismuth molybdate catalyst.

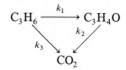

The corresponding stoichiometric reactions are

$$C_3H_6 + O_2 \xrightarrow{k_1} C_3H_4O + H_2O$$

$$C_3H_4O + 3.50_2 \xrightarrow{k_2} 3CO_2 + 2H_2O$$

$$C_3H_6 + 4.50_2 \xrightarrow{k_3} 3CO_2 + 3H_2O$$

The following ratios of the rate constants at 460°C were reported by Adams et al.: $k_3/k_1 = 0.10$, $k_2/k_1 = 0.025$.

(a) Compute the moles of acrolein produced per mole of propylene reacted versus the fraction of propylene reacted for plug-flow and perfectly mixed flow reactors. Plot these results.

(b) Compute the composition of the reactor effluent. Report these results in tabular form.

4-17. The following reaction network has been proposed by Adams et al. [18] for the oxidation-dehydrogenation of butene over a bismuth molybdate catalyst:

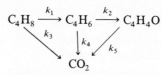

At 460°C, $k_3/k_1 = 0.05$ and $(k_2 + k_4)/k_1 = 0.05$. Determine the moles of butadiene produced per mole of butene reacted when 10% of the butene is reacted in (a) a plug-flow reactor, and (b) a perfectly mixed flow reactor.

Answer: (a) 0.94995 (b) 0.94737.

4-18. Sotelo et al. [23] studied the liquid phase oxidation of cumene initiated by ozone in the presence of sodium cyclohexane carboxylate (sodium naphthenate, NaNap). The experiments were conducted in a 0.5-liter spherical glass reactor that was submerged in a thermostated bath containing silicone oil as a heating medium. The O_2–O_3 mixture was bubbled into the liquid cumene during the initiation period, after which oxygen was fed until the hydroperoxide reached a maximum concentration. A high-performance liquid chromatograph (HPLC) with a C18 μ Bondapack column and an UV detector was used for the analysis of the liquid phase. The following reaction mechanism was proposed to explain the data.

$$RH + O_3 \xrightarrow{k_1} R^{\cdot} + OH^{\cdot} + O_2 \tag{1}$$

$$R^{\cdot} + O_2 \xrightarrow{k_2} RO_2^{\cdot} \tag{2}$$

$$RO_2^{\cdot} + RH \xrightarrow{k_3} ROOH + R^{\cdot} \tag{3}$$

$$2\,ROOH \xrightarrow{k_4} (ROOH)_2 \rightarrow fr \tag{4}$$

$$2\,ROOH + Na^+ \xrightarrow{k_5} (ROOH \cdot ROONa) \rightarrow fr \tag{5}$$

$$ROOH + RO_2^{\cdot} \xrightarrow{k_6} ROH + O_2 + R_2^{\cdot} \rightarrow mp \tag{6}$$

$$2\,RO_2^{\cdot} \xrightarrow{k_7} mp \tag{7}$$

where $\qquad$ $R^{\cdot} =$ isopropylphenyl radical, $C_6H_5—(CH_3)_2C^{\cdot}$

$\qquad$ RH = cumene

$\qquad$ ROOH = cumene hydroperoxide

$\qquad$ ROH = dimethylphenyl carbinol

$\qquad$ fr = free radicals

$\qquad$ $RO_2^{\cdot}$ = radical

$\qquad$ mp = molecular products

For the reaction period from initiation to maximum selectivity of the peroxide, the rate of reaction (6) was small and was thus neglected in the development of the rate expressions.

(a) By use of the pseudo-steady-state assumption, by neglecting reaction (6), and by assuming that reactions (1) through (7) are molecular reactions, show that

$$[RO_2^{\cdot}] = \left(\frac{r_i}{k_7}\right)^{1/2}$$

where $\qquad$ $r_i = [k_4 + k_5 Na^+][ROOH]^2 = k_F[ROOH]^2$

$\qquad$ $r_{ROOH} = k_3[RH][RO_2^{\cdot}] = k_T[RH][ROOH]$

$\qquad$ $k_T = k_F^{1/2}(k_3/k_7^{1/2})$

(b) What assumptions are implied by the following equation:

$$[RH] = [RH]^0 - [ROOH]$$

where $[RH]^0$ denotes the initial cumene concentration.

(c) Show that for the batch reactor

$$\frac{[RH]^0 - [ROOH]}{[ROOH]} = C^0 \exp\left\{ k_T [RH]^0 (t - t_i) \right\}$$

where t_i = initiation time

$$C^0 = \frac{[RH]^0 - [ROOH]_i}{[ROOH]_i}$$

$[ROOH]_i$ = hydroperoxide concentration at t_i

(d) For the following values of the rate constants,

$$\left(\frac{k_3}{k_7^{1/2}} \right) = 229.7 \, e^{-1738/T} \text{ in liters/g mol h}$$

$$k_4 = 8.69 \times 10^{11} \, e^{-13,383/T} \text{ in liters/g mol h}$$

$$k_5 = 4.15 \times 10^4 \, e^{-5387/T} \text{ in (liters)}^2/\text{g mol h}$$

and the reaction conditions

$$T = 115°C, \quad [\text{NaNap}]^0 = [\text{Na}^+] = 0.573 \times 10^{-3} \text{ g mol/liter}$$

$[ROOH]_i = 0.040$ g mol/liter

calculate the conversion of cumene for a reaction time of 16 h.

4-19. Anthony and Singh [24] used the techniques presented in Part III.A to interpret the data for the gas phase conversion of methanol to olefins over small-pore zeolites. The reactions considered in this analysis were as follows:

$$2A \rightleftarrows B + C, \quad K_{p1} = 5, \quad r_A = k_{p1} p_A^2 - k'_{p1} p_B p_C \qquad (1)$$

$$B \rightarrow D + C, \quad r_D = k_{p2} p_B \qquad (2)$$

where A = methanol, B = dimethyl ether, C = water, and D = ethylene. On the basis of the following conditions, generate the data and plot the selectivities of B, D, and C versus the conversion of A for a plug-flow reactor. The selectivities of B, C, and D are defined as moles of B, C, or D produced per mole of A reacted. The conditions are as follows:

(1) Feed is pure methanol.
(2) $K_{p1} = 5$, and repeat for $k_{p1}P/k_{p2} = 4$, 1, and 0.2.
Answers for $k_{p1}P/k_{p2} = 4$

Selectivity	Conversion					
	0	20	40	60	80	95
B	0.5	0.48	0.45	0.38	0.16	0.0035
C	0.5	0.52	0.55	0.61	0.84	0.996
D	0.0	0.016	0.047	0.11	0.34	0.496

4-20. Repeat Problem 4-19 for the following mechanism used by Anthony and Singh [24] in the interpretation of the data for the conversion of methanol to olefins in the gas phase over small-pore zeolites.

$$2A \rightleftharpoons B + C \tag{1}$$

$$2A \rightarrow D + 2C \tag{2}$$

where A, B, C, and D have the same meanings as in Problem 4-20. For an equilibrium constant for reaction (1) of 5 and a feed of pure methanol, generate the data and plot the selectivities of B, C, and D versus the conversion of A for ratios of k_1/k_2 equal to 4, 1, and 0.2 for a plug-flow reactor. Selectivities are defined in Problem 4-19.

4-21. Given the reactions

$$A \underset{k_1'}{\overset{k_2}{\rightleftharpoons}} R, \qquad K_1 = 5 \tag{1}$$

$$R \underset{k_2'}{\overset{k_2}{\rightleftharpoons}} S, \qquad K_2 = 4 \tag{2}$$

and the values of the rate constants over three different catalysts:

Catalyst	k_1 (s^{-1})	k_2/k_1
I	0.243	3
II	0.5	0.3
III	2.7	1.5

(a) On the basis of a feed composed of pure A fed to a plug-flow reactor, calculate the equilibrium composition in mole percent which would be obtained if only reaction (1) occurred.

(b) Repeat part (a) for the case where only reaction (2) occurs and the feed is composed of pure R.

(c) Calculate the equilibrium composition in mole percent for the system $A \rightleftharpoons R \rightleftharpoons S$ on the basis of a feed composed of pure A.

(d) Calculate the values of n_A/n_A^0, n_R/n_A^0, and n_S/n_A^0 as a function of $k_1\theta = k_1 V/v_T$ in a plug-flow reactor for each catalyst and a feed of pure A.

(e) Plot the results found in parts (c) and (d) as a function of $k_1\theta$ for each catalyst.

(f) What effect does the ratio of the rate constants have on n_R/n_A^0 in the neighborhood of the maximum? What would happen in this neighborhood if a catalyst were found that gave a value of $k_2/k_1 = 0.005$?

(g) Can n_R/n_A^0 or n_S/n_A^0 in part (d) ever exceed the equilibrium compositions found in parts (a) and (b)?

4-22. Riley and Anthony [25] investigated the decomposition of decanol in a tubular reactor over each of two catalysts, H-ZSM-5 and Ni-ZSM-5. It may be assumed that the plug-flow assumption is valid. Suppose that the major reactions occurring are as follows:

$$2C_{10}H_{21}OH \rightleftarrows C_{10}H_{21}OC_{10}H_{21} + H_2O \tag{1}$$

$$C_{10}H_{21}OC_{10}H_{21} \rightleftarrows 2C_{10}H_{20} + H_2O \tag{2}$$

Didecylether $\rightleftarrows$ Decene

The following data were reported by Riley and Anthony. Plot the moles of decene and didecylether produced, respectively, per mole of decanol reacted to show that these data support the preceding stoichiometry.

PRIMARY PRODUCT DISTRIBUTION IN WT% FOR 1-DECANOL / H-ZSM-5

Temperature (K)	Conversion[a] (%)	C_{10}[b] (%)	C_{20}[b] (%)	C_9A[b] (%)
572	1.27	0.00	99.61	0.00
596	8.07	31.28	68.49	0.00
598	13.23	58.59	37.98	0.00
599	26.94	74.87	23.83	0.00
600	13.42	50.38	46.50	0.00
618	45.29	83.30	13.66	1.59
645	60.91	84.83	5.63	5.11
666	66.75	83.75	6.28	7.90

[a] Conversions are on a decanal, n-nonane, CO, and CO_2 free basis.
[b] Product distributions are on a decanal, n-nonane, CO, CO_2, and H_2O free basis. C_{10} denotes $C_{10}H_{20}$, C_{20} denotes $C_{10}H_{21}OC_{10}H_{21}$, and C_9A denotes the C_9 aromatics.

NOTATION

A^* = activated molecules or free radical

C_A^* = concentration of the activated molecules

M = concentration of the monomer

R_j = polymer molecule or the concentration of polymer molecules leaving j monomer units

$R^{\cdot}$ = free radical

τ = defined by Equation (4-35) and has the units of (moles)(time)/volume

REFERENCES

1. MacMullin, R. B., *Chem. Eng. Prog.*, 44:183 (1948).

2. Smith, J. M., *Chemical Engineering Kinetics*, McGraw-Hill Book Company, New York (1972).

3. Aris, R., *Introduction to the Analysis of Chemical Reactors*, Prentice-Hall, Inc., Englewood Cliffs, N.J. (1965).

4. Allen, G., G. Gee, and C. Stretch, *J. Polymer Sci.*, 48:189 (1960).

5. Bhattacharyya, D. N., J. Smid, and M. Szwarc, *J. Phys. Chem.*, *69*:624 (1965).

6. Lindermann, F. A., *Trans. Faraday Soc.*, *17*:598 (1922).

7. Hinshelwood, C. N., *Proc. Roy. Soc.*, *A113*:230 (1927).

8. Date, R. D., J. B. Butt, and H. Bliss, *I & EC Fund.*, *8*:687 (1960).

9. Billingsley, D. S., and C. D. Holland, *I & EC Fund.*, *2*:252 (1963).

10. Hong, C. C., F. Lenzi, and W. H. Rapson, *Canadian J. Chem. Eng.*, *45*:349 (1967).

11. Farkas, Adalbert, *Hydrocarbon Processing*, p. 121 (July 1970).

12. Callahan, J. C., R. K. Grasselli, E. C. Milberger, and H. A. Streeker, *Preprints of the Division of Petroleum Chemistry of American Chemical Society 14, No. 3*, C13 (Sept. 1969).

13. Gulyaev, G. V., and L. S. Polak, *Kinetika i Kataliz* (in Russian), *6*:399 (1965); *Kinetics and Catalysis* (English trans.), *6*:352 (1965).

14. Bollard, J. L., *Proc. Roy. Soc.*, *A186*:218 (1946); *Quart. Rev. Chem. Soc.*, *31* (1949).

15. Pacey, P. D., and J. H. Purnell, *I & EC Fund.*, *11*:233 (1972).

16. Alagy, J., L. Asselineau, C. Busson, B. Cha, and H. Sandler, *Hydrocarbon Processing 47:No. 12*, 131 (1968).

17. Allan, D. E., and A. Voorhies, *I & EC Prod. Research Develop.*, *11*:15 (1972).

18. Adams, C. R., H. H. Voge, C. Z. Morgan, and W. E. Armstrong, *J. Catalysis*, *3*:379 (1964).

19. Habelt, W. W., and A. P. Selker, Paper presented at *Central States Section of the Combustion Institute*, Madison, Wisc., March 26–27, 1974.

20. Zeldovich, Ya. B., P. Ya. Sadovnikov, and D. A. Kamenetskii, *Academy of Sciences of the USSR, Institute of Chemical Physics*, Moscow–Leningrad, translated by M. Shelef, 1947, Scientific Research Staff, Ford Motor Co.

21. Baulch, D. L., D. D. Drysdale, D. G. Horne, and A. C. Lloyd, Report No. 4, Department of Physical Chemistry, Leeds University, U.K., December 1969.

22. Jones, F. L., "A Simulation of the Opposed-Jet Diffusion Flame under the Influence of an Electric Field," Ph.D. Thesis, Pennsylvania State University, University Park, 1971.

23. Sotelo, J. L., F. J. Beltram, J. Beltran-Hereida, and M. Gonzalez, *Ind. Eng. Chem. Prod. Res. Dev.*, *24*:650 (1985).

24. Anthony, R. G., and B. B. Singh, *Chem. Eng. Comm.*, *6*:215 (1980).

25. Riley, M., and R. G. Anthony, *J. Catalysis, 103*, 87 (1987).

Thermodynamics of
Chemical Reactions

5

Most chemical reactions are accompanied by the evolution or the absorption of heat. Since the rate of reaction depends strongly on temperature, it is necessary to take the heat of reaction into account in the design of chemical reactors. In Part I, the methods for calculating the standard heats of reaction as provided by chemical thermodynamics are presented. In Part II, the effects of pressure and heats of mixing on the enthalpies of the components in the reacting mixture, which were neglected in Part I, are included. Also needed in the analysis of the feasibility of carrying out chemical reactions is the free energy change for a reaction. The free energy change makes it possible to calculate the maximum conversion which is theoretically attainable. The concept of the free energy change associated with a chemical reaction and equilibrium constants are treated in Part III.

I. USE OF THE CONCEPTS OF CHEMICAL THERMODYNAMICS IN THE CALCULATION OF THE STANDARD HEAT OF REACTION

The fundamental thermodynamic quantities of internal energy U, enthalpy H, Gibbs free energy G, Helmholtz free energy A, and the entropy function S, are related as follows:

$$H = U + Pv \tag{5-1}$$

$$G = H - TS \tag{5-2}$$

$$A = U - TS \tag{5-3}$$

where T is the absolute temperature and Pv is the pressure–volume product. Although the conversion factor for Pv has been omitted, it is supposed that the Pv product has the units of Btu/lb mol, cal/g mol, or joules/kg mol. The quantities H, U, G, and A have the units of Btu/lb, Btu/lb mol, cal/g, cal/g mol, joules/kg, or joules/kg mol and entropy has the units of Btu/(lb mole °R), cal/(g mol K), or joules/(kg mol K).

For a pure component or for a mixture at constant composition, the first and second laws of thermodynamics are contained in the single expression

$$dU = T\,dS - P\,dv \qquad (5\text{-}4)$$

By use of this expression and the relationships given by Equations (5-1) through (5-3), the following differential expressions are obtained:

$$dH = T\,dS + v\,dP \qquad (5\text{-}5)$$

$$dG = -S\,dT + v\,dP \qquad (5\text{-}6)$$

$$dA = -S\,dT - P\,dv \qquad (5\text{-}7)$$

The functions U, A, G, H, and S are called state functions because changes in these functions depend only on the initial and final states of a system, and they are consequently independent of the path taken by a system in passing from the initial to the final state. The fact that the final values of these functions are independent of path is used extensively in the calculation of the heats of reaction and the free energy changes for reactions.

A. Standard Heat of Reaction

The *standard heat of reaction* at the temperature T is defined as the sum of the enthalpies of the products in their respective standard states at the temperature T minus the sum of the enthalpies of the reactants in their respective standard states at the temperature T. The standard states may be selected arbitrarily, and two of the standard states which are commonly used are shown in Table 5-1.

Now consider the reaction

$$aA(g) + bB(g) = cC(s) + dD(g) \qquad (5\text{-}8)$$

where C denotes carbon. To compute the standard heat of reaction at 25°C, it is helpful to visualize each component as being in a separate container at its standard state, as follows:

aH_A°		bH_B°			cH_C°		dH_D°
a moles of A perfect gas at 1 atm and 25°C	$+$	b moles of B perfect gas at 1 atm and 25°C	$a\,\Delta H_{r298}^{\circ}$ $=$		c moles of C pure graphite at 1 atm and 25°C	$+$	d moles of D perfect gas at 1 atm and 25°C

From the definition of the standard heat of reaction, it follows that the heat of

TABLE 5-1
STANDARD STATES

Physical State	Standard State
Solid	The most stable form at 1 atm pressure and at the specified temperature
Gas	Perfect gas at 1 atm pressure and at the specified temperature

Standard States Used Herein	
Substances	Standard State
All elements and compounds except carbon	Perfect gas at 1 atm pressure and at the specified temperature
Carbon	Graphite form at 1 atm pressure and at the specified temperature

reaction per mole of A at the reference temperature of 25°C for the reaction given by Equation (5-8) is given by

$$\Delta H_{r298}^{\circ} = \left(\frac{1}{a}\right)[cH_c^{\circ} + dH_D^{\circ} - aH_A^{\circ} - bH_B^{\circ}] \tag{5-9}$$

When the standard state of a perfect gas at 1 atm pressure is used for all substances except carbon, and the standard state of carbon is taken to be graphite at 1 atm pressure (see Table 5-1), the corresponding enthalpies are identified by the superscript ∘. When these enthalpies are evaluated at the temperature of 25°C (298.15 K), the subscript 298 is used. (In the examples and tables, 298 is used to mean 298.16 K, an old value for 25°C.) The use of separate containers for each of the substances involved in the reaction emphasizes the fact that the standard heat of reaction in no way accounts for any heat of solution effects. The standard heat of reaction is used in the calculation of the heat of reaction at any specified temperature as demonstrated in a subsequent section.

B. Standard Heat of Formation

The *standard heat of formation* ΔH_f° of a compound at the temperature T is the standard heat of reaction when the given compound in its standard state at the temperature T is *formed from its elements* in their standard states at the temperature T. The elements in their standard states are taken as the reference state; that is, they are assigned an enthalpy of zero. In the case of gases such as oxygen and nitrogen, the standard state could be taken as either the atomic or the molecular forms of oxygen and nitrogen. The molecular forms are generally taken as the respective standard states because this choice generally leads to a simplification in the calculation of heats of reaction involving these

TABLE 5-2
TYPICAL HEATS OF FORMATION AT 25°C*
(Taken from [1] and [2])

Substance	ΔH_f^0 kcal / mol	Substance	ΔH_f^0 kcal / mol
H (g)	52.095	1-Butene (g)	− 0.03
O (g)	59.553	Acetylene (g)	54.194
Cl (g)	28.989	Benzene (g)	19.820
CO (g)	− 26.416	Toluene (g)	11.950
CO_2 (g)	− 94.051	o-Xylene (g)	4.540
H_2O (g)	− 57.798	m-Xylene (g)	4.120
H_2O (l)	− 68.315	p-Xylene (g)	− 4.290
Methane (g)	− 17.889	Methanol (l)	− 57.11
Ethane (g)	− 20.236	Ethanol (l)	− 66.20
Propane (g)	− 24.820	Glycine (s)	− 126.330
Ethylene (g)	12.496	Acetic acid (l)	− 116.4
Propylene (g)	4.879	1,3-Butadiene (g)	26.33

*These heats of formation are based on the standard state of carbon in the solid state of graphite at 1 atm and 25°C, and molecular hydrogen, chlorine, and oxygen in the standard state of a perfect gas at 1 atm and 25°C.

gases. Typical heats of formation are given in Table 5-2. The heat of formation of methane per mole of methane formed from its elements by the reaction

$$C(s) + 2H_2(g) = CH_4(g) \qquad (5\text{-}10)$$

may be represented graphically as follows:

H_C^o		$2H_{H_2}^o$			$H_{CH_4}^o$
1 mole of carbon: graphite at 1 atm and 25°C	+	2 moles of H_2: perfect gas at 1 atm and 25°C	ΔH_{r298}^o =	1 mole methane: perfect gas at 1 atm and 25°C

The heat of reaction is given by

$$\Delta H_{f298}^o = \Delta H_{r298}^o = \left[H_{CH_4}^o - H_C^o - 2H_{H_2}^o \right]_{298} \qquad (5\text{-}11)$$

Since $H_C^o = H_{H_2}^o = 0$, it follows that

$$\Delta H_{r298}^o = \Delta H_{f298}^o = H_{CH_4, 298}^o \qquad (5\text{-}12)$$

Thus the standard heat of formation of CH_4 is equal to its enthalpy above its elements in their standard states.

The standard heat of reaction at 25°C for the production of synthesis gas from methane and steam

$$CH_4(g) + H_2O(g) = CO(g) + 3H_2(g) \qquad (5\text{-}13)$$

may be calculated from the heats of formation at 25°C as follows. Addition of the reactions

$$CH_4(g) = C(s) + 2H_2(g), \qquad \Delta H = -\Delta H_f^\circ = 17.889 \text{ kcal}$$

$$H_2O(g) = \tfrac{1}{2}O_2(g) + H_2(g), \qquad \Delta H = -\Delta H_f^\circ = 57.7979 \text{ kcal}$$

$$C(s) + \tfrac{1}{2}O_2(g) = CO(g), \qquad \Delta H = \Delta H_f^\circ = -26.4157 \text{ kcal}$$

yields

$$CH_4(g) + H_2O(g) = CO(g) + 3H_2(g), \qquad \Delta H_{r298}^\circ = 49.2712 \text{ kcal}$$

This calculation is a numerical demonstration of the general formula for the calculation of the heat of reaction from heats of formation; that is, for the general reaction

$$aA + bB = cC + dD \qquad (5\text{-}14)$$

the standard heat of reaction at temperature, T, to give ΔH_{rT}° is given by

$$\Delta H_{rT}^\circ = \frac{1}{a}\left[c\,\Delta H_{fC}^\circ + d\,\Delta H_{fD}^\circ - a\,\Delta H_{fA}^\circ - b\,\Delta H_{fB}^\circ \right]_T \qquad (5\text{-}15)$$

where it is understood that each heat of formation is evaluated at the reference temperature, T. Equation (5-15) may be used to calculate the standard heat of reaction at any temperature T, provided that the heats of formation are available.

    Another method for calculating heats of reaction that is in common practice makes use of the differences

$$H_T^\circ - H_0^\circ \quad \text{or} \quad \frac{H_T^\circ - H_0^\circ}{T}$$

The enthalpy of a compound or element in its standard state at temperature T is denoted by H_T° and the corresponding enthalpy at 0 K is denoted by H_0°. The heat of formation of a compound in its standard state from its elements in their standard states at 0 K is denoted by ΔH_{f0}°. To demonstrate the presentation of these data, an abbreviated set of data is shown in Table 5-3. The use of these data is illustrated by the calculation of the standard heat of formation of methane at 25°C by the reaction

$$C(298.16 \text{ K}) + 2H_2(298.16 \text{ K}) = CH_4(298.16 \text{ K})$$

The standard heat of formation of methane at 25°C is obtained by adding the following reactions:

$C(0 \text{ K}) + 2H_2(0 \text{ K}) = CH_4(0 \text{ K}),$	$\Delta H_{f0}^\circ = -15.987,$	$\Delta H = \Delta H_{f0}^\circ$
$C(298.16 \text{ K}) = C(0 \text{ K}),$	$\Delta H = -0.2516,$	$\Delta H = -(H_T^\circ - H_0^\circ)_C$
$2H_2(298.16 \text{ K}) = 2H_2(0 \text{ K}),$	$\Delta H = -4.0476,$	$\Delta H = -2(H_T^\circ - H_0^\circ)_{H_2}$
$CH_4(0 \text{ K}) = CH_4(298.16 \text{ K}),$	$\Delta H = 2.3970,$	$\Delta H = (H_T^\circ - H_0^\circ)_{CH_4}$

$$\overline{C(298.16 \text{ K}) + 2H_2(298.16 \text{ K}) = CH_4(298.16 \text{ K}), \ \Delta H_f^\circ = -17.889 \text{ kcal/g mol}}$$

$$(5\text{-}16)$$

The general case of the formation of compound C in its standard state from

TABLE 5-3
SELECTED VALUES OF THE ENTHALPY FUNCTION $H_T^\circ - H_0^\circ$ (taken from [1])

Substance	ΔH_{f0}° kcal / mol	$H_T^\circ - H_0^\circ$ in cal / mol at T K indicated					
		298.16	400	600	800	1000	1500
H_2 (g)	0	2023.81	2731.0	4128.6	5537.4	6965.8	10,694.2
O_2 (g)	0	2069.78	2792.4	4279.2	5854.1	7397.0	11,776.4
C (graphite)	0	251.56	502.6	1198.1	2081.7	3074.6	5814
CO (g)	-27.2019	2072.63	2783.8	4209.5	5699.8	7256.5	11,358.8
CO_2 (g)	-93.9686	2238.11	3194.8	5322.4	7689.4	10,222	17,004
H_2O (g)	-57.107	2367.7	3194.0	4882.2	6689.6	8608.0	13,848
Methane (g)	-15.987	2397	3323	5549	8321	11,560	21,130
Ethane (g)	-16.571	2856	4296	8016	12,760	18,280	34,500
Propane (g)	-19.482	3512	5556	10,930	17,760	25,670	48,650
Ethylene (g)	14.522	2525	3711	6732	10,480	14,760	27,100
Acetylene (g)	54.329	2391.5	3541.2	6127	8999	12,090	20,541
Benzene (g)	24.000	3401	5762	12,285	20,612	30,163	57,350
Toluene (g)	17.500	4306	7269	15,334	25,621	37,449	71,250
o-Xylene (g)	11.096	5576	9291	19,070	31,386	45,531	85,960
m-Xylene (g)	10.926	5325	8925	18,563	30,817	44,933	85,330
p-Xylene (g)	11.064	5358	8929	18,499	30,690	44,755	85,080

its elements in their standard states, all at the temperature T, is represented by

$$aA + bB = cC \tag{5-17}$$

where A and B are elements. The formula for the heat of formation of compound C at temperature T is obtained by adding the following equations:

$$aA(0 \text{ K}) + bB(0 \text{ K}) = cC(0 \text{ K}), \qquad \Delta H = c\,\Delta H_{f0}^\circ$$

$$aA(T \text{ K}) = aA(0 \text{ K}), \qquad \Delta H = -a(H_T^\circ - H_0^\circ)_A$$

$$bB(T \text{ K}) = bB(0 \text{ K}), \qquad \Delta H = -b(H_T^\circ - H_0^\circ)_B$$

$$\underline{cC(0 \text{ K}) = cC(T \text{ K}), \qquad \Delta H = c(H_T^\circ - H_0^\circ)_C}$$

$$aA(T \text{ K}) + bB(T \text{ K}) = cC(T \text{ K})$$

$$c\,\Delta H_{fC}^\circ = c\,\Delta H_{f0C}^\circ + c(H_T^\circ - H_0^\circ)_C - \left[a(H_T^\circ - H_0^\circ)_A + b(H_T^\circ - H_0^\circ)_B\right]$$

or

$$\Delta H_{fC}^\circ = \Delta H_{f0C}^\circ + (H_T^\circ - H_0^\circ)_C - \left[\frac{a}{c}(H_T^\circ - H_0^\circ)_A + \frac{b}{c}(H_T^\circ - H_0^\circ)_B\right] \tag{5-18}$$

Now suppose that compound C reacts to give compounds D and E by the equation

$$cC = dD + eE \tag{5-19}$$

and that it is desired to find the heat of reaction at temperature T by use of

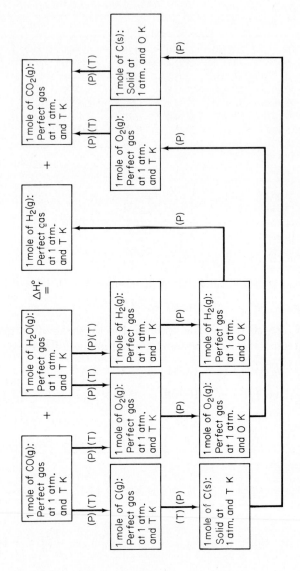

Figure 5-1. Use of an alternate path to relate the standard heat of reaction to the standard heats of formation.

the data in Table 5-3. Since the heat of reaction may be computed from the heats of formation by use of Equations (5-15) and since Equation (5-18) is applicable for each compound, it follows that the heat of reaction per mole of C is given by

$$\Delta H_r^\circ = \frac{1}{c}\left[d\,\Delta H_{f0D}^\circ + e\,\Delta H_{f0E}^\circ - c\,\Delta H_{f0C}^\circ + d(H_T^\circ - H_0^\circ)_D \right.$$

$$\left. + e(H_T^\circ - H_0^\circ)_E - c(H_T^\circ - H_0^\circ)_C \right] \quad (5\text{-}20)$$

Note that when each member involved in the reaction is a compound as in Equation (5-19) the terms involving the elements A and B in Equation (5-18) cancel in the development of Equation (5-20) because the same set of elements required to form the reactants is also required to form the products. This fact is vividly demonstrated when Equation (5-20) is derived by use of a path such as the one shown in Figure 5-1 for the reaction

$$CO(g) + H_2O(g) = H_2(g) + CO_2(g)$$

C. Variation of the Standard Heat of Reaction with Temperature

The heat of reaction at any temperature T may be obtained directly from heat of formation and enthalpy data of the types presented in Tables 5-2 and 5-3 by use of Equations (5-15) and (5-20), respectively, provided, of course, that the data are available at the temperature T of interest. The variation of the heat of reaction with temperature may also be accounted for by use of heat capacity data. For the general reaction given by Equation (5-8) at any temperature T, Equation (5-9) is applicable, where H_C°, H_D°, H_A°, and H_B° are evaluated at the temperature T. Since the heat capacity of a pure component A is defined by

$$\left(\frac{\partial H_A^\circ}{\partial T} \right)_P = C_{pA}^\circ \quad (5\text{-}21)$$

Equation (5-9) may be differentiated to give

$$\left(\frac{\partial \Delta H_r^\circ}{\partial T} \right)_P = \frac{c}{a}\left(\frac{\partial H_C^\circ}{\partial T} \right)_P + \frac{d}{a}\left(\frac{\partial H_D^\circ}{\partial T} \right)_P - \left(\frac{\partial H_A^\circ}{\partial T} \right)_P - \frac{b}{a}\left(\frac{\partial H_B^\circ}{\partial T} \right)_P$$

$$(5\text{-}22)$$

and thus

$$\left(\frac{\partial \Delta H_r^\circ}{\partial T} \right)_P = \frac{c}{a}C_{pC}^\circ + \frac{d}{a}C_{pD}^\circ - C_{pA}^\circ - \frac{b}{a}C_{pB}^\circ = \Delta C_p^\circ \quad (5\text{-}23)$$

which holds for each T at the specified pressure P.

One advantage of the selection of a perfect gas at 1 atm pressure at a given temperature as the standard state is the fact that the heat capacity of any

TABLE 5-4

SELECTED VALUES AT THE INDICATED TEMPERATURE, IN cal/g mole K OF
HEAT CAPACITIES FOR PERFECT GAS STATE (Taken from [1])

Substance	298.16	400	600	800	1000	1500
H_2 (g)	6.892	6.975	7.009	7.080	7.219	7.720
O_2 (g)	7.020	7.196	7.670	8.063	8.336	8.739
C (monatomic) (g)	4.980	4.975	4.971	4.970	4.969	4.975
CO (g)	6.965	7.013	7.276	7.625	7.931	8.416
CO_2 (g)	8.874	9.876	11.310	12.293	12.980	13.954
H_2O (g)	8.025	8.186	8.676	9.245	9.850	11.233
Methane (g)	8.536	9.721	12.55	15.10	17.21	20.71
Ethane (g)	12.58	15.68	21.35	25.83	29.33	34.90
Propane (g)	17.57	22.54	30.88	37.08	41.83	49.26
Ethylene (g)	10.41	12.90	17.10	20.20	22.57	26.36
Acetylene (g)	10.499	11.973	13.728	14.933	15.922	17.704
Benzene (g)	19.52	26.74	37.74	45.06	50.16	57.67
Toluene (g)	24.80	33.25	46.58	55.72	62.19	71.78
o-Xylene (g)	31.85	41.03	55.98	66.64	74.35	85.93
m-Xylene (g)	30.49	40.03	55.51	66.41	74.23	85.89
p-Xylene (g)	30.32	39.70	55.16	66.14	74.02	85.79
1-Butene	20.47	26.04	35.14	41.80	46.82	54.62
1,3-Butadiene	19.01	24.29	31.84	36.84	38.81	46.34

element or compound may be determined spectroscopically at conditions that approximate zero pressure. Since all substances approach perfect gases as the pressure goes to zero, these heat capacities may be used for perfect gases at any pressure because the change of C_p with pressure for a perfect gas is equal to zero [3, 4, 5]. Consequently, the change in ΔH_r° with temperature at any pressure may be computed by use of the heat capacities for the perfect gases. Tabular collections of these heat capacities are available (see Table 5-4); see also Ref. (1), (18), and (20).

Suppose that the heat of reaction is known at some temperature T_0. Then integration of Equation (5-23) over the temperature range from T_0 to T yields

$$\Delta H_r^\circ = \Delta H_r^\circ|_{T_0} + \int_{T_0}^{T} \Delta C_p^\circ \, dT \tag{5-24}$$

This expression may be derived by use of the path shown in Figure 5-2. Since ΔH_r° is independent of path, it may be computed by summing the ΔH's around the path shown in Figure 5-2. Thus, when ΔH_r° is based on one mole of A, it follows from Figure 5-2 that

$$\Delta H_r^\circ = \frac{1}{a} \left[\int_{T}^{T_0} a C_{pA}^\circ \, dT + \int_{T}^{T_0} b C_{pB}^\circ \, dT + a \, \Delta H_r^\circ|_{T_0} \right.$$
$$\left. + \int_{T_0}^{T} c C_{pC}^\circ \, dT + \int_{T_0}^{T} d C_{pD}^\circ \, dT \right] \tag{5-25}$$

which is readily reduced to the result given by Equation (5-24). If the heat capacities of the perfect gases in Equation (5-25) are expressed as functions of

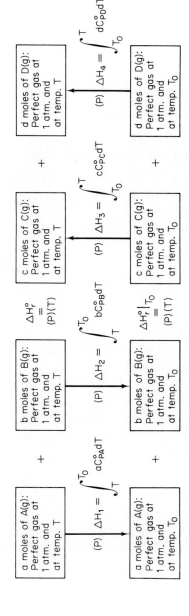

Figure 5-2. Path for calculating ΔH_r° at temperature T from a knowledge of ΔH_r° at T_0 and heat capacity data.

temperature as follows,

$$C_{pi}^\circ = \alpha_i + \beta_i T + \gamma_i T^2 + \delta_i T^3$$

then the integrals in Equation (5-25) may be evaluated and the resulting expression for ΔH_r° may be stated as follows:

$$\Delta H_r^\circ = \Delta H_r^\circ|_{T_0} + \Delta\alpha(T - T_0) + \frac{\Delta\beta}{2}(T^2 - T_0^2)$$

$$+ \frac{\Delta\gamma}{3}(T^3 - T_0^3) + \frac{\Delta\delta}{4}(T^4 - T_0^4) \tag{5-26}$$

where

$$\Delta\alpha = \frac{1}{a}(c\alpha_C + d\alpha_D - a\alpha_A - b\alpha_B)$$

$$\Delta\beta = \frac{1}{a}(c\beta_C + d\beta_D - a\beta_A - b\beta_B)$$

$$\Delta\gamma = \frac{1}{a}(c\gamma_C + d\gamma_D - a\gamma_A - b\gamma_B)$$

$$\Delta\delta = \frac{1}{a}(c\delta_C + d\delta_D - a\delta_A - b\delta_B)$$

Since the heat capacities for perfect gases are independent of pressure, the same numerical value for ΔH is obtained if the specified pressure of 1 atm in Figure 5-2 is replaced by any arbitrary pressure P. Thus, when the standard states of all compounds involved in a reaction are taken to be perfect gases at 1 atm at the temperature of the reaction, the heat of reaction ΔH_r° is independent of pressure; that is,

$$\left(\frac{\partial \Delta H_r^\circ}{\partial P}\right)_T = 0 \tag{5-27}$$

or

$$\Delta H_r^\circ|_{P,T} = \Delta H_r^\circ|_{1,T}$$

If the constituents of the reacting mixture behave as perfect gases,

then

$$\Delta H_r|_{P,T} = \Delta H_r^\circ|_{P,T} = \Delta H_r^\circ|_{1,T} \tag{5-28}$$

If, on the other hand, the behavior of the constituents of the reaction mixture does depend on pressure, the heat of reaction at P and T may be found as shown in Part II.

EXAMPLE 5-1

Calculate the standard heat of reaction for the dehydrogenation of 1-butene to 1,3-butadiene at 900 K by use of the heat of formation data given in Table 5-2 and the heat capacities given in Table 5-4.

1-Butene   1,3-Butadiene

SOLUTION

For convenience, let the reaction be represented as follows:

$$A \rightleftarrows B + C$$

The heat of reaction at 25°C is computed by use of the heats of formation given in Table 5-2 as follows:

$$\Delta H_r^\circ = \Delta H_{fB}^\circ + \Delta H_{fC}^\circ - \Delta H_{fA}^\circ$$

$$= 26.33 + 0.0 - (-0.03) = 26.36 \text{ kcal/g mole}$$

By use of Equation (5-24)

$$\Delta H_{r900}^\circ = 26{,}360 + \int_{298.16}^{900} \left[C_{pB}^\circ + C_{pC}^\circ - C_{pA}^\circ \right] dT$$

The integrand is evaluated numerically by use of the heat capacities (cal/g mol K) given in Table 5-4 as follows:

T K	C_{pC}°	C_{pB}°	C_{pA}°	$\Delta C_p^\circ = C_{pB} + C_{pC}^\circ - C_{pA}^\circ$
298.16	6.892	19.01	20.47	5.432
300	6.895	19.11	20.57	5.435
400	6.975	24.29	26.04	5.225
500	6.994	28.52	30.93	4.584
600	7.009	31.84	35.14	3.709
700	7.036	34.55	38.71	2.876
800	7.080	36.84	41.80	2.120
900	7.142	38.81	44.49	1.462

The integral appearing in the expression for ΔH_r° at 900 K is evaluated by use of the trapezoidal rule. For the first increment in temperature

$$\int_{298.16}^{300} \Delta C_p^\circ \, dT = \left[\Delta C_{p298}^\circ + \Delta C_{p300}^\circ \right] (300 - 298.16) \tfrac{1}{2} = 9.998$$

From 300 to 900 K, equal increments in temperature of 100 K are used, and

$$\int_{300}^{900} \Delta C_p^\circ \, dT = \left[\frac{\Delta C_{p300}^\circ}{2} + \left(\Delta C_{p400}^\circ + \Delta C_{p500}^\circ + \cdots + \Delta C_{p800}^\circ \right) + \frac{\Delta C_{p900}^\circ}{2} \right] (100)$$

$$= 2196.25 \text{ cal/g mol}$$

Thus the heat of reaction at 900 K is given by

$$\Delta H_{r900}^\circ = 26{,}360 + 2196.25 + 9.998$$

$$= 28{,}568 \text{ cal/g mol} \quad \text{or} \quad 28.568 \text{ kcal/g mol}$$

II. EFFECT OF HEATS OF MIXING AND PRESSURE ON THE ENTHALPY OF A MIXTURE

In all the enthalpy changes represented in Part I by various paths, the participating components are shown in separate containers at given values of temperature and pressure. To represent the actual reacting mixture at any point in a reactor, these components must be mixed (or placed in solution). Methods for computing the change of enthalpy that accompanies mixing are developed and demonstrated next.

A. Deviation of the Enthalpy of a Mixture of Gases at a Given Pressure and Temperature from the Enthalpy of a Mixture of the Same Gases in the Perfect Gas State at the Same Temperature

The path shown in Figure 5-3 may be used to compute the change in enthalpy when $n_1, n_2, \ldots, n_c$ (moles or molar flow rates) of components $1, 2, \ldots, c$ in the state of perfect gases at $P = 1$ atm and the temperature T are mixed at P and T to form the solution as it exists in a reactor. In the first step shown in Figure 5-3, the perfect gas is expanded isothermally from $P = 1$ atm to $P = P_0$. In the limit as P_0 goes to zero, state 2 becomes a converter in the sense that a perfect gas enters and an actual gas leaves state 2. Thus, in the limit as P_0 approaches zero,

$$\Delta H_1 = \sum_{i=1}^{c} n_i \left[H_i(0, T) - H_i^\circ(1, T) \right]$$

where n_T is equal to the sum of $n_1, n_2, \ldots, n_c$ moles of components 1 through c. Next the actual gas is compressed isothermally to the pressure P, and for this step

$$\Delta H_2 = \sum_{i=1}^{c} n_i \left[H_i(P, T) - H_i(0, T) \right]$$

In the third step, $n_1, n_2, \ldots, n_c$ moles of components $1, 2, \ldots, c$ are mixed at constant temperature and pressure, and for this step

$$\Delta H_3 = \sum_{i=1}^{c} n_i \left[\overline{H}_i(P, T) - H_i(P, T) \right]$$

where $\overline{H}_i(P, T)$ is the partial molar enthalpy of component i at the temperature, pressure, and composition of the final state shown in Figure 5-3. [In the interest of simplicity, the complete functional representation of the partial molar enthalpy, $\overline{H}_i(P, T, n_1, n_2, \ldots, n_c)$ is abbreviated as $\overline{H}_i(P, T)$ or simply $\overline{H}_i$.] Thus the total change in enthalpy ΔH between the final and initial states

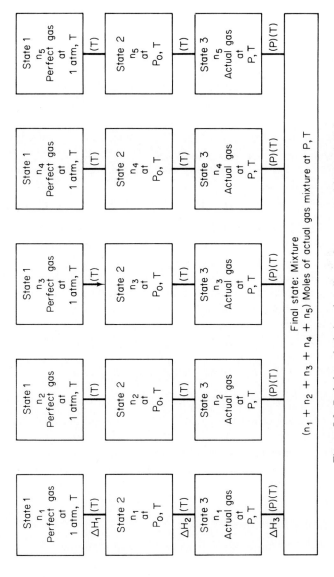

Figure 5-3. Path for calculating the effects of pressure and heats of mixing on the enthalpy of a mixture. [Taken from C. D. Holland, *I & EC Fundamentals*, *16*: No. 1, 143 (1977). Used with permission of the American Chemical Society.]

for the path shown in Figure 5-3 is given by

$$\Delta H = \Delta H_1 + \Delta H_2 + \Delta H_3 = \sum_{i=1}^{c} \left[\overline{H}_i(P, T) - H_i^{\circ}(1, T) \right] n_i$$

$$= n_T H - n_T H^{\circ} \tag{5-29}$$

where
$$n_T H = \sum_{i=1}^{c} n_i \overline{H}_i(P, T)$$

$$n_T H^{\circ} = \sum_{i=1}^{c} n_i H_i^{\circ}(1, T)$$

Let the deviation of the enthalpy of 1 mole of a mixture from the enthalpy of 1 mole of the perfect gas mixture be denoted by Ω. Then the enthalpy H per mole of mixture is equal to the enthalpy of the perfect gas mixture plus the deviation per mole; that is,

$$H = H^{\circ} + \Omega \tag{5-30}$$

Generally, the direct calculation of the partial molar enthalpies from an equation of state is far more difficult than is the calculation of Ω from an equation of state. Fortunately, there exists a set of functions $\{\hat{H}_i\}$ called the *virtual values* of the partial molar enthalpies which is a simple function of Ω and which gives the correct enthalpy of the mixture. These functions were introduced by Holland and Eubank [6]. The virtual value of the partial molar enthalpy is related to Ω as follows:

$$\hat{H}_i = H_i^{\circ} + \Omega \tag{5-31}$$

where
$$H_i^{\circ} = H_i^{\circ}(1, T), \qquad \Omega = \Omega(P, T, n_1, n_2, \ldots, n_c)$$
$$\hat{H}_i = \hat{H}_i(P, T, n_1, n_2, \ldots, n_c)$$

The quantities $\{\hat{H}_i\}$ are called the virtual values of the partial molar enthalpies because they may be used to compute the correct enthalpy of a mixture, but they may differ individually from the true values of the partial molar enthalpies. A proof of this statement follows the development of expressions for the partial molar enthalpies.

B. Partial Molar Enthalpies

The ideal solution relationship for the enthalpy of a mixture

$$n_T H = \sum_{i=1}^{c} n_i H_i(P, T) \tag{5-32}$$

is not adequate for computing the enthalpy of highly nonideal solutions. The enthalpy of a nonideal solution may be expressed in terms of the partial molar enthalpies of the individual components. The origin of these enthalpies is a consequence of the characteristics of homogeneous functions. A function $f(x_1, x_2, \ldots, x_c)$ is said to be homogeneous of order m if

$$f(\lambda x_1, \lambda x_2, \ldots, \lambda x_c) = \lambda^m f(x_1, x_2, \ldots, x_c) \qquad (5\text{-}33)$$

where λ is any arbitrary, real number. By this definition, the total enthalpy of a mixture $\mathscr{H}$ (which is equal to $n_T H$) is homogeneous of degree 1 because the enthalpy of two moles of a given mixture is twice that of 1 mole of the same mixture at a fixed temperature and pressure; that is,

$$\mathscr{H}(\lambda n_1, \lambda n_2, \ldots, \lambda n_c) = \lambda \mathscr{H}(n_1, n_2, \ldots, n_c) \qquad (5\text{-}34)$$

If the function $f(x_1, x_2, \ldots, x_c)$ is homogeneous of degree m in the x_i's and has continuous first partial derivatives, then, by Euler's theorem,

$$x_1 \frac{\partial f}{\partial x_1} + x_2 \frac{\partial f}{\partial x_2} + \cdots + x_c \frac{\partial f}{\partial x_c} = m f(x_1, x_2, \ldots, x_c) \qquad (5\text{-}35)$$

Since the total enthalpy of a mixture is homogeneous of degree 1 in the n_i's at a fixed temperature and pressure, Euler's theorem gives

$$n_1 \left[\frac{\partial (n_T H)}{\partial n_1} \right]_{P,T,n_j} + n_2 \left[\frac{\partial (n_T H)}{\partial n_2} \right]_{P,T,n_j} + \cdots + n_c \left[\frac{\partial (n_T H)}{\partial n_c} \right]_{P,T,n_j} = n_T H$$
$$(5\text{-}36)$$

The subscript n_j is used to denote the fact that the moles of all components are to be held fixed, except the particular one with which the partial derivative is to be taken. The partial derivatives appearing in Equation (5-36) were given the name *partial molar enthalpies* by Gibbs. They are commonly identified by the symbol $\overline{H}_i$; that is,

$$\overline{H}_i = \left[\frac{\partial (n_T H)}{\partial n_i} \right]_{P,T,n_j} \qquad (5\text{-}37)$$

Equation (5-37) may be used to state Equation (5-36) in this more compact form:

$$n_T H = \sum_{i=1}^{c} n_i \overline{H}_i \qquad (5\text{-}38)$$

C. Relationship between the Virtual Values of Partial Molar Enthalpies and the Partial Molar Enthalpies

The virtual value of the partial molar enthalpy [defined by Equation (5-31)] and the partial molar enthalpy [defined by Equation (5-37)] may be related in the following manner. First let each member of Equation (5-30) be multiplied by the total number of moles to give

$$n_T H = n_T H^\circ + n_T \Omega \qquad (5\text{-}39)$$

Then at constant temperature and pressure, partial differentiation of each member of Equation (5-39), with respect to n_i with T, P, and the moles of each component except i held fixed, gives

$$\overline{H}_i = \sum_{k=1}^{c} H_k^{\circ} \left(\frac{\partial n_k}{\partial n_i} \right)_{P,T,n_j} + \Omega \left(\frac{\partial n_T}{\partial n_i} \right)_{P,T,n_j} + n_T \left(\frac{\partial \Omega}{\partial n_i} \right)_{P,T,n_j}$$

or

$$\overline{H}_i = H_i^{\circ} + \Omega + n_T \left(\frac{\partial \Omega}{\partial n_i} \right)_{P,T,n_j} \tag{5-40}$$

The difficulty of computing the partial molar enthalpies lies in the relatively large number of terms required to evaluate the derivative $(\partial \Omega / \partial n_i)_{P,T,n_j}$. For example, one entire page (page 478) of Reference [7] is required to present the formula obtained by Papadopoulos et al. [8] for the partial molar enthalpy when it is based on the Benedict–Webb–Rubin equation of state [9].

Fortunately, it can be shown as follows that the derivative $(\partial \Omega / \partial n_i)_{P,T,n_j}$ may be neglected in the calculation of the enthalpy of a mixture. Multiplication of each term of Equation (5-40) by n_i followed by the summation over all components yields

$$\sum_{i=1}^{c} \overline{H}_i n_i = \sum_{i=1}^{c} H_i^{\circ} n_i + \Omega n_T + n_T \sum_{i=1}^{c} n_i \left(\frac{\partial \Omega}{\partial n_i} \right)_{P,T,n_j} \tag{5-41}$$

By Equation (5-36), the left side of Equation (5-41) is equal to $n_T H$, and by Euler's theorem [Equation (5-35)], it follows that

$$\sum_{i=1}^{c} n_i \left(\frac{\partial \Omega}{\partial n_i} \right)_{P,T,n_j} = 0 \tag{5-42}$$

since Ω is homogeneous of degree 0. Thus Equation (5-41) reduces to Equation (5-39) and division of each term of this expression by n_T gives Equation (5-30). Since the last term of Equation (5-40) disappears in the calculation of the enthalpy of a mixture, it follows that this derivative may be omitted in the expression for $\overline{H}_i$ to give the expression for $\hat{H}_i$, Equation (5-31). Proof of this statement follows. Multiplication of each term of Equation (5-31) by n_i, followed by the summation over all components i, gives

$$\sum_{i=1}^{c} \hat{H}_i n_i = n_T H^{\circ} + n_T \Omega \tag{5-43}$$

Comparison of Equations (5-38), (5-39), and (5-43) shows that

$$n_T H = \sum_{i=1}^{c} \hat{H}_i n_i = \sum_{i=1}^{c} \overline{H}_i n_i \tag{5-44}$$

where $\hat{H}_i$ is defined by Equation (5-31).

Although the virtual values of the partial molar enthalpies give the correct enthalpy of a mixture as shown by Equation (5-44), they do differ individually from the partial molar enthalpies. They are related by use of Equations (5-31) and (5-40) as follows:

$$\overline{H}_i = \hat{H}_i + n_T \left(\frac{\partial \Omega}{\partial n_i} \right)_{P,T,n_j} \tag{5-45}$$

Fortunately, the difference between the values of the partial molar enthalpies $\{ \overline{H}_i \}$ and their corresponding virtual values $\{ \hat{H}_i \}$ has no effect on the value of the enthalpy of the mixture, as shown by Equation (5-44). Thus, in the calculation of the enthalpy of a mixture, the partial molar enthalpies $\{ \overline{H}_i \}$ may be replaced by their respective virtual values $\{ \hat{H}_i \}$ without altering the value of the enthalpy of the mixture. This important result is most useful in reactor design and analysis.

D. Calculation of the Virtual Values of the Partial Molar Enthalpies from an Equation of State

For illustrative purposes, consider the Benedict–Webb–Rubin equation of state [9].

$$P = RT\rho + \left(B_0 RT - A_0 - \frac{C_0}{T^2} \right)\rho^2 + (bRT - a)\rho^3$$

$$+ a\alpha\rho^6 + \frac{c\rho^3}{T^2}(1 + \gamma\rho^2)e^{-\gamma\rho^2} \tag{5-46}$$

The usual mixing rules for this equation of state are

$$B_0 = \sum_{i=1}^{c} X_i B_{0i} \qquad a = \left[\sum_{i=1}^{c} X_i a_i^{1/3} \right]^3$$

$$A_0 = \left[\sum_{i=1}^{c} X_i A_{0i}^{1/2} \right]^2 \qquad c = \left[\sum_{i=1}^{c} X_i c_i^{1/3} \right]^3$$

$$C_0 = \left[\sum_{i=1}^{c} X_i C_{0i}^{1/2} \right]^2 \qquad \alpha = \left[\sum_{i=1}^{c} X_i \alpha_i^{1/3} \right]^3$$

$$b = \left[\sum_{i=1}^{c} X_i b_i^{1/3} \right]^3 \qquad \gamma = \left[\sum_{i=1}^{c} X_i \gamma_i^{1/2} \right]^2$$

where X_i is the mole fraction of component i in either phase. The Lorentz rule or a quadratic form of B_0 of the mixture is sometimes used.

$$B_0 = \sum_{ij} X_i X_j \left[(B_{0i})^{1/3} + (B_{0j})^{1/3} \right] / 8 \quad \text{(Lorentz)}$$

$$B_0 = \left[\sum_{i=1}^{c} X_i B_{0i}^{1/2} \right]^{1/2} \quad \text{(quadratic)}$$

Following others, Orye [10] suggested that the following mixing rule for A_0 be used instead of the one listed previously:

$$A_0 = \sum_{i=1}^{c} X_i^2 A_{0i} + \sum_i \sum_{\substack{j \\ j \neq i \\ j > i}} M_{ij} X_i X_j A_{0i}^{1/2} A_{0j}^{1/2}$$

An examination of all the preceding expressions for the evaluation of the constants A_0, B_0, C_0, a, b, c, α, and γ shows that they are all homogeneous of degree 0.

An abbreviated development of the expressions needed to compute the value of the deviation function for a given mixture at a specified temperature and pressure follows. The enthalpy $H_i(P, T)$ of pure component i is related to its fugacity by the well-known relationship

$$H_i(P, T) - H_i^\circ(1, T) = -RT^2 \left(\frac{\partial \log_e f_i}{\partial T} \right)_P \quad (5\text{-}47)$$

where $H_i^\circ(1, T)$ = the enthalpy of pure component i in the perfect gas state at 1 atm pressure and at temperature T

Fugacity may be expressed in terms of pressure P by eliminating dG from Equations (5-71) and (5-73) to give

$$RT d \log_e f = \frac{1}{\rho} dP \quad (5\text{-}48)$$

Since P is given by the equation of state [Equation (5-46)], Equation (5-48) may be integrated at constant temperature to give

$$RT \log_e f = RT \log_e \rho RT + 2\rho \left(B_0 RT - A_0 - \frac{C_0}{T^2} \right) + \frac{3\rho^2 (bRT - a)}{2}$$
$$+ \frac{6a\alpha\rho^5}{5} + \frac{c\rho^2}{T^2} \left[\frac{1 - e^{-\gamma\rho^2}}{\gamma\rho^2} + \frac{e^{-\gamma\rho^2}}{2} + \gamma\rho^2 e^{-\gamma\rho^2} \right] \quad (5\text{-}49)$$

The fugacity in the vapor phase is obtained by using vapor density, and the fugacity in the liquid phase is obtained by using liquid density.

An expression for the enthalpy of a mixture is obtained by first carrying out the differentiation implied by Equation (5-47) with the aid of Equations

(5-46) and (5-49). Then the constants for the pure component in the expression so obtained are replaced by the constants for the mixture. The result so obtained may be represented as follows:

$$H - H° = -RT^2\left(\frac{\partial \log_e f}{\partial T}\right)_{P, M}$$ (5-50)

The subscript M has been added to emphasize that after the differentiation has been performed the constants for the pure component have been replaced by the constants for the mixture; for example, see Orye [10]. Comparison of Equations (5-30) and (5-49) shows that

$$\Omega = -RT^2\left(\frac{\partial \log_e f}{\partial T}\right)_{P, M}$$ (5-51)

The expression Ω for the Benedict–Webb–Rubin equation is

$$\Omega = \left(B_0 RT - 2A_0 - \frac{4C_0}{T^2}\right)\rho + (2bRT - 3a)\frac{\rho^2}{2} + \frac{6a\alpha\rho^5}{5}$$
$$+ \frac{c\rho^2}{T^2}\left[3\frac{1 - e^{-\gamma\rho^2}}{\gamma\rho^2} - \frac{e^{-\gamma\rho^2}}{2} + \gamma\rho^2 e^{-\gamma\rho^2}\right]$$ (5-52)

For the case where A_0, C_0, b, and γ are taken to be functions of temperature, a formula for $H - H°$ or Ω has been given by Orye [10].

When other equations of state are used to represent a mixture, the deviation Ω of the enthalpy of the actual mixture from that of a perfect gas mixture at the same temperature is computed in a manner analogous to that shown for the Benedict–Webb–Rubin equation of state. After Ω has been obtained, the corresponding set of the virtual values of the partial molar enthalpies is computed by use of the defining equation, Equation (5-31).

A tabulation of the expressions of Ω for many of the commonly used equations of state is given by Holland [11].

E. Calculation of the Virtual Values of the Partial Molar Enthalpies from Correlations Based on the Law of Corresponding States

At a given value of the critical compressibility factor Z_c for the mixture, Lydersen et al. [13] (see also Perry et al. [12] and Hougen et al. [14]) were able to correlate the enthalpies of mixtures as functions of the reduced temperatures and pressures. These correlations were of the general form

$$\frac{H - H°}{RT_c} = \frac{f_1(T_r, P_r)}{f_2(T_r, P_r)} \quad \text{(at a given } Z_c)$$ (5-53)

where H = enthalpy of 1 mol of mixture at a given temperature and pressure

$P_c = \sum\limits_{i=1}^{c} P_{ci} X_i$, pseudocritical pressure of the mixture

$\quad P_{ci}$ = critical pressure of pure component i

$T_c = \sum\limits_{i=1}^{c} T_{ci} X_i$, pseudocritical temperature of the mixture

$\quad T_{ci}$ = critical temperature of pure component i

$Z_c = \sum\limits_{i=1}^{c} Z_{ci} X_i$, pseudocritical compressibility factor for the mixture

$\quad Z_{ci}$ = critical compressibility factor for pure component i

$P_r = P/P_c$, reduced pressure

$T_r = T/T_c$, reduced temperature

The simplified formulas for calculating the enthalpy result from the fact that the reduced temperature and reduced pressure are homogeneous functions of degree 0 in the n_i's, which is shown as follows. Since

$$T_r(n_1, n_2, \ldots, n_c) = \frac{T}{T_c} = \frac{n_T T}{\sum\limits_{i=1}^{c} T_{ci} n_i} \tag{5-54}$$

then $$T_r(\lambda n_1, \lambda n_2, \ldots, \lambda n_c) = \frac{\lambda n_T T}{\sum\limits_{i=1}^{c} T_{ci} \lambda n_i} = T_r(n_1, n_2, \ldots, n_c) \tag{5-55}$$

Since T_r and P_r are homogeneous of degree 0, any function of these variables which has continuous first derivatives with respect to the n_i's at a fixed temperature and pressure is also homogeneous of degree 0. Also, it can be shown that any combinations of a set of homogeneous functions is also homogeneous.

Since T_c, f_1 and f_2 are all homogeneous of degree 0, then $T_c f_1/f_2$ is homogeneous of degree 0, and consequently Equation (5-53) may be stated in the form given by Equation (5-30) by setting

$$\Omega = \frac{RT_c f_1}{f_2} \tag{5-56}$$

Expressions for Ω for other equations of state and correlations such as those proposed by others [2, 15, 16] are developed in a manner analogous to that demonstrated here.

EXAMPLE 5-2

For an equimolar mixture of ethane and ethylene, calculate the virtual value of the enthalpy of each component and the enthalpy of the mixture at 292.5 atm and 588.12 K. The critical temperatures and pressures are as follows:

Component	T (°C)	P_c (atm)
Ethane	32.1	48.8
Ethylene	9.7	50.5

Use the plot of $(H° - H)/T_c$ versus T_r at parameters of P_r in Figure 4-36, page 4-52, of the *Chemical Engineer's Handbook* [12]. This figure is based on data originally published by Lydersen et al. [13]. Use the data in Table 5-3 to compute $H_i°$ for each component at $T = 588.12$ K.

SOLUTION

The reduced temperature and the reduced pressure of the mixture are computed as follows:

$$T_r = \frac{T}{T_c} = \frac{588.12}{0.5(305.26) + 0.5(282.86)} = \frac{588.12}{294.06} = 2.0$$

$$P_r = \frac{P}{P_c} = \frac{292.5}{0.5(48.8) + 0.5(50.5)} = \frac{292.5}{49.65} = 5.89$$

By use of Figure 4-36 on page 4-52 of Reference [12],

$$\frac{H° - H}{T_c} = 2.25$$

Thus

$$\Omega = H - H° = (-2.25)(294.06) = -661.6 \text{ cal}$$

The following data at $T = 588.12$ K were found by interpolation from those stated in Table 5-3. The virtual values of the partial molar enthalpies for ethane and ethylene in 1 g mole of the equal molar mixture at 588.12 K and 292.5 atm are found by use of Equation (5-31) as follows:

Component	$H_i° = H_T° - H_0°$ (cal/g mol) at $T = 588.12$ K	$\hat{H}_i = H_i° + \Omega$ (cal/g mol)
Ethane	7769	7107.4
Ethylene	6538	5876.4

Then the enthalpy of 1 g mole of the mixture is computed by use of the virtual values of the partial molar enthalpies as follows:

Component	$X_i \hat{H}_i$ (cal/g mol)
Ethane	3553.7
Ethylene	2938.2
	$H = 6491.9$

F. Virtual Values of the Partial Molar Heat Capacities

The partial molar heat capacity is defined by

$$\overline{C}_{pi} = \left(\frac{\partial \overline{H}_i}{\partial T} \right)_{P,\,n_i} \tag{5-57}$$

where the subscript n_i means that the moles of all components are held fixed in the differentiation process. Termwise differentiation of Equation (5-40) with respect to T and with P and the moles of all components held fixed yields

$$\overline{C}_{pi} = C_{pi}^{\circ} + \left(\frac{\partial \Omega}{\partial T} \right)_{P,\,n_i} + n_T \frac{\partial}{\partial T} \left[\left(\frac{\partial \Omega}{\partial n_i} \right)_{P,\,T,\,n_j} \right]_{P,\,n_i} \tag{5-58}$$

Since the last term on the right side of Equation (5-58) is homogeneous of degree 0, it will vanish in the calculation of the molar heat capacity of the mixture, and, consequently, this term is omitted in the definition of the virtual value of the partial molar heat capacity,

$$\hat{C}_{pi} = C_{pi}^{\circ} + \left(\frac{\partial \Omega}{\partial T} \right)_{P,\,n_i} \tag{5-59}$$

By multiplication of each member of Equations (5-58) and (5-59) by n_i, followed by the summation over all components, it is evident that

$$\sum_{i=1}^{c} \overline{C}_{pi} n_i = \sum_{i=1}^{c} \hat{C}_{pi} n_i \tag{5-60}$$

Thus the virtual values of the partial molar heat capacities lead to the correct heat capacity of a mixture. As shown later, these virtual values of the partial molar heat capacities arise in the formulation of the enthalpy balance on chemical reactors.

The heat capacity C_p is often calculated in terms of second derivatives of PVT data and the perfect gas value C_p°. Except for a very simple equation of state, the final equation for $C_p - C_p^{\circ}$ is complex for a pure component [17], and the prospect of the calculation of C_p for a mixture by the usual route of the partial molar values, $\overline{C}_{pi}$, is so complex as to discourage most workers. By use of the virtual values of the partial molar heat capacities, however, the evaluation of the heat capacity for a mixture is reduced to the use of the equation for a pure component with the parameters and density for the pure component replaced by those for the mixture.

G. Enthalpy Function for the Nonideal Liquid Phase

It will now be demonstrated that Equation (5-30) also applies for a mixture in the liquid phase, provided that the symbol H in Equation (5-30) represents the enthalpy of 1 mole of the liquid mixture and Ω denotes the deviation of the enthalpy of 1 mole of the liquid mixture from the enthalpy of 1 mole of the mixture in the perfect gas state at 1 atm and the temperature T of the mixture.

A path in which each component is taken from its standard state of a perfect gas at 1 atm and the temperature T to the final state of a liquid mixture is shown in Figure 5-4. When the changes in enthalpy for this process are summed over all steps and components, the following result is obtained in the limit as P_0 is allowed to go to zero:

$$\Delta H_1 + \Delta H_2 + \Delta H_3 + \Delta H_4 + \Delta H_5 = \sum_{i=1}^{c} \left[\bar{h}_i(P, T) - H_i^\circ(1, T)\right] n_i$$

$$= n_T H - n_T H^\circ \tag{5-61}$$

where $\bar{h}_i$ is the partial molar enthalpy of component i in the liquid mixture at

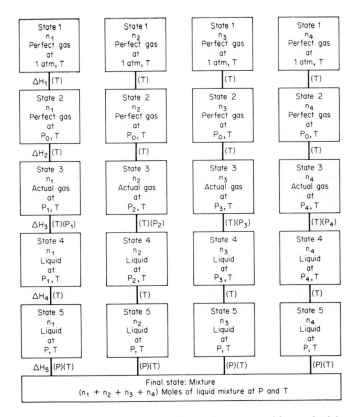

Figure 5-4. Path for computing the effects of pressure and heats of mixing on the enthalpy of a liquid mixture.

P and T, and it is defined by

$$\bar{h}_i = \left[\frac{\partial (n_T H)}{\partial n_i} \right]_{P,T,n_j} \tag{5-62}$$

and H denotes the enthalpy per mole of the liquid mixture. The lowercase letter h is used to emphasize that the partial molar enthalpy is for the liquid state. Thus, when the deviation of the enthalpy of 1 mole of the mixture at P and T from the enthalpy of 1 mole of a perfect gas mixture is set equal to Ω, one again obtains Equation (5-30).

For a liquid mixture, Equation (5-50) takes the following form:

$$H - H^\circ = -RT^2 \left(\frac{\partial \log_e f^L}{\partial T} \right)_{P,M} \tag{5-63}$$

where f^L is the fugacity of the pure component in the liquid phase. Thus

$$\Omega = -RT^2 \left(\frac{\partial \log_e f^L}{\partial T} \right)_{P,M} \tag{5-64}$$

For the special case of an ideal solution which also has the property that its liquid enthalpy is independent of pressure and its vapor behaves as a perfect gas, the enthalpy of the liquid mixture may be computed by setting Ω equal to the negative of the average latent heat of vaporization of the mixture; that is,

$$n_T \Omega = -\sum_{i=1}^{c} n_i \lambda_i(T) \tag{5-65}$$

where $\lambda_i(T)$ is the latent heat of pure component i at the temperature T. Then the enthalpy of such liquid mixtures may be computed as follows:

$$n_T H = \sum_{i=1}^{c} \left[H_i^\circ(1, T) - \lambda_i(T) \right] n_i \tag{5-66}$$

The result given by Equation (5-66) is readily verified as follows. For the case of an ideal solution, the partial molar enthalpy, $\bar{h}_i(P, T)$ is equal to the enthalpy of pure component i, $h_i(P, T)$, in the liquid phase. Thus

$$n_T H = \sum_{i=1}^{c} n_i \bar{h}_i = \sum_{i=1}^{c} n_i h_i$$

and Equation (5-39) may be solved for $n_T \Omega$ to give

$$n_T \Omega = n_T H - n_T H^\circ = \sum_{i=1}^{c} \left[h_i(P, T) - H_i^\circ(1, T) \right] n_i \tag{5-67}$$

Since the effect of pressure on the liquid is negligible, it follows that

$$h_i(P, T) \cong h_i(P_i, T) \tag{5-68}$$

where P_i is the vapor pressure of pure component i. If the vapors behave as perfect gases, then

$$H_i(P_i, T) = H_i^\circ(P_i, T) = H_i^\circ(1, T) \qquad (5\text{-}69)$$

and since the latent heat of vaporization of pure component i is defined by

$$\lambda_i(P_i, T) = H_i(P_i, T) - h_i(P_i, T) \qquad (5\text{-}70)$$

Equation (5-61) reduces to Equation (5-65).

$$n_T \Omega = - \sum_{i=1}^{c} \left[H_i(P_i, T) - h_i(P_i, T) \right] n_i = - \sum_{i=1}^{c} n_i \lambda_i(T)$$

In many instances, data on the behavior of liquid mixtures are meager, and Equation (5-65) will be found useful because many compilations of heats of formation and latent heats are available; for example, see Reference [1].

In conclusion, Equation (5-30) is applicable for both the gas and liquid phases, provided that an appropriate equation of state is used in the evaluation of Ω according to Equations (5-51) and (5-63).

III. ANALYSIS OF CHEMICAL EQUILIBRIA

This analysis is based on the use of the Gibbs free energy function G, which is particularly convenient for expressing how far a chemical reaction is away from equilibrium when the reactants and products are in their standard states. In the developments that follow, any arbitrary standard state may be employed, and a variety of standard states have been used in the past. In the interest of simplicity, only the standard states shown in Table 5-1 are employed. Expressions needed in the analysis of chemical equilibria are developed as follows.

At constant temperature, Equation (5-6) reduces to

$$dG = v \, dP \qquad (5\text{-}71)$$

For 1 mole of a perfect gas ($Pv = RT$), this expression becomes

$$dG = \frac{RT}{P} dP = RT \, d \log_e P \qquad (5\text{-}72)$$

The general form of this expression may be retained for any substance by defining the fugacity f by

$$dG = RT \, d \log_e f \qquad \text{(at constant } T\text{)} \qquad (5\text{-}73)$$

where f is that function required to give the correct free energy change for a process carried out isothermally. For example, for the isothermal process where the pressure on a system is changed from P_1 and P_2 and the corresponding fugacities change from f_1 to f_2, the free energy change is found by

TABLE 5-5
SELECTED VALUES OF THE FREE ENERGY FUNCTION, $(G° - H_0°) / T$ (Taken from [1])
$(G_T° - H_0°)/T$ in cal/(mol K) at indicated T K

Substance	298.16	400	600	800	1000	1500
H_2 (g)	−24.423	−26.422	−29.203	−31.186	−32.738	−35.590
O_2 (g)	−42.061	−44.112	−46.968	−49.044	−50.697	−53.808
C (graphite)	−0.5172	−0.8245	−1.477	−2.138	−2.771	−4.181
CO (g)	−40.350	−42.393	−45.222	−47.254	−48.860	−51.864
CO_2 (g)	−43.555	−45.828	−49.238	−51.895	−54.109	−58.481
H_2O (g)	−37.172	−39.508	−42.768	−45.131	−47.018	−50.622
Methane (g)	−36.46	−38.86	−42.39	−45.21	−47.65	−52.84
Ethane (g)	−45.27	−48.24	−53.08	−57.29	−61.11	−69.46
Propane (g)	−52.73	−56.48	−62.93	−68.74	−74.10	−85.86
Ethylene (g)	−43.98	−46.61	−50.70	−54.19	−57.29	−63.94
Acetylene (g)	−39.976	−42.451	−46.313	−49.400	−52.005	−57.231
Benzene (g)	−52.93	−56.69	−63.70	−70.34	−76.57	−90.45
Toluene (g)	−61.98	−66.74	−75.52	−83.79	−91.53	−108.75
o-Xylene (g)	−65.61	−71.74	−82.81	−93.01	−102.46	123.30
m-Xylene (g)	−67.63	−73.50	−84.20	−94.18	−103.48	−124.11
p-Xylene (g)	−66.25	−72.15	−82.83	−92.76	−102.02	−122.29

integration of Equation (5-73):

$$G(T, P_2) - G(T, P_1) = RT \log_e \frac{f_2}{f_1} \qquad (5\text{-}74)$$

Corresponding to the *standard heat of formation* there is a *standard free energy of formation* $\Delta G_f°$ of a substance, which is defined as the change in free energy that occurs when a substance in its standard state is formed from its elements in their standard states. Thus it is inferred by this definition that $\Delta G_f°$ for any element is zero.

The *standard free energy change* $\Delta G°$ is that free energy change that occurs when products in their standard states are formed from reactants in their standard states. Then, for the reaction given by Equation (5-14), the standard free energy change is stated in terms of the free energies of formation as follows:

$$\Delta G° = c\,\Delta G_{fC}° + d\,\Delta G_{fD}° - a\,\Delta G_{fA}° - b\,\Delta G_{fB}° \qquad (5\text{-}75)$$

Various tables are available that list the free energies of formation of various substances as functions of temperature. As a result of the development of statistical methods, free energy data are frequently tabulated as $(G_T° - H_0°)/T$ as demonstrated by Table 5-5.

At the temperature of absolute zero, it follows from the definition of G [Equation (5-2)] that

$$G_0° = H_0° \qquad (5\text{-}76)$$

$$\Delta G_{f0}° = \Delta H_{f0}° \qquad (5\text{-}77)$$

(This relationship is particularly important because it permits the values for ΔG_{f0}° to be obtained from Table 5-3 in which the values for ΔH_{f0}° are listed.)

Suppose that a compound C in its standard state is formed from its elements in their standard states at the temperature T according to the following reaction:

$$aA(g) + bB(s) = cC(g) \tag{5-78}$$

The free energy change for this reaction can be represented as follows:

a moles of element A: perfect gas at 1 atm and at the temperature T	$+$	b moles of element B: solid at 1 atm and at the temperature T	$c\,\Delta G^\circ$ $=$	c moles of compound C: perfect gas at 1 atm and at the temperature T

The following formula for the free energy of formation for C is developed in a manner analogous to that shown in the development of Equation (5-17).

$$\left(\frac{\Delta G_{fC}^\circ}{T}\right) = \left[\frac{\Delta H_{f0}^\circ}{T} + \frac{G_T^\circ - H_0^\circ}{T}\right]_{\text{compound } C}$$

$$-\frac{a}{c}\left(\frac{G_T^\circ - H_0^\circ}{T}\right)_{\text{element } A} - \frac{b}{c}\left(\frac{G_T^\circ - H_0^\circ}{T}\right)_{\text{element } B} \tag{5-79}$$

Free energies of formation may be used to calculate the standard free energy change ΔG° for a reaction by use of Equation (5-75).

EXAMPLE 5-3

Calculate the free energy of formation for benzene at 800 K by use of Equation (5-79) and the data given in Tables 5-3 and 5-5. The reaction is

$$6C(s) + 3H_2(g) = C_6H_6(g)$$

SOLUTION

Since $\Delta G_{f0}^\circ = \Delta H_{f0}^\circ$, the value of ΔG_{f0}° may be found by use of Table 5-3:

$$\Delta G_{f0}^\circ = \Delta H_{f0}^\circ = 24{,}000 \text{ cal/g mol}$$

From Table 5-5, the following values are obtained:

Substance	$(G_T^\circ - H_0^\circ)/T$(cal/g mol K) at 800 K
C_6H_6	-70.34
H_2	-31.186
C	-2.138

Substitution of these numerical values into Equation (5-79) gives

$$\frac{\Delta G_f^\circ}{800} = \left[\frac{24{,}000}{800} + (-70.34) \right] - 6(-2.138) - 3(-31.186)$$

$$= 66.046 \, (\text{cal/g mol K})$$

Thus the free energy of formation of benzene at 800 K from its elements in their standard states at 800 K is given by

$$\Delta G_f^\circ = (66.046)(800) = 52{,}836.8 \, \text{cal/g mol}$$

1. Relationship between the Standard Free Energy Change ΔG° and the Equilibrium Constant for a Gas Phase Reaction. Suppose compounds C and D are formed from compounds A and B by the following reaction at pressure P and temperature T:

$$aA(g) + bB(g) \rightleftarrows cC(g) + dD(g) \tag{5-80}$$

Since the free energy function G is independent of path, a path such as the one shown in Figure 5-5 may be devised which involves the standard free energy change ΔG° and the equilibrium constant. In this figure, the symbols $\hat{f}_A^V$, $\hat{f}_B^V$, $\hat{f}_C^V$, and $\hat{f}_d^V$ denote the fugacities of components A, B, C, and D, respectively, in the equilibrium gaseous mixture at temperature T and pressure P. The familiar concept of the van't Hoff equilibrium box is used as part of the path. The box is regarded as being so large that a moles of A and b moles of B may be introduced and c moles of C and d moles of D withdrawn without upsetting the equilibrium state of the box, provided, of course, that these compounds are introduced and withdrawn at their respective equilibrium fugacities, $\hat{f}_A^V$, $\hat{f}_B^V$, $\hat{f}_C^V$, and $\hat{f}_D^V$. Addition of the free energy changes around the path yields

$$\Delta G = -RT \log_e \frac{\left(\hat{f}_C^V\right)^c \left(\hat{f}_D^V\right)^d}{\left(\hat{f}_A^V\right)^a \left(\hat{f}_B^V\right)^b} + RT \log_e \frac{\left(f_C^\circ\right)^c \left(f_D^\circ\right)^d}{\left(f_A^\circ\right)^a \left(f_B^\circ\right)^b} \tag{5-81}$$

Equation (5-81) applies regardless of the standard states selected for the pure components. For the special case shown in Figure 5-5 where the standard state of each component (A, B, C, and D) is taken to be a perfect gas at 1 atm at the temperature of the mixture, the second term on the right side of Equation (5-81) vanishes and Equation (5-81) reduces to

$$\Delta G^\circ = -RT \log_e \frac{\left(\hat{f}_C^V\right)^c \left(\hat{f}_D^V\right)^d}{\left(\hat{f}_A^V\right)^a \left(\hat{f}_B^V\right)^b} \tag{5-82}$$

where the superscript $\circ$ has been added to ΔG to denote the particular set of standard states used to compute the free energy change.

For the particular set of standard states shown in Figure 5-5, Equation (5-81) may be stated in terms of the activities a_i and equilibrium constant K

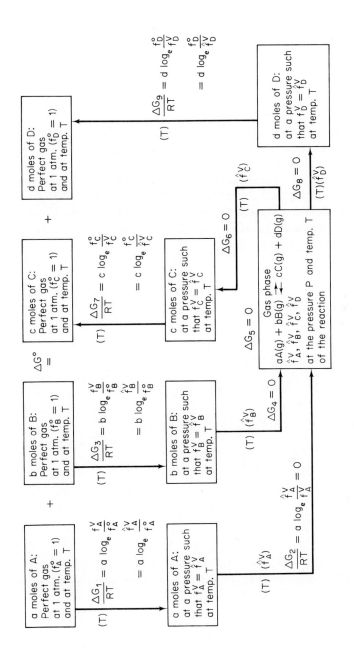

Figure 5-5. Use of a path for relating ΔG^0 and the equilibrium constant for a gas phase reaction.

as follows:

$$\Delta G^\circ = -RT \log_e K \qquad (5\text{-}83)$$

where $\quad K = \dfrac{a_C^c a_D^d}{a_A^a a_B^b}$

$\quad a_i = \hat{f}_i / f_i^\circ = \hat{f}_i$, for the standard states shown in Figure 5-5

For the standard state of a perfect gas at 1 atm at the specified temperature, the activity a_i (a dimensionless variable) is numerically equal to the fugacity $\hat{f}_i$ of component i in the mixture.

The behavior of many fluids may be approximated with good accuracy by use of the concept of an ideal solution. A gaseous mixture is said to form an *ideal* solution at a given temperature and pressure if the fugacity $\hat{f}_i^V$ of each component in the mixture is related to its fugacity in the pure state as follows:

$$\hat{f}_i^V = f_i^V y_i \qquad (5\text{-}84)$$

where $\quad f_i^V =$ fugacity of the pure component i evaluated at the temperature T and pressure P of the mixture

$\quad y_i =$ mole fraction in gas

For such mixtures, Equation (5-82) reduces to

$$\Delta G^\circ = -RT \log_e K_f K_y \qquad (5\text{-}85)$$

where the constants K_f and K_y are defined as follows:

$$K_f = \frac{\left(f_C^V\right)^c \left(f_D^V\right)^d}{\left(f_A^V\right)^a \left(f_B^V\right)^b}, \qquad K_y = \frac{y_C^c y_D^d}{y_A^a y_B^b}$$

Comparison of Equations (5-83) and (5-85) shows that

$$K = K_f K_y \qquad (5\text{-}86)$$

Various types of correlations of fugacities for pure components as functions of temperature and pressure have appeared in the literature.

For the case where each component of the mixture behaves as a perfect gas, it follows that $K = K_f K_y = K_p$, and thus

$$\Delta G^\circ = -RT \log_e K_p = -RT \log_e \left[\frac{p_C^c p_D^d}{p_A^a p_B^b} \right] \qquad (5\text{-}87)$$

EXAMPLE 5-4

Calculate the conversion of pure methanol to dimethyl ether by the reaction

$$2CH_3OH \rightleftarrows (CH_3)_2O + H_2O$$

when pure methanol is allowed to come to equilibrium at 1 atm pressure and at each of the following temperatures: 500 K, 600 K, and 700 K. Assume that at these conditions, the mixture obeys the perfect gas law. The following values of the free energy changes (obtained by use of data given in Reference [18]) may be used.

T K	$\Delta G°$ [kcal/g mol of $(CH_3)_2O$ formed]
500	-2.95
600	-2.56
700	-2.18

SOLUTION

For convenience, let the reaction be represented as follows:

$$2A \rightleftarrows B + C$$

Let x be the moles of A converted per mole of A in the feed. Then the stoichiometry of the reaction requires that

$$n_A = n_A^0 - n_A^0 x$$

$$n_B = n_B^0 + \tfrac{1}{2}n_A^0 x$$

$$n_C = n_C^0 + \tfrac{1}{2}n_A^0 x$$

$$n_T = n_T^0$$

Also, since the feed is to be pure methanol, $n_B^0 = n_C^0 = 0$, and thus $n_T = n_T^0 = n_A^0$.

Since it is given that the mixture obeys the perfect gas law, the expression given by Equation (5-87) is applicable. For the reaction given, the appropriate form of K_p is given by

$$K_p = \frac{p_C p_D}{p_A^2} = \frac{n_C n_D}{n_A^2} = \frac{\left(\tfrac{1}{2}n_A^0 x\right)^2}{\left(n_A^0 - n_A^0 x\right)^2} = \frac{\tfrac{1}{4}x^2}{(1-x)^2}$$

This expression is readily solved for x as a function of K_p to give

$$x = \frac{2\sqrt{K_p}}{1 + 2\sqrt{K_p}}$$

Thus, by use of Equation (5-87) and the preceding expression for the conversion, the desired results are obtained, as summarized in the following table:

T K	$K_p = e^{-\Delta G°/RT}$	$x = 2\sqrt{K_p}/(1 + 2\sqrt{K_p})$
500	19.472	0.898
600	8.560	0.854
700	4.793	0.814

2. Chemical Reactions in the Liquid Phase. Suppose that the reaction given by Equation (5-80) is carried out in the liquid phase at the total pressure P and temperature T; that is,

$$aA(l) + bB(l) = cC(l) + dD(l) \qquad (5\text{-}88)$$

Again the standard state of each component i is taken to be a perfect gas at 1 atm pressure and the temperature T of the reaction. Thus $\Delta G°$ for the liquid phase reaction given by Equation (5-88) is exactly equal to the $\Delta G°$ for the gas phase reaction given by Equation (5-80). For $\Delta G°$ is always computed on the basis of the standard state and is independent of whether the reaction is carried out in the vapor phase, the liquid phase, or in two or more phases.

To relate $\Delta G°$ to the equilibrium constant, it is necessary, however, to take into account the precise manner in which the reaction is carried out. In this case, let the fugacities of components A, B, C, and D in the equilibrium mixture be denoted by $\hat{f}_A^L$, $\hat{f}_B^L$, $\hat{f}_C^L$, and $\hat{f}_D^L$, where the superscript L is used to denote that these fugacities are for components in the liquid phase. The free energy $\Delta G°$ may be related to the equilibrium constant by use of the path shown in Figure 5-6. Consider the path shown for compound A. First compound A is expanded isothermally as a perfect gas to pressure P_0. If the pressure P_0 is allowed to go to zero, the perfect gas may be regarded as an actual substance, because all substances become perfect gases as the pressure goes to zero. After the ΔG's for compound A have been summed, the limit of this sum is taken as P_0 is allowed to go to zero. In the limit as P_0 goes to zero, the gas withdrawn at pressure P_0 may be regarded as an actual gas. Thus this box acts as a "converter" in that a perfect gas enters it at 1 atm and an actual substance leaves it at pressure P_0. Then, for the first step,

$$\Delta G_1 = aRT \log_e \frac{P_0}{1}$$

(The limit as P_0 approaches zero is imposed in a subsequent step.) Next the actual gas is compressed isothermally to a total pressure equal to the vapor pressure P_A of pure compound A at the temperature T. At this pressure, the fugacity of compound A is denoted by f_{A, P_A}^V, and for this step

$$\Delta G_2 = aRT \log_e \frac{f_{A, P_A}^V}{f_{A, P_0}^V}$$

In the third step, compound A is condensed at its vapor pressure P_A (at temperature T). Since the vapor and liquid phases of compound A are in a state of equilibrium at P_A and T, it follows that

$$f_A^V|_{P_A} = f_A^L|_{P_A}$$

and thus $$\Delta G_3 = aRT \log_e \frac{f_{A, P_A}^L}{f_{A, P_A}^V} = 0$$

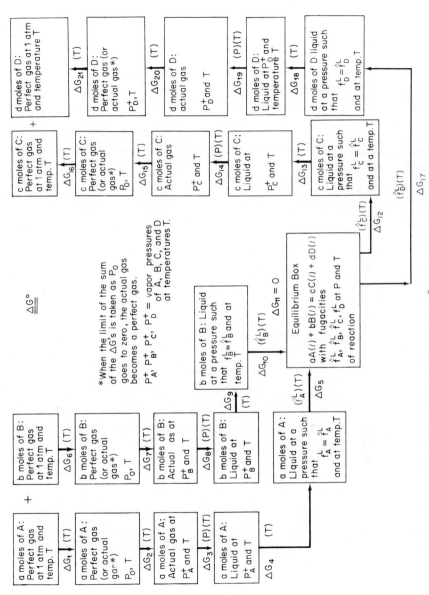

Figure 5-6. Use of a path to relate ΔG^0 and the equilibrium constant for a liquid phase reaction.

Next, compound A is compressed isothermally to a pressure such that it has a fugacity $\hat{f}_A^L$ equal to its fugacity $\hat{f}_A^L$ in the equilibrium box, and for this step

$$\Delta G_4 = aRT \log_e \frac{\hat{f}_A^L}{\hat{f}_{A,\,P_A}^L}$$

For the fifth step, the van't Hoff equilibrium box is used again, where a moles of A and b moles of B are introduced and c moles of C and d moles of D are withdrawn from the equilibrium box at their respective equilibrium fugacities $\hat{f}_A^L$, $\hat{f}_B^L$, $\hat{f}_C^L$, and $\hat{f}_D^L$. For this step,

$$\Delta G_5 = 0$$

Then the change in free energy for compound A around the entire path is given by

$$\Delta G_A^\circ = \lim_{P_0 \to 0} \left[\sum_{j-1}^{5} \Delta G_j \right] = \lim_{P_0 \to 0} \left[aRT \log_e \left(\frac{P_0}{1} \right) \left(\frac{f_{A,\,P_A}^V}{f_{A,\,P_0}^V} \right) \left(\frac{\hat{f}_A^L}{f_{A,\,P_A}^V} \right) \right]$$

$$= aRT \log_e \frac{\hat{f}_A^L}{1}$$

For the entire path

$$\Delta G^\circ = \sum_{i=A,\,B,\,C,\,D} \Delta G_i^\circ = -RT \log_e \frac{\left(\hat{f}_C^L \right)^c \left(\hat{f}_D^L \right)^d}{\left(\hat{f}_A^L \right)^a \left(\hat{f}_B^L \right)^b} \qquad (5\text{-}89)$$

Since

$$a_A = \frac{\hat{f}_A^L}{1}, \qquad a_B = \frac{\hat{f}_B^L}{1}, \qquad a_C = \frac{\hat{f}_C^L}{1}, \qquad a_D = \frac{\hat{f}_D^L}{1}$$

it is possible to state Equation (5-89) in terms of the equilibrium constant K [see Equation (5-88)] for the reaction under consideration.

To state the equilibrium constant K in terms of the mole fractions in the liquid phase, the general expression for the fugacity of a component in a mixture,

$$\hat{f}_i^L = \gamma_i^L f_i^L x_i \qquad (5\text{-}90)$$

is used. In this case, f_i^L is evaluated at the temperature T and pressure P of the reacting mixture, and the activity coefficient γ_i^L is a function of P, T, and the x_i's. Then, for the reaction given by Equation (5-88),

$$K = K_\gamma K_f K_x \qquad (5\text{-}91)$$

The quantities K_γ, K_f, and K_x are defined as follows:

$$K_\gamma = \frac{\gamma_C^c \gamma_D^d}{\gamma_A^a \gamma_B^b}, \qquad K_f = \frac{\left(f_C^L \right)^c \left(f_D^L \right)^d}{\left(f_A^L \right)^a \left(f_B^L \right)^b}, \qquad K_x = \frac{x_C^c x_D^d}{x_A^a x_B^b} \qquad (5\text{-}92)$$

For an ideal solution $\gamma_A = \gamma_B = \gamma_C = \gamma_D = 1$, and thus $K = K_f K_x$. Thus, for an ideal solution,

$$\Delta G^\circ = -RT \log_e K_f K_x \tag{5-93}$$

3. Evaluation of the Fugacities f_A^L, f_B^L, f_C^L, and f_D^L, $P > P_i$ for $i = A$, B, **C, and D.** If the total pressure at which the reaction occurs is greater than the vapor pressures of the respective components at the temperature of the reaction, then f_A^L, f_B^L, f_C^L, and f_D^L at the temperature and pressure of the reaction may be evaluated as indicated by Equation (5-96). The development of this relationship follows. When pure component A in the liquid state is compressed from its vapor pressure at the temperature of the reaction to the total pressure P of the reaction, the following result is obtained by integration of Equation (5-71):

$$\Delta G = \int_{P_A}^{P} v \, dP \cong (P - P_A) v \tag{5-94}$$

Since the effect of pressure on the volume of a liquid is usually negligible, the approximate result given by Equation (5-94) represents the change in free energy with good accuracy. Also, for this same change in pressure at the temperature T of the reaction, Equation (5-74) gives

$$\Delta G = RT \log_e \frac{f_A^L}{f_{A,P_A}^L} = RT \log_e \frac{f_A^L}{f_{A,P_A}^V} \tag{5-95}$$

where f_A^L is the value of the fugacity of component A at the total pressure P and temperature T. After ΔG has been eliminated from Equations (5-94) and (5-95), the following result is obtained upon rearrangement:

$$f_A^L = f_{A,P_A}^V \exp\left[\frac{(P - P_A) v}{RT}\right] \tag{5-96}$$

The fugacity of A at its vapor pressure may be evaluated by use of an appropriate equation of state.

EXAMPLE 5-5

Calculate the liquid fugacity of n-butane in the liquid state at 350 K and 60 atm. The vapor pressure of n-butane at 350 K is 9.35 atm [1]. The molar volume of the liquid at the vapor pressure (or saturation pressure) at 350 K is 0.1072 liter/g mol [12]. From the jb and jb-E tables of Reference [1], the fugacity coefficient ϕ at the vapor pressure at $T = 350$ K is 0.8340.

SOLUTION

The fugacity of the vapor at 350 K is computed as follows:

$$f_{i,P_i}^V = \phi P_i = \left(\frac{f_{i,P_i}^V}{P_i}\right) P_i = 0.8340 \times 9.35 = 7.7979 \text{ atm}$$

The fugacity of the liquid f_i^L at $P = 60$ atm may be estimated by use of Equation (5-96) as follows:

$$f_i^L = f_{i,P_i}^V \exp\left[\frac{(P - P_i)v}{RT}\right]$$

$$f_i^L = 7.7979 \exp\left[\frac{(60 - 9.35)(0.1072)}{(0.08205)(350)}\right] = 9.421 \text{ atm}$$

If the total pressure is less than the vapor pressure of one of the components, the liquid state at the pressure P and temperature T is a hypothetical state since the pure component cannot exist in the liquid state at the pressure P. Consequently, the volume v can no longer be regarded as a constant in the integration, as was done in Equation (5-94). Thus, to evaluate ΔG when one or more of the pure components are in hypothetical states at the temperature T and pressure P of the system, the fugacities of the components in the hypothetical state must be found by extrapolation procedures.

A. Calculation of Fugacities of Components in Hypothetical States

Procedures which may be used to compute the fugacities of components in hypothetical vapor and liquid states are presented next, first for a component in a hypothetical liquid state, and second for a component in a hypothetical vapor state at the temperature T and pressure P of the mixture.

1. Calculation of f_i^L when $P < P_i$ at Temperature T. When the specified value of the total pressure P is less than the vapor pressure P_i of component i at the specified temperature T, component i cannot exist in the liquid phase. Thus the hypothetical liquid state of component i must be found or defined by an extrapolation procedure. The description of this procedure is made easier by use of the fugacity coefficient chart shown in Figure 5-7. The reduced temperature and reduced pressure used in this chart are defined in the usual way,

$$P_r = \frac{P}{P_c}, \qquad T_r = \frac{T}{T_c} \tag{5-97}$$

Suppose, for example, that the specified temperature and pressure for component i gives a reduced pressure $P_{r0} = 0.5$ and a reduced temperature $T_{r0} = 1.0$. This point (P_{r0}, T_{r0}) is seen to lie in the vapor phase above the saturation line in Figure 5-7. The hypothetical liquid state at P_{r0} and T_{r0} may be defined by the extrapolation of the points $(\phi_1, T_{r1}), (\phi_2, T_{r2}), \ldots$, along the isobar P_{r0} to the hypothetical liquid state P_{r0} and T_{r0}. If data for a pure component are available, it is not necessary to use the reduced temperatures and pressures in the construction. An extrapolation procedure for predicting the liquid fugacity of a substance in a hypothetical liquid state is demonstrated by the following numerical example.

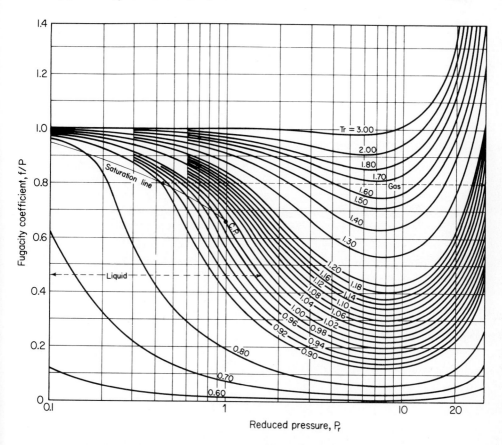

Figure 5-7. Fugacity coefficients of gases and liquids: $Z_c = 0.27$ (From *Chemical Engineer's Handbook*, by Robt. H. Perry, C. H. Chilton, and S. D. Kirkpatrick, 4th ed. pp. 4–53. © 1963 by McGraw-Hill Book Company. Used with permission of McGraw-Hill Book Company.)

EXAMPLE 5-6

In this example the procedure described is used to calculate the liquid fugacity of f_i^L of *n*-butane in a hypothetical liquid state at a pressure below its vapor pressure of 9.35 atm at the given temperature of 350 K. In particular, it is desired to determine the liquid fugacity of *n*-butane at 350 K and 4.52 atm. At 350 K, the molar volume of the saturated liquid is equal to 0.1072 liter/g mol [12]. The following values of the fugacity coefficient were taken from the *jb-E* tables of Reference [1].

T (K)	$\phi\|_{P_i} = f_i^V/P_i$ (at the vapor pressure)	Vapor Pressure P_i (atm)
300	0.9219	2.55
310	0.9066	3.43
320	0.8900	4.52

SOLUTION

To carry out the extrapolation procedure described previously, values of ϕ at different values of the reduced temperature along the isobar $P = 4.52$ are found by use of the data given.

First the given values of ϕ at the vapor pressure are used to compute the fugacity of the vapor at its vapor pressure, f_{i,P_i}^V. These vapor fugacities and Equation (5-96) are then used to compute the fugacities of the liquid in the hypothetical state of $P = 4.52$ atm and $T = 350$ K. Then the desired values of ϕ at $P = 4.52$ atm are computed.

T (K)	$f_{i,P_i}^V = (\phi|_{P_i})P_i$	f_i^L at $p = 4.52$ atm, by Equation (5-96)	ϕ at ($P = 4.52$ atm), $\phi = (f_i^L/4.52)$
300	2.35	2.37	0.524
310	3.11	3.12	0.691
320	4.02	4.02	0.890

The graph of ϕ versus T at $P = 4.52$ atm used to find ϕ at $T = 350$ K is shown in Figure 5-8. The value of ϕ so obtained is

$$\phi = 1.5$$

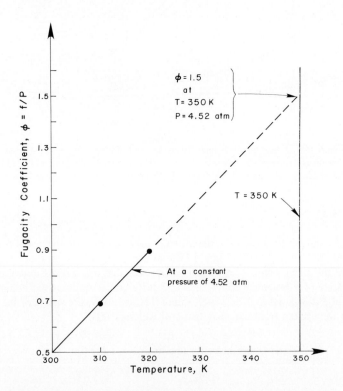

Figure 5-8. Extrapolation procedure for the prediction of the liquid fugacity of n-butane in the hypothetical liquid state at $P = 4.52$ atm and 350 K.

Thus the liquid fugacity of n-butane in the hypothetical liquid state at $T = 350$ K and $P = 4.52$ atm is given by

$$f_i^L = \phi P = \left(\frac{f_i^L}{P} \right) P = (1.5)(4.52) = 6.78 \text{ atm}$$

2. Calculation of f_i^V when $P > P_i$ at Temperature T. In this case, component i cannot exist in the vapor phase because the total pressure is greater than the vapor pressure of component i at temperature T of the mixture. For this component, the state P_{r0} and T_{r0} lies in the liquid region (below the saturation line in Figure 5-7). The fugacity coefficient in the hypothetical vapor state at P_{r0} and T_{r0} is found in a manner analogous to that demonstrated for the hypothetical liquid state. In the present case, however, the points $(\phi_1, T_{r1}), (\phi_2, T_{r2}), \ldots$, are located in the vapor region at the reduced pressure P_{r0}. After these points have been plotted (ϕ versus T_r), the value of ϕ at T_{r0} and P_{r0} is found by extrapolation to T_{r0}. Again, if values of ϕ are available for a pure component, the extrapolation procedure may be carried out in terms of P and T instead of P_r and T_r.

EXAMPLE 5-7

Calculate the vapor fugacity of n-butane in the hypothetical vapor state at 374 K and 20 atm. At 374 K, the vapor pressure of n-butane is 15.4 atm. This vapor pressure was found by interpolation of the data given in the jb and jb-E tables of Reference [1]. From this same reference, the following data were obtained:

T (K)	ϕ at $P = 20$ atm
390	0.7570
392	0.7615
394	0.7659
396	0.7701
400	0.7782

SOLUTION

The data presented are plotted in Figure 5-9. As shown, extrapolation to the temperature of 374 K at $P = 20$ atm gives

$$\phi = 0.725$$

Thus the vapor fugacity of n-butane in the hypothetical vapor state of 374 K and 20 atm is given by

$$f_i^V = (0.725)(20) = 14.5 \text{ atm}$$

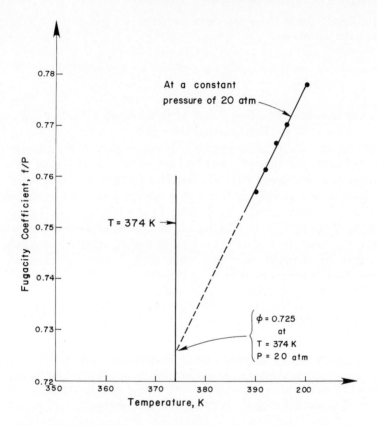

Figure 5-9. Extrapolation procedure for the prediction of the vapor fugacity of n-butane in the hypothetical vapor state at $P = 20$ atm and 374 K.

B. Heterogeneous Chemical Reactions

If components in different phases participate in a chemical reaction, then the reaction is classified as a heterogeneous chemical reaction. To illustrate such a reaction, consider the case where the following reaction is carried out at the total pressure P and temperature T:

$$aA(g) + bB(l) = cC(s) + dD(g) \qquad (5\text{-}98)$$

Let the following standard states be selected.

a moles of A: perfect gas at 1 atm and temperature T		b moles of B: perfect gas at 1 atm and temperature T		c moles of C: solid at 1 atm and temperature T		d moles of D: perfect gas at 1 atm and temperature T
	$+$		$=$		$+$	

If C is an element such as carbon whose standard state is taken to be graphite,

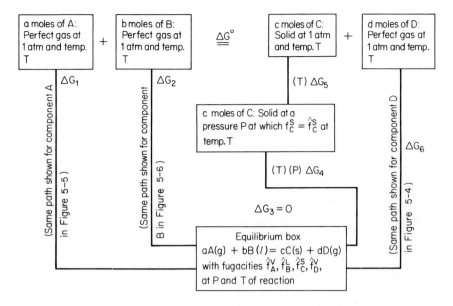

Figure 5-10. Use of a path to relate ΔG^0 and the equilibrium constant for a heterogeneous reaction.

while the standard states of all the other elements are perfect gases at 1 atm at temperature T, then the ΔG computed by use of the alternate path in Figure 5-10 will be equal to ΔG° as computed from the free energies of formation.

For component A, the change in free energy in going from the standard state to the equilibrium box is

$$\Delta G_1 = aRT \log_e \frac{\hat{f}_A^V}{1}$$

as illustrated in Figure 5-5. In the same manner as shown for component B in Figure 5-6,

$$\Delta G_2 = bRT \log_e \frac{\hat{f}_B^L}{1}$$

Since the reaction is in equilibrium, $\Delta G_3 = 0$. For the solid, the change in free energy in going from the equilibrium box to the standard state as shown in Figure 5-10 is given by

$$\Delta G_4 = cRT \log_e \frac{\hat{f}_C^S}{f_C^S} = 0$$

$$\Delta G_5 = cRT \log_e \frac{f_C^\circ}{f_C^S} = -cRT \log_e a_C$$

where f_C° is equal to the fugacity of component C in the standard state

$(P = 1 \text{ atm}, T)$ and a_C is equal to the activity of component C [defined following Equation 5-83)].

The change in free energy of compound D in going from the equilibrium box to its standard state is given by

$$\Delta G_6 = dRT \log_e \frac{1}{\hat{f}_D^V}$$

Since $\Delta G°$ is equal to the sum of the free energy changes around the path shown in Figure 5-10, it follows that

$$\Delta G° = -RT \log_e \frac{(a_C)^c (\hat{f}_D^V)^d}{(\hat{f}_A^V)^a (\hat{f}_B^L)^b} \tag{5-99}$$

Since the liquid phase is composed of pure compound B, it follows that

$$\hat{f}_B^L = f_B^L|_{P,T}$$

where the fugacity of the pure component B may be evaluated by use of Equation (5-95).

To evaluate a_C, another expression is obtained for ΔG_5 by use of Equation (5-94):

$$\Delta G_5 = c \int_P^1 v_s \, dP \cong c(1 - P) v_s$$

where v_s is the volume per mole of C in the solid state. From these two expressions for ΔG_5, the following formula for computing a_C is obtained:

$$a_C = \exp\left[-\left(\frac{1 - P}{RT}\right) v_s\right]$$

Generally, the volume v_s for a solid is so small that its activity can be set equal to unity with good accuracy.

For illustrative purposes, suppose that the approximation $a_C = 1$ is valid and that the vapor phase forms an ideal solution. Then Equation (5-99) reduces to

$$\Delta G° = -RT \log_e \frac{(f_D^V y_D)^d}{(f_A^V y_A)^a (f_B^L)^b}$$

If it is further supposed that the vapor behaves as a perfect gas and that the pressure correction for f_B^L [given by an equation of the form of Equation (5-96)] is insignificant, then the expression for $\Delta G°$ further reduces to

$$\Delta G° = -RT \log_e \frac{(P y_D)^d}{(P y_A)^a (P_B)^b} \tag{5-100}$$

where P_B is the vapor pressure of pure component B at temperature T of the reaction.

EXAMPLE 5-8

The reaction

$$CO_2(g) + C(s) \rightleftarrows 2CO(g)$$

occurs in the reducing zone of a coal gasification reactor. At 800°C, $K_p = 7.64633$ atm, and at 1000°C, $K_p = 149.886$ atm (see Reference [19]). If it is supposed that pure carbon dioxide in the gaseous state is present initially over pure carbon in the solid state, find the equilibrium mole fractions of carbon dioxide and carbon monoxide at each of these temperatures at a pressure of 1 atm.

SOLUTION

Let the reaction be represented by the following symbols:

$$A(g) + B(s) \rightleftarrows 2C(g)$$

Since B is a solid, the activity of B is approximately equal to unity, as discussed previously. Thus assume $a_B = 1$. Then

$$K_p = \frac{p_C^2}{p_A} = \frac{Py_C^2}{y_A}$$

The stoichiometric relationships are as follows:

$$n_A = n_A^0 - n_A^0 x$$

$$n_C = \qquad 2n_A^0 x$$

$$n_T = n_A^0 + n_A^0 x$$

Then

$$y_A = \frac{1-x}{1+x} \quad \text{and} \quad y_C = \frac{2x}{1+x}$$

and

$$K_p = \frac{P(2x)^2(1+x)}{(1+x)^2(1-x)} = \frac{4x^2}{(1+x)(1-x)} = \frac{4x^2}{1-x^2}$$

Thus

$$x^2 = \frac{K_p}{4 + K_p}$$

and at 800°C,

$$x = \left(\frac{7.6433}{4 + 7.6433} \right)^{1/2} = 0.810$$

At 1000°C,

$$x = \left(\frac{149.886}{4 + 149.886} \right)^{1/2}$$

$$= 0.987$$

C. Effect of Temperature, Pressure, and Concentration on Equilibrium Conversion

1. Temperature. The particular choice of the standard state of a perfect gas at 1 atm at the temperature of the reaction makes it possible to express $\Delta G°$ in terms of the specific heats of the reactants and products in the perfect gas state. The variation of $\Delta G°$ or the equilibrium constant with temperature is given by the van't Hoff equation, which relates $\Delta G°$ and $\Delta H_r°$. The van't Hoff equation may be developed in the following manner. Division of each member of Equation (5-2) by T followed by partial differentiation with respect to temperature at constant pressure yields

$$\left[\frac{\partial (G/T)}{\partial T} \right]_P = -\frac{H}{T^2} + \frac{1}{T} \left(\frac{\partial H}{\partial T} \right)_P - \left(\frac{\partial S}{\partial T} \right)_P \qquad (5\text{-}101)$$

From the definition of C_p given by Equation (5-21) and from Equation (5-5), it can be shown that

$$C_p = \left(\frac{\partial H}{\partial T} \right)_P = T \left(\frac{\partial S}{\partial T} \right)_P$$

In view of this result, it is seen that Equation (5-101) reduces to the following form of the van't Hoff equation:

$$\left[\frac{\partial (G/T)}{\partial T} \right]_P = -\frac{H}{T^2} \qquad (5\text{-}102)$$

Now observe that Equation (5-102) is applicable for each component that participates in the reaction, say the reaction given by Equation (5-80). Since

$$\Delta G° = cG_C° + dG_D° - aG_A° - bG_B°$$

and

$$\Delta H_r° = cH_C° + dH_D° - aH_A° - bH_B°$$

and since Equation (5-102) may be stated for each component in its standard state, the van't Hoff relationship may be restated in the form

$$\left[\frac{\partial (\Delta G°/T)}{\partial T} \right]_P = -\frac{\Delta H_r°}{T^2} \qquad (5\text{-}103)$$

Elimination of $\Delta G°$ from Equations (5-83) and (5-103) gives

$$R \left(\frac{\partial \log_e K}{\partial T} \right)_P = \frac{\Delta H_r°}{T^2} \qquad (5\text{-}104)$$

To express the variations of the functions $\Delta G°$ or K with T, let Equation (5-26) be restated in terms of an arbitrary constant of integration C_1 as follows:

$$\Delta H_r° = \Delta \alpha\, T + \frac{\Delta \beta}{2} T^2 + \frac{\Delta \gamma}{3} T^3 + \frac{\Delta \delta}{4} T^4 + C_1 \qquad (5\text{-}105)$$

Substitution of this expression for $\Delta H_r{}^\circ$ into Equation (5-103), followed by integration, yields

$$-\frac{\Delta G^\circ}{T} = \Delta\alpha \log_e T + \frac{\Delta\beta}{2} T + \frac{\Delta\gamma}{6} T^2 + \frac{\Delta\delta}{12} T^3 - \frac{C_1}{T} + C_2 \quad (5\text{-}106)$$

where C_2 is the constant of integration. By use of Equation (5-83), it is possible to restate Equation (5-106) in the alternate form

$$R \log_e K = \Delta\alpha \log_e T + \frac{\Delta\beta}{2} T + \frac{\Delta\gamma}{6} T^2 + \frac{\Delta\delta}{12} T^3 - \frac{C_1}{T} + C_2 \quad (5\text{-}107)$$

By inspection of Equations (5-105) and (5-106), it is evident that the constants of integration C_1 and C_2 may be obtained if $\Delta H_r{}^\circ$ and ΔG° are known either at the same or different temperatures or if ΔG° is known at two different temperatures.

Examination of Equation (5-104) shows that for exothermic reactions the equilibrium constant decreases as the temperature increases. Thus, for exothermic reactions, the equilibrium conversion decreases as the reaction temperature is increased. The converse of these statements may be made for endothermic reactions.

EXAMPLE 5-9

A commercial process for the production of 1,3-butadiene consists of the dehydrogenation of 1-butene in the presence of steam at high temperatures. It is desired to find the maximum conversion of 1-butene to 1,3-butadiene which can be expected at 900 K and 1 atm in a reactor in which the feed consists of 15 mol of steam per mole of 1-butene.

(a) Compute ΔG° and K for the dehydrogenation of 1-butene to 1,3-butadiene (see Example 5-3) by use of Equations (5-105), (5-106), and (5-83).
(b) On the basis of the K found in part (a), calculate the fraction of 1-butene converted to 1,3-butadiene at equilibrium. The following data are available.

Curve Fits of the Heat Capacities (taken from [20]):

$$C_p = \alpha + \beta T + \gamma T^2 + \delta T^3 \ (\text{cal/g mol K})$$

Component	α	$\beta \times 10^2$	$\gamma \times 10^5$	$\delta \times 10^9$
Hydrogen	6.424	0.1039	-0.00780	
1,3-Butadiene	-1.29	8.350	-5.582	14.24
1-Butene	-0.24	8.650	-5.110	12.04

Other Data (taken from [18]).

$$\Delta G_f^\circ \text{ of 1-butene at } 25^\circ C = 17.09 \ (\text{kcal/g mol})$$

$$\Delta G_f^\circ \text{ of 1,3-butadiene at } 25^\circ C = 36.01 \ (\text{kcal/g mol})$$

$$\Delta H_r^\circ \text{ at } 25^\circ C \ (\text{see Example 5-1}) = 26.36 \ (\text{kcal/g mol})$$

SOLUTION

(a) The free energy of formation of 1,3-butadiene by the dehydrogenation of 1-butene by the reaction at 25°C,

$$C_4H_8 \rightleftarrows C_4H_6 + H_2$$

1-Butene   1,3-Butadiene

or
$$A \rightleftarrows B + C$$

is computed as follows:

$$\Delta G° = \Delta G_{fB}° - \Delta G_{fA}° = 36.01 - 17.09 = 18.92 \text{ (kcal/g mol)}$$

Since $\Delta H_r°$ is known at 25°C and the curve fits for the heat capacities are available, the constant C_1 in Equation (5-105) may be evaluated:

$$C_1 = \Delta H_r° - \left[\Delta\alpha T + \frac{\Delta\beta}{2} T^2 + \frac{\Delta\gamma}{3} T^3 + \frac{\Delta\delta}{4} T^4 \right]$$

where
$$\Delta\alpha = \alpha_B + \alpha_C - \alpha_A, \qquad \Delta\beta = \beta_B + \beta_C - \beta_A$$
$$\Delta\gamma = \gamma_B + \gamma_C - \gamma_A, \qquad \Delta\delta = \delta_B + \delta_C - \delta_A$$

By use of the data given,

$$C_1 = 26{,}360 - \left[5.374(298.16) - \frac{0.1961}{2}(10^{-2})(298.16)^2 \right.$$

$$\left. - \frac{0.4798}{3}(10^{-5})(298.16)^3 + \frac{2.17(10^{-9})}{4}(298.16)^4 \right]$$

Thus

$$C_1 = 2.4887 \times 10^4$$

This value of C_1 and the data given at 25°C may be used to compute C_2 in Equation (5-106):

$$C_2 = -\left[\frac{18{,}920}{298.16} + 5.374(\log_e 298.16) - \frac{0.1961}{2}(10^{-2})(298.16) \right.$$

$$\left. - \frac{0.4798(10^{-5})}{6}(298.16)^2 + \frac{2.17}{12}(10^{-9})(298.16)^3 - \frac{2.4884}{298.16}(10^4) \right]$$

$$= -10.2483$$

Equation (5-106) may now be used to evaluate $\Delta G°$ at 900 K.

$$-\frac{\Delta G_{900}°}{900} = 5.374 \log_e 900 - \frac{0.1961}{2}(10^{-2})(900) - \frac{0.4798}{6}(10^{-5})(900)^2$$

$$+ \frac{2.17(10^{-9})(900)^3}{12} - \frac{2.4887(10^4)}{900} - 10.2483$$

$$= -2.743$$

Thus

$$\Delta G_{900}° = (900)(2.735) = 2462 \text{ (cal/g mol)}$$

The equilibrium constant K at 900 K may be computed by use of Equation (5-83):

$$K = \exp\left[-\frac{\Delta G^\circ}{RT}\right] = \exp\left[\frac{-2468}{(1.9872)(900)}\right]$$

$$= 0.252$$

(b) The stoichiometry of the dehydrogenation reaction [see part (a)] may be used to express the molar flow rates in terms of the conversion x as follows:

$$n_A = n_A^0 - n_A^0 x$$

$$n_B = n_A^0 x$$

$$n_C = n_A^0 x$$

$$n_I = n_I^0$$

$$\overline{n_T = n_T^0 + n_A^0 x}$$

where n_I is used to represent the steam. Since it is to be assumed that the mixture behaves as a perfect gas at 900 K and 1 atm, $K = K_f K_y = K_p$,

and

$$K_p = \frac{p_B p_C}{p_A} = \left(\frac{n_B n_C}{n_A n_T}\right) P$$

If n_A, n_B, n_C, and n_T are stated in terms of the conversion x by use of the preceding stoichiometric relationships, a quadratic equation in x is obtained. The solution of this equation for positive values of x is given by

$$x = \frac{-K_p(1 - y_A^0) + \sqrt{\left[(1 - y_A^0)K_p\right]^2 + 4K_p y_A^0 (P + K_p)}}{2 y_A^0 (P + K_p)}$$

where $y_A^0 = n_A^0 / n_T^0$. Since $y_A^0 = \frac{1}{16}$, $K_p = 0.252$, and $P = 1$ atm,

$$x = \frac{(-0.252)(0.9375) + \sqrt{[(0.9375)(0.252)]^2 + 4(0.252)(0.0625)(1.252)}}{2(0.0625)(1.252)}$$

$$= 0.8355$$

For endothermic reactions the equilibrium constant increases as the temperature is increased. Thus the equilibrium conversion for an endothermic reaction increases as the temperature is increased.

2. Effect of Total Pressure and Initial Concentrations. Although the free energy change ΔG° and, hence, the corresponding equilibrium constant K are independent of the pressure at which the reaction is carried out, the equilibrium conversion may depend on the pressure.

For the gas phase reaction,

$$2A \rightleftharpoons R \tag{5-108}$$

$$K = \frac{\hat{f}_R^V}{\left(\hat{f}_A^V\right)^2} \tag{5-109}$$

Suppose that $K = 0.4$, which would correspond to a ΔG° greater than zero.

Furthermore, suppose that R and A are perfect gases so that the fugacities may be replaced by partial pressures. Then

$$K = K_f K_y = K_p = \frac{P_R}{P_A^2} = \frac{y_R P}{y_A^2 P^2} = \frac{y_R}{y_A^2 P} \tag{5-110}$$

Since K is not a function of pressure, the value of y_R can be increased by increasing the pressure. As P increases, y_R/y_A^2 must increase since K is fixed. Thus, for reactions in which the total number of moles decreases as the reaction progresses, an increase in the pressure increases the equilibrium conversion. Conversely, for reactions in which the number of moles increases, a decrease in the pressure produces an increase in the equilibrium conversion.

For the case of an ideal solution, Equation (5-92) shows that at a given temperature and pressure K_x is fixed. Consequently, K_x is independent of conversion. Suppose, for example, that the reaction given by Equation (5-108) occurs in the liquid phase and that the liquid phase is an ideal solution. Then

$$K = K_f K_x = K_f \frac{x_R}{x_A^2} \tag{5-111}$$

where

$$x_A = \frac{n_A}{n_T}, \qquad x_R = \frac{n_R}{n_T}$$

$$n_T = n_A + n_R + n_S$$

$$n_S = \text{moles of solvent}$$

Then

$$K_x = \frac{n_R n_T}{n_A^2} \tag{5-112}$$

From Equation (5-112), it is evident that as the total number of moles n_T is decreased, the equilibrium conversion of A to R increases. Thus, as the relative amount of solvent is decreased, the equilibrium conversion of A to R is increased. Conversely, for reactions in which the total number of moles increase (such as $A \rightleftarrows 2R$), an increase in the relative amount of solvent increases the equilibrium conversion of A to R.

D. Relationship between the Equilibrium Constants K, K_p, K_c, and K_x

If the reactants and products of the reaction given by Equation (5-108) are perfect gases, then

$$K = K_p = \frac{p_R}{p_A^2} \tag{5-113}$$

Since $p_R = C_R RT$ and $p_A = C_A RT$ for a perfect gas, then Equation (5-113) reduces to

$$K_p = \frac{C_R RT}{C_A^2 (RT)^2} = \frac{C_R}{C_A^2}(RT)^{-1}$$

$$= K_c (RT)^{-1} \tag{5-114}$$

In general, $K_p = K_c(RT)^{\Delta n}$, where Δn is equal to the sum of stoichiometric coefficients of the products minus the sum of the stoichiometric coefficients of the reactants. For the liquid phase, consider again the reaction given by Equation (5-108). From Equations (5-91) and (5-111), it follows that

$$K = K_\gamma K_f K_x = K_\gamma K_f \frac{x_R}{x_A^2} \tag{5-115}$$

The mole fractions appearing in the right side of Equation (5-115) may be stated in terms of concentrations as follows:

$$K_x = \frac{(C_T x_R) C_T}{(C_T x_A)^2} = K_c C_T \tag{5-116}$$

where $\qquad C_T =$ molar density of the equilibrium mixture

In general, then,

$$K = K_\gamma K_f K_x = C_T^{-\Delta n} K_\gamma K_f K_c \tag{5-117}$$

where Δn is defined below Equation (5-114).

E. Equilibrium Conversions for Multiple Reactions

Consider the two parallel reactions

$$\begin{aligned} A &\rightleftarrows R \\ A &\rightleftarrows S \end{aligned} \tag{5-118}$$

Sufficient data are generally available [1] for the calculation of $\Delta G°$ for each reaction, and from these values of $\Delta G°$ the corresponding values of K may be computed by use of Equation (5-83). Consider the case where the reactions occur in the gas phase and the gas can be treated as a perfect gas. Then two independent equations at a given temperature T and pressure P may be stated:

$$K_{p1} = \frac{p_R}{p_A} \tag{5-119}$$

$$K_{p2} = \frac{p_S}{p_A} \tag{5-120}$$

For a given set of initial conditions (the moles of A, R, and S initially), these equations may be solved simultaneously.

EXAMPLE 5-10

For the gas phase reactions,

$$A \rightleftarrows R, \qquad K_{p1} = 4$$

$$A \rightleftarrows S, \qquad K_{p2} = 1$$

calculate the equilibrium composition for the case where the products are not present initially.

SOLUTION

(a)
$$K_{p1} = \frac{p_R}{p_A} = \frac{y_R P}{y_A P} = \frac{n_R/n_T}{n_A/n_T}$$

$$K_{p2} = \frac{p_S}{p_A} = \frac{y_S P}{y_A P} = \frac{n_S/n_T}{n_A/n_T}$$

$$n_A = n_A^0 - n_1 - n_2$$

$$n_R = n_R^0 + n_1$$

$$n_S = n_S^0 + n_2$$

Since $n_R^0 = n_S^0 = 0$, then

$$4 = \frac{n_1}{n_A^0 - n_1 - n_2}$$

$$1 = \frac{n_2}{n_A^0 - n_1 - n_2}$$

When these two equations are solved simultaneously for n_1 and n_2, one obtains

$$n_R = n_1 = \tfrac{2}{3} n_A^0$$

$$n_S = n_2 = \tfrac{1}{6} n_A^0$$

The moles of A converted per mole of A in the feed are then given by

$$\frac{n_A^0 - n_A}{n_A^0} = \frac{n_1 + n_2}{n_A^0} = \frac{5}{6}$$

The equilibrium compositions are readily shown to be $y_A = 1/6$, $y_R = 2/3$ and $y_s = 1/6$.

Most problems involving simultaneous reactions are more difficult to solve than the one shown in Example 5-10. Generally, it is necessary to find the equilibrium composition for a set of multiple reactions by use of a trial and error procedure such as the Newton–Raphson method [21], as demonstrated in the next example.

EXAMPLE 5-11

The reactions

$$C(s) + 2H_2 \rightleftarrows CH_4 \tag{1}$$

$$C(s) + CO_2 \rightleftarrows 2CO \tag{2}$$

occur in the reduction of carbon. On the basis of the following data given by Gumz [19], calculate the equilibrium composition at 700°C and 10 atm. The feed consists of 30% CO_2 and 70% hydrogen, and the K_p's for the respective reactions at this temperature are $K_{p1} = 0.13237$ atm^{-1} and $K_{p2} = 1.07266$ atm.

SOLUTION

For convenience, let reactions (1) and (2) be represented as follows:

$$C + 2A \rightleftarrows B \tag{1a}$$

$$C + D \rightleftarrows 2E \tag{2a}$$

Let n_1 and n_2 be equal to the moles of C that react by the first and second reactions, respectively. Then

$$n_A = n_A^0 - 2n_1$$
$$n_B = n_1$$
$$n_D = n_D^0 - n_2$$
$$n_E = 2n_2$$
$$\overline{}$$
$$n_T = n_T^0 - n_1 + n_2$$

Since carbon is a solid at these conditions, its activity may be taken equal to unity, and

$$K_{p1} = \frac{p_B}{p_A^2} = \frac{n_1(n_T^0 - n_1 + n_2)}{(n_A^0 - 2n_1)^2 P}$$

$$K_{p2} = \frac{p_E^2}{p_D} = \frac{(2n_2)^2 P}{(n_T^0 - n_1 + n_2)(n_D^0 - n_2)}$$

To solve for the two unknowns n_1 and n_2, these equations are first restated in functional notation and then solved by use of the Newton–Raphson method [21]. Let

$$f_1(n_1, n_2) = n_1(n_T^0 - n_1 + n_2) - (n_A^0 - 2n_1)^2 PK_{p1}$$
$$f_2(n_1, n_2) = 4n_2^2 P - (n_T^0 - n_1 + n_2)(n_D^0 - n_2)K_{p2}$$

As a basis, take $n_T^0 = 1$. Then $n_A^0 = 0.7$, $n_D^0 = 0.3$, $P = 10$ atm, $K_{p1} = 0.13237$, and $k_{p2} = 1.07266$. The Newton–Raphson method consists of the successive solution of the following equations for Δn_1 and Δn_2:

$$\frac{\partial f_1}{\partial n_1}\Delta n_1 + \frac{\partial f_1}{\partial n_2}\Delta n_2 = -f_1$$

$$\frac{\partial f_2}{\partial n_1}\Delta n_1 + \frac{\partial f_2}{\partial n_2}\Delta n_2 = -f_2$$

where the partial derivatives and the functions are evaluated at $n_1 = n_{1,k}$, $n_2 = n_{2,k}$, and

$$\Delta n_1 = n_{1,k+1} - n_{1,k}$$

$$\Delta n_2 = n_{2,k+1} - n_{2,k}$$

The values assumed to make the given trial are denoted by the subscript k and those to be used for the next trial by the subscript $k + 1$.

The solution values obtained for n_1 and n_2 by this procedure are as follows:

$$n_1 = 0.1773$$

$$n_2 = 0.0739$$

The solution set is, of course, that set of numbers that makes $f_1 = f_2 = 0$. (The values listed were taken to be the solution set because the corresponding values of f_1 and f_2 were approximately zero.)

The equilibrium composition of the gas computed on the basis of the preceding values of n_1 and n_2 is:

Component	Moles	Mole %
H_2	0.3434	38.52
CH_4	0.1773	19.77
CO_2	0.2261	25.22
CO	0.1478	16.49
	0.8966	100.00

Another method which may be used to solve this problem is called *interpolation regula falsi* [21]. For a single-variable function, the interpolation formula for this method is

$$x_3 = \frac{x_1 f(x_2) - x_2 f(x_1)}{f(x_2) - f(x_1)} \tag{5-121}$$

This procedure may be applied to the preceding set of two simultaneous equations as follows. Take either n_1 or n_2 to be the independent variable, say n_1. Then set either f_1 or f_2 equal to zero, say f_1, and solve for the dependent variable n_2. The assumed value of n_1 and the corresponding calculated value of n_2 are then substituted in f_2. If f_2 is equal to zero, the correct value of n_1 was assumed. If f_2 is unequal to zero for the first choice of n_1 (denoted by $n_{1,1}$), it is evaluated from some arbitrarily selected second choice of n_1 (denoted by $n_{1,2}$). Then Equation (5-121) is applied:

$$n_{1,3} = \frac{n_{1,1} f_2(n_{1,2}) - n_{1,2} f_2(n_{1,1})}{f_2(n_{1,2}) - f_2(n_{1,1})} \tag{5-122}$$

After the calculational procedure has been repeated for $n_{1,3}$, the next best value, $n_{1,4}$, is found by interpolating between the points $[n_{1,3}, f_2(n_{1,3})]$ and $[n_{1,2}, f_2(n_{1,2})]$ in a manner analogous to that demonstrated by Equation (5-122) for the points $[n_{1,2}, f_2(n_{1,2})]$ and $[n_{1,1}, f_2(n_{1,1})]$.

To initiate the calculational procedure for the preceding example, two values are selected for n_1, say

$$n_{1,1} = 0.18 \quad \text{and} \quad n_{1,2} = 0.165$$

After f_1 has been set equal to zero and solved for n_2, one obtains

$$n_2 = n_1 - n_T^0 + n_1 \left(\frac{n_A^0}{n_1} - 2 \right)^2 PK_{p1}$$

For $n_{1,1} = 0.18$,

$$n_{2,1} = 0.18 - 1 + 0.18 \left(\frac{0.7}{0.18} - 2 \right)^2 (10)(0.13237) = 0.03011$$

For $n_{1,2} = 0.165$,

$$n_{2,2} = 0.165 - 1 + 0.165 \left(\frac{0.7}{0.165} - 2 \right)^2 (10)(0.13237) = 0.26327$$

When f_2 is evaluated on the basis of $n_{1,1}$ and $n_{2,1}$, one obtains

$$f_2(n_{1,1}) = 4(0.03011)^2(10)$$
$$- [(1.0 - 0.18 + 0.03011) \times (0.3 - 0.03011)(1.07266)]$$
$$= -0.20984$$

Evaluation of f_2 on the basis of $n_{1,2}$ and $n_{2,2}$ gives

$$f_2(n_{1,2}) = 4(0.26327)^2(10)$$
$$- [(1 - 0.165 + 0.26327)(0.3 - 0.26327) \times (1.07266)]$$
$$= 2.72917$$

Then Equation (5-122) is used to obtain the next best value for n_1:

$$n_{1,3} = \frac{0.18(2.72917) - (0.165)(-0.20984)}{2.72917 - (-0.20984)}$$
$$= 0.1789$$

This value of $n_{1,3}$ is used to compute $n_{2,3}$ in the same manner demonstrated for $n_{2,1}$ and $n_{2,2}$. Next f_2 is evaluated on the basis of $n_{1,3}$ and $n_{2,3}$. Then $n_{1,4}$ is found by interpolation between the most recent pair of points as described previously. This process is continued until the value of f_2 is equal to or less than some small preassigned number. The use of search techniques for solving problems involving systems of complex reactions at chemical equilibrium has been presented by Anthony and Himmelblau [22].

The techniques demonstrated for the calculation of the equilibrium constants and equilibrium yields are useful in both the analysis of experimental results and in the design of reactors in which equilibrium conditions prevail.

The techniques demonstrated for the calculation of the heat of reaction are useful in the design of chemical reactors because of the necessity to provide for an appropriate amount of heat transfer.

PROBLEMS

5-1. To determine the effect of the ratio of steam to 1-butene on the equilibrium conversion of 1-butene to 1, 3-butadiene, repeat Example 5-9 for the following steam to 1-butene ratios:
(a) zero moles of steam and 1 mole of 1-butene
(b) 6 moles of steam per mole of 1-butene
(c) 12 moles of steam per mole of 1-butene
     Answer: (a) $x = 0.4486$,   (b) $x = 0.7279$,   (c) $x = 0.811$

5-2. Calculate the equilibrium composition at 900°C and 40 atm for the following reactions:

$$C(s) + 2H_2 \rightleftarrows CH_4 \tag{1}$$

$$C(s) + CO_2 \rightleftarrows 2CO \tag{2}$$

The feed composition is the same as Example 5-11 and the values of K_p taken from Gumz [19] are

$$K_{p1} = 0.019845 \text{ atm}^{-1}$$

$$K_{p2} = 38.6164 \text{ atm}$$

Answer: 40.87, 13.27, 33.94, 11.92 mole % H_2, CH_4, CO, and CO_2 respectively.

5-3. The primary reactions in a coal gasifier are as follows:

$$C(s) + 2H_2 \rightleftarrows CH_4 \tag{1}$$

$$C(s) + CO_2 \rightleftarrows 2CO \tag{2}$$

$$C(s) + H_2O(g) \rightleftarrows H_2 + CO \tag{3}$$

Calculate the equilibrium composition of the gas at 700°C and 20 atm for an initial gas composition of 50% CO_2 and 50% H_2O. The equilibrium constants taken from Gumz [22] are as follows:

$$K_{p1} = 0.13237 \text{ atm}^{-1}$$

$$K_{p2} = 1.07266 \text{ atm}$$

$$K_{p3} = 1.6618 \text{ atm}$$

Answer: 22.1, 3.52, 13.03, 40.22, 20.32 mole % H_2, CH_4, CO, CO_2, and H_2O, respectively.

5-4. In the production of methane from water gas, the following reactions occur:

$$CO + 3H_2 \rightleftarrows CH_4 + H_2O \tag{1}$$

$$CO + H_2O \rightleftarrows CO_2 + H_2 \tag{2}$$

At 550°C, the equilibrium constants given by Gumz [19] are as follows:

$$K_{p1} = 12.460 \text{ atm}^{-2}$$

$$K_{p2} = 3.45381$$

Calculate the equilibrium composition for an initial composition of 4 mol of hydrogen per mole of carbon monoxide and a reactor pressure of 5 atm.
Answer: 30.40, 38.71, 0.4889, 1.272, 29.13 mole % CH_4, H_2, CO, CO_2, and H_2O, respectively.

5-5. For the reactions given in Problem 5-3, calculate ΔH_r°, and K at 1000 K and 1 atm by use of the data given in Tables 5-3 and 5-5.
Answer: (1) $\Delta H_r^\circ = -21.433$ kcal, $K = 0.09829$
 (2) $\Delta H_r^\circ = 40.783$ kcal, $K = 1.029$
 (3) $\Delta H_r^\circ = 32.447$ kcal, $K = 2.488$

5-6. Methanol may be reacted with oxygen over a catalyst to produce formaldehyde at reaction temperatures in the range of 400° to 600°C. The following reactions occur simultaneously:

$$CH_3OH + \tfrac{1}{2}O_2 \rightarrow CH_2O + H_2O \tag{1}$$

$$CH_3OH \rightarrow CH_2O + H_2 \tag{2}$$

(a) On the basis of the following data taken from Reference [18], calculate ΔG° and ΔH_r° at 700 and 800 K for each reaction.

Component	ΔH_f° (kcal/g mol)		ΔG_f° (kcal/g mol)	
	700 K	800 K	700 K	800 K
CH_3OH	-50.88	-51.31	-24.84	-21.10
O_2	0	0	0	0
CH_2O	-29.17	-29.44	-23.59	-22.77
H_2O	-58.71	-58.91	-49.92	-48.65
H_2	0	0	0	0

(b) In part (a) it will be found that the first reaction tends to go to completion and is highly exothermic, whereas the second reaction will be found to have an equilibrium constant of the order of unity and to be highly endothermic. Calculate the equilibrium composition for a feed consisting of 40 mole % CH_3OH and 60 mole % air for a reactor pressure of 1 atm and a reaction temperature of 700 K.

(c) Use the results of part (b) to compute the heat which must be added or removed per mole of feed to the reactor in order to maintain the reaction temperature at 700 K.
Answer: (a) At 700 K, $\Delta H_{r1}^\circ = -37$, $\Delta H_{r2}^\circ = 21.71$, $\Delta G_1^\circ = -48.67$, $\Delta G_2^\circ = 1.25$ kcal/g mol. At 800 K, $\Delta H_{r1}^\circ = -37.04$, $\Delta H_{r2}^\circ = 21.87$, $\Delta G_1^\circ = -50.32$, $\Delta G_2^\circ = -1.67$ kcal/g mol. (b) 4.97, 27.99, 20.76, 7.23, 39.05 mole % CH_3OH, CH_2O, H_2O, H_2, and N_2, respectively. (c) 7.42 kcal must be removed per mole of CH_3OH fed.

5-7. (a) Use the data given in Tables 5-3 and 5-5 to compute $\Delta G°$ and $\Delta H_r°$ for the reaction

$$C_2H_6 \rightleftarrows C_2H_4 + H_2$$

at 1000 K.

(b) Use the value of $\Delta G°$ found in part (a) to compute the maximum conversions which can be achieved at the pressures of 1 atm and 60 torr when pure ethane is fed to the reactor.
Answer: (a) $\Delta H_r° = 34.539$, $\Delta G° = 2.175$ kcal/g mol, (b) $x = 0.5008$ at 1 atm, $x = 0.8996$ at 60 torr.

5-8. In the cracking of ethane, some acetylene is produced. The acetylene produced by this reaction is removed by hydrogenation over a catalyst, which is represented by

$$C_2H_2 + H_2 \rightleftarrows C_2H_4$$

(a) Calculate the equilibrium constants for the hydrogenation of acetylene at 298.15 and 600 K.
(b) For a feed consisting of 90% inerts, 8% hydrogen, and 2% acetylene, calculate the equilibrium conversions of acetylene at 600 K and a pressure of 5 atm.

5-9. A possible reaction for the production of methanol is

$$CO_2 + 3H_2 \rightleftarrows CH_3OH + H_2O$$

Carbon dioxide is available from fossil fuel electric generating plants. Some of the capacity of an electric generating plant could be used to produce hydrogen by the electrolysis of water.

(a) For the reaction given, calculate the heat of reaction, $\Delta H_r°$, at 298.16 and 600 K.
(b) Calculate $\Delta G°$ at 298.16 and at 600 K.
(c) Compute the conversion at 600 K for an operating pressure of 300 atm. Hydrogen and carbon dioxide are fed to the reactor in the ratio of 3 mol of hydrogen per mole of carbon dioxide. In addition to the thermodynamic data given in the tables in this chapter, the following data, which were taken from [18], are available. For methanol:

$$\Delta H_f° = -48.08 \text{ kcal/g mol at } 25°C$$

$$\Delta H_f° = -50.34 \text{ kcal/g mol at } 600 \text{ K}$$

$$\Delta G_f° = -38.8 \text{ kcal/g mol at } 25°C$$

$$\Delta G_f° = -28.52 \text{ kcal/g mol at } 600 \text{ K}$$

Answer: (a) $\Delta H_r° = -11.827$, -14.71 kcal/g mol, (b) $\Delta G° = 0.8221$, 14.769 kcal/g mol.

5-10. Toulene is to be hydrogenated to methylcyclohexane at a reaction temperature of 600 K. Calculate $\Delta H_r°$, $\Delta G°$, and K. For methylcyclohexane, the following data at 600 K were taken from Reference [18].

$$\Delta H_f^\circ = -44.49 \text{ kcal/g mol}$$
$$\Delta G_f^\circ = 69.79 \text{ kcal/g mol}$$

Answer: $\Delta H_r^\circ = -52.4249$, $\Delta G^\circ = 21.432$ kcal/g mol, $K = 1.5586 \times 10^{-8}$

5-11. Methanol may be re-formed to produce methane by the following reaction:

$$4CH_3OH \rightleftarrows 3CH_4 + CO_2 + 2H_2O$$

Calculate ΔH_r° and ΔG° for this reaction at 600 K. For the data needed for methanol, see Problem 5-9.
Answer: $\Delta H_r^\circ = -69.4294$, $\Delta G^\circ = -99.1492$ kcal/g mol

5-12. Ethylene may be produced from carbon monoxide and hydrogen by the reactions

$$2CO + 4H_2 \rightleftarrows C_2H_4 + 2H_2O$$
$$4CO + 2H_2 \rightleftarrows C_2H_4 + 2CO_2$$

Calculate ΔG° and ΔH_r° for each of these reactions at 600 and 800 K.
Answer: At 600 K: $\Delta H_{r1}^\circ = -53.7252$, $\Delta H_{r2}^\circ = -72.324$ kcal/g mol, $\Delta G_1^\circ = -2.6762$, $\Delta G_2^\circ = -10.5416$ kcal/g mol. At 800 K: $\Delta H_{r1}^\circ = -54.978$, $\Delta H_{r2}^\circ = -72.628$ kcal/g mol, $\Delta G_1^\circ = -10.5416$, $\Delta G_2^\circ = 10.1188$ kcal/g mol

5-13. Ethylene may be produced from methanol by the following reaction:

$$2CH_3OH \rightleftarrows C_2H_4 + 2H_2O$$

Calculate ΔG°, ΔH_r°, and K for this reaction at 600 and 800 K. Use the following data, which were taken from Reference [18].

Component	ΔH_f° (kcal/g mol)		ΔG_f° (kcal/g mol)	
	600 K	800 K	600 K	800 K
CH_3OH	-50.34	-51.31	-28.52	-21.10
C_2H_4	10.60	9.77	20.92	24.49
H_2O	-58.91	-58.50	-51.16	-48.65

Answer: At 600 K: $\Delta H_r^\circ = -6.54$ kcal, $\Delta G^\circ = -24.36$ kcal. At 800 K: $\Delta H_r^\circ = -4.61$ kcal, $\Delta G^\circ = -30.61$ kcal

5-14. Styrene is produced by the dehydrogenation of ethylbenzene. In the Styro-Plus technology (Ward et al. [23]), the feed to the dehydrogenation reactors is composed of 98.65% (mol) ethylbenzene, 0.5% styrene, and 0.85% toluene, and a steam to hydrocarbon weight ratio of 1.5. For this feed mixture, calculate the equilibrium composition for an average reactor temperature of 630°C and a pressure of 0.5 atm. By-products are benzene and toluene. Experimental conversion and selectivity were 85.4% and 94.4%, respectively. Calculate the percentage difference between the experimental and the equilibrium values. Selectivity is defined as the moles of styrene produced per mole of ethylbenzene reacted. Consider only the dehydrogenation of ethylbenzene to styrene,

$$C_8H_{10} \rightarrow C_8H_8 + H_2$$

Data (Stull et al. [18])

Styrene

T (K)	C_p° [cal/(mol K)]	$-(G^\circ - H_{298}^\circ)/T$ [cal/(mol K)]	$H^\circ - H_{298}^\circ$ (kcal/mol)	ΔH_f° (kcal/mol)	ΔG_f° (kcal/mol)
900	64.93	100.81	30.32	30.10	87.52
1000	67.92	104.53	36.97	29.83	93.92

Ethylbenzene

T (K)	C_p° [cal/(mol K)]	$-(G^\circ - H_{298}^\circ)/T$ [cal/(mol K)]	$H^\circ - H_{298}^\circ$ (kcal/mol)	ΔH_f° (kcal/mol)	ΔG_f° (kcal/mol)
900	71.25	105.79	32.82	0.27	85.77
1000	74.77	109.82	40.13	-0.05	95.30

5-15. Toluene may be alkylated with methanol over acid catalysts to produce xylenes. When the zeolite ZSM-5 is treated with phosphorus, it is capable of producing *p*-xylenes selectively by alkylating toluene. At high temperatures, toluene disproportionates to produce xylenes and benzene. A secondary reaction which occurs in the system is the aklylation of the xylenes with methanol to produce trimethylbenzenes. The process may be represented by the following reactions:

$$T + M \rightleftarrows X + W$$

$$p\text{-}X \rightleftarrows m\text{-}X \rightleftarrows o\text{-}X$$

$$2T \rightleftarrows X + B$$

$$M + X \rightleftarrows TMB + W$$

where T = toluene, M = methanol, X = xylenes, *o*-X, *m*-X, *p*-X = ortho-, meta-, and para-xylene, B = benzene, TMB = trimethyl benzene, and W = water. Calculate the equilibrium distributions in mole percent at 600 and 900 K on the basis of a feed containing 2 mol of toluene per mole of methanol. (Note: For the equilibrium calculation, X may be used to denote any one of the xylenes, ortho-, meta-, or para-xylene.)

Data (Stull et al. [18])

Toluene

T (K)	C_p° [cal/(mol K)]	$-(G^\circ - H_{298}^\circ)/T$ [cal/(mol K)]	$H^\circ - H_{298}^\circ$ (kcal/mol)	ΔH_f° (kcal/mol)	ΔG_f° (kcal/mol)
600	47.20	89.50	11.13	8.02	48.32
900	60.23	92.88	27.41	6.24	68.93
1000	63.32	96.25	33.60	6.01	75.91

Methanol

T (K)	C_p° [cal/(mol K)]	$-(G^\circ - H_{298}^\circ)/T$ [cal/(mol K)]	$H^\circ - H_{298}^\circ$ (kcal/mol)	ΔH_f° (kcal/mol)	ΔG_f° (kcal/mol)
500	14.22	58.61	2.49	-49.70	-32.11
600	16.02	59.67	4.00	-50.34	-28.52
700	17.62	60.81	5.69	-50.88	-24.84
900	20.29	63.15	9.49	-51.66	-17.30

1,2,3-*Trimethyl benzene*

T (K)	C_p° [cal/(mol K)]	$-(G^\circ - H_{298}^\circ)/T$ [cal/(mol K)]	$H^\circ - H_{298}^\circ$ (kcal/mol)	ΔH_f° (kcal/mol)	ΔG_f° (kcal/mol)
600	64.00	100.91	15.45	−8.00	64.67
900	81.60	114.56	37.51	−10.99	101.76

Dimethyl ether

T (K)	C_p° [cal/(mol K)]	$-(G^\circ - H_{298}^\circ)/T$ [cal/(mol K)]	$H^\circ - H_{298}^\circ$ (kcal/mol)	ΔH_f° (kcal/mol)	ΔG_f° (kcal/mol)
500	22.23	65.84	3.84	−46.24	−14.81
600	25.16	67.48	6.21	−47.10	−8.44

5-16. Fujimoto et al. [24] produced dimethylether by passing a mixture of carbon monoxide and hydrogen over a methanol synthesis catalyst impregnated on γ-alumina. The postulated reaction sequence is

$$CO + 2H_2 \rightleftarrows CH_3OH \tag{1}$$

$$2CH_3OH \rightleftarrows CH_3OCH_3 + H_2O \tag{2}$$

The water gas shift reaction also occurs:

$$CO + H_2O \rightleftarrows CO_2 + H_2 \tag{3}$$

Reactions (1) and (2) can be added to give

$$CO + 4H_2 \rightleftarrows CH_3OCH_3 + H_2O$$

and when reactions (1), (2), and (3) are combined, one obtains

$$3CO + 3H_2 \rightleftarrows CH_3OCH_3 + CO_2$$

(a) Calculate the equilibrium constants for each of the preceding reactions at a pressure of 20 atm and at temperatures of 500 and 600 K.

(b) Calculate the equilibrium constant for each reaction at 550 K by use of the Van't Hoff equation and an average ΔH_r° for each reaction.

(c) Calculate the equilibrium composition in mole percent for the reaction system at 550 K and 20 atm. For convenience, suppose that the feed to the reactor system consists of 2 moles of hydrogen per mole of carbon monoxide. (Note that, even though five reactions are given, only three are independent.)

5-17. The reactions

$$CO + 2H_2 \rightleftarrows CH_3OH$$

and

$$CO_2 + H_2 \rightleftarrows CO + H_2O$$

occur over a zinc chromite catalyst at 600 K and 350 atm or over a copper zinc oxide catalyst at 500 K and 70 atm. Calculate the equilibrium conversion for these two reactions at each set of conditions for a feed consisting of 70% (mole) hydrogen, 20% CO, and 10% CO_2.

5-18. The isomerization reactions of the butylenes are often used in the characterization of catalysts. In many applications, 1-butene is of greater value than the *cis*-
and *trans*-butene-2, and isobutylenes may be more valuable than 1-butene. The
reactions are

$$\text{1-butene} \overset{①}{\rightleftarrows} \text{c-butene-2} \overset{②}{\rightleftarrows} \text{t-butene-2} \overset{③}{\rightleftarrows} \text{i-butene}$$

For the temperature of 555 K and pressure of 5 atm, calculate the following
items on the basis of the data that follow:
(a) Standard heats of reaction
(b) Equilibrium constants for each reaction
(c) Equilibrium composition in mole percent
Clearly state any assumptions made in performing these calculations.

Data (taken from Stull et al. [18])

| T (K) | 500 K | | | 600 K | | |
Thermodynamic Property	C_p°	ΔH_f°	ΔG_f°	C_p°	ΔH_f°	ΔG_f°
1-Butene	30.93	−2.70	29.39	35.14	−3.71	37.91
c-Butene-2	29.39	−4.68	28.42	33.80	−5.82	35.14
t-Butene-2	30.68	−5.33	27.83	34.80	−6.37	34.56
i-Butene	31.24	−6.60	26.78	35.40	−7.58	33.35

C_p° has units of cal/mol K, and ΔH_f° and ΔG_f° have the units of kcal/mol.
Answer: (a) $\Delta H_{r1}^\circ = -2.0515$, $\Delta H_{r2}^\circ = -0.595$, $\Delta H_{r3}^\circ = -1.237$
  (b) $K_1 = 5.914$, $K_2 = 1.699$, $K_3 = 2.540$
  (c) Component   mol %

Component	mol %
1-Butene	2.35
c-Butene-2	13.92
t-Butene-2	23.65
i-Butene	60.07

NOTATION

A = Helmholtz free energy

a_i = activity of component i at the temperature and pressure of the system

C_p° = heat capacity of a perfect gas, defined by Equation (5-21)

C_{pi}° = perfect gas heat capacity of the pure component i

$\overline{C}_{pi}$ = partial molar heat capacity of component i, defined by Equation (5-57)

$\hat{C}_{pi}$ = virtual value of the partial molar heat capacity, defined by Equation (5-59)

f_i = fugacity of pure component i at the temperature and pressure of the system

$\hat{f}_i$ = fugacity of component i in a mixture at the temperature and pressure of the
mixture

f_i° = fugacity of pure component i in its standard state at the temperature of the
system

G = Gibbs free energy

ΔG_f° = standard free energy of formation, which is equal to the change in free energy when a compound in its standard state is formed from its elements in their standard state, all at the temperature T, and the standard states are taken to be as stated in Table 5-1

ΔG_{f0}° = value of ΔG_f° at the temperature 0 K

ΔG° = standard free energy change, which is equal to the change in free energy which occurs when products in their standard states are formed from reactants in their standard states, all at the temperature T

G_T° = free energy of a compound or element in its standard state at the temperature T, and G_0° denotes its free (see Table 5-1) energy at the temperature 0 K

H = enthalpy; also used to denote the enthalpy of 1 mol of mixture

ΔH_r° = standard heat of reaction, which is defined as the sum of the enthalpies of the products in their standard states minus the sum of the enthalpies of the reactants in their standard states (see Table 5-1), all at the temperature T

ΔH_f° = standard heat of formation, which is defined as the standard heat of reaction when a compound in its standard state is formed from its elements in their standard state, all at the temperature T; ΔH_{f0}° is the value of ΔH_f° at 0 K

H_T° = enthalpy of a compound or element in its standard state at the temperature T, and H_0° is the corresponding enthalpy at 0 K

H° = enthalpy of 1 mol of a perfect gas mixture at temperature T

H_i° = perfect gas enthalpy of pure component i at temperature T

$\overline{H}_i$ = partial molar enthalpy of component i, defined by Equation (5-37)

$\hat{H}_i$ = virtual value of the partial molar enthalpy of component i

K = equilibrium constant

K_y, K_p = equilibrium constants for the gas phase

K_x = equilibrium constant for the liquid phase

K_f, K_γ = ratios of fugacities and activities, respectively

n_i = moles of component i or molar flow rate of component i

n_T = total moles of mixture or total molar flow rate

p_i = partial pressure of component i

P = total pressure

P_c = critical pressure

P_i = vapor pressure of component i

P_r = reduced pressure

R = gas constant in consistent units

S = entropy

T = temperature, degrees absolute

T_c = critical temperature, degrees absolute

T_r = reduced temperature

T_0 = temperature at which the heat of reaction is known

U = internal energy

v = molar volume, volume per mole

x_i = mole fraction of component i in the liquid phase

y_i = mole fraction of component i in the gas phase

Greek Letters

γ_i = activity coefficient for component i

ρ = molar density

ϕ = fugacity coefficient

Ω = deviation of 1 mole of mixture from 1 mole of a perfect gas mixture at the same P and T

Superscripts

L = liquid phase

S = solid phase

V = vapor phase

REFERENCES

1. "Selected Values of Physical and Thermodynamic Properties of Hydrocarbons and Related Compounds," API Research Project 44, Thermodynamics Research Center, Texas A & M University, College Station, Texas (1977) (loose-leaf data sheets).

2. Wagman, D. D., *Tables of Selected Values of Chemical Properties*, National Bureau of Standards (July 1, 1973).

3. Denbigh, K. G., *The Principles of Chemical Equilibrium*, Cambridge University Press, New York (1955).

4. Klotz, I. M., *Chemical Thermodynamics*, Prentice-Hall, Inc., Englewood Cliffs, N.J. (1950).

5. Smith, J. M., and H. C. Van Ness, *Introduction to Chemical Engineering Thermodynamics*, McGraw-Hill Book Company, New York (1959).

6. Holland, C. D., and P. T. Eubank, *Hydrocarbon Processing*, *53*: No. *11*, 176 (1974).

7. Holland, C. D., *Multicomponent Distillation*, Prentice-Hall, Inc., Englewood Cliffs, N.J. (1963).

8. Papadopoulos, A., R. L. Pigford, and L. Friend, *Chem. Eng. Prog. Symposium Series*, *49*: No. *7*, 119 (1953).

9. Benedict, M., G. B. Webb, and L. C. Rubin, *J. Chem. Phys.*, *8*:334 (1940), *10*, 747 (1942).

10. Orye, R. V. *I & EC Proc. Design and Develop.*, *8*:579 (1969).

11. Holland, C. D., *Fundamentals of Multicomponent Distillation*, McGraw-Hill Book Company, New York (1981).

12. Perry, John H., Cecil H. Chilton, and Sidney D. Kirkpatrick, *Chemical Engineer's Handbook*, 4th ed. McGraw-Hill Book Company, New York (1963).

13. Lydersen, A. L., R. A. Greenkorn, and O. A. Hougen, Wisc. Univ. Eng. Exp. Sta. Rept., *4* (October 1955).

14. Hougen, O. A., K. M. Watson, and R. A. Ragatz, *Chemical Process Principles*, *Part II*, *Thermodynamics*, 2d ed., John Wiley & Sons, New York (1959).

15. Redlich, O., and J. N. S. Kwong, *Chem. Rev.*, *44*:233 (1949).

16. Yen, L. C., and R. E. Alexander, *AIChEJ.*, *11*:334 (1965).

17. Dowling, D. W., and P. T. Eubank, API Project 44, Report on Investigation on the Calculation of the Compressibility Factor and Thermodynamic Properties of Methane, API Project 44, Thermodynamics Research Center, College Station, Texas (1966).

18. Stull, D. R., E. F. Westrum, Jr., and G. C. Sinke, *The Chemical Thermodynamics of Organic Compounds*, John Wiley & Sons, Inc., New York (1969).

19. Gumz, W., *Gas Producers and Blast Furnaces*, *Theory and Methods of Calculation*, John Wiley & Sons, Inc., New York (1950).

20. Hougen, O. A., K. M. Watson, and R. A. Ragatz, *Chemical Process Principles*, *Part I*, *Material and Energy Balances*, John Wiley & Sons, New York (1959).

21. Carnahan, B., H. A. Luther, and J. O. Wilkes, *Applied Numerical Methods*, John Wiley & Sons, Inc., New York (1969).

22. Anthony, R. G., and D. M. Himmelblau, *J. Physical Chem.* *67*, 1080 (1963).

23. Ward, D. J., S. M. Black, T. Imai, Y. Sato, N. Nakayama, H. Tokano, and K. Egawa, *Hydrocarbon Processing*, *66*:47 (1987).

24. Fujimoto, K., K. Asami, T. Shikada, and H. Tominaga, *Chemistry Letters* (The Chemical Society of Japan): 2051 (1984).

Section II:
Design of Thermal
and Catalytic Reactors

Design of Plug-Flow and Partially Mixed Flow Reactors

6

A necessary part of the design of a reactor is the control of the temperature of the reacting mixture by removing the heat liberated by an exothermic reaction or by supplying heat to compensate for the absorption of heat by an endothermic reaction. The equations needed to take heat effects into account are presented in Part I for plug-flow reactors and in Part II for reactors in which partial mixing occurs. In Part III, the design of tubular reactors is demonstrated by use of several numerical examples.

I. DEVELOPMENT OF ENERGY-BALANCE EQUATIONS FOR PLUG-FLOW REACTORS

The first law of thermodynamics asserts that the energy of the universe is constant. Consequently, the total amount of energy entering minus that leaving the system must be equal to the accumulation of energy within the system. (The word *system* is used to mean that particular part of the universe under consideration.) The following form of the energy balance is readily applied to systems at unsteady-state operation:

$$\int_{t_n}^{t_n + \Delta t} \left[\begin{pmatrix} \text{input of energy} \\ \text{to the system} \\ \text{per unit time} \end{pmatrix} - \begin{pmatrix} \text{output of energy} \\ \text{from the system} \\ \text{per unit time} \end{pmatrix} \right] dt$$

$$= \begin{pmatrix} \text{energy within} \\ \text{the system} \end{pmatrix} \Bigg|_{t_n + \Delta t} - \begin{pmatrix} \text{energy within} \\ \text{the system} \end{pmatrix} \Bigg|_{t_n} \qquad (6\text{-}1)$$

In accounting for all the energy entering and leaving a system, the energy equivalents of the net heat absorbed and the net work done on the system must be taken into account. The terms *heat* and *work* represent energy in the state of transition between the system and its surroundings. A system that has work done on it experiences the conversion of mechanical energy into internal energy. Let the internal energy contained by a unit mass of fluid be denoted by U. The internal energy is the energy associated with the structure and molecular motion of the molecules that make up each unit mass of fluid. Let KE denote the kinetic energy of the mass motion of a unit mass of fluid. The potential energy PE is equal to the work required to raise a unit mass of fluid a given distance above a specified datum level, against the force of gravity. Finally, let the total energy possessed by a unit mass of fluid be denoted by U_T. Then

$$U_T = U + KE + PE \tag{6-2}$$

where all these quantities are stated on a mass basis rather than the mole basis. For convenience, let H_T be defined as follows:

$$H_T = H + KE + PE \tag{6-3}$$

where H = enthalpy in Btu/lb mass

Thus $H_T = U + Pv + KE + PE = U_T + Pv$, where v is volume of fluid per unit mass. In this equation and throughout the development which follows, it is supposed that the Pv product is expressed in energy units (the same as those used for H and U).

Throughout all the developments that follow, it is supposed that the reactor shown in Figure 6-1 is flowing full and that the reacting mixture is in plug flow. Let z_j, z_{j+1}, t_n, and t_{n+1} be arbitrarily selected in the space and

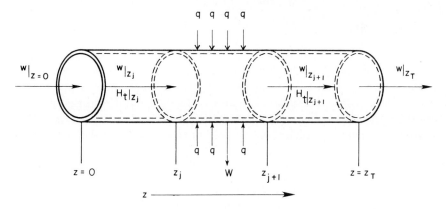

Figure 6-1. Energy balance on the element of volume from z_j to z_{j+1} of a plug-flow reactor. [Taken from Charles D. Holland, *I & EC Fundamentals*, 16:143 (1977). Used with the permission of the American Chemical Society.]

time domains of interest. Then Δz and Δt are fixed as follows:

$$\Delta z = z_{j+1} - z_j$$
$$\Delta t = t_{n+1} - t_n$$

The formulation of the energy balance for the element of volume from z_j to z_{j+1} over the time period from t_n to t_{n+1} follows. The energy entering the element per unit time in the fluid at z_j is given by

$$\left\{ \begin{array}{l} \text{Input of energy} \\ \text{per unit time} \\ \text{by flow} \end{array} \right\} = (wU_T)|_{z_j} \qquad (6\text{-}4)$$

At any time t ($t_n \leq t \leq t_{n+1}$), the work required to force one unit mass of fluid into the element of volume is given by

$$\text{Work per unit mass} = \int_0^v P \, dv \Big|_{z_j} = (Pv)|_{z_j} \qquad (6\text{-}5)$$

Note, this work Pv may vary with time throughout the time period Δt. (In this equation and those which follow, it is understood that the Pv product is to be stated in energy units which are consistent with the remaining terms in the equations.) At any time t,

$$\left\{ \begin{array}{l} \text{Rate at which work} \\ \text{is done on the element} \\ \text{of volume by the entering} \\ \text{fluid} \end{array} \right\} = (wPv)|_{z_j} \qquad (6\text{-}6)$$

Let q [energy per second per meter, say kJ/(s m)] denote the rate of heat transfer to the system from the surroundings at each z ($z_j \leq z \leq z_{j+1}$), where it is understood that q varies in a continuous manner with respect to z at any t. Then

$$\left\{ \begin{array}{l} \text{Heat transferred across} \\ \text{the boundary of the} \\ \text{element of volume} \\ \text{per s} \end{array} \right\} = \int_{z_j}^{z_{j+1}} q \, dz \qquad (6\text{-}7)$$

With regard to the shaft work done by the system on the surroundings, two cases are considered. In the first case, the work leaves the system at a point z lying between z_j and z_{j+1} as shown in Figure 6-1, and the rate at which the system does work on the surroundings is denoted by W (energy/time, say kJ/s). In the second case, it is supposed that work is done by the system in a continuous manner at each point on the boundary, and that the rate at which work is done per unit length is denoted by $\mathcal{W}$ (energy per unit time per meter, say kJ/(s m)). For the second case,

$$\left\{ \begin{array}{l} \text{Shaft work done} \\ \text{by the element} \\ \text{of volume per s} \end{array} \right\} = \int_{z_j}^{z_{j+1}} \mathcal{W} \, dz \qquad (6\text{-}8)$$

The integral difference equation is set up for the first case, from which the final result for the second case is readily obtained. The input terms that appear in

Equation (6-1) are as follows:

$$\begin{pmatrix} \text{Input of energy} \\ \text{to the element} \\ \text{of volume during} \\ \text{the time period } \Delta t \end{pmatrix} = \int_{t_n}^{t_{n+1}} \left\{ (wU_T)|_{z_j} + (wPv)|_{z_j} + \int_{z_j}^{z_{j+1}} q\,dz \right\} dt \quad (6\text{-}9)$$

The output terms are

$$\begin{pmatrix} \text{Output of energy} \\ \text{from the element} \\ \text{of volume during} \\ \text{the time period } \Delta t \end{pmatrix} = \int_{t_n}^{t_{n+1}} \left\{ (wU_T)|_{z_{j+1}} + (wPv)|_{z_{j+1}} + W \right\} dt \quad (6\text{-}10)$$

The accumulation of energy [the right side of Equation (6-1)] is given by

$$\begin{pmatrix} \text{Accumulation of energy} \\ \text{within the element of} \\ \text{volume during the} \\ \text{time period } \Delta t \end{pmatrix} = \int_{z_j}^{z_{j+1}} \rho S U_T \, dz|_{t_{n+1}} - \int_{z_j}^{z_{j+1}} \rho S U_T \, dz|_{t_n} \quad (6\text{-}11)$$

where ρ is the density (mass per unit volume) of the fluid and S is the cross-sectional area of the element volume as shown in Figure 6-1. Since $\rho = 1/v$, and $H_T = U_T + Pv$, it follows from Equations (6-2) and (6-3) that

$$\rho U_T = \rho H_T - \rho(Pv) = \rho H_T - P \quad (6\text{-}12)$$

Then Equation (6-11) may be restated in terms of H_T as follows:

$$\begin{pmatrix} \text{Accumulation of} \\ \text{energy during} \\ \text{time } \Delta t \end{pmatrix}$$

$$= \int_{z_j}^{z_{j+1}} \left[\rho H_T|_{t_{n+1}} - \rho H_T|_{t_n} \right] S \, dz - \int_{z_j}^{z_{j+1}} \left[P|_{t_{n+1}} - P|_{t_n} \right] S \, dz \quad (6\text{-}13)$$

Also, when Equation (6-3) is used to state the inputs and outputs in terms of H_T, the final result for the energy balance is given by

$$\int_{t_n}^{t_{n+1}} \left\{ (wH_T)|_{z_j,t} - (wH_T)|_{z_{j+1},t} + \int_{z_j}^{z_{j+1}} q\,dz - W \right\} dt$$

$$= \int_{z_j}^{z_{j+1}} \left[(\rho H_T)|_{t_{n+1},z} - (\rho H_T)|_{t_n,z} \right] S \, dz$$

$$- \int_{z_j}^{z_{j+1}} \left[P|_{t_{n+1},z} - P|_{t_n,z} \right] S \, dz \quad (6\text{-}14)$$

(Note in the application of this equation that all terms must be expressed in appropriate energy units.) Examination of the second integral on the right side of Equation (6-14) shows that it has the physical significance of being the difference between the amount of work required to sweep out the element of volume at times t_{n+1} and t_n.

If the element of volume does shaft work on the surroundings at each point along the boundary, then W in Equation (6-14) is replaced by the expression given by Equation (6-8).

In dealing with systems involving chemical reactions, it is generally more convenient to work in terms of molar flow rates rather than mass flow rates because enthalpies are generally stated in tables on a mole basis rather than a mass basis. Since mass flow rates are constant for many systems for which the corresponding values of the molar flow rates vary, it might appear simpler to use mass flow rates. However, this is not generally the case, and molar flow rates are used in most of the subsequent developments. Since the same amount of energy is contained in a given amount of material regardless of the units employed, it follows that

$$w \left[\frac{\text{mass}}{\text{time}} \right] H \left[\frac{\text{energy}}{\text{mass}} \right] = n_T \left[\frac{\text{mole}}{\text{time}} \right] H \left[\frac{\text{energy}}{\text{mole}} \right]$$

and

$$S(\text{length}^2) \rho \left[\frac{\text{mass}}{\text{length}^3} \right] H \left[\frac{\text{energy}}{\text{mass}} \right] = S(\text{length}^2) C_T \left[\frac{\text{mole}}{\text{length}^3} \right] H \left[\frac{\text{energy}}{\text{mole}} \right]$$

where the total molar density is denoted by C_T. Suppose that the shaft work is done continuously on the boundary [Equation (6-8)] and that molar flow rates and molar densities are used. Then Equation (6-14) takes the form

$$\int_{t_n}^{t_{n+1}} \left[(n_T H_T)|_{z_j,t} - (n_T H_T)|_{z_{j+1},t} + \int_{z_j}^{z_{j+1}} q \, dz - \int_{z_j}^{z_{j+1}} \mathscr{W} \, dz \right] dt$$

$$= \int_{z_j}^{z_{j+1}} \left[(C_T H_T)|_{t_{n+1},z} - (C_T H_T)|_{t_n,z} \right] S \, dz$$

$$- \int_{z_j}^{z_{j+1}} \left[P|_{t_{n+1},z} - P|_{t_n,z} \right] S \, dz \qquad (6\text{-}15)$$

where H_T is now stated on a mole basis.

After the mean-value theorems have been applied as described in Chapter 1 to Equation (6-15) and the limit of the result so obtained is taken as Δz and Δt approach zero, the following partial differential equation is obtained:

$$-\frac{\partial (n_T H_T)}{\partial z} + q - \mathscr{W} = S \left[\frac{\partial (C_T H_T)}{\partial t} - \frac{1}{j} \frac{\partial P^*}{\partial t} \right], \qquad 0 < z < z_T \quad (6\text{-}16)$$

---

*Although it has become customary to omit the conversion factor required to convert the Pv product in the general energy-balance equation into appropriate energy units, it is not obvious that the corresponding derived term $\partial P / \partial t$ in Equations (6-16) and (6-17) should be preceded by a conversion factor. For convenience, the conversion factor has been included in the text and in the corresponding working equations in Tables 6-1 to 6-4. Also when heat transfer through the fluid in the z direction is taken into account, Equation (6-16) becomes

$$-\frac{\partial (n_T H + S Q_z)}{\partial z} + q - \mathscr{W} = S \left[\frac{\partial (C_T H)}{\partial t} - \frac{1}{j} \frac{\partial P^*}{\partial t} \right]$$

where $Q_z = -k \, \partial T / \partial z$ and k is the thermal conductivity. The generalization for heat transfer by conduction in three-dimensional space is shown in Part II.

If the kinetic and potential effects are negligible ($\Delta PE = 0$, $\Delta KE = 0$), then H_T in Equation (6-16) is replaced by H. Also, note that if the shaft work is done at some point between z_j and z_{j+1} then the integral $\int_{z_j}^{z_{j+1}} \mathscr{W} \, dz$ appearing in Equation (6-15) is replaced by W, the point value of the work done at z_w, and the following result is obtained instead of Equation (6-16):

$$-\frac{\partial(n_T H)}{\partial z} + q = S\left[\frac{\partial(C_T H)}{\partial t} - \frac{1}{j}\frac{\partial P}{\partial t}\right], \qquad 0 < z < z_T, \, z \neq z_w \quad (6\text{-}17)$$

where the kinetic and potential effects are negligible. If the work W or $\mathscr{W}$ is negligible, then the condition $z \neq z_w$ may be removed in Equations (6-17) and (6-19).

For steady-state operation, Equations (6-16) and (6-17) reduce to the following differential equations, respectively:

$$-\frac{d(n_T H_T)}{dz} + q - \mathscr{W} = 0, \qquad 0 < z < z_T \quad (6\text{-}18)$$

and

$$-\frac{d(n_T H)}{dz} + q = 0, \qquad 0 < z < z_T, \, z \neq z_w \quad (6\text{-}19)$$

In most cases W or $\mathscr{W}$ can be neglected because the energy introduced into the reacting mixture by any device such as a stirrer is generally small relative to the heat of reaction, and, similarly, ΔKE and ΔPE are generally negligible. Thus, the approximations $\Delta KE = 0$, $\Delta PE = 0$, $W = 0$, or $\mathscr{W} = 0$ are made in all subsequent development.

A. Enthalpy of a Reacting Mixture

The thermodynamic functions such as enthalpy which have the mathematical property of being independent of path may be expressed relative to any reference state. Although the choice of the reference state is an arbitrary one, not all choices are equally convenient to apply. The use of the elements of a compound at a specified temperature and pressure as the reference state in the calculation of the enthalpy of a compound offers many advantages which become evident in the analysis of complex reactions and systems at unsteady-state operation. In the analysis which follows, the enthalpy of component i above this reference state, denoted by H_{fi}°, is defined as the enthalpy of component i at a given temperature T in the perfect gas state at 1 atm above its elements in their standard states at some arbitrary datum temperature. For example, consider the case where the elements A and B form compound C, and the datum temperature for the elements is taken to be 0 K. Then the enthalpy of compound C is given by

$$H_{f,C}^{\circ} = \Delta H_{f0,C}^{\circ} + (H_T^{\circ} - H_0^{\circ})_C \quad (6\text{-}20)$$

Values for $\Delta H_{f0,C}^{\circ}$ and $(H_T^{\circ} - H_0^{\circ})$ are available from standard thermodynamic tables (see, for example, Table 5-3). For the case of element A, the enthalpy at temperature T is given by

$$H_{f,A}^{\circ} = \left(H_T^{\circ} - H_0^{\circ}\right)_A$$

For the case where the datum temperature of the elements is taken equal to the temperature of compound C, the enthalpy becomes equal to the heat of formation:

$$H_{f,C}^{\circ} = \Delta H_{f,C}^{\circ}$$

When the heats of mixing and the effect of pressure on enthalpy are negligible, then the enthalpy of each component i in the reacting mixture at the temperature T and pressure P can be taken equal to its enthalpy in the perfect gas state at the temperature T and a pressure of 1 atm above its elements in their standard states at a given reference temperature.

Thus, when enthalpies are calculated as outlined previously, the enthalpy $n_T H$ of a perfect gas mixture at any z in the reactor is given by

$$n_T H = \sum_{i=1}^{c} n_i H_{fi}^{\circ} \tag{6-21}$$

Similarly, if the reaction occurs in the liquid phase, then the enthalpy, $n_T H$, of the liquid mixture at any z in the reactor is given by

$$n_T H = \sum_{i=1}^{c} n_i h_{fi}^{\circ} \tag{6-22}$$

where h_{fi}° is the enthalpy of formation of component i in the liquid state at temperature T from its elements in their standard states at the datum temperature. The enthalpies H_{fi}° and h_{fi}° differ by $\lambda_i(T)$, the latent heat of vaporization of component i at the temperature T; that is,

$$h_{fi}^{\circ} = H_{fi}^{\circ} - \lambda_i(T) \tag{6-23}$$

B. Enthalpy of a Reacting Mixture in a Plug-Flow Reactor

For a nonideal solution, the enthalpy H of 1 mole of mixture at a given pressure and temperature may be regarded as a function of either the mole fractions $\{x_i\}$, molar flow rates $\{n_i\}$, or concentrations $\{C_i\}$ of the reacting mixture, since

$$x_i = \frac{n_i}{\sum_{i=1}^{c} n_i} = \frac{n_i}{n_T} = \frac{C_i}{\sum_{i=1}^{c} C_i} = \frac{C_i}{C_T}$$

Thus,

$$H = H(x_1, x_2, \ldots, x_c, P, T) = H(n_1, n_2, \ldots, n_c, P, T)$$
$$= H(C_1, C_2, \ldots, C_c, P, T)$$

or, more concisely, let the arguments $x_1, x_2 \ldots x_c$; $n_1, n_2 \ldots n_c$; and $C_1, C_2 \ldots C_c$ in these functions be represented by the vectors $\mathbf{x}$, $\mathbf{n}$, and $\mathbf{C}$, respectively, to give

$$H = H(\mathbf{x}, P, T) = H(\mathbf{n}, P, T) = H(\mathbf{C}, P, T)$$

Since H is homogeneous of degree 0 and n_T is homogeneous of degree 1 in the $\{n_i\}$, it follows that $n_T H$ is homogeneous of degree 1 in the $\{n_i\}$ at constant temperature and pressure. Thus, by Euler's theorem,

$$n_T H = \sum_{i=1}^{c} n_i \frac{\partial (n_T H)}{\partial n_i} = \sum_{i=1}^{c} n_i \overline{H}_{n_i} = \sum_{i=1}^{c} n_i \hat{H}_{n_i} \qquad (6\text{-}24)$$

where

$$\overline{H}_{n_i} = \left(\frac{\partial (n_T H)}{\partial n_i} \right)_{P, T, n_j}$$

$$\hat{H}_{n_i} = H_{fi}^{\circ} + \Omega(\mathbf{n}, P, T), \qquad (\text{see Chapter 5}).$$

The subscript n_j in the definition of $\overline{H}_{n_i}$ means that all n_j $(n_1, n_2, \ldots, n_c)$ except for n_i are held fixed in taking the partial derivative. Similarly, $C_T H$ is homogeneous of degree 1 in the $\{C_i\}$, and

$$C_T H = \sum_{i=1}^{c} C_i \frac{\partial (C_T H)}{\partial C_i} = \sum_{i=1}^{c} C_i \overline{H}_{C_i} = \sum_{i=1}^{c} C_i \hat{H}_{C_i} \qquad (6\text{-}25)$$

where

$$\overline{H}_{C_i} = \left(\frac{\partial (C_T H)}{\partial C_i} \right)_{P, T, C_j}$$

$$\hat{H}_{C_i} = H_{fi}^{\circ} + \Omega(\mathbf{C}, P, T)$$

Since $\Omega(\mathbf{x}, P, T) = \Omega(\mathbf{n}, P, T) = \Omega(\mathbf{C}, P, T)$, it follows that

$$\hat{H}_{n_i} = \hat{H}_{C_i} = \hat{H}_i = H_{fi}^{\circ} + \Omega \qquad (6\text{-}26)$$

When stated in terms of Ω, the expressions for the partial molar enthalpies $\overline{H}_{n_i}$ and $\overline{H}_{C_i}$ have the following form:

$$\overline{H}_{n_i} = H_{fi}^{\circ} + \Omega + n_T \frac{\partial \Omega}{\partial n_i} = H_{fi}^{\circ} + \frac{\partial (n_T \Omega)}{\partial n_i}$$

$$\overline{H}_{C_i} = H_{fi}^{\circ} + \Omega + C_T \frac{\partial \Omega}{\partial C_i} = H_{fi}^{\circ} + \frac{\partial (C_T \Omega)}{\partial C_i} \qquad (6\text{-}27)$$

To prove that $\overline{H}_{n_i} = \overline{H}_{C_i}$, it is necessary to show that the last terms on the right sides of the expressions given by Equation (6-27) are equal. This proof follows immediately where the total volumetric flow rate of v_T is constant, but the following approach is required when $v_T = v_T(\mathbf{x}, P, T) = v_T(\mathbf{n}, P, T) = v_T(\mathbf{C}, P, T)$. First observe that

$$\frac{\partial x_i}{\partial n_i} = \frac{\partial \left[n_i / \sum_{j=1}^{c} n_j \right]}{\partial n_i} = \frac{\sum_{j=1}^{c} n_j - n_i}{\left[\sum_{j=1}^{c} n_j \right]^2}$$

and, since $C_i = n_i/v_T$,

$$\frac{\partial x_i}{\partial C_i} = \frac{\partial \left[C_i/\sum_{j=1}^{c} C_j \right]}{\partial C_i} = \frac{\sum_{j=1}^{c} C_j - C_i}{\left[\sum_{j=1}^{c} C_j \right]^2} = v_T \frac{\left[\sum_{j=1}^{c} n_j - n_i \right]}{\left[\sum_{j=1}^{c} n_j \right]^2}$$

Thus,

$$\frac{\partial x_i}{\partial C_i} = v_T \frac{\partial x_i}{\partial n_i}$$

Since $\Omega = \Omega(\mathbf{x}, P, T) = \Omega(\mathbf{n}, P, T) = \Omega(\mathbf{C}, P, T)$, it follows that

$$\frac{\partial \Omega}{\partial n_i} = \frac{\partial \Omega}{\partial x_i} \frac{\partial x_i}{\partial n_i}$$

and

$$\frac{\partial \Omega}{\partial C_i} = \frac{\partial \Omega}{\partial x_i} \frac{\partial x_i}{\partial C_i} = v_T \frac{\partial \Omega}{\partial x_i} \frac{\partial x_i}{\partial n_i}$$

Thus

$$\frac{\partial \Omega}{\partial C_i} = v_T \frac{\partial \Omega}{\partial n_i}$$

and multiplication by $C_T = n_T/v_T$ gives

$$C_T \frac{\partial \Omega}{\partial C_i} = n_T \frac{\partial \Omega}{\partial n_i} \tag{6-28}$$

It follows therefore from Equations (6-27) and (6-28) that

$$\overline{H}_{n_i} = \overline{H}_{C_i} = \overline{H}_i \tag{6-29}$$

Although the energy balances are developed later for mixtures which form nonideal solutions, they also apply for mixtures which form ideal solutions. As shown in standard thermodynamic texts, one property of an ideal solution is that the enthalpy of each component in a mixture is equal to its pure component value. That is, the partial molar enthalpy for each component in the mixture is equal to the respective pure component enthalpy,

$$\overline{H}_i(\mathbf{x}, P, T) = H_i(P, T) \tag{6-30}$$

and

$$n_T H = \sum_{i=1}^{c} n_i \overline{H}_i = \sum_{i=1}^{c} n_i H_i \tag{6-31}$$

Since $n_T H = n_T H_f^{\circ} + n_T \Omega$, it follows for an ideal solution that

$$\sum_{i=1}^{c} n_i H_i = \sum_{i=1}^{c} n_i H_{fi}^{\circ} + n_T \Omega$$

Thus,

$$n_T \Omega = \sum_{i=1}^{c} n_i \left(H_i - H_{fi}^{\circ} \right) \tag{6-32}$$

Partial differentiation of each member of Equation (6-32) with respect to n_i, followed by rearrangement, yields the following result for an ideal solution:

$$\frac{\partial (n_T \Omega)}{\partial n_i} + H_{fi}^{\circ}(T) = H_i(P,T) \tag{6-33}$$

Similarly,

$$\frac{\partial \Omega}{\partial T} = \sum_{i=1}^{c} \frac{n_i}{n_T} \frac{\partial H_i}{\partial T} - \sum_{i=1}^{c} \frac{n_i}{n_T} \frac{\partial H_{fi}^{\circ}}{\partial T} = C_p - C_p^{\circ} \tag{6-34}$$

where C_p denotes the molar heat capacity of the ideal solution and C_p° the molar heat capacity of a perfect gas mixture. Likewise,

$$\frac{\partial \Omega}{\partial P} = \sum_{i=1}^{c} \frac{n_i}{n_T} \frac{\partial H_i}{\partial P} \tag{6-35}$$

C. Energy Balances for Plug-Flow Reactors at Steady-State and Unsteady-State Operation

The application of Equations (6-17) and (6-19) is facilitated by restating them in terms of the heats of reaction and the molar heat capacities through the use of component material balances and the virtual values of the partial molar enthalpies.

At steady state, the rate of reaction r_i for each component i is for convenience stated in terms of disappearance. Thus, the component material balance [Equation (2-8)] becomes

$$Sr_i = -\frac{dn_i}{dz} \tag{6-36}$$

The reactor volume in Equation (2-8) has been replaced by its equivalent $V = Sz$, where the cross-sectional area S is assumed to remain constant over all z. Similarly, the component material balance given by Equation (2-104) for unsteady-state operation may be restated in the following form:

$$-\frac{\partial n_i}{\partial z} - Sr_i = S\frac{\partial C_i}{\partial t} \tag{6-37}$$

The energy balance for the general case of unsteady-state operation and nonideal solutions [Equation (6-17)] is first restated in terms of the variables r_i, C_p, and Ω, and all other cases such as the energy balance for steady-state operation are deduced from this result. By use of the relationships given by Equations (6-24) and (6-26), Equation (6-17) is readily restated in the following form:

$$-\frac{\partial \left[\sum_{i=1}^{c} n_i \hat{H}_i\right]}{\partial z} + q = S\left[\frac{\partial \left[\sum_{i=1}^{c} C_i \hat{H}_i\right]}{\partial t} - \frac{1}{j}\frac{\partial P}{\partial t}\right] \tag{6-38}$$

Consider first the left side of Equation (6-38) and carry out the implied partial differentiation to give

$$-\frac{\partial\left[\sum_{i=1}^{c} n_i \hat{H}_i\right]}{\partial z} + q = -\sum_{i=1}^{c} n_i \frac{\partial \hat{H}_i}{\partial z} - \sum_{i=1}^{c} \hat{H}_i \frac{\partial n_i}{\partial z} + q \qquad (6\text{-}39)$$

The first term on the right side of this equation is evaluated by first expanding $\partial \hat{H}_i / \partial z$ by use of the chain rule, as follows:

$$\frac{\partial \hat{H}_i}{\partial z} = \frac{\partial\left(H_{fi}^{\circ} + \Omega\right)}{\partial z} = \left[C_{pi}^{\circ} + \frac{\partial \Omega}{\partial T}\right]\frac{\partial T}{\partial z} + \frac{\partial \Omega}{\partial P}\frac{\partial P}{\partial z} + \sum_{i=1}^{c} \frac{\partial \Omega}{\partial n_i}\frac{\partial n_i}{\partial z}$$

where $C_{pi}^{\circ} = \partial H_{fi}^{\circ}/\partial T$, the molar heat capacity for pure component i in the perfect gas state. Thus,

$$\sum_{i=1}^{c} n_i \frac{\partial \hat{H}_i}{\partial z} = n_T C_p \frac{\partial T}{\partial z} + n_T \frac{\partial \Omega}{\partial P}\frac{\partial P}{\partial z} + \sum_{i=1}^{c} n_T \frac{\partial \Omega}{\partial n_i}\frac{\partial n_i}{\partial z}$$

where

$$C_p = \sum_{i=1}^{c} \frac{n_i}{n_T}\left(C_{pi}^{\circ} + \frac{\partial \Omega}{\partial T}\right)$$

the molar heat capacity for a nonideal solution mixture. Substitution of this result into Equation (6-39) gives

$$-\frac{\partial\left[\sum_{i=1}^{c} n_i \hat{H}_i\right]}{\partial z} + q = -n_T C_p \frac{\partial T}{\partial z} - n_T \frac{\partial \Omega}{\partial P}\frac{\partial P}{\partial z} - \sum_{i=1}^{c}\left(\hat{H}_i + n_T \frac{\partial \Omega}{\partial n_i}\right)\frac{\partial n_i}{\partial z} + q$$

$$(6\text{-}40)$$

Consider next the right side of Equation (6-38) and carry out the implied partial differentiation to give

$$S\left[\frac{\left[\partial \sum_{i=1}^{c} C_i \hat{H}_i\right]}{\partial t} - \frac{1}{j}\frac{\partial P}{\partial t}\right] = S\left[\sum_{i=1}^{c} C_i \frac{\partial \hat{H}_i}{\partial t} + \sum_{i=1}^{c} \hat{H}_i \frac{\partial C_i}{\partial t} - \frac{1}{j}\frac{\partial P}{\partial t}\right]$$

When the first two summations on the right side of this expression are treated in a manner analogous to that shown for the first two terms on the right side of Equation (6-39), the following result is obtained:

$$S\left[\frac{\partial\left[\sum_{i=1}^{c} C_i \hat{H}_i\right]}{\partial t} - \frac{1}{j}\frac{\partial P}{\partial t}\right] = S\left[C_T C_p \frac{\partial T}{\partial t} + C_T \frac{\partial \Omega}{\partial P}\frac{\partial P}{\partial t}\right.$$

$$\left. + \sum_{i=1}^{c}\left(\hat{H}_i + C_T \frac{\partial \Omega}{\partial C_i}\right)\frac{\partial C_i}{\partial t} - \frac{1}{j}\frac{\partial P}{\partial t}\right]$$

where

$$C_p = \sum_{i=1}^{c} \frac{C_i}{C_T}\left[C_{pi}^{\circ} + \frac{\partial \Omega}{\partial T}\right]$$

Elimination of $\partial C_i/\partial t$ from this equation by use of the component material balance yields

$$S\left[\frac{\partial\left[\sum_{i=1}^{c} C_i\hat{H}_i\right]}{\partial t} - \frac{\partial P}{\partial t}\right] = SC_TC_p\frac{\partial T}{\partial t} + SC_T\frac{\partial\Omega}{\partial P}\frac{\partial P}{\partial t}$$

$$- S\sum_{i=1}^{c}\left(\hat{H}_i + C_T\frac{\partial\Omega}{\partial C_i}\right)r_i$$

$$- \sum_{i=1}^{c}\left(\hat{H}_i + C_T\frac{\partial\Omega}{\partial C_i}\right)\frac{\partial n_i}{\partial z} - \frac{S}{j}\frac{\partial P}{\partial t} \quad (6\text{-}41)$$

Substitution of the results given by Equations (6-41) and (6-40) into (6-38) yields the following result upon recognizing that the last summation on the right side of Equation (6-41) cancels with the last summation on the right side of Equation (6-40), since $C_T(\partial\Omega/\partial C_i) = n_T(\partial\Omega/\partial n_i)$ [Equation (6-28)].

$$-n_TC_p\frac{\partial T}{\partial z} - n_T\frac{\partial\Omega}{\partial P}\frac{\partial P}{\partial z} + S\sum_{i=1}^{c}\bar{H}_ir_i + q$$

$$= S\left[C_TC_p\frac{\partial T}{\partial t} + C_T\frac{\partial\Omega}{\partial P}\frac{\partial P}{\partial t} - \frac{1}{j}\frac{\partial P}{\partial t}\right] \quad (6\text{-}42)$$

where $\bar{H}_i$ is defined by Equations (6-27) and (6-29).

The term containing the summation of the $\bar{H}_ir_i$'s in Equation (6-42) constitutes the general expression containing the heats of reaction for any number of reactions. For the case of the single reaction

$$aA + bB = cC + dD \quad (6\text{-}43)$$

$$r_A = \frac{b}{a}r_B = -\frac{c}{a}r_C = -\frac{d}{a}r_D$$

and

$$\sum_{i=1}^{c}\bar{H}_ir_i = \sum_{i=1}^{c}H_{fi}^{\circ}r_i + \sum_{i=1}^{c}\frac{\partial(n_T\Omega)}{\partial n_i}r_i$$

For the reaction given by Equation (6-43), the first sum reduces to

$$\sum_{i=1}^{c}H_{fi}^{\circ}r_i = -r_A\left[\frac{c}{a}H_{fC}^{\circ} + \frac{d}{a}H_{fD}^{\circ} - H_{fA}^{\circ} - \frac{b}{a}H_{fB}^{\circ}\right]$$

$$= -r_A\Delta H_r^{\circ}$$

where ΔH_r° is the standard heat of reaction per mole of A reacted at the temperature T; the superscript $\circ$ denotes the fact that each member of the mixture is taken to be in its standard state of a perfect gas except for the element carbon, which, if present, is taken to be the solid graphite. The elements $[\partial(n_T\Omega)/\partial n_i]r_i$ of the preceding summations are the corrections for

TABLE 6-1
SUMMARY OF ENERGY BALANCES FOR PLUG-FLOW REACTORS WITH PERFECT MIXING IN THE
RADIAL DIRECTION AND NO MIXING IN THE AXIAL DIRECTION

Unsteady State: Nonideal Solution

$$-n_T C_p \frac{\partial T}{\partial z} - n_T \frac{\partial \Omega}{\partial \rho} \frac{\partial P}{\partial z} + S \sum_{i=1}^{c} \bar{H}_i r_i + q = S\left[C_T C_p \frac{\partial T}{\partial t} + C_T \frac{\partial \Omega}{\partial P} - \frac{1}{j} \frac{\partial P^*}{\partial t} \right]$$

$$\bar{H}_i = H_{fi}^\circ + \frac{\partial (n_T \Omega)}{\partial n_i}$$

Unsteady State: Ideal Solutions

$$-n_T C_p \frac{\partial T}{\partial z} - n_T \frac{\partial \Omega}{\partial P} \frac{\partial P}{\partial z} + S \sum_{i=1}^{c} H_i r_i + q = S\left[C_T C_p \frac{\partial T}{\partial t} + C_T \frac{\partial \Omega}{\partial P} \frac{\partial P}{\partial t} - \frac{1}{j} \frac{\partial P}{\partial t} \right]$$

Steady State: Nonideal Solution

$$-n_T C_p \frac{dT}{dz} - n_T \frac{\partial \Omega}{\partial P} \frac{dP}{dz} + S \sum_{i=1}^{c} \bar{H}_i r_i + q = 0$$

Steady State: Perfect Gas Mixture

$$-n_T C_p^\circ \frac{dT}{dz} + S \sum_{i=1}^{c} H_{fi}^\circ r_i + q = 0$$

---

*The conversion factor j, the mechanical equivalent of heat in an appropriate set of units (say $j = 778$ ft-lb$_f$/Btu or 1054 N m/Btu or 1 N m/J), has been included to emphasize that the last term in these equations, $\partial P / \partial t$, must be expressed in the appropriate energy units. Also when heat transfer by conduction in the z direction is included, the term $Sk\ \partial^2 T/\partial z^2$ must be added to the left-hand sides of the above unsteady-state balances, and $Sk\ d^2 T/dz^2$ must be added to the left-hand sides of the steady-state balances.

the deviation of the mixture from a perfect gas mixture. Thus,

$$\sum_{i=1}^{c} \bar{H}_i r_i = -r_A \left[\Delta H_r^\circ + \frac{c}{a} \frac{\partial (n_T \Omega)}{\partial n_C} + \frac{d}{a} \frac{\partial (n_T \Omega)}{\partial n_D} \right.$$

$$\left. - \frac{\partial (n_T \Omega)}{\partial n_A} - \frac{b}{a} \frac{\partial (n_T \Omega)}{\partial n_B} \right] = -r_A \Delta H_r$$

Use of this result gives the following well-known expression for the case where a single reaction [Equation (6-43)] occurs in a plug-flow reactor with negligible pressure drop at steady-state operation:

$$-n_T C_p \frac{dT}{dz} - S r_A \Delta H_r + q = 0 \tag{6-44}$$

For a perfect gas mixture, C_p and ΔH_r are replaced by C_p° and ΔH_r°.

A summary of the energy balances deduced from Equation (6-42) for different types of operation and different types of mixtures is presented in Table 6-1.

II. DEVELOPMENT OF ENERGY-BALANCE EQUATIONS FOR PARTIALLY MIXED REACTORS

The energy balances are developed for the three general types of systems: (1) partial mixing in the axial direction z and perfect mixing in the radial direction r, (2) partial mixing in the generalized three-dimensional rectangular space x, y, z, and (3) partial mixing in the generalized cylindrical coordinate space r, z.

A. Partial Mixing in the Axial Direction z and Perfect Mixing in the Radial Direction r

Mixing occurs in reactors by the mechanisms of molecular diffusion and turbulent motion. The combination of the two mechanisms is commonly called *dispersion*. The net rate of transfer of component i in the positive direction of z is customarily represented by the following modified form of *Fick's first law of diffusion*:

$$J_i = -D_z \frac{\partial C_i}{\partial z} \tag{6-45}$$

where J_i = net molar rate of transfer of component i in the positive

direction of z per unit of cross-sectional area perpendicu-

lar to z, molar flux of component i, $\text{mol}\,(\text{time})^{-1}\,(\text{length})^{-2}$

D_z = dispersion coefficient, $(\text{time})^{-1}(\text{length})^2$

In mixtures, in which dispersion occurs, the various chemically identifiable components i move at different velocities. A number of different types of velocities may be used to express this behavior (Bird et al. [1]). Of these, only the molar average velocity is needed in the developments presented here. Let u_i denote the velocity of component i (the arithmetic average of the velocities of all molecules i in a small element of volume) relative to a fixed coordinate system. The local molar average velocity is defined by

$$u = \frac{\sum_{i=1}^c u_i C_i}{\sum_{i=1}^c C_i} = \frac{\sum_{i=1}^c u_i C_i}{C_T} \tag{6-46}$$

The rate at which moles pass through the cross-sectional area S is equal to $uSC_T = v_T C_T$, where v_T is the total volumetric rate of flow through the area S.

To relate the u_i's, J_i's, and u, it is convenient to define the *diffusion velocity* of component i relative to u as $(u_i - u)$, and the molar flux F_i of component i relative to a fixed coordinate system as

$$F_i = u_i C_i \tag{6-47}$$

Then the molar dispersion of component i relative to the molar average velocity u is given by

$$J_i = C_i(u_i - u) \tag{6-48}$$

Thus, relative to the fixed coordinate system, the molar flux is given by

$$F_i = C_i u + J_i \tag{6-49}$$

These definitions lead to the useful result that the sum of the dispersion fluxes J_i is equal to zero. For, by Equation (6-48),

$$\sum_{i=1}^{c} J_i = \sum_{i=1}^{c} C_i u_i - \left(\sum_{i=1}^{c} C_i \right) u = C_T u - C_T u = 0 \tag{6-50}$$

and thus

$$F_T = \sum_{i=1}^{c} F_i = \left(\sum_{i=1}^{c} C_i \right) u = C_T u \tag{6-51}$$

Next, the question arises as to what set of concentrations should be used in the calculation of the enthalpy of the fluid flowing through the cross-sectional area S at location z at any time t for a system at either steady-state or unsteady-state operation. This question is easily answered by observing that the moles of component i in the total volume Su passing through S at a given z at any time t is given by SuC_i. Thus, over the time period from t_n to $t_n + \Delta t$, the total moles of component i passing through S at z is given by

$$\int_{t_n}^{t_n + \Delta t} SuC_i \, dt$$

and by the *mean-value theorem* of differential calculus,

$$\int_{t_n}^{t_n + \Delta t} SuC_i \, dt = (SuC_i)_m \Delta t$$

where $(SuC_i)_m$ is the mean value of SuC_i which is to be evaluated at time $t_n + \alpha \Delta t$, $0 \leq \alpha \leq 1$. Thus, at any time t, the mole fraction x_i of component i in the fluid passing through S at z is given by

$$x_i = \lim_{\Delta t \to 0} \frac{(SuC_i)_m \Delta t}{\sum_{i=1}^{c} (SuC_i)_m \Delta t} = \frac{C_i}{\sum_{i=1}^{c} C_i} = \frac{C_i}{C_T} \tag{6-52}$$

Thus, at either steady state or unsteady state, the enthalpy at any given position can be expressed as a function of $\{x_i\}$ or $\{C_i\}$ at a given T and P.

The component material balances and the energy balance are formulated for a partially mixed reactor in the same manner as that shown for the plug-flow reactor with the role of the component flow rate n_i and total flow rate n_T being played by the molar fluxes, F_i and F_T, respectively. The component material balance for a tubular reactor such as the one shown in

Figure 6-1 is given by

$$-\frac{\partial F_i}{\partial z} - r_i = \frac{\partial C_i}{\partial t}, \qquad 0 < z < z_T, t > 0 \tag{6-53}$$

and the energy balance by

$$-S\frac{\partial (F_T H)}{\partial z} + q = S\left[\frac{\partial (C_T H)}{\partial t} - \frac{1}{j}\frac{\partial P}{\partial t}\right], \qquad 0 < z < z_T, t > 0 \tag{6-54}$$

where it has been assumed that the cross-sectional area S is independent of time and position. Since $F_T = uC_T = u(\mathbf{C})C_T(\mathbf{C})$, it follows from the definitions of u and C_T that $F_T H$ is homogeneous of degree 1 in the $\{C_i\}$ at a given P and T. Thus, by Euler's theorem,

$$F_T H = \sum_{i=1}^{c} C_i \overline{H}_{Fi} \tag{6-55}$$

where

$$\overline{H}_{Fi} = \left[\frac{\partial (F_T H)}{\partial C_i}\right]_{P,T,C_j}$$

By making use of the fact that u is homogeneous of degree 0 in the $\{C_i\}$, it follows from Euler's theorem that

$$F_T H = \sum_{i=1}^{c} C_i \overline{H}_{Fi} = u\sum_{i=1}^{c} C_i \frac{\partial C_T H}{\partial C_i} + C_T H \sum_{i=1}^{c} C_i \frac{\partial u}{\partial C_i}$$

$$= u\sum_{i=1}^{c} C_i \frac{\partial C_T H}{\partial C_i} = u\sum_{i=1}^{c} C_i \overline{H}_i \tag{6-56}$$

Statement of the partial derivatives of $F_T H$ and $C_T H$ in terms of the molar heat capacities, the pressure effects, and the heats of reaction is facilitated by first expressing the function $F_T H$ in terms of the virtual values of the partial molar enthalpies. Again,

$$H(\mathbf{x}, P, T) = H_f^\circ(\mathbf{x}, T) + \Omega(\mathbf{x}, P, T)$$

In view of Equations (6-24) and (6-26), it follows that

$$H(\mathbf{C}, P, T) = H_f^\circ(\mathbf{C}, T) + \Omega(\mathbf{C}, P, T) \tag{6-57}$$

and

$$uC_T H(\mathbf{C}, P, T) = uC_T H_f^\circ(\mathbf{C}, T) + uC_T \Omega(\mathbf{C}, P, T)$$

Thus,

$$F_T H = F_T H_f^\circ + F_T \Omega = u\sum_{i=1}^{c} C_i H_{fi}^\circ + u\sum_{i=1}^{c} C_i \Omega$$

or

$$F_T H = u\sum_{i=1}^{c} C_i \left(H_{fi}^\circ + \Omega\right) = u\sum_{i=1}^{c} C_i \hat{H}_i \tag{6-58}$$

TABLE 6-2
SUMMARY OF ENERGY BALANCES FOR TUBULAR REACTORS WITH PERFECT MIXING IN THE
RADIAL DIRECTION AND PARTIAL MIXING IN THE AXIAL DIRECTION

1. *Unsteady State: Nonideal Solutions*

$$-F_T C_p \frac{\partial T}{\partial z} - F_T \frac{\partial \Omega}{\partial P}\frac{\partial P}{\partial z} + \sum_{i=1}^{c} \overline{H}_i(r_i + \frac{\partial J_i}{\partial z}) + \frac{q}{S} = C_T C_p \frac{\partial T}{\partial t} + C_T \frac{\partial \Omega}{\partial P}\frac{\partial P}{\partial t} - \frac{1}{j}\frac{\partial P^*}{\partial t}$$

$$\overline{H}_i = H_{fi}^\circ + \frac{\partial(C_T \Omega)}{\partial C_i}$$

2. *Steady State: Nonideal Solutions*

$$-F_T C_p \frac{dT}{dz} - F_T \frac{\partial \Omega}{\partial P}\frac{dP}{dz} + \sum_{i=1}^{c} \overline{H}_i\left(r_i + \frac{dJ_i}{dz}\right) + \frac{q}{S} = 0$$

3. *Steady State: Ideal Solutions*

$$-F_T C_p \frac{dT}{dz} - F_T \frac{\partial \Omega}{\partial P}\frac{dP}{dz} + \sum_{i=1}^{c} H_i\left(r_i + \frac{dJ_i}{dz}\right) + \frac{q}{S} = 0$$

4. *Steady State: Perfect Gas Mixture*

$$-F_T C_p^\circ \frac{dT}{dz} + \sum_{i=1}^{c} H_{fi}^\circ\left(r_i + \frac{dJ_i}{dz}\right) + \frac{q}{S} = 0$$

*For the definition of j, see Table 6-1.

By use of the preceding relationships and the same approach used to obtain Equation (6-42) from (6-17), the following form of Equation (6-54) is obtained:

$$-F_T C_p \frac{\partial T}{\partial z} - F_T \frac{\partial \Omega}{\partial P}\frac{\partial P}{\partial z} + \sum_{i=1}^{c} \overline{H}_i\left(r_i + \frac{\partial J_i}{\partial z}\right) + \frac{q}{S}$$

$$= C_T C_p \frac{\partial T}{\partial t} + C_T \frac{\partial \Omega}{\partial P}\frac{\partial P}{\partial t} - \frac{1}{j}\frac{\partial P}{\partial t}, \qquad 0 < z < z_T > 0 \quad (6\text{-}59)$$

where

$$C_p = \sum_{i=1}^{c} \frac{C_i}{C_T} C_{pi} + \frac{\partial \Omega}{\partial T}$$

$$\overline{H}_i = H_{fi}^\circ + \frac{\partial(C_T \Omega)}{\partial C_i}$$

The forms taken by this equation for several special cases are listed in Table 6-2.

B. Generalized Partial Mixing and Flow in Three Directions x, y, z

The development of the component material balance and the energy balance for three-dimensional flow accompanied by partial mixing in each direction is carried out in a manner similar to that demonstrated for plug flow and partial mixing in the axial or z direction. A summary of the corresponding definitions and associated relationships follows:

$$u_x = \frac{\sum_{i=1}^{c} u_{xi}C_i}{C_T} \qquad u_y = \frac{\sum_{i=1}^{c} u_{yi}C_i}{C_T} \qquad u_z = \frac{\sum_{i=1}^{c} u_{zi}C_i}{C_T}$$

$$J_{xi} = -D_x\frac{\partial C_i}{\partial x} \qquad J_{yi} = -D_y\frac{\partial C_i}{\partial y} \qquad J_{zi} = -D_z\frac{\partial C_i}{\partial z}$$

$$J_{xi} = C_i(u_{xi} - u_x) \qquad J_{yi} = C_i(u_{yi} - u_y) \qquad J_{zi} = C_i(u_{zi} - u_z)$$

$$J_{xT} = \sum_{i=1}^{c} J_{xi} = 0 \qquad J_{yT} = \sum_{i=1}^{c} J_{yi} = 0 \qquad J_{zT} = \sum_{i=1}^{c} J_{zi} = 0$$

$$F_{xi} = C_i u_{xi} \qquad F_{yi} = C_i u_{yi} \qquad F_{zi} = C_i u_{zi}$$

$$F_{xi} = C_i u_x + J_{xi} \qquad F_{yi} = C_i u_y + J_{yi} \qquad F_{zi} = C_i u_z + J_{zi}$$

$$F_{xT} = C_T u_x \qquad F_{yT} = C_T u_y \qquad F_{zT} = C_T u_z$$

For flow and partial mixing in the x, y, and z directions, the following expression is obtained for the component material balance for any component i:

$$-\frac{\partial F_{xi}}{\partial x} - \frac{\partial F_{yi}}{\partial y} - \frac{\partial F_{zi}}{\partial z} - r_i = \frac{\partial C_i}{\partial t},$$

$$0 \le x \le x_T, 0 \le y \le y_T, 0 \le z \le z_T, t > 0 \quad (6\text{-}60)$$

and the energy balance is given by*

$$-\frac{\partial(F_{xT}H)}{\partial x} - \frac{\partial(F_{yT}H)}{\partial y} - \frac{\partial(F_{zT}H)}{\partial z} = \frac{\partial(C_T H)}{\partial t} - \frac{1}{j}\frac{\partial P}{\partial t} \quad (6\text{-}61)$$

Again, by using the same approach used to obtain Equation (6-58), it is readily shown that

$$F_{xT}H = u_x C_T H = \sum_{i=1}^{c} C_i\overline{H}_{xi} - u_x\sum_{i=1}^{c} C_i\overline{H}_i = u_x\sum_{i=1}^{c} C_i\hat{H}_i = u_x\sum_{i=1}^{c} C_i\left(H_{fi}^\circ + \Omega\right)$$

$$F_{yT}H = u_y C_T H = \sum_{i=1}^{c} C_i\overline{H}_{yi} = u_y\sum_{i=1}^{c} C_i\overline{H}_i = u_y\sum_{i=1}^{c} C_i\hat{H}_i = u_y\sum_{i=1}^{c} C_i\left(H_{fi}^\circ + \Omega\right)$$

$$F_{zT}H = u_z C_T H = \sum_{i=1}^{c} C_i\overline{H}_{zi} = u_z\sum_{i=1}^{c} C_i\overline{H}_i = u_z\sum_{i=1}^{c} C_i\hat{H}_i = u_z\sum_{i=1}^{c} C_i\left(H_{fi}^\circ + \Omega\right)$$

*For the general case where heat transfer by conduction is included, Equation (6-61) becomes

$$-\frac{\partial(F_{xT}H + Q_x)}{\partial x} - \frac{\partial(F_{yT}H + Q_y)}{\partial y} - \frac{\partial(F_{zT}H + Q_z)}{\partial z} = \frac{\partial(C_T H)}{\partial t} - \frac{1}{j}\frac{\partial P}{\partial t}$$

where

$$Q_x = -k\frac{\partial T}{\partial x}, \qquad Q_y = -k\frac{\partial T}{\partial y}, \qquad Q_z = -k\frac{\partial T}{\partial z},$$

and k is the thermal conductivity of the reacting medium.

where

$$\overline{H}_{xi} = \left[\frac{\partial(F_{xT}H)}{\partial C_i} \right]_{P,T,C_j}, \qquad \overline{H}_{zi} = \left[\frac{\partial(F_{zT}H)}{\partial C_i} \right]_{P,T,C_j}$$

$$\overline{H}_{yi} = \left[\frac{\partial(F_{yT}H)}{\partial C_i} \right]_{P,T,C_j}$$

By use of the same general approach shown for stating Equation (6-17) in the form given by Equation (6-42), it is readily shown that Equation (6-61) may be restated as follows:

$$-C_T C_p \left(u_x \frac{\partial T}{\partial x} + u_y \frac{\partial T}{\partial y} + u_z \frac{\partial T}{\partial z} \right) - C_T \frac{\partial \Omega}{\partial P} \left(u_x \frac{\partial P}{\partial x} + u_y \frac{\partial P}{\partial y} + u_z \frac{\partial P}{\partial z} \right)$$

$$+ \sum_{i=1}^{c} \overline{H}_i \left(r_i + \frac{\partial J_{xi}}{\partial x} + \frac{\partial J_{yi}}{\partial y} + \frac{\partial J_{zi}}{\partial z} \right) = C_T C_p \frac{\partial T}{\partial t} + C_T \frac{\partial \Omega}{\partial P} \frac{\partial \Omega}{\partial t} - \frac{1}{j} \frac{\partial P}{\partial t},$$

$$0 < x < x_T, 0 < y < y_T, 0 < z < z_T, t > 0 \quad (6\text{-}62)$$

where again, $\overline{H}_i = H_{fi}^\circ + \partial(C_T \Omega)/\partial C_i$. The forms taken by this equation for several special cases are shown in Table 6-3.

C. Generalized Partial Mixing in the Radial and Axial Directions

To develop the component material balances and the enthalpy balance for this case, the definitions and relationships needed for the derivation are of the same form as those shown previously, and those for the r direction are obtained from these by replacing the subscript z wherever it appears by the subscript r.

For flow and partial mixing in the radial and axial directions, the following component material balance is obtained in a manner analogous to that shown for Equation (2-101):

$$-\frac{\partial(rF_{zi})}{\partial z} - \frac{\partial(rF_{ri})}{\partial r} - r_i r = r \frac{\partial C_i}{\partial t}$$

or

$$-\frac{\partial F_{zi}}{\partial z} - \frac{1}{r} \frac{\partial(rF_{ri})}{\partial r} - r_i = \frac{\partial C_i}{\partial t}, \qquad 0 < z < z_T, 0 < r < r_T,$$

$$t > 0 \quad (6\text{-}63)$$

TABLE 6-3

SUMMARY OF ENERGY-BALANCE EQUATIONS FOR THE GENERALIZED CASE OF
THREE-DIMENSIONAL FLOW WITH PARTIAL MIXING

1. Unsteady State: Nonideal Solution

$$-C_T C_p \left(u_x \frac{\partial T}{\partial x} + u_y \frac{\partial T}{\partial y} + u_z \frac{\partial T}{\partial z} \right) - C_T \frac{\partial \Omega}{\partial P} \left(u_x \frac{\partial P}{\partial x} + u_y \frac{\partial P}{\partial y} + u_z \frac{\partial P}{\partial z} \right)$$

$$+ \sum_{i=1}^{c} \overline{H}_i \left(r_i + \frac{\partial J_{xi}}{\partial x} + \frac{\partial J_{yi}}{\partial y} + \frac{\partial J_{zi}}{\partial z} \right) = C_T C_p \frac{\partial T}{\partial t} + C_T \frac{\partial \Omega}{\partial P} \frac{\partial \Omega}{\partial t} - \frac{1}{j} \frac{\partial P^*}{\partial t}$$

2. Steady State: Nonideal Solutions

$$-C_T C_p \left(u_x \frac{\partial T}{\partial x} + u_y \frac{\partial T}{\partial y} + u_z \frac{\partial T}{\partial z} \right) - C_T \frac{\partial \Omega}{\partial P} \left(u_x \frac{\partial P}{\partial x} + u_y \frac{\partial P}{\partial y} + u_z \frac{\partial P}{\partial z} \right)$$

$$+ \sum_{i=1}^{c} \overline{H}_i \left(r_i + \frac{\partial J_{xi}}{\partial x} + \frac{\partial J_{yi}}{\partial y} + \frac{\partial J_{zi}}{\partial z} \right) = 0$$

3. Steady State: Nonideal Solution

$$-C_T C_p \left(u_x \frac{\partial T}{\partial x} + u_y \frac{\partial T}{\partial y} + u_z \frac{\partial T}{\partial z} \right) - C_T \frac{\partial \Omega}{\partial P} \left(u_x \frac{\partial P}{\partial x} + u_y \frac{\partial P}{\partial y} + u_z \frac{\partial P}{\partial z} \right)$$

$$+ \sum_{i=1}^{c} \overline{H}_i \left(r_i + \frac{\partial J_{xi}}{\partial x} + \frac{\partial J_{yi}}{\partial y} + \frac{\partial J_{zi}}{\partial z} \right) = 0$$

4. Steady State: Ideal Solutions

$$-C_T C_p \left(u_x \frac{\partial T}{\partial x} + u_y \frac{\partial T}{\partial y} + u_z \frac{\partial T}{\partial z} \right) - C_T \frac{\partial \Omega}{\partial P} \left(u_x \frac{\partial P}{\partial x} + u_y \frac{\partial P}{\partial y} + u_z \frac{\partial P}{\partial z} \right)$$

$$+ \sum_{i=1}^{c} H_i \left(r_i + \frac{\partial J_{xi}}{\partial x} + \frac{\partial J_{yi}}{\partial y} + \frac{\partial J_{zi}}{\partial z} \right) = 0$$

5. Steady State: Perfect Gas Mixture

$$-C_T C_p^\circ \left(u_x \frac{\partial T}{\partial x} + u_y \frac{\partial T}{\partial y} + u_z \frac{\partial T}{\partial z} \right) + \sum_{i=1}^{c} H_{fi}^\circ \left(r_i + \frac{\partial J_{xi}}{\partial x} + \frac{\partial J_{yi}}{\partial y} + \frac{\partial J_{zi}}{\partial z} \right) = 0$$

*For the definition of j, see Table 6-1.

The energy balance is given by

$$-\frac{\partial (F_{zT} H)}{\partial z} - \frac{1}{r} \frac{\partial (r F_{rT} H)}{\partial r} = \frac{\partial (C_T H)}{\partial t} - \frac{1}{j} \frac{\partial P}{\partial t},$$

$$0 < z < z_T, 0 < r < r_T, t > 0 \quad (6\text{-}64)$$

An expression for $F_{rT} H$ is obtained by following the same approach used to obtain the expression for $F_{zT} H$.

By following the same general approach shown for restating Equation (6-17) in the form given by Equation (6-42), Equation (6-64) is readily restated in the following form.

$$-C_T C_p \left(u_z \frac{\partial T}{\partial z} + u_r \frac{\partial T}{\partial r} \right) - C_T \frac{\partial \Omega}{\partial P} \left(u_z \frac{\partial P}{\partial z} + u_r \frac{\partial P}{\partial r} \right)$$

$$+ \sum_{i=1}^{c} \overline{H}_i \left(r_i + \frac{\partial J_{zi}}{\partial z} + \frac{1}{r} \frac{\partial (r J_{ri})}{\partial r} \right) = C_T C_p \frac{\partial T}{\partial t} + C_T \frac{\partial \Omega}{\partial P} \frac{\partial P}{\partial t} - \frac{1}{j} \frac{\partial P}{\partial t}$$

$$(6\text{-}65)$$

TABLE 6-4

SUMMARY OF THE ENERGY-BALANCE EQUATIONS FOR FLOW AND PARTIAL MIXING IN THE
RADIAL AND AXIAL DIRECTIONS

1. *Unsteady State: Nonideal Solution*

$$-C_T C_p \left(u_z \frac{\partial T}{\partial z} + u_r \frac{\partial T}{\partial r} \right) - C_T \frac{\partial \Omega}{\partial P} \left(u_z \frac{\partial P}{\partial z} + u_r \frac{\partial P}{\partial r} \right)$$

$$+ \sum_{i=1}^{c} \overline{H}_i \left(r_i + \frac{\partial J_{zi}}{\partial z} + \frac{1}{r} \frac{\partial (r J_r)}{\partial r} \right) = C_T C_p \frac{\partial T}{\partial t} + C_T \frac{\partial \Omega}{\partial P} \frac{\partial P}{\partial t} - \frac{1}{j} \frac{\partial P^*}{\partial t}$$

2. *Unsteady State: Nonideal Solution, $u_r = 0$ (tubular reactor)*

$$-u_z C_T C_p \frac{\partial T}{\partial z} - u_z C_T \frac{\partial \Omega}{\partial P} \frac{\partial P}{\partial z} + \sum_{i=1}^{c} \overline{H}_i \left(r_i + \frac{\partial J_{zi}}{\partial z} + \frac{1}{r} \frac{\partial (r J_{ri})}{\partial r} \right)$$

$$= C_T C_p \frac{\partial T}{\partial t} + C_T \frac{\partial \Omega}{\partial P} \frac{\partial P}{\partial t} - \frac{1}{j} \frac{\partial P}{\partial t}$$

3. *Steady State: Nonideal Solution, $u_r = 0$*

$$-u_z C_T C_p \frac{\partial T}{\partial z} - u_z C_T \frac{\partial \Omega}{\partial P} \frac{\partial P}{\partial z} + \sum_{i=1}^{c} \overline{H}_i \left(r_i + \frac{\partial J_{zi}}{\partial z} + \frac{1}{r} \frac{\partial (r J_{ri})}{\partial r} \right) = 0$$

4. *Steady State: Ideal Solutions, $u_r = 0$*

$$-u_z C_T C_p \frac{\partial T}{\partial z} - u_z C_T \frac{\partial \Omega}{\partial P} \frac{\partial P}{\partial z} + \sum_{i=1}^{c} H_i \left(r_i + \frac{\partial J_{zi}}{\partial z} + \frac{1}{r} \frac{\partial (r J_{ri})}{\partial r} \right) = 0$$

5. *Steady State: Perfect Gas Mixture, $u_r = 0$*

$$-u_z C_T C_p^\circ \frac{\partial T}{\partial z} + \sum_{i=1}^{c} H_{fi}^\circ \left(r_i + \frac{\partial J_{zi}}{\partial z} + \frac{1}{r} \frac{\partial (r J_{ri})}{\partial r} \right) = 0$$

*For the definition of j, see Table 6-1.

Again $\overline{H}_i = H_{fi}^\circ + \partial (C_T \Omega)/\partial C_i$, and the form of Equation (6-65) for several special cases is shown in Table 6-4. The special case where $u_r = 0$ corresponds to the tubular reactor shown in Figure 2-18.

To solve the equations listed in Tables 6-1 through 6-4, a knowledge of both the inital conditions and the boundary conditions is required for the unsteady-state equations, and of the boundary conditions for the steady-state equations (1). In general, the energy balance equations must be solved simultaneously with the remaining independent equations that are required to describe the system. The general approach is illustrated in the following section on the design of thermal reactors.

III. DESIGN OF THERMAL REACTORS

Thermal reactors belong to that class of reactors in which the energy of activation required for the reaction to take place at an appreciable rate is provided by heating the reaction mixture. The transfer of the appropriate amount of heat to and/or from the reacting mixture is one of the major problems to be solved in the design of thermal reactors.

The design of thermal reactors is demonstrated by solving several relatively simple examples. Since the analysis of rate data is simplified by carrying out a reaction isothermally, many experimental reactors are designed for this type operation. Most industrial reactors are not operated isothermally, and the design of these reactors involves the simultaneous solution of the material- and energy-balance equations. Numerical examples are presented for the purpose of demonstrating the design of tubular reactors.

To begin the design of a reactor, answers must be found to several questions, such as: What types of materials of construction are inert to the reacting system at the temperatures and pressures being considered? What method is to be used to transfer heat to and from the reacting mixture? What pressure drop across the reactor should be permitted? What volume (or time) is required to effect the desired conversion? Will the load-bearing properties of the ground or building support the reactor? Many of the questions which must be answered are the same questions which must be answered in the design of any structure. Of these questions, only the ones concerning heat transfer, fluid flow, and chemical reactions are considered in the following analysis.

To effect the necessary heat transfer, several types of reactors have been developed. Cracking reactions, such as the cracking of propane and ethane to ethylene, are commonly carried out in direct-fired reactors. In these reactors, the reactants are passed through tubes which are surrounded by a direct-fired combustion furnace. In these furnaces, heat is transferred to the heated side of the wall at approximately the same rate at all points.

In another class of reactors, the latent heat of vaporization of the heat-transfer medium is either transferred to or removed from the reacting mixture. An example of this type of reactor is a steam-jacketed reactor in which heat is supplied to the reacting mixture by the condensing steam in the jacket.

A third type of reactor is a heat-exchanger reactor in which the reactants pass through the reactor either in the tubes or on the outside of the tubes. The heat-transfer medium passes through the reactor in a direction countercurrent or parallel (cocurrent) with respect to the reacting mixture. When the reacting mixture is used to preheat the feed, the reactor is sometimes called an autothermal reactor.

A summary of the equations needed in the design of the tubular reactors considered herein follows. These equations are applicable when the reactants and the heat-transfer medium make a single pass through the reactor, the fluids are in plug flow, and when the reactor is at the condition of steady-state operation.

Material balance: $\quad r_A = -\dfrac{dn_A}{dV} = n_A^0 \dfrac{dx}{dV} = \dfrac{n_A^0}{S}\dfrac{dx}{dz}$ $\qquad$ (6-66)

Energy balance on reactants, plug flow: $\quad wC_p\dfrac{dT}{dz} + r_A S\,\Delta H_r - q = 0$ $\quad$ (6-67)

Energy balances on heat-transfer medium

1. *Cocurrent, plug flow:* $w_c C_{pc} \dfrac{dT_c}{dz} + q = 0$ (6-68)

2. *Countercurrent, plug flow:* $-w_c C_{pc} \dfrac{dT_c}{dz} + q = 0$ (6-69)

In these equations, the positive direction of z is taken to be the direction of flow of the reactants. For cocurrent, countercurrent, and steam-heated reactors, the heat-transfer rate q (Btu per hour per foot of length) is given by

$$q = U_h a (T_c - T)$$ (6-70)

For direct-fired reactors, only the first two of the preceding equations are needed, and q is usually given as an estimated numerical value in Btu per hour per square foot of heat-transfer area. Also, it should be pointed out that the length variable z may be replaced by the corresponding volume V at any z, since $V = Sz$.

In addition to the preceding equations, the Fanning equation [2] is needed to calculate the pressure drop:

$$-\frac{dP}{dz} = \frac{2fG^2}{D\rho g_c} = \frac{2fu^2\rho}{Dg_c}$$ (6-71)

where
P = pressure in lb_f/ft^2

D = inside diameter of the tube, ft

G = mass flux, lb_m/ft^2 s

g_c = 32.17 $(lb_m \; ft)/lb_f \; s^2$

f = friction factor; dimensionless

z = direction of flow, ft

ρ = density, lb_m/ft^3

u = average velocity, ft/s

(In these definitions, the symbol lb_m is used to mean pounds mass, and the symbol lb_f is used to mean pounds force. In all other equations, only mass units are involved, and consequently the subscript m is omitted in the interest of simplicity.)

If an appropriate correspondence exists between the operating variables in a batch reactor and in a tubular reactor, then the material-balance equations for tubular reactors may be obtained from those for a batch reactor. For example, if both reactors are operated isothermally, the fluid density is constant, the batch reactor is perfectly mixed, and plug flow prevails throughout the length of the tubular reactor, then the variable t in the equations for the batch reactor may be replaced by the variable θ (the true residence time,

$\theta = V/v_T$) to give the material-balance equations for the tubular reactor. The residence time $\theta = V/v_T$ required to achieve a specified conversion may be found by use of the material-balance equations. Even though the residence time gives no explicit information about the tube diameter, the computed value of the residence time does depend on the assumption of plug flow. To achieve plug flow, an appropriate number of tubes having a given diameter must be selected. The selection of the number of tubes of a given diameter must also be made in a manner such that the heat-transfer requirements and the pressure-drop limitations are all satisfied simultaneously. The interrelationships of the variables involved in the design of a reactor are demonstrated by the following example.

EXAMPLE 6-1

It is desired to carry out the first-order unidirectional reaction $A \rightarrow R$ in the liquid phase in a tubular reactor at a temperature of 120°F. The reaction is exothermic, and a heat-exchanger reactor is to be used. The reacting medium is to be passed through the tubes and the cooling medium through the shell side of the exchanger. Water is to be used as the heat-transfer medium, and the transfer of heat from the outer shell to the surroundings may be neglected. The following data are available:

$C_p = 8$ Btu/lb mol °F, for the reacting mixture

$C_{pc} = 1.0$ Btu/lb °F for the cooling water

$C_A^0 = 0.0245$ lb mol/ft^3, initial concentration of the feed

$k = 0.128$ s^{-1}, rate constant

$v_T = 360$ ft^3/h, flow rate of the feed to the reactor

$U_h = 150$ Btu/ft^2 h °F

$\Delta H_r = -30,600$ Btu/lb mol of A reacted

$\mu = 0.44$ centipoise or 1.06 lb/ft h, viscosity of the reacting mixture

$\rho = 56.16$ lb/ft^3

(a) Calculate the reactor volume required to obtain a 90% conversion of A per pass.
(b) If 1-in. (20 BWG) reactor tubes are used, find the number and length of the tubes required to satisfy the following conditions:

1. The reactor volume found in part (a).
2. The pressure drop across the reactor is not to exceed 20 lb/in.^{2}.
3. To approximate plug flow, the Reynolds number in each tube is not to be less than 10^4.

(c) Should parallel or countercurrent flow be used? Why? Find the inlet temperature and the flow rate of the cooling water required for isothermal operation.
(d) Construct a graph of the temperature of the cooling water versus the residence time.

SOLUTION

(a) Since the reaction is first order and isothermal, Equations (2-13), (2-16), and (2-19) may be used to compute first θ and then the reactor volume.

$$\theta = -\frac{1}{k} \log_e (1 - x)$$

$$\theta = -\frac{1}{0.128} \log_e (1 - 0.9) = 18 \text{ s}$$

$$V = \theta v_T = (18)(0.1 \text{ ft}^3/\text{s}) = 1.8 \text{ ft}^3$$

(b) From page 11-11 of Reference [2], the following data were obtained for 20 BWG tubes:

$$D \text{ (inside diameter)} = 0.93 \text{ in.} = 0.0775 \text{ ft}$$

$$S \text{ (internal cross-sectional area)} = 4.717 \times 10^{-3} \text{ ft}^2$$

$$a \text{ (internal heat-transfer area per foot of tube)} = 0.2435 \text{ ft}^2/\text{ft}$$

The conditions stated may be satisfied by use of three reactor tubes with a length of 127.2 ft. For

$$V = NSL = (3)(4.717 \times 10^{-3})(127.2) = 1.8 \text{ ft}^3$$

The velocity u per tube is given by

$$u = \frac{L}{\theta} = \frac{127.2}{18} = 7.07 \text{ ft/s}$$

Then

$$N_{Re} = \frac{Du\rho}{\mu} = \frac{(0.0775)(7.07)(56.16)}{(1.06)(1/3600)} = 1.045 \times 10^5$$

From Figure 5-25 of Reference [2], the Fanning friction factor $f = 0.0044$. Then, by use of the integrated form of Equation (6-71),

$$(P)_{\text{inlet}} - (P)_{\text{outlet}} = \frac{(2)(0.0044)(127.2)(7.07)^2(56.16)}{(0.0775)(32.2)(144)} = 8.74 \text{ psi}$$

Thus the reactor should be constructed as shown in Figure 6-2 with three 180° bends. The pressure drop should not exceed the allowable drop of 20 psi, since the pressure drop resulting from entrance effects and return bends should be less than 7 psi. Consequently, the total pressure drop should not exceed 16 psi, which is less than the permissible drop of 20 psi.

(c) Since the rate of reaction has its maximum value at the inlet of the reactor, the maximum rate of heat transfer must occur at the inlet of the reactor. The condition of isothermal operation can be achieved when the coolant flows parallel with respect to the reactants. Isothermal operation is also possible by use of parallel flow for the case of a first-order reaction which takes place in the gas phase at constant density, whereas, isothermal operation is impossible (unless $\Delta H_r = 0$) when countercurrent flow is used (see Problem 6-9).

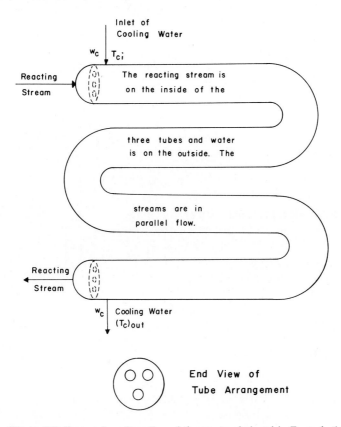

Figure 6-2. Proposed configuration of the reactor designed in Example 6-1.

The equations needed to demonstrate that isothermal operation is possible when parallel flow is used and to compute the required inlet temperature and flow rate of the coolant are one and the same. The equations needed may be developed in the following manner. Since $dT/dz = 0$, Equation (6-58) reduces to

$$q = r_A S \Delta H_r \tag{A}$$

When q is replaced by its equivalent as given by Equation (6-70), the following expression is obtained upon rearrangement:

$$T_c - T = \frac{r_A S \Delta H_r}{U_h a} \tag{B}$$

Since

$$r_A = kC_A = kC_A^0 e^{-k\theta}$$

it is possible to state Equation (B) in terms of θ as follows:

$$T_c - T = \frac{kS C_A^0 \Delta H_r e^{-k\theta}}{U_h a} \tag{C}$$

Since

$$T_{ci} - T = \frac{kSC_A^0\,\Delta H_r}{U_h a} \tag{D}$$

at $\theta = 0$, Equation (C) may be restated as follows:

$$T_c - T = (T_{ci} - T)e^{-k\theta} \tag{E}$$

When Equation (6-68) is integrated on the basis of the assumption that the temperature T of the reaction does not vary with position, the result so obtained may be rearranged and stated in the form

$$T_c - T = (T_{ci} - T)\exp\left[-\left(\frac{U_h a}{w_c C_{pc}}\right)z\right]$$

Since $\theta = V/v_T = Sz/v_T$, it follows that

$$T_c - T = (T_{ci} - T)\exp\left[-\left(\frac{U_h a}{w_c C_{pc}}\right)\left(\frac{v_T}{S}\right)\theta\right] \tag{F}$$

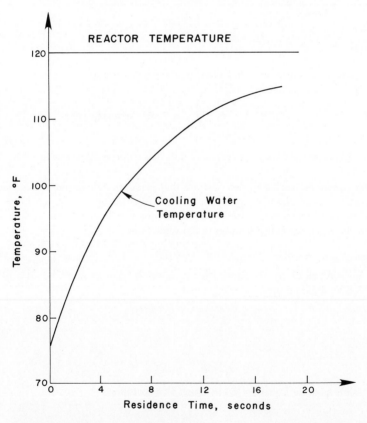

Figure 6-3. Isothermal operation of the reactor in Example 6-1 by parallel flow of the reacting stream and the coolant.

Comparison of Equations (E) and (F) shows that isothermal operation is possible provided that the following relationship exists between the rate constant k and the other variables:

$$k = \left(\frac{U_h a}{C_{pc} S}\right)\left(\frac{v_T}{w_c}\right) = \left(\frac{4U_h}{C_{pc} D}\right)\left(\frac{v_T}{w_c}\right) \tag{G}$$

The inlet temperature of the cooling water may be computed by use of Equation (D).

$$T_{ci} = T + \frac{kDC_A^0 \,\Delta H_r}{4U_h}$$

$$T_{ci} = 120 + \frac{(0.128)(3600)(0.0775)(0.0245)(-30,600)}{(4)(150)} = 75.38°F$$

The flow rate w_c of the cooling water may be computed by solving Equation (G) for w_c to give

$$w_c = \left(\frac{4U_h}{C_{pc} D}\right)\left(\frac{v_T}{k}\right) = \frac{(4)(150)(0.1)}{(1)(0.0775)(0.128)} = 6048 \text{ lb/h}$$

Since all the variables (k, T, and T_{ci}) in Equation (E) are now known, it may be used to compute the temperature T_c at any θ. A graph of these results is shown in Figure 6-3.

1. Use of Condensing Vapors as the Heat-Transfer Medium. The use of condensing steam to supply the necessary heat to carry out an endothermic reaction is illustrated by the numerical example solved by Billingsley et al. [3] and by Logan [4]. Typical steam-heated reactors are depicted in Figure 6-4.

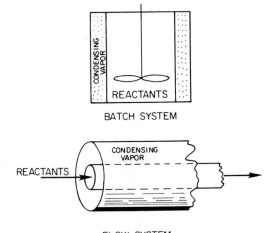

Figure 6-4. Sketches of reactors heated by use of condensing vapors. [Taken from D. S. Billingsley, W. S. McLaughin, Jr., N. E. Welch and C. D. Holland, *Ind. Eng. Chem.*, 50:741 (1958). Used with permission of the American Chemical Society.]

EXAMPLE 6-2

The reaction $A \rightarrow$ products is to be carried out in a steam-heated heat-exchanger reactor. The charge rate is 10,000 lb/h per tube. The feed, pure A, enters the reactor at a temperature of 70°F. It is desired to determine the volume of reactor required for 90% conversion of A. The internal diameter of the reactor tubes may be taken as 1 in. Steam at 240°F is available for heating purposes. The rate of the reaction may be taken to be first order with respect to A, and the rate constant may be taken as $k(1/h) = 203 \times 10^{15} e^{-46,846/RT}$, where T is in °R. The physical properties of the system, which may be assumed to remain constant throughout the reactor, are as follows:

$$\rho = 80 \text{ lb/ft}^3$$

$M_w = 200$ lb/lb mol, molecular weight of the reacting steam

$U_h = 400$ Btu/(h ft^2 °F) based on the inside diameter of the reactor tube

$C_p = 0.75$ Btu/(lb °F)

$\Delta H_r = 40,000$ Btu/lb mol

$E = 46,846$ Btu/lb mol

$A = 203 \times 10^{15}$ (1/h)

SOLUTION

Since the reaction is first order, the rate expression may be stated in the form

$$r_A = kC_A = k\frac{n_A}{v_T} = \frac{kn_A^0}{v_T}(1 - x) \tag{A}$$

Also, since the variation of k with T is given by the Arrhenius expression ($k = Ae^{-ERT}$), it is possible to express k at any T in terms of k_1 (the value of k at $T = T_1$) as follows:

$$k = k_1 e^{b(1 - 1/\phi)} \tag{B}$$

where

$$b = \frac{E}{RT_1}$$

$$\phi = \frac{T}{T_1}$$

Elimination of r_A and k from Equations (A), (B), and (6-66), followed by arrangement, yields

$$\frac{dx}{dV} = \frac{k_1 e^{b(1 - 1/\phi)}(1 - x)}{v_T}$$

TABLE 6-5
SYMBOLS IN EQUATIONS (C) AND (E) OF EXAMPLE 6-2
FOR VARIOUS TYPES OF REACTORS

Type and Arrangement	s	m/s	c/s	Z
Flow				
Reactants inside tubes	$\dfrac{v_T}{k_1}$	$\dfrac{\rho C_p S k_1}{U_h a}$	$\dfrac{n_A^0 \Delta H_r S k_1}{v_T U_h a T_1}$	$\dfrac{k_1 V}{v_T}$
Reactants outside tubes	$\dfrac{v_T}{k_1}$	$\dfrac{\rho C_p S_o k_1}{U_h a}$	$\dfrac{n_A^0 \Delta H_r S_o k_1}{v_T U_h a T_1}$	$\dfrac{k_1 V}{v_T}$
Batch				
Reactants inside tubes	$\dfrac{1}{k_1}$	$\dfrac{\rho C_v S k_1}{U_h a}$	$\dfrac{N_A^0 \Delta U_r S k_1}{V U_h a T_1}$	$k_1 t$
Reactants outside tubes	$\dfrac{1}{k_1}$	$\dfrac{\rho C_v S_o k_1}{U_h a}$	$\dfrac{N_A^0 \Delta U_r S_o k_1}{V_o U_h a T_1}$	$k_1 t$

where S_o, V_o = cross-sectional area and volume, respectively, between the outside of the tubes and the shell of the heat-exchanger reactor

or

$$\frac{dx}{dZ} = e^{b(1 - 1/\phi)}(1 - x) \tag{C}$$

where

$$Z = \frac{k_1 V}{v_T}, \quad \text{dimensionless}$$

Since $a/S = N\pi D/(N\pi D^2/4)$, it is evident that Equations (6-67) and (6-70) may be combined and restated as follows:

$$-wC_p \frac{dT}{dV} + \frac{4U_h}{D}(T_c - T) - r_A \Delta H_r = 0 \tag{D}$$

where it is, of course, understood that a, S, and U are all based on the same tube diameter. When r_A in Equation (D) is replaced by its equivalent as given by Equations (A) and (C), and the resulting expression is restated in terms of the dimensionless variables ϕ, Z, and Φ (where $\Phi = T_c/T_1$), the following result is obtained upon rearrangement:

$$\phi + \frac{m}{s}\frac{d\phi}{dZ} + \frac{c}{s}\frac{dx}{dZ} = \Phi \tag{E}$$

The symbols m/s and c/s are defined in Table 6-5 for various types of reactors. The selection of the temperature T_1 is arbitrary. In the problem under consideration, T_1 was taken equal to the steam temperature, 700°R, which for this problem is also the maximum possible temperature.

Several numerical methods are available for solving nonlinear differential equations [5]. The system under consideration may be solved by the Euler and Runge and

Kutta predictor methods, which are available in many software packages. However, since more powerful corrector methods are needed to solve systems of stiff differentials, it was elected to solve Equations (C) and (E) by use of a combination of the trapezoidal corrector [5] and the Newton–Raphson method [5]. The trapezoidal method is a special case of the more general implicit method [5] in which the weight factor μ is taken equal to $\frac{1}{2}$; that is,

$$\int_{Z_j}^{Z_j+\Delta Z_j} f(x)\, dZ = \left[\mu f(x_{j+1}) + (1-\mu)f(x_j)\right]\Delta Z_j \qquad (6\text{-}72)$$

where x_j is the value of x at $Z = Z_j$ and x_{j+1} is the value of x at Z_{j+1}. For $\mu = \frac{1}{2}$, the implicit method reduces to the trapezoidal rule:

$$\int_{Z_j}^{Z_j+\Delta Z_j} f(x)\, dZ = \frac{\Delta Z_j}{2}\left[f(x_{j+1}) + f(x_j)\right] \qquad (6\text{-}73)$$

The first step of the suggested numerical procedure for solving the problem at hand consists of integrating Equations (C) and (E) with respect to Z from Z_j to $Z_j + \Delta Z_j$ (where $\Delta Z_j = Z_{j+1} - Z_j$ is fixed at some preselected value) as follows:

$$0 = \frac{\Delta Z_j}{2}\left[e^{b(1-1/\phi)}(1-x) + e^{b(1-1/\phi_j)}(1-x_j)\right] - (x - x_j) \qquad (\text{F})$$

$$0 = \frac{m}{s}(\phi - \phi_j) + \frac{c}{s}(x - x_j) + \frac{\Delta Z_j}{2}(\phi + \phi_j) - \Phi\Delta Z_j \qquad (\text{G})$$

In the interest of simplicity, the subscript $j + 1$ has been omitted on the unknown values of x and ϕ at Z_{j+1}.

On the basis of known values of ϕ_j and x_j at Z_j, it is desired to find the positive values of ϕ and x which satisfy Equations (F) and (G). To find this solution set of values of ϕ and x by the Newton–Raphson method, the first step is to replace the zeros on the left sides of Equations (F) and (G) by $F_1(\phi, x)$ and $F_2(\phi, x)$. Then the problem is to find the positive values of ϕ and x which make

$$F_1(\phi, x) = 0$$
$$F_2(\phi, x) = 0$$

simultaneously. The Newton–Raphson method consists of the repeated use of the linear terms of the Taylor series expansion of the functions F_1 and F_2 about an assumed set of values ϕ_n and x_n.

$$0 = F_1(\phi_n, x_n) + \frac{\partial F_1(\phi_n, x_n)}{\partial \phi}\Delta\phi + \frac{\partial F_1(\phi_n, x_n)}{\partial x}\Delta x \qquad (\text{H})$$

$$0 = F_2(\phi_n, x_n) + \frac{\partial F_2(\phi_n, x_n)}{\partial \phi}\Delta\phi + \frac{\partial F_2(\phi_n, x_n)}{\partial x}\Delta x \qquad (\text{I})$$

where

$$\Delta\phi = \phi_{n+1} - \phi_n$$
$$\Delta x = x_{n+1} - x_n$$

After these equations have been solved simultaneously for $\Delta\phi$ and Δx, the next set of values to be assumed are found by use of the definition of $\Delta\phi$ and Δx.

To demonstrate the use of this method, the calculations are shown for the first increment at the entrance of the reactor. The parameters are evaluated as follows:

$$b = \frac{E}{RT_1} = \frac{46{,}846}{(1{,}9872)(700)} = 33.677$$

$$k_1 = Ae^{-E/RT_1} = Ae^{-b} = 203 \times 10^{15}e^{-33.677} = 480.6 \ \text{h}^{-1}$$

$$Z = \frac{k_1 V}{v_T} = \frac{(480.6V)(80 \ \text{lb/ft}^3)}{(10{,}000 \ \text{lb/h})} = 3.84V$$

where V is in cubic feet.

$$\frac{m}{s} = \frac{\rho C_p S k_1}{U_h a} = \frac{\rho C_p D k_1}{4U_h} = \frac{(80)(0.75)\left(\frac{1}{12}\right)(480.6)}{(4)(400)} = 1.502$$

$$\frac{c}{s} = \frac{n_A^0 \, \Delta H_r \, S k_1}{v_T U_h a T_1} = \frac{n_A^0 \, \Delta H_r \, D k_1}{4 v_T U_h T_1} = \frac{(10{,}000/200)(40{,}000)\left(\frac{1}{12}\right)(480.6)}{4(10{,}000/80)(400)(700)} = 0.572$$

$$\Phi = \frac{T_c}{T_1} = \frac{700}{700} = 1.0$$

For the first increment, let $\Delta Z_j = 1$. At $Z = 0$, $T = T_0 = 530°\text{R}$, and $x = x_0 = 0$. Thus

$$\phi_0 = \frac{530}{700} = 0.757$$

$$x_0 = 0$$

For the first trial, assume $T = 615°\text{R}$ at Z_1. Then

$$\phi_1 = \frac{615}{700} = 0.8786$$

Assume

$$x_1 = 0.005$$

For the first increment,

$$F_1(\phi, x) = \frac{\Delta Z_1}{2}\left[e^{b(1-1/\phi)}(1-x) + e^{b(1-1/\phi_0)}(1-x_0)\right] - (x - x_0)$$

Then

$$F_1(\phi_1, x_1) = 0.5\left[e^{33.677(1-1/0.8786)}(1-0.005) + e^{33.677(1-1/0.757)}(1)\right] - 0.005$$

$$= -2.5 \times 10^{-4}$$

Similarly, for the first increment,

$$F_2(\phi, x) = \frac{m}{s}(\phi - \phi_0) + \frac{c}{s}(x - x_0) + \frac{\Delta Z_1}{2}(\phi + \phi_0) - \Phi \Delta Z_1$$

$$F_2(\phi, x) = 1.502(0.8786 - 0.757) + (0.572)(0.005) + 0.5(0.8786 + 0.757) - (1)(1)$$

$$= 0.0033.$$

The partial derivatives appearing in the Newton–Raphson equations are evaluated as follows:

$$\frac{\partial F_1}{\partial \phi} = \frac{\Delta Z_1}{2}\left[\frac{b}{\phi^2}e^{b(1-1/\phi)}(1-x)\right]$$

$$= 0.5\left[\frac{33.677}{(0.8786)^2}e^{33.677(1-1/0.8786)}(1-0.005)\right] = 0.2068$$

$$\frac{\partial F_1}{\partial x} = -\frac{\Delta Z_1}{2}\left[e^{b(1-1/\phi)}\right]-1$$

$$= -0.5\left[e^{33.677(1-1/0.8786)}\right]-1 = -1.005$$

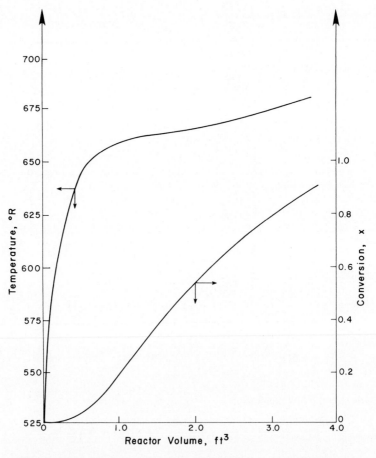

Figure 6-5. Temperature and conversion profiles in the steam-heated flow reactor designed in Example 6-2.

Next

$$\frac{\partial F_2}{\partial \phi} = \frac{m}{s} + \frac{\Delta Z_1}{2} = 1.502 + 0.5 = 2.002$$

$$\frac{\partial F_2}{\partial x} = \frac{c}{s} = 0.572$$

When the values of the functions F_1, F_2 and their partial derivatives at the point ϕ_1, x_1 are substituted in the Newton–Raphson equations [Equations (H) and (I)], the following set of linear equations in the two unknowns $\Delta\phi$ and Δx is obtained:

$$0 = -2.5 \times 10^{-4} + 0.2068\,\Delta\phi - 1.005\,\Delta x$$

$$0 = 33.0 \times 10^{-4} + 2.002\,\Delta\phi + 0.572\,\Delta x$$

These two equations may be solved simultaneously for $\Delta\phi$ and Δx to give

$$\Delta\phi = -0.001489$$

$$\Delta x = -0.0005552$$

Thus the values of ϕ and x to be assumed for the next trial for the first increment ΔZ are given by the expressions for any trial n:

$$\phi_{n+1} = \phi_n + \Delta\phi = 0.8786 - 0.001489 = 0.8771$$

$$x_{n+1} = x_n + \Delta x = 0.005 - 0.0005552 = 0.00445$$

This procedure is repeated for the first increment ΔZ_1 until the differences between the assumed and calculated values of ϕ and x are less than some small preassigned number. When the solution set satisfying this criterion has been found, the Newton–Raphson equations are applied to the second increment, and the quantities ϕ_j and x_j in Equations (H) and (I) are replaced by solution set (ϕ_1, x_1) at Z_1.

The solution sets (ϕ, x) or (T, x) versus the reactor volume obtained by the procedure described are displayed in Figure 6-5. It is to be observed that very little of A is converted in the first 0.43 ft^3 of the reactor. In this section of the reactor, the reactants are being heated to a temperature at which the rate constant k becomes significant. As in any numerical method, the choice of the increment size ΔZ_j has some effect on the numerical results [5] as demonstrated in Table 6-6.

TABLE 6-6
EFFECT OF INCREMENT SIZE ΔZ ON THE SOLUTION OF EXAMPLE 6-2 [4]

Volume (ft^3)	Conversion			Temperature (°R)		
	$\Delta Z = 0.1$	0.01	0.001	$\Delta Z = 0.1$	0.01	0.001
0.264	0.005	0.002	0.002	614.9	613.1	613.0
0.792	0.132	0.123	0.123	657.9	656.7	656.7
1.320	0.326	0.316	0.315	660.9	660.7	660.7
1.848	0.503	0.493	0.492	664.4	664.2	664.2
2.376	0.657	0.649	0.648	668.6	668.3	668.2
2.903	0.786	0.779	0.778	673.4	673.1	673.0
3.431	0.885	0.880	0.879	679.1	678.8	678.7
3.56	0.903	0.900	0.900	680.7	680.3	680.2

2. Design of a Direct-Fired Tubular Reactor for Carrying out a Gas Phase Reaction. One of the best known examples of the use of direct-fired tubular reactors is the production of ethylene and propylene by the pyrolysis of hydrocarbon feedstocks. To achieve an improved yield of olefins, contact times in these reactors have been reduced while the outlet temperatures have been increased. A new design of these reactors was used in the plant shown in Figure 6-6, which contained six cracking furnaces. By use of reaction temperatures of 1600° to 1700°F and contact times less than 0.1 s, 33% (by weight) of ethylene may be produced from naphtha in this furnace [6]. (The contact time may be approximated by dividing the holdup volume of the reactor by the log mean of the volumetric flow rate, evaluated at the inlet and outlet conditions of the reactor.) While it has been established that pyrolysis proceeds by interacting free radical mechanisms [7, 8], the pyrolysis reactions may be approximated with good accuracy by a first-order system [6].

To illustrate the design of direct-fired tubular reactors, the problem solved by Murdoch and Holland [9, 10] will be used. Although Murdoch and Holland presented a design method based on an approximate analytical solution to the equations describing the reactor, a numerical method will be used here.

As the term is used here, a direct-fired tubular reactor consists essentially of straight tubes connected in series by return bends. Most but not all of the straight portion is heated on its outside by flue gas. The rate of heat transfer to

Figure 6-6. Idemitsu's 246,500-metric-ton-a-year olefins plant at Chiba, Japan, is the first such facility in which all furnaces are based on the advanced, patented Millisecond technology of the M. W. Kellogg Company. Olefin yields are substantially increased due to the short residence time in the furnaces, five-hundredths to one-tenth of a second (courtesy of the M. W. Kellogg Company).

the tubes is assumed to be constant, at different levels, in the radiant and convection sections. The number of straight lengths of tubes is equal to the number of return bends. The unit upon which the calculations will be based is therefore a hook-shaped section consisting of one tube and one 180-degree return bend. The actual reactor consists of n of these hook-shaped sections in series. Thus the reactor is equal in length to a long, straight pipe of length $n(L_s + L_b)$, where L_s is equal to the length of straight pipe and L_b is equal to the length of the bend. Since the bends result in a greater pressure drop than an equal length of straight pipe, the friction factor must be modified accordingly. This may be done by defining a modified friction factor f for the Fanning equation [Equation (6-71)] as follows:

$$f = \frac{Bf_0}{N_{Re}^{0.2}} \tag{6-74}$$

where
$$B = \frac{L_{ep}}{L_s + L_b}$$

f_0 = coefficient for a straight pipe for which $B = 1$

L_{ep} = equivalent length of pipe for pressure drop

$N_{Re} = DG/\mu$, Reynolds number

For the case of direct-fired tubes, the heat flux in Btu per hour per square foot of heating surface is generally known, rather than the heat transfer rate per foot of tube. To distinguish between these two rates of heat transfer, a script $\mathcal{q}$ will be used to represent the heat flux.*

Suppose that the heat-input rate $\mathcal{q}_a$ (Btu/h ft^2 of inside area) is constant over the heated length L_h. The actual input rate $\mathcal{q}$ over the entire length of pipe may be computed as follows:

$$\mathcal{q} = \mathcal{q}_a \left(\frac{L_h}{L_s + L_b} \right) \tag{6-75}$$

It is customary to take
$$L_b = 4D \tag{6-76}$$

and
$$L_{ep} = L_s + 60D \tag{6-77}$$

where D denotes the inside diameter of the pipe and where consistent units are employed. Therefore,
$$B = \frac{L_s + 60D}{L_s + 4D}$$

---

*This script $\mathcal{q}$ (Btu/h ft^2 of surface area) differs from the q (Btu/h ft of length) used in other sections of this book. The two rates are related as follows:

$$\int_{z_j}^{z_j + \Delta z} q \, dz = \int_{z_j}^{z_j + \Delta z} \mathcal{q}a \, dz = \int_{z_j}^{z_j + \Delta z} \mathcal{q} \left(\frac{4S}{D} \right) dz = \int_{V_j}^{V_j + \Delta V} \left(\frac{4\mathcal{q}}{D} \right) dV \quad \text{or} \quad q = \frac{4S\mathcal{q}}{D}$$

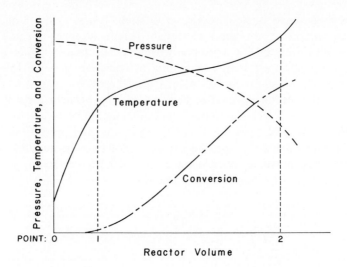

Figure 6-7. Pressure, temperature, and conversion profiles in a fired reactor with a constant heat input rate. [Taken from P. G. Murdoch and C. D. Holland, *Petroleum Refiner*, 33:159 (1954). Used with permission of the Gulf Publishing Company.]

Let an endothermic reaction occur in the reactor and suppose, first, that q is constant throughout. Then the variation, along the tube, of temperature, pressure, and conversion will be qualitatively as indicated in Figure 6-7. The feed enters at a relatively low temperature such that the reaction rate is very small. In the first portion of the reactor the feed is preheated to a temperature high enough to cause the reaction to occur at a small, but measurable rate. This is normally referred to as the *preheat section*; it is bounded by points 0 and 1 of Figure 6-7.

As the reaction commences, the rate of temperature rise diminishes because some of the heat input is utilized as heat of reaction, but unless the heat of reaction is very large, the temperature rise continues until a predetermined outlet value is reached, at point 2. The portion of the reactor between points 1 and 2 will be called the *reaction section*. The reactor may be ended at any desired length, so point 2 may in a specific case be either nearer to or farther from point 1 than indicated by the figure.

For the case where a single reaction occurs in the gas phase, three equations in the three unknowns P, T, and x are required to describe the reacting system. The pressure at any point along the reactor may be computed by replacing f in the Fanning equation [Equation (6-71)] by its equivalent given by Equation (6-74) to give

$$-\frac{dP}{dz} = \frac{2Bf_0 G^2}{DN_{Re}^{0.2} g_c \rho} \tag{6-78}$$

By use of the definition of Reynolds number and the gas law ($Pv = zRT$), it is possible to restate this equation in the following form:

$$-P\frac{dP}{dV} = \beta T\frac{n_T}{n_T^o} \tag{6-79}$$

where P = absolute pressure, $lb/in.^2$

$\quad\quad T$ = temperature of the reacting mixture, °R

$\quad\quad V$ = internal volume of the reactor tube; the internal volume of the straight lengths plus the return bends, ft^3

$\quad\quad \dfrac{n_T}{n_T^o} = 1 + \delta y_A^o x$ for the single reaction

$\quad\quad y_A^o$ = mole fraction of the base component A in the feed; $y_A^o = n_A^o/n_T^o$

$\quad\quad x = (n_A^o - n_A)/n_A^o$

$\quad\quad \beta = \left(\dfrac{41.2}{M_w}\right)\left[\dfrac{z\,Bf_0}{(6D)^{6.8}}\right]\left[\dfrac{w}{1000}\right]^{1.8}\mu^{0.2}$, $(psi)^2/(ft^3)(°R)$

$\quad\quad \delta$ = change in the number of moles per mole of A reacted [for the reaction $aA + bB \rightleftarrows cC + dD$, $\delta = (c + d - a - b)/a$]

$\quad\quad M_w$ = molecular weight of the feed

$\quad\quad z$ = compressibility factor for the reacting mixture

$\quad\quad D$ = internal diameter of a reactor tube, ft

$\quad\quad w$ = feed rate to the reactor, $w = n_T^o M_w$, lb/h

$\quad\quad \mu$ = viscosity, centipoise

For the case where the reaction is first order with respect to A,

$$r_A = kp_A \tag{6-80}$$

Elimination of r_A from the material balance (6-66) by use of Equation (6-80), followed by the restatement of the expression so obtained in terms of pressure and conversion, gives

$$n_T^o\frac{dx}{dV} = \frac{k(1-x)P}{(1+\delta y_A^o x)} \tag{6-81}$$

The enthalpy balance on the reactants [Equation (6-67)] may be restated in the following form:

$$\frac{dT}{dV} = m - s\frac{dx}{dV} \tag{6-82}$$

where

$$m = \frac{4q}{DwC_p}$$

$$s = \frac{y_A^0 \Delta H_r}{M_w C_p}$$

To design the reactor, it is necessary to solve Equations (6-79), (6-81), and (6-82) simultaneously. When the equations are integrated with respect to volume by use of the trapezoidal rule (where needed) and the expressions so obtained are restated in functional form, the following results are obtained:

$$F_1(P, T, x) = \frac{\Delta V}{2}\left[\beta T\left(1 + \delta y_A^0 x\right) + \beta T_j\left(1 + \delta y_A^0 x_j\right)\right] + \frac{1}{2}\left[P^2 - P_j^2\right]$$

(6-83)

$$F_2(P, T, x) = \frac{\Delta V}{2}\left[\frac{k(1-x)P}{1 + \delta y_A^0 x} + \frac{k_j(1-x_j)P_j}{1 + \delta y_A^0 x_j}\right] - n_T^0(x - x_j) \quad (6-84)$$

$$F_3(P, T, x) = m\,\Delta V - s(x - x_j) - (T - T_j) \tag{6-85}$$

For each element of reactor tube, Equations (6-83) through (6-85) may be solved simultaneously for values of P, T, and x at the end of the increment by use of the Newton–Raphson method in a manner analogous to that demonstrated in Example 6-2. This procedure is repeated for each increment until the desired conversion has been obtained.

In computer programs of the Newton–Raphson method, the functions and all variables are customarily normalized. Although matrix methods exist for performing this operation [11], the following algebraic procedure is generally satisfactory for relatively small problems. The variables of temperature and pressure may be normalized by stating them in terms of dimensionless variables T/T_b and P/P_b, where T_b and P_b are arbitrarily selected base values. After the variables have been restated in dimensionless form, the functions (or the original Taylor series expansions) should be restated such that the individual terms have values lying within the interval 0 to 1 or as close as possible to it. For example, instead of using the function F_2 as formulated previously, a new function F_2/n_T^0 would be more suitable. In the interest of keeping all the actual variables in view, the functions and variables in Equations (6-83) through (6-85) are used in the solution of the following illustrative example.

EXAMPLE 6-3

Many cracking reactions such as those shown in Example 2-5 may be represented by a single reaction of the form

$$A \rightarrow 2B$$

Suppose that such a reaction is to be carried out in a fired tubular reactor which may be represented by the plug-flow model. The rate of reaction is first order with respect to A,

$$r_A = kp_A$$

The feed contains 95% of reactant A on a molar basis ($y_A^0 = 0.95$). The charge rate w is 8000 lb/h per tube. The inlet temperature of the reacting stream is 600°F. The inlet pressure is 100 lb/in.^{2} (absolute). For the preheat section, the heat input rate is 6000 Btu/ft^2 h; for the reactor section, the heat input rate is 11,000 Btu/ft^2 h. The internal diameter of the tube is 0.28 ft. (The preheat section is defined for this problem as that reactor volume required to achieve a 1% conversion of the base reactant A.)

Determine the reactor volume required to convert 94.3% of A. Also, determine the outlet temperature and pressure of the reacting mixture. The physical properties are as follows:

$C_p = 0.75$ Btu/lb °R, heat capacity of the reacting mixture

$k = 4 \times 10^{11} e^{-46,500/T}$ $\left[\text{lb mol}/(\text{h})(\text{ft}^3)(\text{lb/in.}^2) \right]$, where T is in °R

$M_w = 60$, molecular weight of the reacting mixture

$\Delta H_r = 36,000$ Btu/lb mol of A reacted

$\mu = 0.02$ centipoise, viscosity of the reacting mixture

$z Bf_0 = 0.072$

SOLUTION

Since the problem may be solved by use of the same numerical procedure used to solve Example 6-2, an abbreviated presentation of the trial procedure for the first increment of reactor volume is given. For the preheat section,

$$m = \frac{4q}{DwC_p} = \frac{(4)(6,000)}{(0.28)(8000)(0.75)} = 14.286°R$$

For the reaction section,

$$m = \frac{11}{6}(14.286) = 26.191°R$$

and

$$s = \frac{y_A^0 \Delta H_r}{M_w C_p} = \frac{(0.95)(36,000)}{(60)(0.75)} = 760°R$$

$$\beta = \left(\frac{41.2}{M_w} \right) \left[\frac{z Bf_0}{(6D)^{6.8}} \right] \left[\frac{w}{1000} \right]^{1.8} \mu^{0.2} = 0.02804$$

$$\delta = \frac{2-1}{1} = 1$$

Take the first increment of volume $\Delta V = 2.0$ ft^3. For the first trial for this increment of volume, assume

$$P_1 = 99.1, \qquad T_1 = 1102, \quad \text{and} \quad x_1 = 2 \times 10^{-7}$$

Then the functions F_1, F_2, and F_3 have the values

$$F_1(P, T, x) = -28.97$$
$$F_2(P, T, x) = 8.95 \times 10^{-6}$$
$$F_3(P, T, x) = -13.43$$

The partial derivatives needed in the Newton–Raphson method are evaluated at P_1, T_1, and x_1 as follows:

$$\frac{\partial F_1}{\partial P} = P = 99.1$$

$$\frac{\partial F_1}{\partial T} = \frac{\Delta V}{2}\left[\beta\left(1 + \delta y_A^0 x\right)\right] = 0.02804$$

$$\frac{\partial F_1}{\partial x} = \frac{\Delta V}{2}\left[\beta T \delta y_A^0\right] = 29.355$$

$$\frac{\partial F_2}{\partial P} = \frac{\Delta V}{2}\left[\frac{k(1 - x)}{1 + \delta y_A^0 x}\right] = 1.89 \times 10^{-7}$$

$$\frac{\partial F_2}{\partial T} = \frac{\Delta V}{2}\left[\frac{(1 - x)P}{1 + \delta y_A^0 x}\right]\frac{\partial k}{\partial T}, \qquad \text{where} \qquad \frac{\partial k}{\partial T} = \frac{46{,}500}{T^2}k$$

$$= 7.17 \times 10^{-7}$$

$$\frac{\partial F_2}{\partial x} = \frac{\Delta V}{2}\left[\frac{-k(1 - x)P \delta y_A^0}{\left[1 + \delta y_A^0 x\right]^2} - \frac{kP}{\left[1 + \delta y_A^0 x\right]}\right] - n_T^0 = -133.3$$

$$\frac{\partial F_3}{\partial P} = 0$$

$$\frac{\partial F_3}{\partial T} = 1$$

$$\frac{\partial F_3}{\partial x} = -760$$

The Newton–Raphson equations for the first trial reduce to

$$0 = -28.97 + 99.1\,\Delta P + 0.028\,\Delta T + 29.4\,\Delta x$$

$$0 = -8.95 \times 10^{-6} + 1.89 \times 10^{-7}\Delta P + 7.17 \times 10^{-7}\Delta T - 133.3\,\Delta x$$

$$0 = -13.43 + 0.0\,\Delta P - 1.0\,\Delta T - 760\,\Delta x$$

Solution of these equations for ΔP, ΔT, and Δx gives

$$\Delta P = 0.296$$
$$\Delta T = -13.43$$
$$\Delta x \cong 0$$

The trial values for the second trial for the first increment are then computed as follows:

$$P_2 = P_1 + \Delta P = 99.40$$
$$T_2 = T_1 + \Delta T = 1088.6$$
$$x_2 = x_1 + \Delta x = 2 \times 10^{-7}$$

After the solution set of values for P, T, and x at V_1 has been found, the same procedure is repeated for the second and all subsequent elements. The conversion, pressure, and temperature profiles obtained on the basis of the specified values of the tube diameter (0.28 ft) and the heat input rates are tabulated in Table 6-7.

TABLE 6-7

CONVERSION, TEMPERATURE, AND PRESSURE PROFILES FOR THE FIRED
TUBULAR REACTOR DESIGNED IN EXAMPLE 6-3

Volume (ft^3)	Conversion	Temperature $(°R)$	Pressure $(psia)$	Heat Flux $(Btu / h\,ft^2)$
2.000	0.0000001	1089.998	99.365	6000.00
4.000	0.0000004	1118.567	87.740	6000.00
6.000	0.0000013	1147.136	98.094	6000.00
8.000	0.0000038	1175.705	97.428	6000.00
10.000	0.0000103	1204.269	96.740	6000.00
12.000	0.0000261	1232.826	96.031	6000.00
14.000	0.0000638	1261.366	95.300	6000.00
16.000	0.0001497	1289.871	94.546	6000.00
18.000	0.0003381	1318.297	93.769	6000.00
20.000	0.0007359	1346.564	92.968	6000.00
22.000	0.0015447	1374.519	92.143	6000.00
24.000	0.0031224	1401.889	91.292	6000.00
26.000	0.0060570	1428.227	90.415	6000.00
28.000	0.0112678	1457.598	89.510	11000.00
30.000	0.0232743	1500.852	88.563	11000.00
32.000	0.0483613	1534.164	87.564	11000.00
34.000	0.0881695	1556.288	86.504	11000.00
36.000	0.1385463	1570.380	85.374	11000.00
38.000	0.1942797	1580.402	84.167	11000.00
40.000	0.2521445	1588.803	82.878	11000.00
42.000	0.3106262	1596.735	81.501	11000.00
44.000	0.3690593	1604.703	80.030	11000.00
46.000	0.4271251	1612.952	78.461	11000.00
48.000	0.4846281	1621.627	76.787	11000.00
50.000	0.5414043	1630.855	74.999	11000.00
52.000	0.5972794	1640.769	73.090	11000.00
54.000	0.6520435	1651.526	71.049	11000.00
56.000	0.7054300	1663.330	68.864	11000.00
58.000	0.7570879	1676.449	66.519	11000.00
60.000	0.8065413	1691.242	63.997	11000.00
62.000	0.8531300	1708.213	61.296	11000.00
64.000	0.8959184	1728.072	58.326	11000.00
66.000	0.9335703	1751.834	55.111	11000.00
66.800	0.9451888	1761.336	53.913	11000.00

The problem stated as Example 6-3 is easier to solve than is the problem which commonly confronts the designer in which the diameter and the heat input rates to the preheat and reaction sections are to be picked for a specified conversion such that the outlet temperature and the pressure drop are equal to or less than the design specifications [9, 10]. One way to solve a problem of this type is as follows. Values of D and q are assumed and the volume required to achieve the specified conversion is found by the numerical procedure illustrated in the example. If the calculated values of the outlet temperature and pressure do not satisfy the specifications on the outlet temperature and pressure, another set of values for q and D are assumed and the corresponding outlet temperature and pressure at the required conversion are computed. Picking the tube diameter is simplified by the fact that the pressure drop is inversely proportional to the tube diameter raised to the 6.8 power. Thus a slight change in D has a significant effect on ΔP. Obviously, the picking of q and D may be formulated as a problem in optimization [12]. Alternately, the approximate method of Murdoch and Holland [9, 10] may be used.

PROBLEMS

6-1. For a plug-flow reactor at unsteady-state operation wherein c components are involved in chemical reactions, the corresponding energy balance is given by Equation (6-42) and the first expression in Table 6-1. Show that when the conditions stated in this table [(2) *Steady State: Nonideal Solutions*; (3) *Steady State: Perfect Gas Mixture*] are imposed on Equation (6-42), the resulting energy balances are those given in the table.

6-2. For an adiabatic reactor, $q = 0$ for all z. Then for a plug-flow reactor the enthalpy balance enclosing the entrance and any position z is given by

$$n_T H|_{z=0} - n_T H|_z = 0$$

For the case where a single reaction occurs in the perfect gas phase, show that the conversion x and temperature T at the position z are related as follows:

$$x = -\frac{\int_{T_0}^{T} (C_p^\circ)_{\text{feed}} \, dT}{y_A^0 \, \Delta H_r^\circ}$$

where T_0 is the temperature of the feed, y_A^0 is the mole fraction of A in the feed, and the ΔH_r° is the heat of reaction at the temperature T per mole of A reacted.

6-3. Suppose that the following reactions occur in the gas phase in a plug-flow reactor:

$$A \rightarrow 2C \tag{1}$$

$$2A \rightarrow D \tag{2}$$

Let n_1 denote the moles of A per unit time that have reacted at a given z by the first reaction, and let n_2 denote the moles of A per unit time that have reacted by the second reaction at a given z. Let r_{A1} denote the rate of disappearance of A by the first reaction and r_{A2} the rate of disappearance of A by the second reaction. Then show that

$$r_{A1} = -\tfrac{1}{2} r_C, \qquad r_{A1} + r_{A2} = r_A$$

$$r_{A2} = -2 r_D$$

where r_C and r_D are the rates of disappearance of C and D. Then show that Equation (6-42) may be restated as follows for steady-state operation and a perfect gas mixture:

$$-n_T C_p^\circ \frac{dT}{dz} - S[r_{A1} \Delta H_{r1}^\circ + r_{A2} \Delta H_{r2}^\circ] + q = 0$$

where $\qquad \Delta H_{r1}^\circ = 2H_{fC}^\circ - H_{fA}^\circ, \qquad \Delta H_{r2}^\circ = \tfrac{1}{2} H_{fD}^\circ - H_{fA}^\circ$

6-4. For a flow reactor with partial axial mixing in which the reacting mixture forms a nonideal solution, show that when the enthalpies in Equation (6-54) are evaluated by use of the virtual values of the partial molar enthalpies and the set $\{\partial C_i/\partial t\}$ is eliminated from the expression so obtained by use of Equation (6-53), the expression in Table 6-2 for the energy balance is obtained.

6-5. Obtain the partial differential equation given by Equation (6-17) by use of the mean-value theorems and the limiting processes as demonstrated in Chapter 1.

6-6. The *heat-source method* for making enthalpy balances is based on the separation of the enthalpy of the mixture into a sensible heat term and a heat of reaction term. In this case, let H represent the sensible heat per unit mass of the reacting mixture at any z in the reactor. Let Q_r° denote the heat evolved at temperature T per mole of A reacted; that is, let

$$Q_r^\circ = -\Delta H_r^\circ \qquad\qquad (A)$$

(a) Show that the energy balance on the element of volume $S(z_{j+1} - z_j)$ of the tubular reactor shown in Figure 6-1 is given by

$$wH|_{z_j} - wH|_{z_{j+1}} + \int_{z_j}^{z_{j+1}} r_A Q_r^\circ S \, dz + \int_{z_j}^{z_{j+1}} q \, dz = 0 \qquad (B)$$

where r_A is defined in terms of the disappearance of A.

(b) Although the sensible heat of the reacting mixture depends on both conversion and temperature, the separation of the sensible heat effects from the heat of reaction effects makes it necessary to define the derivative of the sensible heat with respect to z as follows:

$$\frac{dH}{dz} = C_p^\circ \frac{dT}{dz} \qquad\qquad (C)$$

where C_p° is the heat capacity of the mixture on a mass basis. Show that after the mean-value theorems have been applied to Equation (B) as demonstrated in Chapter 2 and after the limit has been taken as Δz goes to zero the following result is obtained:

$$-wC_p^\circ \frac{dT}{dz} + r_A Q_r^\circ S + q = 0 \qquad (D)$$

(c) Since

$$w\left[\frac{mass}{time}\right]C_p^\circ\left[\frac{energy}{(mass)(degree\ of\ temp)}\right]$$

$$= n_T\left[\frac{moles}{time}\right]C_p^\circ\left[\frac{energy}{(moles)(degree\ of\ temp)}\right]$$

and in view of the definition of Q_r° given by Equation (A), show that Equation (D) may be restated in the form given by Equation (6-44) for a perfect gas. [The definition given by Equation (C) can lead to confusion in the treatment of nonideal solutions, where the enthalpies of the components in the reacting mixture depend on composition.]

6-7. For the case where the reactions are carried out in the liquid phase in a plug-flow reactor, use Equation (6-22) in the evaluation of $n_T H$ in the energy balance given by Equation (6-19) to show that

$$-n_T c_p^\circ \frac{dT}{dz} + S\sum_{i=1}^{c} h_{fi}^\circ r_i + q = 0$$

where $h_{fi}^\circ = H_{fi}^\circ - \lambda_i$, enthalpy of the liquid at the temperature T

$$c_p^\circ = \sum_{i=1}^{c}\frac{n_i}{n_T}\frac{\partial h_{fi}^\circ}{\partial T} = \sum_{i=1}^{c}\frac{n_i}{n_T}\left(\frac{\partial H_{fi}^\circ}{\partial T} - \frac{\partial \lambda_i}{\partial T}\right)$$

6-8. Repeat Problem 6-7 for the case where the single reaction given by Equation (6-43) occurs in the liquid phase, and show that the corresponding energy balance may be stated in the following form:

$$-n_T c_p^\circ \frac{dT}{dz} - r_A(\Delta H_r^\circ)_{liq} S + q = 0$$

where $(\Delta H_r^\circ)_{liq} = \frac{c}{a}h_{fC}^\circ + \frac{d}{a}h_{fD}^\circ - h_{fA}^\circ - \frac{b}{a}h_{fB}^\circ$

6-9. In Example 6-1, it was shown that for the case of an exothermic, first-order reaction, the reaction stream could be maintained at constant temperature in a heat-exchanger reactor by passing an appropriate cooling medium through the exchanger in the cocurrent direction.

(a) For the first-order irreversible reaction, show that isothermal operation is impossible if countercurrent flow is used in the heat-exchanger reactor, where $\Delta H_r \neq 0$.

(b) For second- and higher-order irreversible reactions, show that isothermal operation is impossible in a heat-exchanger reactor, where $\Delta H_r \neq 0$.

6-10. (a) Show that the energy balance for a flow reactor with both partial radial and partial axial mixing over a time period Δt and distances Δz and Δr reduces to Equation (6-64), upon application of the mean-value theorems and the limiting process wherein Δt, Δr, and Δz are allowed to approach zero.

(b) Show that when the enthalpies appearing in Equation (6-64) are evaluated by use of the virtual values of the partial molar enthalpies ($\hat{H}_i = H_{fi}^\circ + \Omega$), the resulting expression for the energy balance given by Equation (6-65) and in Table 6-4 is obtained.

6-11. (a) Show that for the general case of three-dimensional flow accompanied by partial mixing in each of the directions the component material balance expression given by Equation (6-60) is obtained in the limit as Δx, Δy, Δz, and Δt are allowed to go to zero for a balance on the element of volume $\Delta x \Delta y \Delta z$ over the time period Δt. (*Hint:* Apply the mean-value theorems to the component material balance by beginning with the intermost differences of the integrands.)

(b) Repeat part (a) for an energy balance and obtain the expression given by Equation (6-61).

(c) Show that, upon evaluation of the derivatives in Equation (6-61) and making use of the component material balance [Equation (6-60)], the resulting expression given by Equation (6-63) and in Table 6-3 for the energy balance is obtained.

6-12. Vanscoy [13] modeled a commercial ethane-cracking furnace which had 4-in. (internal diameter) tubes. The feed consisted of 5304 lb/h of ethane and 1500 lb/h of steam. The inlet temperature and pressure of the feed was 1250°F and 75 psia. Plug flow throughout the cracking furnace was assumed. On the basis of the following reaction mechanism and the data listed, calculate the conversion, temperature, and pressure profiles for this reactor, as well as the selectivity for the production of ethylene, where,

$$\text{Selectivity} = \frac{\text{moles } C_2H_4 \text{ produced}}{\text{moles } C_2H_6 \text{ reacted}}$$

The proposed reaction mechanism follows:

$$C_2H_6 \underset{k_2}{\overset{k_1}{\rightleftharpoons}} 2CH_3^\bullet \tag{1}$$

$$CH_3^\bullet + C_2H_6 \underset{k_4}{\overset{k_3}{\rightleftharpoons}} CH_4 + C_2H_5^\bullet \tag{2}$$

$$C_2H_5^\bullet \underset{k_6}{\overset{k_5}{\rightleftharpoons}} C_2H_4 + H^\bullet \tag{3}$$

$$C_2H_6 + H^\bullet \overset{k_7}{\longrightarrow} H_2 + C_2H_5^\bullet \tag{4}$$

$$2C_2H_5^\bullet \overset{k_8}{\longrightarrow} n\text{-}C_4H_{10} \tag{5}$$

$$2C_2H_5^\bullet \overset{k_9}{\longrightarrow} C_2H_4 + C_2H_6 \tag{6}$$

$$CH_3^\bullet + C_2H_5^\bullet \overset{k_{10}}{\longrightarrow} C_3H_8 \tag{7}$$

$$C_2H_5^\bullet + C_2H_4 \overset{k_{11}}{\longrightarrow} C_3H_6 + CH_3^\bullet \tag{8}$$

Expressions for the variations of the rate constants with temperature follow:

$$k_i = A_i e^{-E_i/RT}, \quad T \text{ in K}$$

k_i	A_i	E_i, kcal/mol
k_1	2.81×10^{17}	93
k_2	1.34×10^{13}	0
k_3	2.5×10^{11}	10
k_4	1×10^{12}	16
k_5	3.98×10^{13}	38
k_6	5.4×10^{13}	5.4
k_7	3.78×10^{12}	6.0
k_8	2.0×10^{13}	0
k_9	5.0×10^{12}	0
k_{10}	7×10^{13}	0
k_{11}	3.16×10^{12}	18

For the first-order reactions, k_i and A_i have the units of reciprocal seconds or s^{-1}, and for second-order reactions k_i and A_i have the units of $cm^3/(s\text{-g mol})$.

The following table gives the perfect gas heat capacities and heats of formation. To simplify this problem, take the heat capacity of each component to be a constant.

$$\Delta H_f \text{ at } 1520°F$$

Component	Btu/lb mol	Btu/°F C_p lb mol
Hydrogen	0	7.62
Carbon	0	5.42
Methane	−38,867	18.10
Ethylene	16,206	23.54
Ethane	−45,795	18.68
Propylene	−323	36.02
Propane	−55,844	43.74
n-Butane	−67,233	56.62
steam	—	9.10

The following expression for the viscosity as a function of temperature and conversion may be used:

$$\mu = (3.1922 - 0.3417\chi)10^{-4}T^{2/3}$$

where μ is in centipoise, T is in Kelvin, and

χ = conversion of ethane, moles of ethane reacted per mole of ethane fed

The heat flux on the exterior of the tubes may be represented by the following equations:

$$q = 21,000 - 130z + 0.38z^2 \quad 0 \le z \le 160 \text{ ft}$$

and

$$q = 9928, \quad 160 \le z \le 480 \text{ ft}$$

where $z = 0$ at the inlet and 480 ft at the exit of the reactor and q has the units of Btu/(h ft^2).

Hints and Assumptions:

1. Use the pseudo-steady state assumption for the free radicals. (Otherwise, the differential equations are stiff and an excessive amount of computing time is required.)

2. The rate constants are based on the molecular components and not on the free radicals. For example,

$$(r_{ethane})_{dis.} = r_1 - r_2 + r_3 - r_4 + r_7 - r_9$$

where

$$r_1 = k_1 C_{C_2H_6} = k_1 C_E$$

$$r_2 = k_2 C_{CH_3}^2 = k_2 C_M^2.$$

3. When calculating the concentration of the ethyl radical, assume that

$$k_1 C_E \gg k_2 C_M^2.$$

$$2[(k_8 + k_9)(k_1 C_E)] \gg k_{10} C_M.$$

so that

$$C_{E.} = \left[\frac{k_1 C_E}{k_8 + k_9} \right]^{1/2}$$

These assumptions eliminate the iterative calculations required to determine the concentrations of the free radicals at each point z in the reactor tube.

4. Instead of calculating the heat of reaction for each reaction, use heats of formation and rates for each molecular component and neglect the contributions of the free radicals; that is, $r_i H_{fi}^\circ \cong 0$ for each free radical [see Equation (6-42)].

5. Assume that the reacting fluid can be treated as a perfect gas and that $z B f_0 = 0.072$ in Equation (6-79).

6-13. Utilizing the reaction and viscosity data from Problem 6-12, a feed temperature of 1560°F, a feed velocity of 200 ft/s, a 1-in. I.D. by 20-ft reactor tube, and a heat flux equal to 40,000 Btu/(h ft^2), calculate the ethane conversion, temperature, and pressure profiles. Also calculate the selectivity of ethylene.

6-14. Verify the results presented in Table 6-5 for the plug-flow reactor.

6-15. The reactor in Example 6-2 is to be used for the liquid phase reaction

$$A + B \rightarrow C, \qquad r_A = k_C C_A C_B$$

The initial concentration of A, $C_A^0 = 1.2$ g mol/liter and $C_B^0/C_A^0 = 2$. The charge rate per tube is 10,000 lb/h. The feed, consisting of A, B, and a solvent, enter the reactor at 75°F. Steam at 250°F is available for heating purposes. The physical properties of the system, which we may be assumed to be constant throughout the reactor, are as follows:

$$\mu = 0.40 \text{ cp} = 0.004 \text{ g/(cm s)}$$

$$\rho = 55 \text{ lb/ft}^3$$

$$Mw = 85 \text{ lb/lb mol (of the mixture)}$$

$$U_h = 400 \text{ Btu/(h ft}^2 \text{ °F) based on inside diameter of the tubes}$$

$$C_p = 0.70 \text{ Btu/(lb °F) of reaction mixture}$$

$$\Delta H_r = 35,000 \text{ Btu/lb mol}$$

$$E = 46,000 \text{ Btu/lb mol}$$

$$A = 210 \times 10^{15} \text{ ft}^3/(\text{lb mol h})$$

molar density = 10.37 g mol/liter

(a) What dimensionless groups control this reaction?
(b) Calculate:
1. Temperature profile and conversion profile in the reactor.
2. Reactor volume required for 90% and 99.9% conversion of A.
3. Pressure drop for the reactor.

Runge–Kutta (see Chapter 3), Euler's, or the implicit method illustrated in Example 6-2 may be used to solve the equations describing the reactor.

NOTATION

(See also Chapters 1 through 5)

a = surface area for heat transfer per foot of length of pipe; for a single pipe $a = \pi D$, and for N pipes connected in parallel, $a = N\pi D$

B = factor used in the definition of a modified friction factor for the Fanning equation [see Equation (6-74)]

c = parameter used in the description of heat-exchanger reactors (see Table 6-5)

C_i = concentration of component i, moles per unit volume

$\mathbf{C}$ = vector of concentrations

C_T = total molar density

C_p = molar heat capacity, cal/g mol °C or Btu/lb mol °F

D = inside diameter; also used to denote the dispersion coefficient

D_r, D_z = dispersion coefficients for the r and z directions, cm^2/s

D_x, D_y = dispersion coefficients in the x and y directions, cm^2/s

f = Fanning friction factor [see Equation (6-74)]

F_i = molar flux relative to a fixed coordinate system (g mol)/(s cm^2)

F_T = total flux

g_c = 32.174 (lb$_m$ ft)/(lb$_f$ s^2); the symbol lb$_m$ represents pounds mass and the symbol lb$_f$ represents pounds force

G = mass flux, $lb_m/s\ ft^2$

h_i = enthalpy of pure component i in the liquid state, cal/g mol or Btu/lb mol

h_{fi}° = enthalpy of formation of component i in the liquid state at the temperature T from its elements in their standard states at the datum temperature

H_i = enthalpy of pure component i in the vapor state, cal/g mol or Btu/lb mol

H_{fi}° = enthalpy of pure component i at a given temperature T in the perfect gas state at 1 atm above its elements in their standard states at some arbitrary datum temperature

$\overline{H}_i$ = partial molar enthalpy of component i; partial derivative of the function $n_T H$ with respect to n_i with all other n_i's and P and T held fixed, cal/g mol

$\hat{H}_i$ = virtual value of the partial molar enthalpy

H_T = enthalpy of fluid plus the kinetic and potential energy contributions; defined by Equation (6-3)

H_c = enthalpy of the heat-transfer medium

L_b = length of bend

L_s = length of straight pipe

L_{ep} = equivalent length of pipe for pressure drop

m = a parameter; for heat-exchanger reactors, m is defined in Table 6-5, and for direct-fired reactors, m is defined following Equation (6-82)

M_w = molecular weight

M_c = mass of heat-transfer medium in the jacket or holdup space

n = number of hook-shaped sections connected in a series in a direct-fired reactor; N = number of tubes connected in parallel

n_i = molar flow rate of component i

n_T = total molar flow rate

$\mathbf{n}$ = vector of component molar flow rates

N_{Re} = Reynolds number; $N_{Re} = DG/\mu$

P = absolute pressure, in consistent units

q = rate of heat transfer per unit length in consistent units, Btu/h ft [see Equation (6-7)]

$\mathscr{q}$ = rate of heat transfer per unit of heat-transfer area in consistent units, Btu/(h ft^2) [see Equation (6-75)]

r_i = rate of disappearance of component i by reaction, moles per unit time per unit volume of the reacting mixture

s = a parameter; for heat-exchanger reactors, s is defined in Table 6-5, and for direct-fired reactors s is defined following Equation (6-87)

S = cross-sectional area based on the inside diameter; S_o is the cross-sectional area based on the outside area

S_c = cross-sectional area of the space occupied by the heat-transfer medium

t = time

T = temperature of the reacting mixture

T_c = temperature of the heat-transfer medium; T_{ci} is the inlet, and $(T_c)_{out}$ is the outlet temperature of the medium

U_T = internal energy plus the kinetic and potential energy contributions; defined by Equation (6-2)

U_h = overall heat-transfer coefficient in consistent units, Btu/h ft^2 °F

v = volume per mole

v_T = total volumetric flow rate

w = mass flow rate

w_c = mass flow rate of the heat-transfer medium in consistent units, g/s, lb/h

y_A^0 = mole fraction of A in the feed

z = length of the path followed by the reacting mixture, and the positive direction of z is taken to be the direction of flow of the reacting mixture; if the reacting mixture is perfectly mixed in the reactor, then the positive direction of z is taken to be the direction of flow of the heat-transfer medium

Z = dimensionless parameter used in the description of heat-exchanger reactors (see Table 6-5)

z = compressibility factor (for any actual gas, $z = Pv/RT$)

Greek Letters

δ = change in the number of moles per mole of the base reactant A every time the reaction goes one time

$\lambda_i(T)$ = latent heat of vaporization as defined by Equation (6-23); since the effect of pressure on enthalpy is neglected in the text, $\lambda_i(T)$ can be taken equal to the latent heat of vaporization at the saturation temperature and pressure

μ = viscosity in consistent units, centipoise

ρ = mass density in consistent units, g/cm^3, lb/ft^3

ϕ = dimensionless variable used in the design of heat-exchanger reactors, $\phi = T/T_i$

$\Phi = T_c/T_i$

REFERENCES

1. Bird, R. B., W. E. Stewart, and E. N. Lightfoot, *Transport Phenomena*, John Wiley & Sons, Inc., New York (1962).

2. Perry, R. H., C. H. Chilton, and S. D. Kirkpatrick, editors, *Chemical Engineering Handbook*, 4th ed., McGraw-Hill Book Company, New York (1963).

3. Billingsley, D. S., W. S. McLaughlin, Jr., N. E. Welch, and C. D. Holland, *I & EC* 50:741 (1958).

4. Logan, G. F., Master of Engineering Report, Texas A & M University (Aug. 1970).

5. Carnahan, B., H. A. Luther, and J. O. Wilkes, *Applied Numerical Methods*, John Wiley & Sons, Inc., New York (1969).

6. Ennis, B. P., H. B. Boyd, and R. Orriss, *Chemtech*, 693 (Nov. 1975).

7. Bukens, A. B., and G. F. Froment, *Ind. Eng. Chem. Proc. Des. Develop.* 7:435 (1968).

8. Purnell, J. H., and D. A. Leathard, *Proc. Roy. Soc.*, *A*305:517 (1968).

9. Murdoch, P. G. and C. D. Holland, *Petroleum Refiner*, 33:159 (1954).

10. Murdoch, P. G. and C. D. Holland, *Petroleum Refiner*, 33:163 (1954).

11. Holland, C. D., *Fundamentals and Modeling of Separation Processes, Absorptions, Distillation, Evaporation, and Extraction*, Prentice-Hall, Inc., Englewoods Cliffs, N.J. (1975).

12. Wilde, D. J., *Optimum Seeking Methods*, Prentice-Hall, Inc., Englewood Cliffs, N.J. (1964).

13. Vanscoy, D. A., "A Computer Model for Ethane Pyrolysis," Master of Engineering Report, Texas A & M University, 1978.

Batch Reactors and Perfectly Mixed Flow Reactors

7

Historically, the chemical industry began with batch reactors, which are still widely used to carry out many reactions such as those involving the production of specialty chemicals. The sketch of a typical batch reactor is shown in Figure 7-1. A variation of batch reactors, called *perfectly mixed flow* reactors or *continuously stirred tank* reactors, could be regarded as a batch reactor which has been converted into a continuous reactor. Also, a perfectly mixed flow reactor could be regarded as a variation of a plug-flow reactor in which the mixing becomes perfect in both the radial and axial directions. The sketch of a typical perfectly mixed flow reactor is shown in Figure 7-2.

The design of batch and perfectly mixed flow reactors is considered in Parts I and II, respectively. In Part III, the general behavior of selected design equations is demonstrated through the use of the reaction phase plane.

I. BATCH REACTORS

This treatment is divided into two parts. In the first part, the general energy-balance equations needed in the design of batch reactors are developed, and in the second part the design of batch reactors is demonstrated by use of several numerical examples.

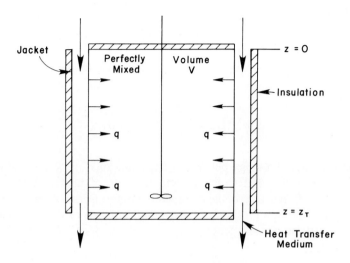

Figure 7-1. A typical batch reactor.

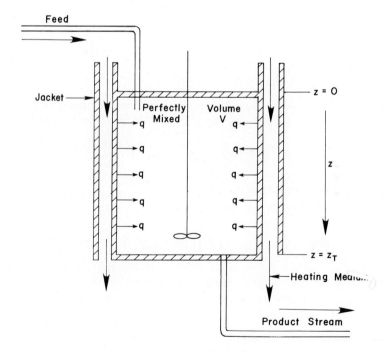

Figure 7-2. Sketch of a perfectly mixed flow reactor.

A. Development of Energy-Balance Equations for Batch Reactors

A typical batch reactor is shown in Figure 7-1. Generally, the kinetic energy effects as well as the work done by the stirrer on the reacting mixture are negligible relative to the heat effects associated with the reaction. Thus the energy balance on the reacting mixture of volume V over the time period from t_n to t_{n+1} is given by

$$\int_{t_n}^{t_{n+1}} Q\, dt = (C_T V U)|_{t_{n+1}} - (C_T V U)|_{t_n} \tag{7-1}$$

where U is the internal energy per mole of reacting mixture, C_T is the molar density of the mixture, and Q is equal to the rate of heat transfer to the reacting mixture. If the rate of heat transfer differs at each point along the boundary, Q in Equation (7-1) may be expressed in terms of the point value q (energy per unit time per unit length) as

$$Q = \int_0^{z_T} q\, dz \tag{7-2}$$

Since enthalpies are more frequently tabulated than are internal energies, it is desirable to restate $C_T V U$ in terms of the enthalpy as

$$C_T V U = C_T V \left(H - \frac{1}{j} P v \right) = C_T V H - \frac{1}{j} P V \tag{7-3}$$

where j is the mechanical equivalent of heat, defined in Table 6-1. Since the molar density $C_T = N_T/V = 1/v$ (where N_T is the total number of moles in the reacting mixture at any time t), Equation (7-3) may be restated as follows:

$$C_T V U = N_T H - \frac{1}{j} P V \tag{7-4}$$

Thus, when $C_T V U$ in Equation (7-1) is replaced by its equivalent as given by Equation (7-4), one obtains

$$\int_{t_n}^{t_{n+1}} Q\, dt = \left(N_T H - \frac{1}{j} P V \right)\Bigg|_{t_{n+1}} - \left(N_T H - \frac{1}{j} P V \right)\Bigg|_{t_n} \tag{7-5}$$

This expression for the energy balance may be used to obtain the following differential equation by use of the mean-value theorems and limiting processes as demonstrated in Chapter 1.

$$Q = \frac{d(N_T H)}{dt} - \frac{1}{j} \frac{d(P V)}{dt} \tag{7-6}$$

When Equation (7-6) is restated in terms of the virtual values of the partial molar enthalpies ($\hat{H}_i = H_{fi}^\circ + \Omega$), one obtains

$$Q = \frac{d\left(\sum_{i=1}^c N_i \hat{H}_i\right)}{dt} - \frac{1}{j}\frac{d(PV)}{dt} \tag{7-7}$$

and thus

$$Q = \sum_{i=1}^c N_i \frac{d\hat{H}_i}{dt} + \sum_{i=1}^c \hat{H}_i \frac{dN_i}{dt} - \frac{1}{j}\frac{d(PV)}{dt} \tag{7-8}$$

The first summation on the right side of Equation (7-8) is evaluated in a manner analogous to that demonstrated in Chapter 6, and the set $\{dN_i/dt\}$ in the second summation is eliminated by use of the component material balance

$$r_i = -\frac{1}{V}\frac{dN_i}{dt} \tag{7-9}$$

After these operations have been performed, one obtains

$$N_T C_p \frac{dT}{dt} + N_T \frac{\partial \Omega}{\partial P}\frac{dP}{dt} - \frac{1}{j}\frac{d(PV)}{dt} - V\sum_{i=1}^c \overline{H}_i r_i - Q = 0 \tag{7-10}$$

where, as shown in Chapter 6,

$$\overline{H}_i = H_{fi}^\circ + \frac{\partial(N_T \Omega)}{\partial N_i}$$

$$N_T C_p = \sum_{i=1}^c N_i\left(C_{pi}^\circ + \frac{\partial \Omega}{\partial T}\right)$$

For the case where the single reaction ($aA + bB = cC + dD$) occurs, Equation (7-10) reduces to

$$N_T C_p \frac{dT}{dt} + N_T \frac{\partial \Omega}{\partial P}\frac{dP}{dt} - \frac{1}{j}\frac{d(PV)}{dt} + \Delta H_r V r_A - Q = 0 \tag{7-11}$$

where the expression for ΔH_r in Equation (7-11) may be obtained by replacing the n_i's and n_T in the expression preceding Equation (6-44) by the N_i's and N_T.

If the change of pressure and volume with time is negligible, Equation (7-11) reduces to

$$N_T C_p \frac{dT}{dt} + \Delta H_r V r_A - Q = 0 \tag{7-12}$$

It should be noted that the energy balances could have been formulated in terms of the internal energy functions rather than the enthalpy functions. The decision to use the enthalpy functions was made because of the desirability of stating the energy balances for all types of reactors in terms of the same type of energy functions and because thermodynamic data are, perhaps, more commonly stated in terms of enthalpy, rather than internal energy functions.

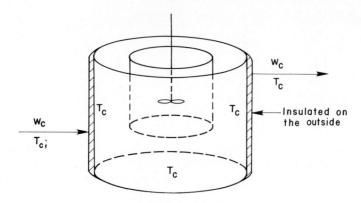

Figure 7-3. The heat-transfer medium is perfectly mixed.

B. Design of Batch Reactors

To maintain the reacting mixture at the desired temperature throughout the course of the reaction cycle, it is necessary to provide for an appropriate amount of heat transfer either to or from the reacting mixture. The design of batch reactors in which various mechanical arrangements are used for the transfer of heat from (or to) the reacting mixture is shown by the solution of several illustrative examples. To demonstrate that a given mechanical and geometrical arrangement provides the desired amount of heat transfer, it is necessary to solve the material and energy balances for each time t of the reaction cycle.

To reduce the calculations involved in the design of reactors, only one reaction is assumed to occur in the illustrative examples which follow.

For the case where the heat-transfer medium is best described as being perfectly mixed in the space provided for it (for example, in the jacket shown in Figure 7-3), then

$$Q = U_h A (T_c - T) \tag{7-13}$$

where A is the total area of the heating surface in contact with the reacting mixture. If the heat-transfer medium is passed through any number of tubes in parallel which are in contact with the reacting mixture (see Figures 7-4 and 7-5), then

$$Q = \int_0^L q \, dz = \int_0^L U_h a (T_c - T) \, dz \tag{7-14}$$

where a is equal to the total surface of heat-transfer area per foot of length; that is, $a = N\pi D$, where N is the number of tubes of diameter D which are connected in parallel. When the reactants are perfectly mixed, the length z is measured in the direction of flow of the heat-transfer medium. The total length of the tubes or coils exposed to the reacting mixture is denoted by L. The assumption that the heat-transfer medium is in plug flow is generally regarded

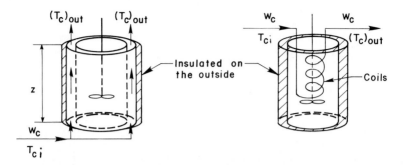

Figure 7-4. Heat-transfer medium passes through the jacket.

Figure 7-5. Heat-transfer medium passes through the coils.

as being valid, provided that the Reynolds number for the fluid is equal to or greater than 10^4.

An abbreviated development of the energy-balance equations for the heat-transfer medium is presented next for each of two cases: (1) it is supposed that the heat-transfer medium is perfectly mixed, and (2) it is supposed that the heat-transfer medium is in plug flow.

Consider first the case where the heat-transfer medium is perfectly mixed as indicated in Figure 7-1. The energy balance on the perfectly mixed heat-transfer medium of mass $\mathcal{M}_c$ which is contained in the jacket space over the time period from time t_n to $t_n + \Delta t$ is given by

$$\int_{t_n}^{t_n + \Delta t} \left[w_c (H_{ci} - H_c) - Q \right] dt = \mathcal{M}_c H_c |_{t_n + \Delta t} - \mathcal{M}_c H_c |_{t_n} \qquad (7\text{-}15)$$

where H_{ci} = enthalpy of the heat-transfer medium entering the jacket at any time t, Btu/lb

H_c = enthalpy of the heat-transfer medium in the jacket and also leaving the jacket at any time t, Btu/lb

w_c = mass flow rate of the heat-transfer medium as it enters and leaves the jacket, lb/h

Equation (7-15) applies for fluids that form either ideal or nonideal solutions. By use of the mean-value theorems followed by a limiting process as demonstrated in Chapter 1, the integral-difference Equation (7-15) may be reduced to the following differential equation:

Energy balance on heat-transfer medium, perfectly mixed: $\qquad w_c(H_{ci} - H_c) - Q = \mathcal{M}_c C_{pc} \dfrac{dT_c}{dt} \qquad (7\text{-}16)$

The appropriate expression for Q is given by Equation (7-13).

Consider next the case where the heat-transfer medium passes in plug flow through one or more tubes in parallel or one or more coils in parallel as shown in Figure 7-3. The energy balance on the heat-transfer medium contained in all the parallel tubes in the element Δz over the time period from t_n to $t_n + \Delta t$ (see Figure 7-4) is given by

$$\int_{t_n}^{t_n+\Delta t}\left[w_cH_c|_{z_j,t} - w_cH_c|_{z_j+\Delta z,t} - \int_{z_j}^{z_j+\Delta z}q\,dz\right]dt$$

$$= \int_{z_j}^{z_j+\Delta z}\left[S_c\rho_cH_c|_{t_n+\Delta t,z} - S_c\rho_cH_c|_{t_n,z}\right]dz \qquad (7\text{-}17)$$

where w_c is the total mass flow rate of the heat-transfer medium through the parallel tubes, S_c is the total cross-sectional area of all parallel tubes, ρ_c is the mass density of the heat-transfer medium, q is defined by Equation (7-14), and z is positive in the direction of flow of the heat-transfer medium. Again by use of the procedure demonstrated in Chapter 1, Equation (7-17) may be reduced to the following partial differential equation:

Energy balance on heat-transfer
medium, plug flow:
$$-w_cC_{pc}\frac{\partial T_c}{\partial z} - q = S_c\rho_cC_{pc}\frac{\partial T_c}{\partial t} \qquad (7\text{-}18)$$

Equation (7-18) is also based on the assumption that the axial transfer of heat through the flowing stream is negligible. In addition to the preceding equations, the Fanning equation [Equation (6-71)] is needed to calculate pressure drop.

In summary, the design equations required to describe the batch reactors considered herein are given by Equations (7-9), (7-11), (7-13), (7-14), (7-16), and (6-71). The design of batch reactors is demonstrated by the numerical examples which follow.

1. Isothermal Operation: The Heat-Transfer Medium Is Perfectly Mixed. To illustrate the design of reactors of this type, two design problems are solved. In Example 7-1, styrene is to be polymerized isothermally in a batch reactor. Water is to be used as the coolant. Reactors in which both the reactants and the coolant are perfectly mixed are described by Equations (7-9), (7-12), (7-13), and (7-16). In addition to the assumptions listed in the development of these equations, it is also assumed that the heat capacities, the overall heat-transfer coefficient, and the pressure remain constant throughout the course of the reaction. The assumption of constant heat capacities permits the first term of Equation (7-16) to be restated as follows:

$$w_c(H_{ci} - H_c) = w_cC_{pc}(T_{ci} - T_c) \qquad (7\text{-}19)$$

Since the reaction is to be carried out isothermally, $dT/dt = 0$, and Equations (7-12) and (7-13) may be combined to give

$$U_hA(T_c - T) = r_AV\Delta H_r \qquad (7\text{-}20)$$

Since r_A is generally a function of time [for example, for a first-order reaction,

$r_A = k_c C_A^0 \exp(-k_c t)$], it is evident from Equation (7-20) that T_c must vary with time in order for the reaction temperature to remain constant. By inspection of Equations (7-16), (7-19), and (7-20), it is evident that the outlet temperature T_c of the coolant also depends on its inlet temperature T_{ci}, its mass flow rate w_c, and other variables. When Equation (7-20) is used to eliminate T_c from Equation (7-16), the following expression is obtained:

$$w_c C_{pc}(T_{ci} - T) - \left[\frac{w_c C_{pc}}{U_h A} + 1\right] r_A V \Delta H_r = \left[\frac{\mathcal{M}_c C_{pc} V \Delta H_r}{U_h A}\right] \frac{dr_A}{dt} \quad (7\text{-}21)$$

This expression may be used to compute the flow rates w_c required to maintain the temperature of the reactants constant at a specified variation of the inlet temperature, T_{ci}, with time. Alternately, the inlet temperatures T_{ci} required to maintain T constant at specified values of w_c may be computed by use of Equation (7-21).

EXAMPLE 7-1

Styrene is to be polymerized isothermally in a batch reactor at a reaction temperature of 20°C. An alkyllithium initiator is to be used, and the corresponding rate expression and rate constant k_c, which were determined by Worsfold and Bywater [1], are to be used. Water is to be used as the heat-transfer medium. It may be assumed that the water in the jacket surrounding the reaction vessel is perfectly mixed. On the basis of the data given, find the inlet temperature of the cooling water T_{ci}, the flow rate w_c, and the outlet temperature required to carry the reaction out isothermally.

Given:

C = total polymer concentration (g mol/liter)

C_A^0 = 2 g mol/liter, initial concentration of styrene in the solvent benzene

V = 400 ml, volume of the reactor occupied by the reacting mixture

C_p = 0.42 cal/g °C, heat capacity of the reacting mixture

C_{pc} = 1 cal/g °C, heat capacity to be used for water

k_c = 0.387 $(\text{liter/g mol})^{1/2}$/min

U_h = 25 Btu/(h ft² °F), based on the area A defined next

A = 0.35 ft², the heat-transfer area based on the inside diameter of the reactor

ΔH_r = -16.6 kcal/g mol of styrene polymerized

$r_A = k_c C^{1/2} C_A$, g mol of styrene reacted per liter per minute

C_A = concentration of styrene (g mol/liter) at any time t

$\mathcal{M}_c$ = 100 g, mass of coolant in jacket

Initial concentration of the initiator = 0.01 g mol/liter

Assumptions:

1. Assume that the reaction of the initiator (*sec*-butyllithium) with the monomer styrene is exceedingly fast, relative to the polymerization of styrene, and that the initiator reaction goes to completion [2]. Thus $C = 0.01$ g mol/liter for all time t.
2. The change in density of the reacting mixture and of the coolant may be neglected.
3. $\Delta H_r \cong \Delta U_r$ for this liquid phase reaction.

SOLUTION

For a batch reactor in which the density is constant, it was shown in Chapter 1 that Equation (7-9) reduces to $r_A = -dC_A/dt$. Thus

$$-\frac{dC_A}{dt} = k_c C^{1/2} C_A = (0.387)(0.01)^{1/2} C_A = 0.0387 C_A$$

Integration of this expression followed by rearrangement gives

$$C_A = C_A^0 e^{-0.0387t} = 2 e^{-0.0387t}$$

Thus

$$r_A = (0.0774) e^{-0.0387t} \tag{A}$$

The numerical value of $U_h A$, which is needed in subsequent calculations, is found as follows:

$$U_h A = \left(\frac{25 \text{ Btu}}{\text{ft}^2 \text{ h }^\circ\text{F}}\right)(0.35 \text{ ft}^2)\left(\frac{1.8^\circ\text{F}}{^\circ\text{C}}\right)\left(\frac{252 \text{ cal}}{\text{Btu}}\right)\left(\frac{1 \text{ h}}{60 \text{ min}}\right) = 66.15 \text{ cal}/^\circ\text{C min} \tag{B}$$

By use of this result and the data given, the variation of T_c with time may be found by use of Equation (7-20):

$$T_c = T + \frac{r_A V \Delta H_r}{U_h A} = 20 + \frac{(0.0774 e^{-0.0387t})(0.4)(-16,600)}{66.15}$$

Thus

$$T_c = 20 - 7.77 e^{-0.0387t} \tag{C}$$

This expression for T_c was used to prepare the following table.

Time (min)	0	5	10	15	20	30	40	50	60
T_c (°C)	12.2	13.6	14.7	15.7	16.4	17.6	18.3	18.9	19.2

For the outlet temperature to take on the values listed in this table, it is necessary to pick values of the flow rate w_c and inlet temperature T_{ci} which satisfy Equation (7-21). To evaluate dr_A/dt, which appears in Equation (7-21), the expression given by Equation (A) is differentiated to give

$$\frac{dr_A}{dt} = -(0.0774)(0.0387) e^{-0.0387t} \tag{D}$$

With the aid of this result and other data presented, it is possible to express the flow rate w_c as a function of time by use of Equation (7-21) as follows:

$$w_c = \frac{V\Delta H_r\left[(r_A/C_{pc}) + (\mathcal{M}_c/U_h A)(dr_A/dt)\right]}{(T_{ci} - T) - \dfrac{r_A V\Delta H_r}{U_h A}}$$

Thus

$$w_c = \frac{-484 e^{-0.0387t}}{(T_{ci} - 20) + 7.77 e^{-0.0387t}} \tag{E}$$

Since the flow rate of the coolant is positive, the inlet temperature T_{ci} must be selected such that

$$0 < \left(T_{ci} + 7.77 e^{-0.0387t}\right) < 20$$

If the value $T_{ci} = 10°C$ is selected, then the corresponding flow rates w_c required to achieve isothermal operation are as follows:

Time (min)	0	5	10	15	20	30	40	50	60
w_c (g/min)	217	124	70	48	35	20	12	8	5

In the design of an isothermal reactor, careful consideration should be given to the specification of the initial concentration of the reactants. To achieve the variation in flow rates computed in this example would require the use of a very fast acting controller on the flow rate w_c of the water. By carrying this reaction out at a lower temperature and at a lower initial concentration of the monomer, $(T_c - T)$ is smaller than in this example and isothermal operation is easier to achieve (see Problem 7-9).

2. Isothermal Operation: Use of Cooling Coils. To obtain a greater heat-transfer area than is available in a jacketed reactor, cooling coils are often mounted inside batch reactors. The operation and design of such reactors are described by Equations (7-9), (7-11), (7-14), and (7-18). In many instances, the accumulation term on the right side of Equation (7-18) is insignificant relative to the other terms and can be neglected. For such reactors, Equation (7-18) becomes

$$w_c C_{pc}\frac{dT_c}{dz} + q = 0 \tag{7-22}$$

which is recognized as the dynamic form of the steady-state equation for the transfer of heat from a flowing stream to the wall of the pipe. This differential equation may be solved by first eliminating q by use of the relationship $q = U_h a(T_c - T)$, followed by the separation of variables, and then integrating from $T_c = T_{ci}$ at $z = 0$ to $(T_c)_{out}$ at $z = z_T$.

$$(T_c)_{out} - T = (T_{ci} - T)e^{-\alpha} \tag{7-23}$$

where it has been assumed that the ratio $\alpha = U_h A / w_c C_{pc}$ is independent of the position z. Also, Equation (7-22) may be integrated termwise with respect to z as follows:

$$\int_0^L w_c C_{pc} \frac{dT_c}{dz} dz = -\int_0^L q \, dz = -Q$$

to give

$$Q = w_c C_{pc} \left[T_{ci} - (T_c)_{out} \right] \tag{7-24}$$

Equations (7-23) and (7-24) may be combined and rearranged to give the well-known expression

$$Q = U_h A \, \Delta T_{lm} \tag{7-25}$$

where

$$\Delta T_{lm} = \frac{[(T_c)_{out} - T] - [T_{ci} - T]}{\log_e \left\{ [(T_c)_{out} - T] / [T_{ci} - T] \right\}}$$

An equivalent but perhaps a more convenient expression to use for Q in the example that follows may be obtained by eliminating $[(T_c)_{out} - T_{ci}]$ from Equations (7-23) and (7-24) to give

$$Q = w_c C_{pc} (T_{ci} - T)(1 - e^{-\alpha}) \tag{7-26}$$

When the pressure and volume are constant and when the reaction is carried out isothermally, $dT/dt = 0$, and Equation (7-11) reduces to

$$Q = r_A V \Delta H_r \tag{7-27}$$

Since r_A can be expressed as a function of time, Equation (7-27) serves to give the total rate of heat transfer through the coil at any time t. The use of the preceding equations in the design of a reactor is illustrated by the following example.

EXAMPLE 7-2

Instead of using an external jacket on the reactor in Example 7-1, suppose that a cooling coil is to be used and that the outside of the reactor is to be insulated. The insertion of the cooling coil in the reaction space (the volume V) reduces the volume available for the reactants by 25 cm^3. The overall heat-transfer coefficient for the coil is estimated to be 100 Btu/h ft^2 °F. (This coefficient is based on the outside surface area of the tube). The tubing has an outside diameter of $\frac{3}{16}$ in. and a wall thickness of 0.032 in. The total length of the coil is 33 in. For a cooling water flow rate of 450 ml/min, calculate the inlet and outlet temperatures of the cooling water required to maintain the reaction temperature at 20°C throughout the course of the reaction. Assume that the heat loss from the outer shell of the reactor to the surroundings is negligible.

SOLUTION

When the expression given by Equation (A) of Example 7-1 is substituted into Equation (7-27), one obtains

$$Q = \left[(0.0774) e^{-0.0387t} \right] V \Delta H_r$$
$$= (0.0774 e^{-0.0387t})(0.375)(-16,600)$$
$$= -482 e^{-0.0387t} \ (\text{cal/min}) \tag{A}$$

When this expression for Q is substituted into Equation (7-26), an expression for computing the inlet temperature of the cooling water at any time t is obtained:

$$T_{ci} = T - \frac{482 e^{-0.0387t}}{w_c C_{pc}(1 - e^{-\alpha})} \tag{B}$$

where $\alpha = \dfrac{U_h A}{w_c C_{pc}} = \dfrac{\left(100 \dfrac{\text{Btu}}{\text{h ft}^2 \ ^\circ\text{F}}\right)\left(\dfrac{1 \ \text{h}}{60 \ \text{min}}\right)(0.1349 \ \text{ft}^2)\left(1.8 \dfrac{^\circ\text{F}}{^\circ\text{C}}\right)\left(252 \dfrac{\text{cal}}{\text{Btu}}\right)}{(450 \ \text{g/min})(1 \ \text{cal/g} \ ^\circ\text{C})}$

$$\alpha = 0.227$$

and $\qquad\qquad\qquad 1 - e^{-\alpha} = 0.203$

Thus Equation (B) reduces to

$$T_{ci} = 20 - \frac{482 e^{-0.0387t}}{(450)(0.203)} = 20 - 5.28 e^{-0.0387t} \tag{C}$$

To find the outlet temperature of the coolant, Equation (7-24) is first solved for $(T_c)_{\text{out}}$ to give

$$(T_c)_{\text{out}} = T_{ci} - \frac{Q}{w_c C_{pc}}$$

Elimination of Q by use of Equation (A) and substituting for $w_c C_{pc}$ gives

$$(T_c)_{\text{out}} = T_{ci} + \frac{482 e^{-0.0387t}}{450} = T_{ci} + 1.071 e^{-0.0387t} \tag{D}$$

Values of the inlet and outlet temperatures computed by the use of Equations (C) and (D) are presented in the following table:

t (min)	0	5	10	15	20	30	40	50	60
T_{ci} (°C)	14.72	15.65	16.41	17.05	17.57	18.35	18.88	19.24	19.48
$(T_c)_{\text{out}}$ (°C)	15.79	16.53	17.14	17.65	18.08	18.69	19.11	19.39	19.59

II. PERFECTLY MIXED FLOW REACTORS

The sketch of the typical perfectly mixed flow reactor shown in Figure 7-2 is used in the following development of the steady-state and unsteady-state energy-balance expressions.

A. Steady-State Operation

Suppose that the reactor shown in Figure 7-2 is operated at steady state and that any number of reactions occur within the reactor. Generally, the changes in the kinetic and potential energies of the entering and leaving streams are negligible, as well as the energy equivalent of the work done by the stirrer on the reacting mixture. Under these conditions, the energy balance on the reacting mixture contained in the reactor at steady-state operation is given by

$$n_T^\circ H(T_0) - n_T H(T) + \int_0^{z_T} q \, dz = 0 \qquad (7\text{-}28)$$

where $q = Ua(T_c - T)$, T_c is the temperature of the heat-transfer medium at any z

$H(T_0)$ = enthalpy of the reactants, evaluated at the feed composition and at the feed temperature T_0 and pressure P_0 at the inlet to the reacting mixture

For the case where any number of reactions occur and the nonideal enthalpies are expressed in terms of the virtual values of the partial molar enthalpies, Equation (7-28) becomes

$$\sum_{i=1}^{c} n_i^0 \hat{H}_i(T_0) - \sum_{i=1}^{c} n_i \hat{H}_i(T) + \int_0^{z_T} q \, dz = 0 \qquad (7\text{-}29)$$

where

$$\hat{H}_i(T) = H_{fi}^\circ(T) + \Omega(\mathbf{n}, P, T)$$

$$\hat{H}_i(T_0) = H_{fi}^\circ(T_0) + \Omega(\mathbf{n}^0, P_0, T_0)$$

Use of the component material balance expression

$$V r_i = n_i^0 - n_i \qquad (7\text{-}30)$$

(where r_i is defined in terms of disappearance for all components) to eliminate the n_i's from Equation (7-29) gives

$$\sum_{i=1}^{c} n_i^0 \left[\hat{H}_i(T_0) - \hat{H}_i(T) \right] + V \sum_{i=1}^{c} \hat{H}_i(T) r_i + \int_0^{z_T} q \, dz = 0 \qquad (7\text{-}31)$$

B. Unsteady-State Operation Nonideal Solutions

If it is supposed that the kinetic and potential energies as well as work contributions are negligible, then the energy balance on the reacting mixture in the perfectly mixed flow reactor over the time period from t_n to t_{n+1} may be represented as follows:

$$\int_{t_n}^{t_{n+1}}\left[n_T^0 H(T_0) - n_T H(T) + \int_0^{z_T} q\, dz\right] dt = (C_T VU)|_{t_{n+1}} - (C_T VU)|_{t_n}$$

(7-32)

where $H(T_0)$ and $H(T)$ are used to represent $H(\mathbf{n}_0, T_0, P_0)$ and $H(\mathbf{n}, T, P)$, respectively.

This expression may be reduced to the following integral-difference equation by use of the mean-value theorems and limiting processes as demonstrated in Chapter 1.

$$n_T^0 H(T_0) - n_T H(T) + \int_0^{z_T} q\, dz = \frac{d(C_T VU)}{dt}$$

(7-33)

$$n_T^0 H(T_0) - n_T H(T) + \int_0^{z_T} q\, dz = \frac{d(VC_T H)}{dt} - \frac{1}{j}\frac{d(PV)}{dt}$$

(7-34)

where the relationship $U = H - P/C_T$ was used to eliminate U from Equation (7-33). Observe that

$$VC_T H = \sum_{i=1}^{c} VC_i \hat{H}_i = \sum_{i=1}^{c} V\left(\frac{N_i}{V}\right)\hat{H}_i = \sum_{i=1}^{c} N_i \hat{H}_i$$

where N_i is the total moles of component i in the volume V. Thus, when stated in terms of the virtual values of the partial molar enthalpies, Equation (7-34) becomes

$$\sum_{i=1}^{c}\left[n_i^0 \hat{H}_i(T_0) - n_i \hat{H}_i(T)\right] + \int_0^{z_T} q\, dz = \frac{d\left[\sum_{i=1}^{c} N_i \hat{H}_i\right]}{dt} - \frac{1}{j}\frac{d(PV)}{dt}$$

(7-35)

The corresponding component material balance given by Equation (2-117) takes the following form for any component i:

$$n_i^0 - n_i - r_i V = \frac{dN_i}{dt}$$

(7-36)

When this component material balance is used in the evaluation of the summation appearing on the right side of Equation (7-35) in a manner analogous to that demonstrated in Chapter 6, the following expression is obtained upon rearrangement:

$$\sum_{i=1}^{c} n_i^0\left[\bar{H}_i(T_0) - \bar{H}_i(T)\right] + V\sum_{i=1}^{c}\bar{H}_i r_i + \int_0^{z_T} q\, dz$$

$$= N_T C_p \frac{dT}{dt} + N_T \frac{\partial\Omega}{\partial P}\frac{dP}{dt} - \frac{1}{j}\frac{d(PV)}{dt}$$

(7-37)

The energy-balance expressions for all types of mixtures and modes of operation are readily deduced from Equation (7-37). For example, for the case where the single reaction $aA + bB = cC + dD$ forms an ideal solution $\bar{H}_i = \hat{H}_i = H_i$ and is carried out at steady-state operation, Equation (7-37) reduces to

$$\sum_{i=1}^{c} n_i^0 \big[\hat{H}_i(T_0) - \hat{H}_i(T) \big] - r_A V \Delta H_r + \int_0^{z_T} q \, dz = 0$$

or
(7-38)

$$-n_T^0 \int_{T_0}^{T} (C_p)_{\text{feed}} \, dT - r_A V \Delta H_r + \int_0^{z_T} q \, dz = 0$$

where
$$n_T^0 (C_p)_{\text{feed}} = \sum_{i=1}^{c} n_i^0 \frac{dH_i}{dT}$$

and ΔH_r is defined in Chapter 6 just above Equation (6-44).

C. Design of Perfectly Mixed Flow Reactors

Typical reactors of this type are shown in Figures 2-6 and 2-7. A summary of the equations needed to describe reactors of this type follow for the case where the single reaction $aA + bB = cC + dD$ is carried out at steady-state operation, and it is supposed that the reacting mixture forms an ideal solution.

$$\textit{Material balance:} \quad r_A = \frac{n_A^0 - n_A}{V} \tag{7-39}$$

Energy balance on the reactants:

$$-n_T^0 \int_{T_0}^{T} (C_p)_{\text{feed}} \, dT - r_A V \Delta H_r + Q = 0 \tag{7-40}$$

where $(C_p)_{\text{feed}}$ is evaluated on the basis of the composition of the feed.
Energy balance on the heat-transfer medium:

$$\text{1.} \quad \textit{Perfectly mixed:} \quad w_c (H_{ci} - H_c) + Q = 0 \tag{7-41}$$

$$\text{2.} \quad \textit{Plug flow:} \quad w_c C_{pc} \frac{dT_c}{dz} + q = 0 \tag{7-42}$$

The total rate of heat transfer (Btu/h) from the heat-transfer medium to the reacting mixture is computed by use of Equation (7-13) for the case where the heat-transfer medium is perfectly mixed and by Equation (7-14) for the case where the heat-transfer medium is in plug flow.

In the first problem presented in this section, a perfectly mixed flow reactor is designed to operate isothermally. For such a reactor, "isothermal operation" is used to mean that the temperatures of the reacting mixture entering and leaving the reactor are equal.

EXAMPLE 7-3

A batch reactor is sometimes converted to a perfectly mixed flow reactor by introducing the feed continuously and removing the product continuously. To illustrate this type of operation, suppose that the batch reactor described in Example 7-1 is to be operated at steady state as a perfectly mixed flow reactor. The initial concentrations of styrene and initiator are the same as was stated in Example 7-1. Also, as in Example 7-1, the rate of reaction of styrene may again be represented by the pseudo first-order rate expression

$$r_A = \bar{k}C_A = 0.0387C_A$$

If the reaction is to be carried out isothermally ($T_0 = T$) at 20°C and a conversion of 80% is to be obtained, find the volumetric flow rate of the feed required to achieve the desired conversion in a reactor having a volume of 400 ml.

Cooling water at 15°C is available for use as the heat-transfer medium. Find the cooling water rate w_c for isothermal operation of the reactor at the specific conversion of 80%.

SOLUTION

Elimination of r_A from the material-balance expression [Equation (7-39)] and the experimental expression for the rate of reaction yields

$$\frac{n_A^0 x}{V} = \frac{n_A^0 - n_A}{V} = \bar{k}C_A = \bar{k}C_A^0(1 - x) \tag{A}$$

Elimination of n_A^0 from Equation (A) by use of the relationship

$$n_A^0 = C_A^0 v_T$$

followed by rearrangement yields

$$v_T = \frac{\bar{k}(1 - x)V}{x} \tag{B}$$

Thus

$$v_T = \frac{0.0387(1 - 0.8)(0.4)}{0.8}$$

$$= 0.00387 \text{ liter/min} = 3.87 \text{ cm}^3/\text{min}$$

Since the reaction is to be carried out isothermally, $T_0 = T$, Equation (7-40) reduces to

$$Q = r_A V \Delta H_r = x n_A^0 \, \Delta H_r = x C_A^0 v_T \, \Delta H_r$$

Thus

$$Q = \left(\frac{2 \text{ g mol}}{\text{liter}}\right)\left(\frac{0.00387 \text{ liter}}{\text{min}}\right)(0.8)\left(\frac{-16,600 \text{ cal}}{\text{g mol}}\right)$$

$$= -102.79 \text{ cal/min}$$

The temperature T_c of the cooling water may be computed by use of Equation (7-13) as follows:

$$T_c = T + \frac{Q}{U_h A}$$

Since $U_h A = 66.15$ cal/°C min [see Equation (B) of Example 7-1], it follows that

$$T_c = 20°C + \left[\frac{-102.79 \text{ cal/min}}{66.15 \text{ cal/min °C}} \right] = 18.45°C$$

The flow rate w_c and the inlet temperature of T_{ci} of the cooling water are related by Equation (7-41). Since $w_c(H_{ci} - H_c) = w_c C_{pc}(T_{ci} - T_c)$, Equation (7-41) may be rearranged to give

$$w_c = \frac{Q}{C_{pc}(T_{ci} - T_c)}$$

Thus

$$w_c = \frac{-102.79}{15 - 18.45} = 29.8 \text{ g/min} = 29.8 \text{ cm}^3/\text{min}$$

Generally, exothermic reactions are more easily controlled in a perfectly mixed flow reactor than in other types of reactors because, in the former, the reaction rate is proportional to the smallest (the final) concentration of the reactant.

EXAMPLE 7-4

Suppose that the reactor described in Example 7-1 is to be operated adiabatically at steady state as a perfectly mixed flow reactor. The inlet temperature of the reacting mixture is 25°C. Styrene is to be polymerized to a conversion of 80%. The concentration of styrene in the feed is 0.65 g mol/liter and that of the initiator is 0.01 g mol/liter. The remainder of the feed is the solvent, cyclohexane, which has a concentration of 8.58 g mol/liter. (Again it may be assumed as in Example 7-1 that the initiator has been completely consumed by reaction by the time the feed enters the reactor and that the total polymer concentration C remains constant at 0.01 g mol/liter.) The rate of polymerization of styrene is given by the following rate equation [1]:

$$r_A = k_c C^{1/2} C_A$$

where

$$k_c = 1.72 \times 10^{13} e^{-18,788/RT} \left(\frac{1}{\text{min}} \right) \left[\frac{\text{liter}}{\text{g mol}} \right]^{1/2}$$

Since the polymer remains relatively dilute in the reacting mixture throughout the course of the reaction, the density of the mixture may be assumed to remain constant. The molecular weight M_w of the feed is 85.4 g/g mol, the heat capacity of the feed is 0.42 cal/g °C, the molar density of the feed is 9.23 g mol/liter, and the mole fraction y_A^0 of styrene in the feed is 0.0704.

Calculate the outlet temperature of the reacting stream and the volumetric flow rate of the feed required to achieve the desired conversion in a reactor having a volume of 400 ml.

SOLUTION

The outlet temperature T of the reacting mixture may be found by first solving Equation (7-40) for T to give

$$T = T_0 - \frac{r_A V \Delta H_r}{w(C_p)_{feed}} \tag{A}$$

where $(C_p)_{feed}$ has the units of cal/g °C. By use of Equation (7-39), $r_A V = n_A^0 - n_A = n_A^0 x$, and the fact that $w = n_T^0 M_w$, Equation (A) may be restated in the form

$$T = T_0 - \frac{y_A^0 x \Delta H_r}{M_w(C_p)_{feed}} \tag{B}$$

where $y_A^0 = n_A^0/n_T^0$.

$$T = 25 - \frac{(0.0704)(0.8)(-16,600)}{(85.4)(0.42)} = 51.1°C \tag{C}$$

The total volumetric flow rate of the feed may be computed by use of Equation (B) of Example 7-3:

$$v_T = k\left(\frac{1 - x}{x}\right)V$$

At 51.1°C, the pseudo first-order rate constant k has the following value:

$$k = kC^{1/2} = (1.72 \times 10^{13})(0.01)^{1/2}\exp\left[\frac{-18,778}{(1.9872)(324.1)}\right]$$

$$= 0.374 \text{ min}^{-1}$$

Then

$$v_T = (0.374)\left[\frac{1 - 0.8}{0.8}\right](0.4) = 0.0374 \text{ liter/min}$$

$$= 37.4 \text{ cm}^3/\text{min}$$

and

$$\theta = \frac{V}{v_T} = \frac{400}{37.4} = 10.7 \text{ min}$$

D. Design of a Reactor in Which Any Number of Reactions Occur

In the preceding design examples, only one reaction occurred. In most commercial reactors, however, more than one reaction usually occurs. The term "reaction" is used to represent either a reversible or an irreversible reaction. For each reaction which occurs, there exists one independent component material balance. Thus, instead of only one material-balance equation, such as the ones shown for batch and perfectly mixed flow reactors, there exists instead one material-balance equation for each reaction. Likewise, in the energy-balance equation for the reacting stream, either the enthalpies for each component must be included or a heat of reaction term must be included for each reaction. The design of reactors in which multiple reactions occur is illustrated by use of the following specific example.

EXAMPLE 7-5

Methane is to be chlorinated in a brick-lined reactor which is operated adiabatically as a perfectly mixed flow reactor. The composition of the feed is 80% methane and 20% chlorine on a mole basis. The feed enters the reactor at 25°C. The reactor is to be operated at a residence time such that essentially all the chlorine in the feed reacts to form one of the following chlorinated products:

$$CH_4 + Cl_2 \rightarrow CH_3Cl + HCl \tag{1}$$

$$CH_3Cl + Cl_2 \rightarrow CH_2Cl_2 + HCl \tag{2}$$

$$CH_2Cl_2 + Cl_2 \rightarrow CHCl_3 + HCl \tag{3}$$

$$CHCl_3 + Cl_2 \rightarrow CCl_4 + HCl \tag{4}$$

By use of the equations for a perfectly mixed flow reactor, calculate the temperature of the reacting mixture leaving the reactor and the moles of each chlorinated product produced per mole of methane fed.

It may be assumed that the mixture is a perfect gas and that the reactions are unidirectional. The respective rates of disappearance of chlorine by each of the reactions are as follows:

$$r_1 = k_1 p_{CH_4} p_{Cl_2}, \qquad r_2 = k_2 p_{CH_3Cl} p_{Cl_2},$$

$$r_3 = k_3 p_{CH_2Cl_2} p_{Cl_2}, \qquad r_4 = k_4 p_{CHCl_3} p_{Cl_2}$$

The ratios of the rate constants are as follows [3]:

$$\frac{k_2}{k_1} = 0.424 e^{1960/RT}$$

$$\frac{k_3}{k_1} = 1.0$$

$$\frac{k_4}{k_1} = 2.2 e^{-2000/RT}$$

where $R = 1.9872$ and T is in Kelvin.

The heats of reactions per mole of chlorine reacted at 25°C are as follows:

$$\Delta H_{r1} = -24,740 \text{ cal/g mol}$$

$$\Delta H_{r2} = -24,170 \text{ cal/g mol}$$

$$\Delta H_{r3} = -23,400 \text{ cal/g mol}$$

$$\Delta H_{r4} = -21,300 \text{ cal/g mol}$$

These values were computed by use of heats of formation data given in Reference [4].

The following curve fits for the heat capacities were taken from Hougen et al. [5] and the heats of formation from Reference [4].

Component	ΔH_f° at 298K	α	$\beta \times 10^2$	$\gamma \times 10^5$	$\delta \times 10^9$
CH_4	-17.89	4.75	1.200	0.303	-2.630
CH_3Cl	-20.63	3.05	2.596	-1.244	2.300
CH_2Cl	-22.80	4.20	3.419	-2.350	6.068
$CHCl_3$	-24.20	7.61	3.461	-2.668	7.344
CCl_4	-24.00	12.24	3.4	-2.995	8.828
HCl	-22.00	7.244	-0.1824	0.3170	-1.036
Cl_2	0.0	6.8214	0.57095	-0.5107	1.547

$C_p = \alpha + \beta T + \gamma T^2 + \delta T^3$, T is in K, and C_p has the units of cal/g mol K.

SOLUTION

For convenience, let the four reactions and their rates be represented as follows:

$$(1) \quad A + B \rightarrow C + D, \qquad r_1 = \frac{n_1}{V} = k_1 p_A p_B$$

$$(2) \quad C + B \rightarrow E + D, \qquad r_2 = \frac{n_2}{V} = k_2 p_C p_B$$

$$(3) \quad E + B \rightarrow F + D, \qquad r_3 = \frac{n_3}{V} = k_3 p_E p_B \qquad \text{(A)}$$

$$(4) \quad F + B \rightarrow G + D, \qquad r_4 = \frac{n_4}{V} = k_4 p_F p_B$$

where $A = CH_4$, $B = Cl_2$, $C = CH_3Cl$, $D = HCl$, $E = CH_2Cl_2$, $F = CHCl_3$, and $G = CCl_4$. The moles of chlorine reacted per unit time by reactions (1), (2), (3), and (4) are denoted by n_1, n_2, n_3, and n_4, respectively. The flow rates n_A, n_B, n_C, n_D, n_E, n_F, and n_G may be expressed in terms of n_1, n_2, n_3, and n_4 by use of the stoichiometry of the four reactions as follows:

$$n_A = n_A^0 - n_1$$

$$n_B = n_B^0 - \sum_{k=1}^4 n_k$$

$$n_C = n_1 - n_2$$

$$n_D = \sum_{k=1}^4 n_k \qquad \text{(B)}$$

$$n_E = n_2 - n_3$$

$$n_F = n_3 - n_4$$

$$n_G = n_4$$

$$\overline{n_T = n_T^0}$$

Since a reactor volume is to be used such that essentially all the chlorine is to be consumed by the reaction,

$$n_B \cong 0$$

Thus the problem to be solved consists of finding the values of T, n_A, n_C, n_D, n_E, n_F, and n_G which satisfy (1) the rate expressions, (2) the stoichiometric relationships, and (3) the energy-balance equation. This problem will be formulated in terms of two equations in two unknowns.

For a perfectly mixed flow reactor which is operated adiabatically, Equation (7-31) reduces to

$$\sum_{i=1}^{c} n_i^0 [H_i(T_0) - H_i(T)] + V \sum_{i=1}^{c} H_i(T) r_i = 0 \tag{C}$$

for ideal solution behavior, where $i = A, B, C, D, E, F$, and G.

First consider the second sum in Equation (C). To restate this sum in terms of the heats of reaction, the following relationships between the net rates of disappearance of $A, B, \ldots, G$ (that is, $r_A, r_B, \ldots, r_G$) and the rates of disappearance of B by each reaction (that is, r_1, r_2, r_3, r_4) are needed. The stoichiometry of the four reactions requires that

$$r_A = r_1, \qquad r_D = - \sum_{k=1}^{4} r_k$$

$$r_B = \sum_{k=1}^{4} r_k, \qquad r_E = r_3 - r_2 \tag{D}$$

$$r_C = r_2 - r_1, \qquad r_F = r_4 - r_3$$

$$r_G = - r_4$$

Since

$$V \sum_{i=1}^{c} H_i(T) r_i = V[H_A r_A + H_B r_B + H_C r_C + H_D r_D + H_E r_E + H_F r_F + H_G r_G]$$

the following result is obtained upon replacing $r_A, r_B, \ldots, r_G$ by their equivalents as given by Equation (D).

$$V \sum_{i=1}^{c} H_i(T) r_i = - V \sum_{k=1}^{4} \Delta H_{rk} \, r_k$$

where

$$\Delta H_{r1} = H_C + H_D - H_A - H_B$$

$$\Delta H_{r2} = H_E + H_D - H_C - H_B$$

$$\Delta H_{r3} = H_F + H_D - H_E - H_B$$

$$\Delta H_{r4} = H_G + H_D - H_F - H_B$$

By use of the definition of heat capacity [Equation (5-21)], it is possible to restate the first summation of Equation (C) in the following form:

$$\sum_{i=1}^{c} n_i^0 [H_i(T_0) - H_i(T)] = - n_T^0 \int_{T_0}^{T} (C_p)_{\text{feed}} \, dT \tag{E}$$

Since the feed contains components A and B alone, the heat capacity of the feed is given by

$$(C_p)_{feed} = \frac{1}{n_T^0} \left[n_A^0 C_{pA} + n_B^0 C_{pB} \right]$$

Thus Equation (C) may be stated in the following equivalent form:

$$n_T^0 \int_{T_0}^{T} (C_p)_{feed} \, dT + V \sum_{k=1}^{4} \Delta H_{rk} \, r_k = 0 \qquad (F)$$

In the solution of the problem at hand, it was elected to replace $r_k V$ in Equation (F) by its equivalent n_k [see Equation (A)]. The corresponding functional expression may be formulated:

$$F = \int_{T_0}^{T} (C_p)_{feed} \, dT + \frac{1}{n_T^0} \sum_{k=1}^{4} \Delta H_{rk} \, n_k \qquad (G)$$

where each ΔH_{rk} is evaluated at the temperature T. The heat of reactions as a function of temperature for each reaction is found by use of Equation (5-24). For example, for the first reaction,

$$\Delta H_{r1} = \Delta H_{r1}|_{T_0} + \int_{T_0}^{T} \Delta C_p \, dT$$

where

$$\Delta C_{p1} = C_{pC} + C_{pD} - C_{pA} - C_{pB}$$

By use of the heat capacities given, the integration implied in this expression may be carried out. Use of the resulting expression so obtained together with the value of ΔH_{r1} at T_0 yields the desired functional expression [Equation (5-105)]:

$$\Delta H_{rk} = C_{1k} + \Delta \alpha_k T + \frac{\Delta \beta_k}{2} T^2 + \frac{\Delta \gamma_k}{3} T^3, \qquad k = 1,2,3,4 \qquad (H)$$

where ΔH_{rk} is in cal per g mol, T is in Kelvin, and

Reaction No.	C_{1k}	$\Delta \alpha_k$	$\Delta \beta_k \times 10^2$	$\Delta \gamma_k \times 10^5$	$\Delta \delta_k \times 10^9$
1	$-24{,}459$	-1.2774	0.6426	0.7193	2.347
2	$-24{,}647$	1.5726	0.06965	-0.2783	1.185
3	$-24{,}269$	3.8326	-0.7114	0.5097	-1.307
4	$-23{,}346$	5.0526	-0.8144	0.5007	-1.102

After the integration implied in the first term of Equation (G) has been carried out, the following result is obtained:

$$\int_{T_0}^{T} (C_p)_{feed} \, dT = 5.16428 + 0.5371(10^{-2}) T^2 + 0.04675(10^{-5}) T^3$$

$$-0.44865(10^{-9}) T^4 - 2024.7 \qquad (I)$$

where T is in Kelvin.

The moles of B (chlorine) reacted by each reaction may be expressed in terms of the ratio n_A/n_1 and the temperature T as follows. From the stoichiometric relationship given by Equation (B),

$$n_1 = \frac{n_A^0}{1 + n_A/n_1} \tag{J}$$

Since perfect gas behavior is assumed, the following expressions are obtained by division of r_2, r_3, and r_4 by r_1 [see Equation (A)]:

$$\frac{k_2 n_C}{k_1 n_A} = \frac{n_2}{n_1}$$

$$\frac{k_3 n_E}{k_1 n_A} = \frac{n_3}{n_1} \tag{K}$$

$$\frac{k_4 n_F}{k_1 n_A} = \frac{n_4}{n_1}$$

By use of these results and the stoichiometric relationships of Equation (B), the following expressions for n_2, n_3, and n_4 as a function of n_A/n_1 and T are obtained:

$$n_2 = \frac{n_1}{[1 + (k_1/k_2)(n_A/n_1)]} \tag{L}$$

$$n_3 = \frac{n_2}{[1 + (k_1/k_3)(n_A/n_1)]} \tag{M}$$

$$n_4 = \frac{n_3}{[1 + (k_1/k_4)(n_A/n_1)]} \tag{N}$$

Thus the problem reduces to finding a set of values n_A/n_1 and T such that $F = 0$ and such that the sum of n_1, n_2, n_3, and n_4 is equal to n_B^0, that is, such that $n_B = 0$. Thus the second function to be satisfied may be stated as follows:

$$G = 1 - \frac{1}{n_B^0}(n_1 + n_2 + n_3 + n_4) \tag{O}$$

This problem may be solved by use of an interpolation procedure wherein T is regarded as the independent variable. For each choice of T, the value of n_A/n_1 that makes $G = 0$ is found. Then an improved choice of T is found by interpolation on the function F. On the basis of 100 g mol of feed, the calculations involved in making the first trial follow:

Trial 1: Assume $T_1 = 725$K. Since $n_T^0 = 100$ g mol and the feed consists of 80% methane and 20% chlorine, it follows that

$$n_A^0 = 80, \qquad n_B^0 = 20$$

At 725K,

$$\frac{k_1}{k_2} = 0.605, \qquad \frac{k_1}{k_3} = 1, \qquad \frac{k_1}{k_4} = 1.822$$

For the first trial on G, take

$$\left(\frac{n_A}{n_1}\right)_1 = 4.5$$

Then

$$n_1 = \frac{n_A^0}{1 + n_A^0/n_1} = \frac{80}{1 + 4.5} = 14.5455$$

$$n_2 = \frac{n_1}{1 + (k_1/k_2)(n_A/n_1)} = \frac{14.5455}{1 + (0.605)(4.5)} = 3.9075$$

$$n_3 = \frac{n_2}{1 + (k_1/k_3)(n_A/n_1)} = \frac{3.9075}{1 + (1)(4.5)} = 0.7105$$

$$n_4 = \frac{n_3}{1 + (k_1/k_4)(n_A/n_1)} = \frac{0.7105}{1 + (1.822)(4.5)} = 0.0772$$

and thus $\quad G_1 = 1 - \frac{1}{20}(14.5455 + 3.9075 + 0.7105 + 0.0772)$

$\qquad\qquad G_1 = 0.03797$

For the second trial on G, take

$$\left(\frac{n_A}{n_1}\right)_2 = 4.0$$

By repeating the same procedure, the following results are obtained:

$$n_1 = 16.000, \qquad n_2 = 4.6784, \qquad n_3 = 0.9357, \qquad n_4 = 0.1129$$

and

$$G_2 = -0.08635$$

By use of the method of interpolation *regula falsi* [6], the next best value of (n_A/n_1) is found:

$$\left(\frac{n_A}{n_1}\right)_3 = \frac{(4.0)(0.03799) - (-0.08635)(4.5)}{0.03799 - (-0.08635)} = 4.347$$

A summary of the iterations on G at $T = 725K$ follows:

Trial No.	n_A/n_k	n_1	n_2	n_3	n_4	G_k
1	4.5	14.5454	3.9074	0.7104	0.0772	0.0379
2	4.0	16.0000	4.6784	0.9357	0.1129	-0.0864
3	4.347	14.9617	4.1218	0.7709	0.0864	0.0029
4	4.335	14.9950	4.1393	0.7759	0.0872	0.0001

Next the value of F at $T = 725K$ and $n_A/n_1 = 4.335$ is computed. The first term of the function F [Equation (G)] is evaluated by use of Equation (I):

$$\int_{298.16}^{725} (C_p)_{\text{feed}} \, dT = 4596$$

The second term $(1/n_T^0)\sum_{k=1}^{4} \Delta H_{rk} n_k$ of the function F is evaluated by use of

Equation (H) and the solution set of n_k's at $n_A/n_1 = 4.335$ (given previously):

Reaction No.	ΔH_{rk}	$1/n_T^0(n_k \, \Delta H_{rk})$
1	$-24,448$	-3666
2	$-23,595$	-977
3	$-22,803$	-177
4	$-21,262$	-18
		-4838

Thus

$$F_1(725) = 4596 - 4838 = -242$$

Trial 2: Assume $T_2 = 750K$. At 750K,

$$\frac{k_1}{k_2} = 0.6331, \qquad \frac{k_1}{k_3} = 1, \qquad \frac{k_1}{k_4} = 1.739$$

At $T = 750K$, the value of n_A/n_1 needed to make $G = 0$ was found to be $n_A^0/n_1 = 4.3$. The iterative procedure demonstrated in trial 1 was used. In summary, at $T = 750K$ and $n_A/n_1 = 4.3$, the following result was obtained:

$$n_1 = 15.0943, \qquad n_2 = 4.0556, \qquad n_3 = 0.7652, \qquad n_4 = 0.0903$$

and

$$G = -0.0002$$

The function F is now evaluated at $T = 750K$ and $n_A/n_1 = 4.3$. First, by Equation (I),

$$\int_{298.16}^{750} \left(C_p \right)_{\text{feed}} dT = 4925$$

By use of the preceding results and Equation (H),

Reaction No.	ΔH_{rk}	$1/n_T^0(n_k \, \Delta H_{rk})$
1	$-24,439$	-3689
2	$-23,576$	-956
3	$-22,792$	-174
4	$-21,239$	-19
		-4838

Then

$$F_2(75) = 4925 - 4838 = 87$$

Interpolation on the functional values F_1 and F_2 at the corresponding values T_1 and T_2 gives T_3, the value of T to be assumed for the third trial:

$$T_3 = \frac{725(87) - 750(-242)}{87 - (-242)} = 743.4$$

Trial 3: Assume $T_3 = 743.4$. At this temperature,

$$\frac{k_1}{k_2} = 0.6257, \qquad \frac{k_1}{k_2} = 1, \qquad \frac{k_1}{k_4} = 1.76$$

and at $(n_A/n_1) = 4.31$, $G \cong 0$. More precisely at $T_3 = 743.4$ and $(n_A/n_1) = 4.31$,

$$n_1 = 15.066, \qquad n_2 = 4.0754, \qquad n_3 = 0.7675, \qquad n_4 = 0.0894$$

and

$$G = 0.00009$$

Next F is evaluated. First

$$\int_{298.16}^{743.4} (C_p)_{\text{feed}} \, dT = 4838$$

and in a manner analogous to that described previously, the following results are obtained:

Reaction No.	ΔH_{rk}	$1/n_T^0(n_k \, \Delta H_{rk})$
1	$-24,439$	-3682
2	$-23,576$	-981
3	$-22,792$	-175
4	$-21,239$	-19
		-4837

Thus

$$F_3 = 4838 - 4837 = 1$$

Since this difference is within the accuracy of the calculations, $F_3 \cong 0$. The solution set of values of the variables follows:

$$T = 743.4\text{K}, \qquad \frac{n_A}{n_1} = 4.31$$

$$n_A = 64.934$$

$$n_C = n_1 - n_2 = 10.991$$

$$n_E = n_2 - n_3 = 3.3079$$

$$n_F = n_3 - n_4 = 0.6781$$

$$n_G = n_4 = 0.0894$$

$$n_D = \sum_{k=1}^{4} n_k = 19.9983$$

$$n_B = 20 - \sum_{k=1}^{4} n_k = 0.0077$$

and thus

$$n_C/n_A^0 = 0.1374,$$

$$n_E/n_A^0 = 0.413,$$

$$n_F/n_A^0 = 0.008476,$$

$$n_G/n_A^0 = 0.00106.$$

III. THE REACTION PHASE PLANE

This section was included in order to demonstrate that, for certain types of reactor problems, more than one solution may exist. In particular, it is shown that, for the design of an adiabatic perfectly mixed flow reactor in which a first-order reversible, exothermic reaction is to be carried out, more than one steady-state solution to the material and energy balances may exist. Thus the designer or operator of a reactor in which such an exothermic reaction occurs could obtain different conversions at the same residence time. Researchers van Heerden [7] and Bilous and Amundson [8] were among the first to point out the importance of analyses of the following type.

To provide some further insight into the design of reactors, an analysis of the material- and energy-balance equations for the case of a single reversible reaction is presented. The rate of reaction r for the case where a single reaction occurs may be represented as a function of conversion and temperature; that is, $r = r(x, T)$. The conversion–temperature plane (the x–T plane) is defined as the *reaction phase plane*. For example, consider the reaction

$$A \rightleftarrows S \tag{7-43}$$

The rate of reaction of A is given by

$$r = k_1 C_A - k_2 C_S \tag{7-44}$$

where
$$k_1 = A_1 e^{-E_1/RT}, \qquad k_2 = A_2 e^{-E_2/RT}$$

The notation appearing in these expressions is a simplification of that introduced in Chapter 1 in that k_1 and k_2 are used to represent k_{c1} and k_{c2}, E_1 and E_2 are used to represent E_{A1} and E_{A2}, and r is used to represent r_A.

The rate of reaction r may be expressed as a function of temperature T and conversion x as follows:

$$r = C_A^0 \left[k_1(1 - x) - k_2 x \right] \tag{7-45}$$

From Figure 12-1 of Chapter 12, it is evident that exothermic and endothermic reactions are characterized by the following relationships between the activation energies.

$$E_1 < E_2 \qquad \text{(exothermic)} \tag{7-46}$$

$$E_1 > E_2 \qquad \text{(endothermic)} \tag{7-47}$$

Suppose that the initial concentrations of A and S are such that the reaction takes place from left to right. That is,

$$r(0, T) = k_1 C_A^0 - k_2 C_S^0 > 0 \tag{7-48}$$

At equilibrium, $r = 0$ and $x = x_e$ (the equilibrium conversion), and Equation (7-45) may be solved for the equilibrium conversion:

$$x_e = \frac{1}{1 + k_2/k_1} \tag{7-49}$$

Then

$$\frac{dx_e}{dT} = -\left[\frac{1}{(1 + k_2/k_1)^2}\right]\left[\left(\frac{a_2}{RT^2}\right)\frac{k_2}{k_1}\right] \tag{7-50}$$

where $$a_2 = E_2 - E_1$$

These relationships may be used to show that the limiting values of x_e and dx_e/dT for an exothermic reaction are as follows:

$$\lim_{T \to 0} x_e = 1$$

$$\lim_{T \to \infty} x_e = \frac{1}{1 + A_2/A_1}$$

$$\lim_{T \to 0} \frac{dx_e}{dT} = 0 \tag{7-51}$$

$$\lim_{T \to \infty} \frac{dx_e}{dT} = 0$$

For an exothermic reaction, $a_2 = E_2 - E_1 > 0$, and consequently, by Equation (7-50), $dx_e/dT < 0$. A typical curve of x_e versus T is shown in Figure 7-6 for a first-order reversible, exothermic reaction. For a first-order reversible, endothermic reaction, $E_2 - E_1 < 0$. A typical graph of x_e versus T for these reactions is shown in Figure 7-7.

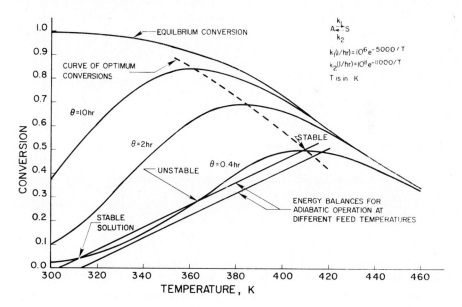

Figure 7-6. An exothermic, first-order reversible reaction in a perfectly mixed flow reactor.

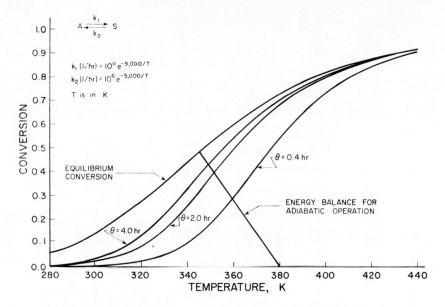

Figure 7-7. An endothermic, first-order reversible reaction in a perfectly mixed flow reactor.

Suppose that the reaction represented by Equation (7-43) is carried out in a perfectly mixed flow reactor. Then the material balance given by Equation (7-39) may be stated in the form

$$x = \mathcal{R}\theta \tag{7-52}$$

where

$$\mathcal{R} = \frac{r}{C_A^0}$$

$$\theta = \frac{V}{v_T}$$

For the special case where the rate of reaction r is given by Equation (7-45), the following expression for x as a function of θ and T is readily obtained by eliminating $\mathcal{R}$ from Equations (7-45) and (7-52) to give

$$x = \frac{k_1}{k_1 + k_2 + 1/\theta} \tag{7-53}$$

Although the shapes of the curves of x versus T at parameters of θ may be deduced directly by use of Equation (7-53), a more general method is presented for deducing the shapes of these curves. Instead of choosing a specific reaction for making the analysis, the general case is considered in which the rate of reaction is given by any arbitrary continuous function which may be represented symbolically as $\mathcal{R} = \mathcal{R}(x, T)$. The conversion x which may be achieved in a perfectly mixed flow reactor is known from experiment to depend on the residence time θ and the temperature T of the reacting mixture; that is,

$$x = x(\theta, T)$$

This dependency of x on θ and T is also implied by Equation (7-52). Thus the total differential of $\mathcal{R}$ may be represented by

$$d\mathcal{R} = \left(\frac{\partial \mathcal{R}}{\partial x}\right)_T dx + \left(\frac{\partial \mathcal{R}}{\partial T}\right)_x dT$$

The partial derivative of $\mathcal{R}$ with respect to T with θ held fixed may be computed by use of the expression

$$\left(\frac{\partial \mathcal{R}}{\partial T}\right)_\theta = \left(\frac{\partial \mathcal{R}}{\partial x}\right)_T \left(\frac{\partial x}{\partial T}\right)_\theta + \left(\frac{\partial \mathcal{R}}{\partial T}\right)_x \tag{7-54}$$

Partial differentiation of each member of Equation (7-52) with respect to T with θ held fixed gives

$$\left(\frac{\partial \mathcal{R}}{\partial T}\right)_\theta = \frac{1}{\theta}\left(\frac{\partial x}{\partial T}\right)_\theta \tag{7-55}$$

Elimination of $(\partial \mathcal{R}/\partial T)_\theta$ from Equations (7-54) and (7-55), followed by rearrangement, gives

$$\frac{\partial x}{\partial T} = \frac{\theta(\partial \mathcal{R}/\partial T)_x}{1 - \theta(\partial \mathcal{R}/\partial x)_T} \tag{7-56}$$

By application of the chain rule to the members of the Equation (7-54), the following expression for the second derivative of $\mathcal{R}$ with respect to T with θ held fixed is obtained:

$$\left(\frac{\partial^2 \mathcal{R}}{\partial T^2}\right)_\theta = \left(\frac{\partial^2 x}{\partial T^2}\right)_\theta \left(\frac{\partial \mathcal{R}}{\partial x}\right)_T + \left(\frac{\partial x}{\partial T}\right)_\theta^2 \left(\frac{\partial^2 \mathcal{R}}{\partial x^2}\right)_T$$
$$+ 2\left(\frac{\partial x}{\partial T}\right)_\theta \left(\frac{\partial^2 \mathcal{R}}{\partial x \partial T}\right) + \left(\frac{\partial^2 \mathcal{R}}{\partial T^2}\right)_x \tag{7-57}$$

But, by Equation (7-55),

$$\left(\frac{\partial^2 \mathcal{R}}{\partial T^2}\right)_\theta = \frac{1}{\theta}\left(\frac{\partial^2 x}{\partial T^2}\right)_\theta \tag{7-58}$$

From Equations (7-57) and (7-58), the following expression for the second derivative of x with respect to T and θ held fixed is obtained:

$$\left(\frac{\partial^2 x}{\partial T^2}\right)_\theta = \frac{\theta\left[\left(\frac{\partial x}{\partial T}\right)_\theta^2 \left(\frac{\partial^2 \mathcal{R}}{\partial x^2}\right)_T + 2\left(\frac{\partial x}{\partial T}\right)_\theta \left(\frac{\partial^2 \mathcal{R}}{\partial x \partial T}\right) + \left(\frac{\partial^2 \mathcal{R}}{\partial T^2}\right)_x\right]}{1 - \theta(\partial \mathcal{R}/\partial x)_T} \tag{7-59}$$

For the special case of the first-order reversible reaction given by Equation (7-43), the following expressions for the partial derivatives of $\mathcal{R}$ appearing in Equation (7-59) may be found by use of Equation (7-45):

$$\left(\frac{\partial \mathcal{R}}{\partial x}\right)_T = -(k_1 + k_2)$$

$$\left(\frac{\partial^2 \mathcal{R}}{\partial x^2}\right)_T = 0$$

$$\left(\frac{\partial \mathcal{R}}{\partial T}\right)_x = \frac{1}{RT^2}\left[E_1 k_1(1 - x) - E_2 k_2 x\right]$$

$$\left(\frac{\partial^2 \mathcal{R}}{\partial T^2}\right)_x = -\frac{2}{RT^3}\left[E_1 k_1(1 - x) - E_2 k_2 x\right] \qquad (7\text{-}60)$$

$$\qquad\qquad\qquad + \frac{1}{RT^4}\left[E_1^2 k_1(1 - x) - E_2^2 k_2 x\right]$$

$$\left(\frac{\partial^2 \mathcal{R}}{\partial x\, \partial T}\right) = -\frac{1}{RT^2}\left(E_1 k_1 + E_2 k_2\right)$$

Consider first the case of an exothermic, first-order reversible reaction for which $E_1 < E_2$. Since $(\partial \mathcal{R}/\partial x)_T < 0$, as shown, it follows that the denominator of the expression for $(\partial x/\partial T)_\theta$ [Equation (7-56)] is always positive. However, for an exothermic reaction, the derivative $(\partial \mathcal{R}/\partial T)_x$ which appears in the numerator of Equation (7-56) may change signs. Let T_k denote the temperature at which a supposed change in signs occurs, that is, at $T = T_k$,

$$\left(\frac{\partial \mathcal{R}}{\partial T}\right)_x = 0$$

and thus

$$\left(\frac{\partial x}{\partial T}\right) = 0 \qquad (7\text{-}61)$$

At $(\partial \mathcal{R}/\partial T)_x = 0$,

$$E_k k_1(1 - x) = E_2 k_2 x$$

and thus

$$\left(\frac{\partial^2 \mathcal{R}}{\partial T^2}\right)_x = \frac{E_2 k_2 x}{RT^4}\left(E_1 - E_2\right) \qquad (7\text{-}62)$$

Since $E_1 - E_2 < 0$, it follows that, at $T = T_k$,

$$\left(\frac{\partial^2 \mathcal{R}}{\partial T^2}\right)_x < 0 \qquad (7\text{-}63)$$

By use of these results at $T = T_k$ and the fact that $(\partial^2 \mathcal{R}/\partial T^2)_T = 0$ for all T, it follows from Equation (7-57) that, at $T = T_k$,

$$\left(\frac{\partial^2 x}{\partial T^2}\right)_\theta < 0 \tag{7-64}$$

From the calculus, it is known that, if the second derivative of a function is negative at the value of the independent variable for which the first derivative is equal to zero, then the function passes through a maximum. Thus, for each choice of θ, x passes through a maximum as shown in Figure 7-6. Since a maximum value of x exists for each θ for the first-order reversible, exothermic reaction, there exists an optimum operating line (or optimum curve) which passes through the maximums of all x versus T curves as indicated by the broken line in Figure 7-6.

For the case of an endothermic, reversible reaction for which $E_1 > E_2$, it follows from Equation (7-60) that, for all $x < x_e$,

$$\left(\frac{\partial \mathcal{R}}{\partial T}\right)_x > \frac{\mathcal{R}}{RT^2} \tag{7-65}$$

Since the initial conditions were selected such that $\mathcal{R}(0, T) > 0$, then $\mathcal{R}(x, T) > 0$ for all $x < x_e$ at any given T. Then it is evident that for the endothermic, reversible reaction

$$\frac{\partial \mathcal{R}}{\partial T} > 0 \tag{7-66}$$

For all $x < x_e$ at any given T, and for these reactions Equations (7-56) and (7-60) lead to the result

$$\frac{\partial x}{\partial T} > 0 \tag{7-67}$$

Typical curves of x versus T at fixed values of θ are shown in Figure 7-7.

From an analysis similar to that outlined, the curves for irreversible reactions in perfectly mixed flow reactors may be deduced. Curves for an exothermic reaction are shown in Figure 7-8.

The graph of the equation for the energy balance on the reaction phase diagram is sometimes called the *reaction path*. For purposes of illustration, suppose that the perfectly mixed flow reactor shown in Figure 7-2 is to be operated adiabatically. For this type of operation, Equation (7-40) may be solved for x to give

$$x = -\frac{n_T^0 C_{pm}(T - T_0)}{n_A^0 \, \Delta H_r} \tag{7-68}$$

where C_{pm} is the mean value of the heat capacity of the feed over the temperature interval between T_0 and T. If, however, the variation of the ratio $n_T^0 C_{Pm}/n_A^0 \, \Delta H_r$ with temperature is neglected, then Equation (7-68) is a straight line with a positive slope for an exothermic reaction and a negative slope for an endothermic reaction (see Figures 7-6 through 7-8). For the

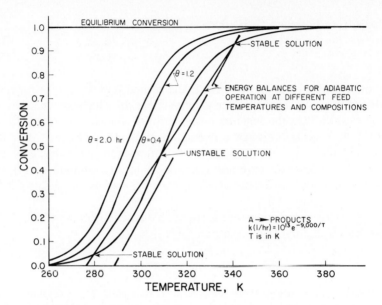

Figure 7-8. An exothermic, first-order irreversible reaction in a perfectly mixed flow reactor.

exothermic, reversible reaction considered, the three intersections of the energy balance [Equation (7-68)] with the material balance [Equations (7-52) and (7-46)] shown in Figure 7-8 suggest that as many as three possible solutions (three sets of values of x and T at a given θ) exist. Whether or not one would obtain these solutions for a specified θ would depend on several factors, such as the inlet temperature of the feed and the shapes of the curves for the material- and energy-balance equations.

The "unstable solution" shown in Figures 7-6 and 7-8 is sometimes called the *ignition point* because at this point a slight increase in the temperature leads to an increase in the rate and the conversion, which results in a further increase in the temperature until the upper solution point is reached. Similarly, if at the unstable solution point the temperature is decreased, the conversion and temperature will tend to decrease until the lower solution point is reached. Thus a minor upset of a reactor operating at the unstable solution point would cause it to move to either the upper or lower solution point. If, on the other hand, the reactor is being operated at either the upper or lower solution point, it tends to return to these respective points after minor upsets. The behavior of reactors for other types of reactions may be deduced by means of an analysis similar to the one shown for the first-order reversible reaction.

For a first-order reversible, endothermic reaction, the energy- and material-balance lines intersect at only one point for any given θ, and reactors in which endothermic reactions are carried out are sometimes referred to as being inherently stable.

A summary of reactor models and the conditions under which multiple solutions exist has been presented by Hlavecek and Votruba [9]. In particular, the plug-flow model does not exhibit multiple solutions at steady-state operation.

PROBLEMS

7-1. The reactor described in Example 7-1 is to be used to produce the substance R in the following reaction:

$$A \rightarrow R, \qquad r_A = k_1 C_A^2$$

where $k_1 = 0.04$ liter/(g mol min) at 20°C and $C_A^0 = 2$ g mol/liter. The heat of the reaction is -16.6 kcal/g mol. All other data are the same as stated in Example 7-1. The reactor is to be operated isothermally at 20°C.
(a) At the selected times of $t = 0, 5, 10, 15, 20$, and 25 min, calculate the temperature that the cooling water must possess in order for isothermal operation to be achieved. Assume that the cooling water in the jacket is perfectly mixed.
(b) Cooling water at an inlet temperature of 0°C is available. At the selected times of $t = 0, 5, 10, 15, 20$, and 25 min, calculate the flow rate of the cooling water required for isothermal operation. What problem do you envision in the operation of the reactor?
Answer: (a) $T_c = 3.94, 11.81, 15.04, 16.68, 17.62, 18.22°C$
        (b) $w_c = 204.65, 37.98, 18.87, 11.71, 8.09, 5.96$ g/min

7-2. The reactor described in Example 7-1 is to be operated adiabatically to polymerize styrene. The initial temperature is 5°C. Cyclohexane is to be used as the solvent. Find the times at which the reacting mixture takes on the temperatures of 10°, 20°, 30°, 40°, and 50°C, given

$$r_A = k_1 C^{1/2} C_A$$

C_A = concentration of styrene, C = concentration of polystyryllithium

$C = 0.005$ g mol/liter, $C_A^0 = 1$ mol/liter, $M_w = 86.2$ g/(g mol)

The variation of k_1 with temperature is presented in Example 7-4. The molar density of the feed is 9.58 g mol/liter, and the specific heat of the feed is 0.42 cal/g °C. Neglect the variation of N_T, C_V, and ΔU_r with temperature throughout the course of the reaction.
Answer: $t \cong 40, 81, 99, 109$, and 119 min

7-3. Obtain the partial differential equation given by Equation (7-10) from Equations (7-8) and (7-9).

7-4. Verify the results presented in Table 6-5 for the batch reactor.

7-5. Begin with the functional relationships

$$\mathcal{R} = \mathcal{R}(x, T)$$
$$x = x(\theta, T)$$

and Equation (7-52),

$$x = \mathscr{R}\theta$$

and develop the expression for $(\partial^2 x/\partial T^2)_\theta$, which is given by Equation (7-59).

7-6. Begin with Equations (7-35) and (7-36) and apply the chain rule as required to obtain the final result given by Equation (7-37).

7-7. In studying reactions in the laboratory, it is common practice to immerse the reactor in a large, perfectly mixed, constant-temperature bath. Suppose the reaction presented in Example 7-1 is to be carried out in a 400-cm³ reactor with a heat-transfer area of 0.35 ft². The initial temperature of the reacting mixture is 20°C. The initial concentrations and the heat-transfer coefficient are the same as those given in Example 7-1. The reactor is immersed in the bath, which is maintained at 20°C throughout the course of the reaction. Compute the reaction temperature and the conversion at selected time intervals. The variation of the reaction rate constant k with temperature is given in Problem 1-6.
Answer: At $t = 4,\ 10,\ 14,\ 18$ min, $T = 28.16,\ 28.87,\ 25.96,\ 23.92°C$, and $x = 0.185,\ 0.48,\ 0.596,$ and 0.671

7-8. If the rate of reaction for the first-order exothermic, reversible reaction $A \rightleftarrows R$ is given by

$$r = C_A^0 \left[A_1 (1 - x) e^{-E_1/RT} - A_2 x e^{-E_2/RT} \right]$$

show that the maximum conversion at a given residence time θ for a perfectly mixed flow reactor (see Figure 2-7) occurs at that temperature given by the following expression.

$$T = \frac{-E_2}{R \log_e \left[E_1/A_2\theta (E_2 - E_1) \right]}$$

7-9. Rework Example 7-1 for the case where the initial concentrations are taken to be the following set:

$$C_A^0 = 0.5 \text{ g mol/liter}$$
$$C = 0.001 \text{ g mol/liter}$$

Answer:

Time (min)	0	5	10	20	30	40	50	60
T_c (°C)	19.39	19.42	19.46	19.52	19.575	19.62	19.67	19.71
w_c (g/min)	3.988	3.98	3.73	3.28	2.88	2.54	2.24	2.01

7-10. For the liquid phase reaction

$$A \rightleftarrows R$$
$$K_c = 5$$
$$\Delta H_r = 15,000 \text{ cal/g mol}$$
$$r_A = k_c (C_A - C_R/K_c)$$
$$k_c = 0.043 \text{ min}^{-1}$$

Pure A is charged to a batch reactor that is operated isothermally at 575 K and 10 atm. The reactor is jacketed and hot oil is circulated through the jacket as required to maintain the reaction temperature at 575 K. The heat capacity of the oil is 0.6 cal/g °C and the mass of oil in the jacket is 100 g. The volume of the reactor is 400 cm³. Additional data are

$C_A^0 = 2$ mol/liter and the charge to the reactor is a solvent with the reactant A

$C_p = 0.5$ cal/g °C, for the reacting mixture

$U_h A = 50$ cal/°C min

Assume that the oil in the jacket is perfectly mixed and that the accumulation term [the right side of Equation (7-16)] is negligible relative to the other terms in this equation. Calculate:
(a) The equilibrium conversion and the time required to reach 90% of this conversion.
(b) The temperature which the hot oil in the jacket must possess at the beginning of the reaction to achieve isothermal operation.
(c) The temperature that the hot oil must possess at 90% of the equilibrium for isothermal operation.
(d) The flow rate of the hot oil at time $t = 0$ and at the time corresponding to a conversion equal to 90% of the equilibrium conversion. The inlet temperature of the oil is 590 K.
Answer: (a) $X_{\text{equil}} = 0.833$, $t = 44.6$ min, (b) $T_c = 585.32$ K, (c) $T_c = 576.32$ K, (d) $w_c \cong 184$ g/min and 6.14 g/min

7-11. The hydrodeoxygenation of 1-naphthol catalyzed by a sulfided N_i Mo/γ–Al$_2$O$_3$ catalyst was investigated by Li et al. [10] in a batch reactor. The initial charge consisted of 1-naphthol dissolved in a large excess of *n*-hexadecane. This reactant solution was saturated with hydrogen, which provided a large stoichiometric excess. The catalyst was crushed and sieved to give particle sizes ranging from 10 to 150 mesh. The catalyst concentration was 1.82 kg/m³. Reaction temperatures and pressures ranged from 473 to 623 K and 22 to 34 atm. The following reaction network was postulated and the corresponding pseudo first-order reaction rate constants and activation energies given were obtained.

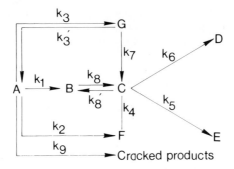

$A =$ 1-naphthol

$B =$ naphthalene

$C =$ tetralin (1,2,3,4-tetrahydro naphthaline)

$D =$ *trans*-decahydronaphthaline
(*trans*-decalin)

$E =$ *cis*-decalin

$F =$ 1-tetralone

$G =$ 5,6,7,8,-tetrahydro-1-naphthol

Reaction No.	k_i or k_i' at 473K (m^3/s kg catalyst)	Activation Energies (kJ/mol)
1	1.64×10^{-6}	139 ± 32
2	4.19×10^{-6}	100 ± 23
3	1.85×10^{-5}	44 ± 20
3'	2.60×10^{-5}	—
4	1.16×10^{-6}	132 ± 38
5	6.5×10^{-7}	—
6	1.87×10^{-6}	—
7	4.64×10^{-4}	—
8	1.34×10^{-4}	—
8'	9.83×10^{-6}	—
9	1.40×10^{-5}	77 ± 23

Since Li et al. [10] were unable to determine the activation energies for reactions 3', 5, 6, 7, 8, and 8', take the activation energies for these reactions to be equal to zero.

Li et al. found that at low temperatures the aromatic ring hydrogenated before hydrodeoxygenation occurred and that at high temperatures hydrodeoxygenation is occurring. Minimum hydrogen consumption thus occurs at the higher temperature. Assume that no hydrogen is produced or consumed in reaction 9, which produces the cracked products. (Hydrogen is probably consumed, but because of the lack of data, use the preceding assumption.)

(a) Calculate the moles of hydrogen consumed per mole of 1-naphthol reacted after a reaction time of 24,000 s for a reaction temperature of 473 K. (*Hint:* Calculate the product distribution of the compounds in the reaction network and then make a hydrogen balance.)

(b) At the conversion of 1-naphthol determined in part (a), calculate the moles of hydrogen consumed per mole of 1-naphthol reacted and the reaction time required to achieve this conversion at 623 K. Do the results found in parts (a) and (b) substantiate the observations of Li et al.?

(c) To demonstrate the effect of temperature on product distribution, calculate and plot the moles of i produced per mole of 1-naphthol reacted versus the conversion of 1-naphthol at 473 and at 623 K.

Answer: (a) 1.164 moles H_2 consumed per mol 1-naphthol consumed, (b) 0.266 moles H_2 consumed per mol 1-naphthol consumed, and $x = 0.804$.

7-12. Kladko [11] reports on the application of kinetics and reaction engineering to a "real engineering design problem." The product, B, was to be produced by the isomerization of chemical A. Only 4000 lb of B were to be produced. Laboratory studies in small-scale reactors indicated no problems in producing B from A in a batch reactor with a 12-h reaction time. However, on conducting the reaction in a 750-gal glass-lined reactor, a rapid temperature rise occurred, control of the temperature was lost, a rapid pressure increase occurred, and the reactor contents were rapidly vented through a pressure safety valve. This type of occurence is referred to as a *runaway* reaction. Thermochemical calculations using bond energies were then performed, and the reaction was found to be exothermic instead of endothermic, as initially expected, with a heat of reaction of -83 cal/g. Batch kinetic experiments were conducted and the following rate equation was found to be applicable at 163°C.

$$r_A = k_c(1 - b)C_A$$

where

b = weight fraction of the aromatic solvent that was used

r_A = lb of A consumed/(gal h)

C_A = lb of A/gal

$k_c = 0.80 \exp[-14{,}570(1/T - 1/436)]h^{-1}$ with T in Kelvin.

Side reactions occur if the temperature of the reaction exceeds 170°C. Therefore, the reaction temperature must be less than 170°C. The heat capacity of the reaction mixture was 1.3 cal/(g °C), and density was 1.2 g/cm³.

(a) A 1000-gal glass-lined reactor, which can be heated or cooled with oil, is available to produce 4000 lb of B. The reaction is to be conducted at 163°C. To accomplish this objective, 1500 lb of solvent is added to the reactor, which is enough material to cover the thermocouple in the reactor. To this, 4200 lb of A is added, and the final conversion of A is to be 97%. Several simulations are conducted, and a feed schedule is used whereby the feed rate of A is increased in a series of step increments over a period of 8 h, held constant for 3 h, and then decreased over a series of step increments over a period of 5 h.

The reaction is continued for another 3 or 4 h until a 97% conversion of the A charged to the reactor is obtained. For this problem, the feed temperature is 20°C, and the feed schedule is approximated as follows:

$$F_A = 25t + 125 \text{ lb/h for the first 8 h}$$

$$F_A = 400 \text{ lb/h for the next 3 h}$$

$$F_A = 424 - 16t \text{ for } 11 < t \leq 17$$

$$F_A = 0 \text{ for } t > 17$$

The reaction then proceeds until 97% conversion of A is obtained. Calculate the conversion of A, the weight fraction of A in the reactor, the mass and volume of fluid in the reactor, and the heat flux as a function of time that is required to maintain the reaction at 163°C. Since pure A is fed to the reactor over a portion of the reaction period and no material is withdrawn from the reactor, this type of reactor is referred to as a semibatch reactor. If the value of UA is equal to 1500 Btu/(h °F), calculate the hot oil temperature required to maintain isothermal operation. Assume all heat is transferred through the jacket of the reactor and that the contents of the jacket are perfectly mixed.

(b) A series of three equal-volume continuous-stirred tank reactors is considered for the production of 2 million pounds per year of B. The reactors are to operate with a pure A feed to the first reactor and a reaction temperature of 163°C. For a feed rate of A of 30 gal/h and a conversion of A of 97% for the system, calculate the residence time and volume of each reactor, weight fraction of A in each reactor, and the heat flux required for each reactor, which is operating at 163°C. How many hours per year of operation will be required to produce 2 million lb/yr of B?

NOTATION

(See also Chapters 1 to 6)

a = heat transfer area per unit length

C_p = molar heat capacity

C_p° = molar heat capacity of a perfect gas

C_T = total molar density

H_{fi}° = enthalpy of component i in the gaseous state of a perfect gas at temperature T and at a pressure of 1 atm above the elements in their standard states at a given datum temperature

n_i = molar flow rate of component i

n_T = total molar flow rate

N_i = moles of component i

N_T = total moles of mixture

Q = total rate of heat transfer; energy per unit time

q = point value of the rate of heat transfer; energy transferred per unit time per unit of length

r_i = rate of disappearance of component i by reaction, moles per unit time per unit volume of reacting mixture

$\mathscr{R}$ = rate of reaction used in the reaction phase plane analysis: $\mathscr{R} = r_A/C_A^0$

S = cross-sectional area

S_c = cross-sectional area occupied by the heat-transfer medium

t = time

T = temperature of the reacting mixture

T_c = temperature of the heat-transfer medium

U = internal energy

U_h = overall heat-transfer coefficient

v = molar volume of the reacting mixture

v_T = total volumetric flow rate

V = volume of the reactor filled by the reacting mixture

w = mass flow rate

w_c = mass flow rate of the heat-transfer medium in consistent units

z = length of the path followed by the reacting mixture, and the positive direction of z is taken to be the direction of flow of the reacting mixture; if the reacting mixture is perfectly mixed in the reactor, then the positive direction is taken to be the direction of flow of the heat-transfer medium

Greek Letters

$\alpha = U_h A/w_c C_{p_c}$

ρ = mass density

REFERENCES

1. Worsfold, D. J., and S. Bywater, *Canadian J. Chem. Eng.*, 38:1891 (1960).

2. Hsieh, H. J., *Polymer Sci.*, A3:153 (1965).

3. Johnson, P. R., S. L. Parsons, and J. B. Roberts, *Ind. Eng. Chem.*, 51:499 (1959).

4. Stull, D. R., E. F. Westrum, Jr., and G. C. Sinke, *The Chemical Thermodynamics of Organic Compounds*, John Wiley & Sons, Inc., New York (1969).

5. Hougen, O. A., K. M. Watson, and R. A. Ragatz, *Chemical Process Principles*, Material and Energy Balances, Vol. 1, John Wiley & Sons, Inc., New York (1943).

6. Holland, C. D., *Fundamentals and Modeling of Separation Processes, Adsorption, Distillation, Evaporation, and Extraction*, Prentice-Hall, Inc., Englewood Cliffs, N.J. (1975).

7. van Heerden, C., *Ind. Eng. Chem.*, 45:1242 (1953); *Chem. Eng. Sci.*, 8:113 (1958).

8. Bilous, O., and N. R. Amundson, *AIChE J.*, 1:513 (1955).

9. V. Hlavacek and J. Votruba, Ch. 6, pg. 314, *Chemical Reactor Theory, A Review*, Leon Lapidus and Neal R. Amundson, eds., Prentice-Hall, Inc., Englewood Cliffs, N.J. (1977).

10. Li, C.-L., Z.-R. Xu, Z.-A. Cao, and B. C. Gates, *AIChE J.*, 31, (No. 1):171 (1985).

11. Kladko, M., *Chemtech*, p. 1 (March 1971).

Fundamentals of Heterogeneous Catalysis

8

A catalyst is generally defined as a substance which alters the speed of a reaction but does not itself undergo any chemical change. In reality, most substances classified as catalysts are eventually destroyed by their subsequent combination with products produced by the reaction in which the catalyst participates. Frost and Pearson [1] suggest that, from a practical point of view, a catalyst should be defined as a substance which alters the speed of a given reaction, regardless of the final fate of the catalyst.

Since most catalysts are employed for the purpose of increasing the rate of a particular reaction, it follows that the presence of the catalyst leads to a mechanism having a lower energy barrier than the barrier which exists when the reaction occurs in the absence of the catalyst. The primary role of the catalyst is to change the mechanism to one with a lower activation energy. It follows that if the forward rate of a given reaction is increased the reverse rate must be increased by the same proportion because the thermodynamic equilibrium is unaltered by the presence of the catalyst.

The mechanisms of heterogeneous catalysis are very complex. Most catalysts are prepared by one of two methods. The first is the impregnation of an active or inactive support with a material which is catalytic. Some supports are activated alumina, silica gels, silica aluminas, Fuller's earth, and activated charcoal. The support material is used because of its high internal surface area per unit mass. In the second method, catalyst supports are not used; instead the solid chemicals are slurried, mixed, pelletted, and then dried.

A certain amount of heat treatment of the catalyst pellet is required in both methods. Also, catalyst preparation may involve a pelletizing step which affects the final pore structure of the catalyst. The interior pore structure is commonly subdivided into the micropore and macropore structure, or small and large pores. Two parameters used to characterize the interior of a catalyst are surface area and pore size distribution. Other parameters which have been considered are the crystal structure, the electronic potential, and the semiconductivity and photoconductivity of the catalyst. Instruments used in catalytic research are electron microscopes, nuclear magnetic resonance (NMR), gas chromatography, mass spectroscopy, electron spin resonance, and others.

This chapter is devoted to the development of rate equations for catalytic reactor design. The development is based on the postulate that the following steps are required for a chemical reaction to occur.

1. Mass transfer by film diffusion of the reactants from the flowing stream of the reacting fluid to the exterior surface of the catalyst.
2. Diffusion of the molecules into the interior of the catalyst (pore diffusion).
3. Adsorption of the reactants on the surface of the catalyst.
4. Reaction between adsorbed molecules (or between an adsorbed molecule and a molecule in the gas phase). Once the reactants have been adsorbed, surface migration may occur. The effect or extent of surface migration (or surface diffusion) is difficult to characterize.
5. Desorption of product molecules from the surface of the catalyst.
6. Diffusion of the products from the interior to the exterior of the catalyst surface (pore diffusion).
7. Mass transfer by film diffusion of the products from the exterior of the catalyst surface to the flowing stream.

These seven steps are seen to occur in series, and for a flow process to be at steady-state operation, the individual rates must be equal, provided that each rate is stated on the same basis. Thus it might appear possible to develop rate expressions involving the individual rate constants (or resistances) and the overall potential (or driving force) in a manner analogous to that used to develop the heat-transfer equation: $Q = UA \, \Delta T_{lm}$. However, it is generally impossible to obtain such an expression because of the impossibility of stating the individual rate expressions in the explicit forms needed to derive such an expression. Because of the complexity of the process, working expressions for the reaction rates are commonly based on the rate expression for the *rate controlling step*, the step having the smallest rate constant. (This step is sometimes called the *slowest rate* in spite of the fact that the rates of the individual steps are equal.)

Rate expressions for steps 3 through 5 are developed in this chapter. The rate expressions presently used for chemical adsorption have grown out of the theory of physical adsorption. Also, experimental methods that give a measure of the effectiveness of a catalyst or a catalyst support, such as the measurement of surface area, have grown out of the theory of physical adsorption. Thus Part I is devoted primarily to the development of the theory of physical adsorption. Part II considers chemical adsorption, reaction, and desorption, and Part III is concerned with reactor design. The effect of pore diffusion and mass transfer to and from the reacting stream and the interface at the surface are treated in Chapter 9.

I. PHYSICAL ADSORPTION

When a gas is brought into contact with an evacuated solid (such as charcoal) and part of it is taken up by the solid, any one of several processes may have occurred. If the gas molecules occupy the void space within the solid, the process is called *absorption*. If the molecules become attached to the surface of the adsorbent, the process is known as *adsorption*. If the interactions between the gas and the solid are weak, similar to those involved in condensation, the process is called *physical adsorption*, and if the interaction between the adsorbed molecules and the surface is strong, similar to chemical bonding, the process is called *chemical adsorption*. Physical adsorption is also called *van der Waals adsorption*, which implies that the van der Waals forces which are active in condensation are also involved in physical adsorption.

All physical adsorption and most chemical adsorption processes are *exothermic*, a result which has been verified experimentally time and again. However, in the case of chemical adsorption, two processes have been found to be endothermic: the chemisorption of molecular hydrogen on glass and the adsorption of molecular hydrogen (followed by dissociation) on iron surfaces which have been contaminated with sulfur. That one should expect physical adsorption to be exothermic can be deduced from thermodynamic considerations in the following manner. An atom of the adsorbent that is on the surface of the adsorbent is subjected to an internal pull that is greater than the outward forces, which results in a tendency to decrease the surface. When a molecule is adsorbed, a balance of some of the inward forces is achieved, which in turn decreases the surface tension. Thus the surface energy (the product of surface tension and surface area) is decreased by adsorption. Consequently, adsorption results in a decrease in free energy for the system; $\Delta G < 0$. Prior to adsorption, the gas molecules are free to move in three-dimensional space, and following adsorption they are free to move in at most the two-dimensional space along the surface. Thus adsorption is accompanied by a decrease in entropy for the system; $\Delta S < 0$. Since adsorption is accompanied both by a decrease in free energy and a decrease in entropy, it follows that the process is always exothermic, $\Delta H < 0$, since $\Delta H = \Delta G + T\Delta S < 0$. The heats of physical adsorption are of the same order of magnitude as the heats of condensation of gases, and the heats of chemical adsorption are of the same order of magnitude as those for chemical reactions.

The amount of a gas adsorbed by physical adsorption at a given pressure increases as the saturation temperature is approached. At a given temperature and pressure, the amount of a gas adsorbed increases with the normal boiling

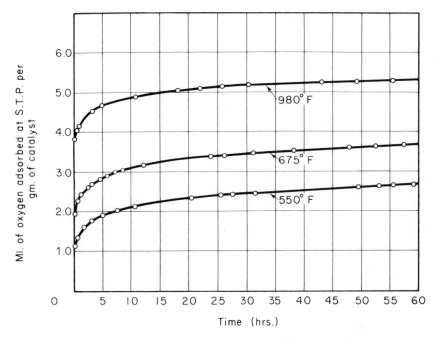

Figure 8-1. Chemical adsorption of oxygen on a chromium catalyst supported on aluminum oxide. [Taken from C. D. Holland and P. G. Murdoch, *AIChE J*. 3:386 (1957). Used with permission of the American Institute of Chemical Engineers.]

point of the gas or with the critical temperature. Chemical adsorption, on the other hand, does not generally occur at relatively low temperatures, which is demonstrated by data taken by Holland and Murdoch [2] and presented in Figure 8-1. Also, the initial amounts adsorbed increase as the temperature is increased. The rate of chemical adsorption is seen to be relatively fast at first and then very slow. An equilibrium was never achieved at the temperatures shown in Figure 8-1.

Physical adsorption is readily reversible with respect to temperature and pressure, whereas chemically adsorbed gases are difficult to remove even by evacuation and heating. In fact, the only way the oxygen adsorbed (see Figure 8-1) could be removed was to react it with hydrogen [2].

Furthermore, the amount of gas chemically adsorbed does not alter the amount which can be physically adsorbed. For example, the same amount of nitrogen was adsorbed in surface area determinations on the reduced adsorbent as was adsorbed by the adsorbent in each of the three states of oxidation shown in Figure 8-1. Furthermore, the adsorbed nitrogen could be easily removed by heating and evacuating, which did not, however, remove the adsorbed oxygen. In effect, the chemically adsorbed oxygen becomes a part of the surface and is removable only by chemical reaction.

A. Types of Physical Adsorption

Since the initial observations of C. K. Scheele in 1773 of the adsorption of gases by solids, a wide variety of both adsorbents and adsorbates has been investigated. The amount of a pure gas adsorbed by a unit mass of a given adsorbent is a function of temperature and pressure alone,

$$v_s = f(P, T) \tag{8-1}$$

where $v_s = cm^3$ of gas adsorbed at STP (0°C and 1 atm) per gram of adsorbent. The results of adsorption experiments are most commonly presented in the form of *adsorption isotherms*, volume adsorbed as a function of pressure at constant temperature.

$$v_s = f(P), \qquad \text{at constant } T \tag{8-2}$$

Sometimes the results are presented in the form of *adsorption isobars*, the volume of gas adsorbed as a function of temperature at constant pressure.

$$v_s = f(T), \qquad \text{at constant } P \tag{8-3}$$

Occasionally, results are presented as *adsorption isosteres*, equilibrium pressure as a function of temperature at a fixed amount of gas adsorbed.

$$P = f(T), \qquad \text{at constant } v_s \tag{8-4}$$

A variety of shapes of curves have been observed by the various investigators who have studied the adsorption of many different gases by many different types of adsorbents. Brunauer [3] suggested the classification of these results according to the five types of isotherms shown in Figure 8-2. Adsorption isotherms of type I are generally attributed to unimolecular adsorption. These curves are also referred to as Langmuir isotherms because they are described by the model proposed by Langmuir [4, 5, 6], which is presented in a subsequent section. The *S-shaped* or *sigmoid* isotherms, type II, are generally regarded as being descriptive of multimolecular adsorption. Brunauer [3] suggests that type III isotherms represent the formation of multimolecular layers before a unimolecular layer has been adsorbed. This behavior is attributed to the fact that the attraction between the adsorbent surface and the gas is less than the attraction between liquid molecules. Brunauer [3] showed that types IV and V could be represented by equations that took into account the effect of capillary condensation.

B. Adsorption Models

In an attempt to explain the wide variety of experimental results characterized by Figure 8-2, many theories and models have been proposed. Of these, only those needed to provide continuity in the development of the expressions employed in the description of adsorption equilibria of multicomponent mixtures are presented.

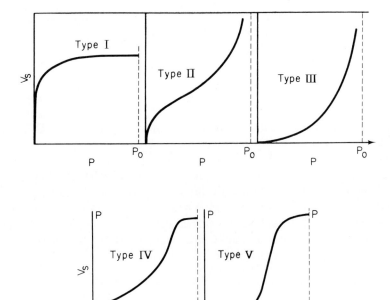

Figure 8-2. The five types of physical adsorption as proposed by Brunauer. [Taken from Stephen Brunauer, L. S. Deming, W. E. Deming, and Edward Teller, *J. Amer. Chem. Soc.*, 52:1723 (1940). Used with permission of the American Chemical Society.]

1. Henry's Law. Frequently, for adsorptions carried out isothermally at low pressures, the amount of gas adsorbed can be described by Henry's law,

$$v = k_0 P \qquad (8\text{-}5)$$

where $\qquad\qquad k_0 =$ Henry's law constant

$\qquad\qquad\quad P =$ adsorption equilibrium pressure

At high pressures, the volume adsorbed in many instances becomes independent of pressure [7]; that is,

$$v = k_0' \qquad (8\text{-}6)$$

Thus the power of the pressure for some systems ranges from 0 to 1. For intermediate pressures, the Freundlich equation is commonly used.

2. The Freundlich Equation. This empirical expression is of the form

$$v = k_1 P^{1/n} \qquad (8\text{-}7)$$

where $n > 1$. More recently, this expression has been developed theoretically on the basis of a logarithmic distribution of adsorption sites [8, 9]. This expression often gives the best fit for isotherms of type I [10].

3. The Langmuir Equation. Brunauer [3] considered the Langmuir equation to be, perhaps, the most important single equation in the field of adsorption. Three derivations of this equation have been presented: the original kinetic derivation of Langmuir [4, 5, 6], the thermodynamic derivation of Volmer [11], and statistical derivations by Fowler [12] and others. Of these, only the kinetic derivation is presented here.

When a molecule strikes the surface of an adsorbent, it may be either elastically reflected from the surface without any energy exchange taking place or it may be elastically adsorbed with the release of energy. Most frequently, the collision is inelastic, and the molecule remains in contact with the surface. Langmuir [4, 5, 6] attributed the phenomenon of adsorption to the average time that a molecule resides on the surface of an adsorbent. In summary, the postulates of Langmuir are as follows:

1. Of the molecules striking the surface, only those that strike a bare surface are candidates for adsorption. That is, molecules that strike an adsorbed molecule are elastically reflected.
2. The probability of desorption of a molecule from the surface is the same whether or not the neighboring positions on the surface are empty or filled by other molecules. This amounts to assuming that the interaction between adsorbed molecules is negligible.

Let μ be equal to the number of molecules striking a unit of surface area per unit time. Let θ be equal to the fraction of the surface covered by adsorbed molecules. Then the candidates for adsorption by postulate 1 are given by $(1 - \theta)\mu$. Let α denote the fraction of the molecules striking the bare surface which are condensed. Then

$$\text{Rate of adsorption} \left(\frac{\text{molecules}}{\text{s cm}^2} \right) = \alpha(1 - \theta)\mu \qquad (8\text{-}8)$$

Let ν denote the rate of evaporation (or desorption) from a completely covered surface. Then, by postulate 2,

$$\text{Rate of desorption} \left(\frac{\text{molecules}}{\text{s cm}^2} \right) = \nu\theta \qquad (8\text{-}9)$$

Then, at equilibrium, the fraction of surface covered by adsorbed molecules is given by

$$\alpha(1 - \theta)\mu = \nu\theta \qquad (8\text{-}10)$$

or

$$\theta = \frac{(\alpha/\nu)\mu}{1 + (\alpha/\nu)\mu} \qquad (8\text{-}11)$$

The following expression for μ, the number of molecules striking 1 cm^2 of surface per second, is given by the kinetic theory of gases [13],

$$\mu = \frac{P}{(2\pi mkT)^{1/2}} \qquad (8\text{-}12)$$

where

m = mass of a molecule

k = Boltzmann constant

T = temperature of the gas, K

An expression for ν may be deduced from the concepts of the kinetic theory. Let q denote the heat given off when a molecule is adsorbed. Then, to be desorbed, a molecule must possess an energy equal to or greater than q. If it is supposed that the adsorbed molecules possess a Maxwellian energy distribution in two degrees of freedom, then it may be shown that the fraction of molecules having an energy equal to or greater than q is given by $e^{-q/kT}$. Thus Equation (8-11) may be restated as follows:

$$\theta = \frac{bP}{1 + bP} \qquad (8\text{-}13)$$

The dependence of b on temperature is given by

$$b = \frac{\alpha e^{q/kT}}{(2\pi mkT)^{1/2}} \qquad (8\text{-}14)$$

Since θ may be expressed as the ratio of the volume of gas adsorbed at pressure P to the volume required to form a unimolecular layer at the same temperature T, Equation (8-13) is commonly stated as

$$\frac{v}{v_m} = \frac{bP}{1 + bP} \qquad (8\text{-}15)$$

where v = volume of gas adsorbed

v_m = volume of gas adsorbed when the adsorbent is covered with a complete unimolecular layer

If $bP \ll 1$, then Equation (8-15) reduces to the form of Henry's law given by Equation (8-5), and if $bP \gg 1$, the form of Henry's law given by Equation (8-6) is obtained.

When the adsorption consists of a partially filled unimolecular layer, it is generally possible to obtain a satisfactory fit with Equation (8-15), the Langmuir equation. Also, this equation has been used with some success in the interpretation of chemical adsorption data. However, the development of the Langmuir equation was presented here, primarily, because it represents the starting point for the development of the BET method, the theory of multimolecular adsorption as proposed by Brunauer, Emmett, and Teller [14].

4. The BET Equation. The multimolecular adsorption theory proposed by Brunauer, Emmett, and Teller [14] was the first attempt to present a unified theory of physical adsorption of pure gases. The equations resulting from this theory may be used to correlate the five types of adsorption shown in Figure 8-2. The multimolecular theory of adsorption constitutes a generalization of Langmuir's theory in that the restriction of unimolecular adsorption is removed. In addition, the following postulates are made.

1. Let $s_0, s_1, s_2, \ldots, s_n, \ldots$ represent the surface area covered by $0, 1, 2, \ldots, n, \ldots$ layers of adsorbed molecules. Postulate that the rate of condensation on s_n is equal to the rate of evaporation from surface s_{n+1} ($n = 0, 1, 2, \ldots$).
2. Postulate that the evaporation–condensation properties of the molecules in the second and higher adsorbed layers are the same as those of the lower layers.

By postulate 1, the rate of adsorption on the bare surface is equal to the rate of evaporation from the first layer,

$$a_1 P s_0 = b_1 s_1 e^{-E_1/RT} \tag{8-16}$$

where E_1 is the heat of adsorption for the first layer, and a_1 and b_1 are constants. The relationship between the theory for multimolecular adsorption and the theory of Langmuir's adsorption is shown as follows. The rate of adsorption on the bare surface is equal to $a_1 P s_0$, which is proportional to $\alpha(1 - \theta)\mu$. The rate of desorption from the surface, which is covered by one layer of adsorbed molecules, is equal to $b_1 s_1 \exp(-E_1/RT)$, which is proportional to $\nu\theta$. Thus Equation (8-16) consists of an alternate statement of Langmuir's equation for unimolecular adsorption, and involves the assumption that a_1, b_1, and E_1 are independent of adsorbed molecules already present in the first layer. Similarly, the rate of condensation on top of the first layer is equal to the rate of evaporation from the second layer,

$$a_2 P s_1 = b_2 s_2 e^{-E_2/RT}$$

Continuation of this argument yields

$$a_3 P s_2 = b_3 s_3 e^{-E_3/RT}$$

$$\vdots \qquad \vdots$$

$$a_j P s_{j-1} = b_j s_j e^{-E_j/RT} \tag{8-17}$$

$$\vdots \qquad \vdots$$

$$a_n P s_{n-1} = b_n s_n e^{-E_n/RT}$$

(It is of interest to note that, in spite of the fact that Langmuir's name is commonly associated with unimolecular adsorption, he also formulated a set of equations in 1918 [6] for the case where the adsorption spaces may hold more than one adsorbed molecule. These equations were of the same form as those given by Equation (8-17), but their summation was handled in a manner different from that proposed by Brunauer, Emmett, and Teller [14].) The summation procedure used by Brunauer, Emmett, and Teller follows. The total surface of an adsorbent is given by

$$A = \sum_{j=0}^{n} s_j \tag{8-18}$$

where n is equal to the total number of adsorbed layers. The total volume adsorbed is given by

$$v = v_0 \sum_{j=0}^{n} js_j \tag{8-19}$$

where v_0 is equal to the volume adsorbed per unit area when the adsorbent surface is covered with a unimolecular layer. Then it follows that

$$\frac{v}{Av_0} = \frac{v}{v_m} = \frac{\sum_{j=0}^{n} js_j}{\sum_{j=0}^{n} s_j} \tag{8-20}$$

where v_m is equal to the volume adsorbed when the adsorbent is covered with a complete unimolecular layer. By postulate 2,

$$E_2 = E_3 = \cdots = E_n = E_L$$

$$\frac{b_2}{a_2} = \frac{b_3}{a_3} = \cdots = \frac{b_n}{a_n} = g \tag{8-21}$$

where g is a constant and E_L is equal to the heat of condensation. Each of the areas $s_1, s_2, \ldots, s_j, \ldots$ may be expressed in terms of s_0 in the following manner. Let

$$y = \left(\frac{a_1}{b_1}\right) Pe^{E_1/RT} \quad \text{and} \quad x = \left(\frac{P}{g}\right) e^{E_L/RT} \tag{8-22}$$

Then

$$s_1 = ys_0$$

$$s_2 = xs_1$$

$$s_3 = xs_2 = x^2 s_1$$

$$\vdots \quad \vdots \qquad \vdots \tag{8-23}$$

$$s_j = xs_{j-1} = x^{j-1} s_1 = yx^{j-1} s_0 = cx^j s_0$$

where

$$c = \frac{y}{x} = \left(\frac{a_1 g}{b_1}\right) e^{(E_1 - E_L)/RT}$$

This result may be used to compute v/v_m as given by Equation (8-20) as follows:

$$\frac{v}{v_m} = \frac{cs_0 \sum_{j=0}^{n} jx^j}{s_0 \left[1 + c\sum_{j=1}^{n} x^j\right]} \tag{8-24}$$

The sum of the geometric series appearing in the denominator is given by

$$\sum_{j=1}^{n} x^j = \frac{x - x^{n+1}}{1 - x} \tag{8-25}$$

The summation in the numerator of Equation (8-24) is evaluated in the following manner. First observe that

$$\sum_{j=1}^{n} jx^j = x \frac{d\left[\sum_{j=1}^{n} x^j\right]}{dx} \tag{8-26}$$

Then, by use of the result obtained from Equation (8-25), it follows that

$$\sum_{j=1}^{n} jx^j = x \frac{d\left[(x - x^{n+1})/(1 - x)\right]}{dx}$$

Thus

$$\sum_{j=1}^{n} jx^j = \frac{x}{(1 - x)^2}\left[1 - (n + 1)x^n + nx^{n+1}\right] \tag{8-27}$$

Substitution of the results given by Equations (8-25) and (8-27) into Equation (8-24), followed by rearrangement, yields

$$\frac{v}{v_m} = \left(\frac{cx}{1 - x}\right)\left[\frac{1 - (n + 1)x^n + nx^{n+1}}{1 + (c - 1)x - cx^{n+1}}\right] \tag{8-28}$$

Brunauer, Emmett, and Teller [14] suggest that the width of the pores, cracks, and capillaries of the adsorbent sets a limit on the maximum number of layers that can be adsorbed.

A physical interpretation of x is obtained from the following considerations. At $P \geq P_0$ (the saturation pressure of the given pure component at the temperature T of the adsorption), any amount of vapor may be condensed. The condensed vapor will appear to have been adsorbed. Consequently, the volume adsorbed becomes unbounded at the saturation pressure. As x approaches unity, these same characteristics are exhibited by Equation (8-28). By repeated application of l'Hôpital's rule, the following result is obtained:

$$\lim_{x \to 1} \frac{v}{v_m} = \frac{c(n^2 + n)}{2(1 + cn)} \tag{8-29}$$

Then, at $x = 1$, it follows that v/v_m is unbounded as n increases without bound. Agreement between the physical considerations and Equation (8-28) is achieved by the condition that

$$\text{at} \quad P = P_0, \qquad x = 1 \tag{8-30}$$

From Equation (8-22),

$$\frac{x}{x_0} = \frac{P}{P_0} \tag{8-31}$$

where

$$x_0 = \frac{P_0}{g} e^{E_L/RT}$$

Comparison of Equations (8-30) and (8-31) shows that $x_0 = 1$, or

$$x = \frac{P}{P_0} \tag{8-32}$$

For the special case where $n = 1$, Equation (8-28) reduces to

$$\frac{v}{v_m} = \frac{cx}{1 + cx}$$

which is recognized as Langmuir's equation:

$$\frac{v}{v_m} = \frac{c(P/P_0)}{1 + c(P/P_0)} \tag{8-33}$$

For $x < 1$, it is readily shown that the limit of v/v_m [given by Equation (8-28)] as the number of layers n is allowed to increase without bound is given by the following relationship:

$$\frac{v}{v_m} = \frac{cx}{(1 - x)(1 - x + cx)} \tag{8-34}$$

or

$$\frac{v}{v_m} = \frac{cP}{(P_0 - P)[1 + (c - 1)(P/P_0)]} \tag{8-35}$$

This expression is called the BET equation (after Brunauer–Emmett–Teller [14]). In the application of this expression for the determination of the surface area on an adsorbent, the following linear form may be employed.

$$\frac{P}{v(P_0 - P)} = \frac{c - 1}{v_m c}\frac{P}{P_0} + \frac{1}{v_m c} \tag{8-36}$$

Then for multimolecular adsorption of the type described by Equation (8-36), a plot of $P/v(P_0 - P)$ versus P/P_0 (where $P/P_0 < 1$) should yield a straight line having a slope of $(c - 1)/v_m c$ and an intercept of $1/v_m c$. From the values obtained for the slope and intercept, the volume v_m required to cover the surface with a unimolecular layer is computed. After v_m has been determined, the surface area of the adsorbent may be calculated as demonstrated in Example 8-1.

It is now fairly common practice to use nitrogen isotherms for the determination of the surface areas of catalysts (or adsorbents). These isotherms are determined at a temperature at or near the normal boiling point of nitrogen. Between relative pressures P/P_0 of 0.05 to 0.35, the plots of Equation (8-36) are approximately linear for type II isotherms. For other types of isotherms, either the general expression given by Equation (8-28) or a similar one [3] may be employed. For catalysts having low surface areas (< 3 m^2/g), krypton is frequently used in the determination of surface areas.

Although the theoretical basis of the BET equation has been questioned by several authors, it is nevertheless generally accepted as giving reliable values for the surface area of an adsorbent. A typical surface area determination is demonstrated by the following example.

EXAMPLE 8-1

The following data were taken from run 183 of Reference [2]. These data were obtained for a chromium catalyst (which was supported on aluminum oxide) by use of a low-temperature adsorption apparatus which was similar to the one recommended by Brunauer [3]. Evaluate the surface area.

Data: The temperature of the nitrogen adsorption is $-319.4°$F, and at this temperature the vapor pressure P_0 of nitrogen is 81.11 mm of Hg.

Adsorption Pressure, P*	Volume of Nitrogen Adsorbed, v[†]
1.145	11.38
18.130	18.17
32.870	23.07
41.640	26.85
46.900	29.88

*P is in cm of Hg absolute.
[†]v is in milliliters of nitrogen (at 0°C and 1 atm) adsorbed per gram of adsorbent.

To estimate the area of surface covered by an adsorbed molecule of nitrogen, the formula used by Brunauer et al. [14] is available:

$$\sigma_m = 1.091\left(\frac{M}{N\rho}\right)^{2/3} \tag{A}$$

where M = molecular weight of nitrogen

        N = Avogadro's number

        ρ = density of liquified nitrogen in grams per cubic centimeter = 0.808

        σ_m = area occupied per molecule in square centimeters

SOLUTION

A plot of these data in the form of Equation (8-36) is shown in Figure 8-3. The slope was found to be

$$\text{Slope} = 0.072 = \frac{c - 1}{v_m c}$$

Since the intercept $1/v_m c$ is very nearly equal to zero for this example, it follows that

$$\frac{c - 1}{v_m c} = \frac{1}{v_m} - \frac{1}{v_m c} \cong \frac{1}{v_m}$$

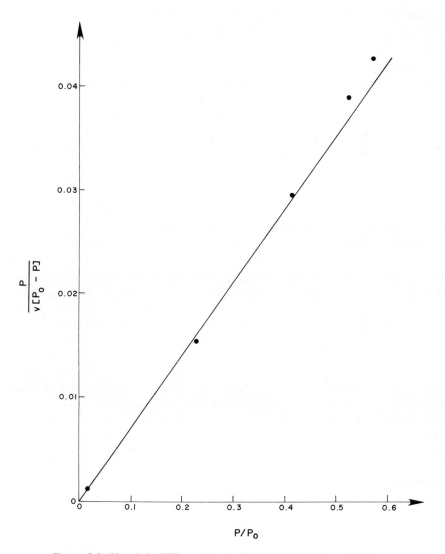

Figure 8-3. Plot of the BET equation for the data given in Example 8-1.

and thus $v_m = 13.9$ ml (at 0°C and 1 atm) per gram of adsorbent. The area per molecule may be estimated by substituting $M = 28$, $N = 6.023 \times 10^{23}$, and $\rho = 0.808$ g cm^3 in Equation (A). The result so obtained is $\sigma_m = 16.2 \times 10^{-16}$ (cm^2 per molecule). By converting v_m to molecules adsorbed per gram of adsorbent and multiplying this result by σ_m, the surface area A_s in square centimeters per gram of catalyst is given.

$$A_s = \left(13.9\frac{cm^3}{g}\right)\left(\frac{g\ mol}{22.4 \times 10^3\ cm^3}\right)\left(\frac{6.023 \times 10^{23}\ molecules}{g\ mol}\right)\left(\frac{16.2 \times 10^{-16}\ cm^2}{molecule}\right)$$

$$= 6.05 \times 10^5\ cm^2/g$$

or
$$A_s = 60.5\ m^2/g$$

Livingston [15] recommended that a $\sigma_m = 15.4 \times 10^{-16}$ (cm^2 per molecule) be used for nitrogen instead of the value computed above.

The large surface areas exhibited by many adsorbents suggests that an adsorbent is made up of an intricate network of pores and that most of the surface area lies in the interior of the adsorbent. Since the diameters of these pores may be of the same order of magnitude as that of molecules involved in a chemical reaction, considerable attention has been given to the determination of pore volume in recent years [16, 17, 18]. An extension of the BET theory of multimolecular adsorption of pure gases to include the adsorption of mixtures of gases was recently presented by Gonzales and Holland [19].

II. CHEMICAL ADSORPTION, REACTION, AND DESORPTION

In contrast to physical adsorption, chemical bonds are involved in chemical adsorption [20]. A further division of this type of adsorption into chemisorption and activated adsorption is common practice. Although the precise distinctions vary somewhat among authors, the general distinction usually made is that activated adsorption is a slower process and has a higher degree of reversibility than chemisorption. For convenience, all adsorption other than physical is referred to herein as *chemical adsorption*. Most of the past experimentation involving this type of adsorption has been pointed toward showing either the heterogeneity or homogeneity of catalyst surfaces.

Taylor and coworkers [21, 22, 23] have interpreted their investigations, which involved the adsorption of hydrogen by industrial catalysts, as showing the heterogeneity of industrial catalyst surfaces. On the other hand, Roberts [24], Beeck [7], and Eucken [25] have investigated pure metal catalysts, usually in the form of films. They found that all adsorption sites of these films behaved in essentially the same manner and concluded that such surfaces possessed a high degree of homogeneity. However, such films rarely make good commercial catalysts [23].

In the development of rate expressions for catalytic reactions, it is supposed that most of the chemical reaction which occurs in a flow system at steady-state operation can be accounted for on the basis of only those sites which have extremely fast rates relative to the other sites. These sites were called *active centers* by Taylor [23]. The concept of active centers was applied to industrial catalytic reactions by Hougen and Watson [26]. More recently, a classification of catalytic reactions was presented by Burwell [27]. For the oxygen adsorption shown in Figure 8-1, the sites which could be classified as the active sites would be among those that appear to have instantaneous rates in a batch system.

A. Formulation of the Rates of Chemical Adsorption, Desorption, and Reaction

Rate expressions for the three rate processes, adsorption, reaction, and desorption (steps 3, 4, and 5 of the introduction of this chapter), are developed for the special case of heterogeneous systems consisting of a gas phase and a solid catalyst.

1. Monomolecular Surface Reactions. Consider first the simple case where compound A is transformed to compound B in the presence of a catalyst:

$$A \rightleftarrows B \tag{8-37}$$

One mechanism for such a reaction is the following:

$$A(g) + S \underset{k_1'}{\overset{k_1}{\rightleftarrows}} AS \tag{8-38}$$

$$AS \underset{k_2'}{\overset{k_2}{\rightleftarrows}} BS \tag{8-39}$$

$$BS \underset{k_3'}{\overset{k_3}{\rightleftarrows}} B(g) + S \tag{8-40}$$

where
$$S = \text{active sites}$$
$$AS = \text{chemically adsorbed } A$$
$$BS = \text{chemically adsorbed } B$$

In the first step of the mechanism [Equation (8-38)], molecule A is adsorbed on the catalyst. In the second step of the mechanism [Equation (8-39)], the adsorbed molecule A obtains sufficient energy to cause it to form the product molecule B, all of which occurs on the single site. In the final step, molecule B is desorbed [Equation (8-40)] to the gas phase. Some of the ways in which molecules adsorb and react on catalytic surfaces are illustrated in Table 8-1.

The rate of chemical adsorption is formulated on the same basis as Langmuir's theory of physical adsorption. Thus the rate of chemical adsorp-

TABLE 8-1

RESTRICTED CLASSIFICATION OF ELEMENTARY REACTIONS

(Taken from R. L. Burwell Jr., *Chem. Eng. News* (Aug. 22, 1966). Used with permission of the American Chemical Society.)

Reaction	Broken Bonds	Bonds Made
1. Adsorption and its reverse, desorption		
(a) $S + NH_3(g) \rightleftarrows H_3^+NS^-$	None	$A-S$
(b) $S + H(g) \rightleftarrows HS$	None	$A-S$
2. Dissociative adsorption and its reverse, associative desorption		
(a) $2S + H_2(g) \rightleftarrows 2HS$	$A-A$	$2(A-S)$
(b) $2S + CH_4(g) \rightleftarrows CH_3S + HS$	$A-B$	$A-S, B-S$
(c) $2S + CH_2=CH_2(g) \rightleftarrows SCH_2CH_2S$	$A-A$	$2(A-S)$
3. Dissociative surface reaction and its reverse, associative surface reaction		
(a) $2S + C_2H_5S \rightleftarrows HS + SCH_2CH_2S$	$A-B-S$	$A-S, B-S$
(b) $2S + SCH_2CH_2CH_2CH_2S \rightleftarrows 2(SCH_2CH_2S)$	$S-A-A-S$	$2(A-S)$
4. Reactive adsorption and its reverse, reactive desorption		
(a) $HS + C_2H_4(g) \rightleftarrows C_2H_5S$	$A-S, B-B$	$A-BS, B-S$
(b) $H_2C=CH_2 + D_2(g) + S \rightleftarrows SCH_2CH_2D + DS$	$A-A$	$A-BS, B-S, A-S$
(c) $SCH_2CH_2S + H_2(g) + S \rightleftarrows SC_2H_5 + S + HS$	$B-S, A-A$	$A-BS, A-S$
5. Quasisorbed reactions		
(a) $H(g) + HS \rightleftarrows H_2(g) + S$	$A-S$	$A-A$
(b) $2HS + C_2H_4(g) \rightleftarrows 2S + C_2H_6(g)$	$2(A-S), B-B$	$2(A-B)$
(c) $HS + D_2(g) + S \rightleftarrows S + HD(g) + DS$	$AS, B-C, B-B$	$E-S, A-B, B-S$
(d) $DS + H_2C=CHCH_3(g) \rightleftarrows DH_2CCH=CH_2(g) + HS$	$D-E$	$C-D$

tion is proportional to the number of molecules of type A that strike the surface per unit time times the fraction of the surface that contains vacant active sites. By Equation (8-12), the total number of molecules striking a unit surface in unit time is proportional to the total pressure. Consequently, the number of molecules of type A that strike a unit of surface in unit time is proportional to p_A, the partial pressure of A. Let the fraction of the total number of active sites occupied by molecules of type A be denoted by θ_A and the fraction of those covered by molecules of type B by θ_B. Then $1 - \theta_A - \theta_B$ represents that fraction of the total number of sites which are vacant. If the condensation coefficient in Equation (8-8) is replaced by the proportionality or rate constant k_1, then the rate of chemical adsorption may be expressed as follows:

$$(r_1)_{\text{forward}} = k_1 p_A (1 - \theta_A - \theta_B) \tag{8-41}$$

where p_A is the partial pressure of component A at any point in the pore of a catalyst particle at any given location in a reactor. Again, as in physical desorption, the rate of chemical desorption A is taken to be proportional to θ_A, the fraction of the surface covered by A:

$$(r_1)_{\text{reverse}} = k_1' \theta_A \tag{8-42}$$

Then the net rate r_1 (molecules of A adsorbed per unit time per unit of surface area) is given by

$$r_1 = k_1 p_A (1 - \theta_A - \theta_B) - k_1' \theta_A \tag{8-43}$$

It has become customary because of convenience in the analysis of processes to express the rate r_1 in the units of

$$\frac{\text{moles of } A \text{ adsorbed}}{(\text{unit time})(\text{unit mass of catalyst})} \tag{8-44}$$

instead of those used in the previous analysis. From a theoretical point of view, r_1 could have been based either on a unit of reactor volume, on a unit of surface area, or on the number of active sites (as discussed below), rather than on a unit mass of catalyst.

The forward rate at which the surface reaction proceeds may be taken proportional to the amount of A in the adsorbed state, AS, or the fraction of the active sites covered by adsorbed molecules of A,

$$(r_2)_{\text{forward}} = k_2 \theta_A \tag{8-45}$$

and the reverse rate by

$$(r_2)_{\text{reverse}} = k_2' \theta_B \tag{8-46}$$

Thus the net rate is given by

$$r_2 = k_2 \theta_A - k_2' \theta_B \tag{8-47}$$

The net rate of reaction for the desorption of the product molecule B from the surface is given by

$$r_3 = k_3\theta_B - k_3'p_B(1 - \theta_A - \theta_B) \tag{8-48}$$

Since θ_A and θ_B are difficult to determine experimentally, it is desirable to eliminate these variables from the rate expressions. Elimination of these variables may be effected by making use of the fact that at steady-state operation any set of rate processes that are in series must occur at the same rate. For if it is supposed that the rates were not equal, then there would be an accumulation or depletion with respect to time, which is contrary to the steady-state assumption. Thus

$$r_1 = r_2 = r_3 = r \tag{8-49}$$

and Equations (8-43), (8-47), and (8-48) may be regarded as three equations in the three unknowns θ_A, θ_B, and r. Since these equations are linear in these variables, they may be stated in matrix notation as follows:

$$\begin{bmatrix} K_2 & -1 & -1/k_2' \\ K_1p_A + 1 & K_1p_A & 1/k_1' \\ p_B/K_3 & 1 + p_B/K_3 & -1/k_3 \end{bmatrix} \begin{bmatrix} \theta_A \\ \theta_B \\ r \end{bmatrix} = \begin{bmatrix} 0 \\ K_1p_A \\ p_B/K_3 \end{bmatrix}$$

where $K_1 = k_1/k_1'$, $K_2 = k_2/k_2'$, and $K_3 = k_3/k_3'$. This matrix equation may be solved for θ_A, θ_B, and r by Gaussian elimination. The rate expression so obtained follows:

$$r = \frac{k_1p_A - f_1(k_1'p_B/K_3)}{1 - (k_1p_A/k_2') + f_1f_2} \tag{8-50}$$

The functions f_1 and f_2 are defined as follows:

$$f_1 = \frac{1 + K_1p_A(1 + K_2)}{K_2(1 + p_B/K_3) + p_B/K_3}, \qquad f_2 = \left[\frac{k_1'}{k_3} + \left(1 + \frac{p_B}{K_3}\right)\left(\frac{k_1'}{k_2'}\right) \right] \tag{8-51}$$

The expression given by Equation (8-50) is readily reduced to the following form:

$$r = \frac{k[p_A - p_B/K]}{A_0 + B_0p_A + C_0p_B} \tag{8-52}$$

where the constants k, K, A_0, B_0, and C_0 are defined as follows:

$$A_0 = K_2 + \frac{k_1'}{k_3} + \frac{k_1'}{k_2'}$$

$$B_0 = \frac{k_1}{k_3}\left(1 + K_2 + \frac{k_3}{k_2'}\right)$$

$$C_0 = \frac{1}{K_3}\left(1 + K_2 + \frac{k_1'}{k_2'}\right) \tag{8-53}$$

$$k = k_1K_2$$

$$K = K_1K_2K_3$$

The use of the notation θ_A and θ_B to represent the fractions of the active sites covered by molecules of types A and B, respectively, follows the original notation suggested by Langmuir [4, 5, 6]. More recently, this notation was employed by Walas [28]. However, instead of this notation, concentration-type units are most commonly employed in the chemical engineering literature to represent the number of active sites covered by molecules of a given type.

The formulation of the concept of concentration of active sites may be effected by regarding a gram mole simply as the number 6.023×10^{23}. Thus 1 g mol of active sites is equal to 6.023×10^{23} sites, and 1 lb mol of sites is equal to $454 \times 6.023 \times 10^{23}$ sites. For metal catalysts, Boudart [29] suggests the use of 10^{15} sites per square centimeter. The area referred to is the BET area obtained by nitrogen adsorption. This number of sites corresponds to the number of metal atoms on the surface of the catalyst. For acidic catalysts, such as those used in cracking reactions, there are approximately 10^{13} sites per square centimeter [30]. On the basis of the numbers of sites per square centimeter and the BET surface area, the number of sites per unit mass of catalyst may be computed. The mass of catalyst required to obtain 1 mol of sites may then be computed. Then the total concentration of sites C_T may be defined as follows:

$$C_T = \frac{\text{total moles of sites}}{\text{unit mass of catalyst}} \tag{8-54}$$

Obviously, the concentration could have been defined on the basis of surface area or a unit volume of catalyst bed rather than on the basis of a unit mass of catalyst. However, because of convenience, not only the concentrations but also the rates of adsorption, desorption, and reaction are defined on the basis of a unit mass of catalyst. Thus θ_A and θ_B may be stated in terms of a ratio of concentration units as follows;

$$\theta_A = \frac{C_{AS}}{C_T}, \qquad \theta_B = \frac{C_{BS}}{C_T} \tag{8-55}$$

The quantities C_{AS} and C_{BS} are defined as follows:

$$C_{AS} = \frac{\text{moles of } A \text{ adsorbed}}{\text{unit mass of catalyst}} \tag{8-56}$$

$$C_{BS} = \frac{\text{moles of } B \text{ adsorbed}}{\text{unit mass of catalyst}} \tag{8-57}$$

The definitions of C_{AS}, C_{BS}, and C_T suppose that only one molecule is adsorbed per active site.

Equations (8-43), (8-47), and (8-48) may be restated in terms of concentration units as follows:

$$r_1 = k_1 p_A C_S - k_1' C_{AS} \tag{8-58}$$

$$r_2 = k_2 C_{AS} - k_2' C_{BS} \tag{8-59}$$

$$r_3 = k_3 C_{BS} - k_3' p_B C_S \tag{8-60}$$

where $C_S = C_T - C_{AS} - C_{BS}$, the concentration of vacant sites. The rates r_1, r_2, r_3 are defined as the moles of A adsorbed, the moles of A reacted, and the moles of B desorbed, respectively, per unit time per unit mass of catalyst. As a matter of convenience, the same symbols were employed for the rate constants in Equations (8-43), (8-47), (8-48), and (8-58) through (8-60). Again $r_1 = r_2 = r_3 = r$ and Equations (8-58), (8-59), and (8-60) may be solved to give

$$r = \frac{k[p_A - p_B/K]}{A_0 + B_0 p_A + C_0 p_B} \tag{8-61}$$

where k, A_0, B_0, and C_0 have the same formal definitions given below Equation (8-52).

An alternative definition of the rate of reaction, called the turnover number, used by catalytic chemists is defined as the number of molecules (or moles) of a given chemical species which reacts or is adsorbed per second per active catalytic site (or per mole of active sites).

B. Approximate Rate Expressions

In the same manner as demonstrated in the treatment of the mechanisms of homogeneous reactions, the rate equations can be simplified by use of the concept of the rate controlling step. In the development that follows, which follows that of Hougen and Watson [26], one reaction is regarded as the rate controlling step and the remaining reactions are taken to be in equilibrium.

To illustrate this method, suppose first that the adsorption of A given by Equation (8-38) is the rate controlling step of the mechanism. By this statement it is meant that k_1 and k_1' are small relative to the other rate constants, and that almost as soon as A is adsorbed, it reacts to form compound B, which is desorbed immediately. That is, suppose the rate of reaction is given by r_1 [Equation (8-58)] and that for all practical purposes, $r_2/k_2 \cong 0$, $r_3/k_3 \cong 0$, which is usually written simply as

$$r_2 \cong r_3 \cong 0 \tag{8-62}$$

Thus it follows from Equations (8-59), (8-60), and (8-62) that

$$C_{AS} = \frac{C_{BS}}{K_2} \tag{8-63}$$

$$C_{BS} = \frac{p_B C_S}{K_3} \tag{8-64}$$

Elimination of C_{BS} from these two expressions yields

$$C_{AS} = \frac{p_B C_S}{K_2 K_3} \tag{8-65}$$

Equations (8-64) and (8-65) are used to eliminate C_{AS} and C_{BS} in the following expression for the total number of sites.

$$C_T = C_S + C_{AS} + C_{BS} \tag{8-66}$$

After rearrangement, one obtains

$$C_S = \frac{C_T}{1 + (p_B/K_2K_3) + p_B/K_3} \tag{8-67}$$

In view of the result given by Equation (8-65), it follows that Equation (8-58) may be restated in the form

$$r = k_1 C_S \left(p_A - \frac{p_B}{K_1K_2K_3} \right) \tag{8-68}$$

Thus, by combining Equations (8-67) and (8-68),

$$r = \frac{C_T k_1 (p_A - p_B/K)}{1 + (p_B/K_2K_3) + p_B/K_3} \tag{8-69}$$

where

$$K = K_1K_2K_3$$

For the case where the surface rate r_2 is controlling, $r_1 \cong r_3 \cong 0$, and thus Equation (8-58) reduces to

$$C_{AS} = K_1 p_A C_S \tag{8-70}$$

and Equation (8-60) reduces to

$$C_{BS} = \frac{p_B C_S}{K_3} \tag{8-71}$$

Substitution of these results into Equation (8-66), followed by rearrangement, yields

$$C_S = \frac{C_T}{1 + K_1 p_A + p_B/K_3} \tag{8-72}$$

Elimination of C_{AS} and C_{BS} from Equation (8-59) by use of Equations (8-70) and (8-71), followed by elimination of C_S through the use of Equation (8-72), gives

$$r = \frac{C_T k (p_A - p_B/K)}{1 + K_1 p_A + p_B/K_3} \tag{8-73}$$

where

$$K = K_1K_2K_3 \quad \text{and} \quad k = k_2K_1$$

Finally, for the case where the rate controlling step is r_3 (the desorption of B), the resulting rate expression is given by

$$r = \frac{C_T k (p_A - p_B/K)}{1 + K_1(1 + K_2) p_A} \tag{8-74}$$

where
$$k = k_3 K_1 K_2$$
$$K = K_1 K_2 K_3$$

1. Bimolecular Surface Reactions. The following mechanism proposed by Hougen and Watson [26] for the reaction $A \rightleftarrows B$ may be regarded as a bimolecular reaction because by this mechanism it is supposed that an adsorbed molecule A becomes sufficiently energized to form an activated complex with an adjacent active center, which then decomposes to form an adsorbed product molecule B. Thus, instead of Equation (8-39), the surface reaction becomes

$$AS + S \underset{k_2'}{\overset{k_2}{\rightleftarrows}} BS + S \tag{8-75}$$

Then, by application of the law of mass action, the net rate at which the surface reaction proceeds is given by the difference of the forward and reverse second-order reaction rates,

$$r_2 = k_2 C_{AS} C_S - k_2' C_{BS} C_S \tag{8-76}$$

Hougen and Watson [26] present the following elegant justification for taking the concentration of adsorbed A and adjacent empty sites as being equal to the product of the concentration C_{AS} and C_S. Consider a single adsorbed molecule A, and let s be the number of adjacent sites which are equidistant from A. If θ_S is the fraction of the sites which are empty, then an average adsorbed molecule A will have $s\theta_S$ sites adjacent to it which are empty. If each adsorbed A is so surrounded, then for C_{AS} moles of adsorbed A, it follows that

$$\left.\begin{array}{l}\text{Product of the}\\ \text{concentration}\\ \text{of adsorbed } A\\ \text{and adjacent}\\ \text{vacant sites}\end{array}\right\} = C_{AS} s \theta_S = \frac{s}{C_T} C_{AS} C_S \tag{8-77}$$

Since s and C_T are constants, they may be incorporated into the rate constants to give the forward rate shown [see Equation (8-76)].

If it is supposed that the surface reaction is the rate controlling step of the mechanism given by Equations (8-38), (8-40), and (8-75), and whose rate expressions are given, respectively, by Equations (8-58), (8-60), and (8-76), then the rate of reaction is given by

$$r = \frac{k[p_A - p_B/K]}{\left[1 + K_1 p_A + (p_B/K_3)\right]^2} \tag{8-78}$$

where
$$K = K_1 K_2 K_3$$
$$k = k_1 K_1 C_T^2$$

The rate expressions corresponding to suppositions of other rate controlling steps are developed in a manner analogous to that described previously. It is of interest to note that, although r_2 is second order with respect to concentrations, the numerator of the final expression is first order with respect to the partial pressures. However, for the following reaction and proposed mechanism, both the surface reaction rate and the numerator of the rate expression which is obtained by taking the surface reaction to be the rate controlling step are second order.

Consider the bimolecular reaction

$$A + B \rightleftarrows C + D \qquad (8\text{-}79)$$

which is carried out in the presence of a catalyst. One possible mechanism is:

$$A(g) + S \underset{k_1'}{\overset{k_1}{\rightleftarrows}} AS \qquad (8\text{-}80)$$

$$B(g) + S \underset{k_2'}{\overset{k_2}{\rightleftarrows}} BS \qquad (8\text{-}81)$$

$$AS + BS \underset{k_3'}{\overset{k_3}{\rightleftarrows}} CS + DS \qquad (8\text{-}82)$$

$$CS \underset{k_4'}{\overset{k_4}{\rightleftarrows}} C(g) + S \qquad (8\text{-}83)$$

$$DS \underset{k_5'}{\overset{k_5}{\rightleftarrows}} D(g) + S \qquad (8\text{-}84)$$

Equation (8-82) represents the surface mechanism wherein adjacently adsorbed molecules A and B become sufficiently energized and react, forming molecules C and D, which are adsorbed on the same sites originally occupied by A and B.

For the case where the surface reaction is the rate controlling step, the expression obtained for the reaction rate is as follows:

$$r = \frac{K'k_3[p_A p_B - (p_C p_D/K)]}{[1 + K_1 p_A + K_2 p_B + (p_C/K_4) + (p_D/K_5)]^2} \qquad (8\text{-}85)$$

where
$$K' = K_1 K_2 C_T^2$$
$$K = K_1 K_2 K_3 K_4 K_5$$

Rate expressions corresponding to the cases where one of the adsorption or desorption steps is taken to be the rate controlling step are developed in a manner analogous to that demonstrated for the set of reactions given by Equations (8-38) through (8-40) (see Problem 8-2).

An alternate mechanism for the reaction given by Equation (8-79) consists of the reaction of A and B in the gas phase with catalysts sites to produce molecules C and D in the gas phase. More precisely, suppose

$$A(g) + S_1 \underset{k_1'}{\overset{k_1}{\rightleftarrows}} C(g) + S_2$$

$$B(g) + S_2 \underset{k_2'}{\overset{k_2}{\rightleftarrows}} D(g) + S_1 \tag{8-86}$$

and that the concentrations of sites S_1 and S_2 are small. Furthermore, suppose that these sites are formed at approximately the same rate at which they react (the pseudo steady-state assumption). The mechanism is seen to consist of two reactions instead of the five proposed previously. The net rate of disappearance of A is given by

$$r_A = k_1 p_A C_{S_1} - k_1' p_C C_{S_2} \tag{8-87}$$

On the basis of the pseudo steady-state assumption, the net rates of disappearance of sites S_1 and S_2 are approximately zero. Thus

$$r_{S_2} = -r_{S_1} = k_1 p_A C_{S_1} - k_1' p_C C_{S_2} - k_2 p_B C_{S_2} + k_2' p_D C_{S_1} \cong 0$$

and

$$C_{S_2} = C_{S_1} \left[\frac{k_1 p_A + k_2' p_D}{k_2 p_B + k_1' p_C} \right] \tag{8-88}$$

Elimination of C_{S_2} from Equation (8-88) and the expression for the conservation of sites

$$C_T = C_{S_1} + C_{S_2}$$

gives

$$C_{S_1} = \frac{C_T(k_2 p_B + k_1' p_C)}{k_1 p_A + k_2 p_B + k_1' p_C + k_2' p_D} \tag{8-89}$$

Use of Equations (8-88) and (8-89) to eliminate C_{S_1} and C_{S_2} from Equation (8-87) gives the desired expression for r_A:

$$r_A = \frac{C_T k_1 k_2 (p_A p_B - p_C p_D / K)}{k_1 p_A + k_2 p_B + k_1' p_C + k_2' p_D} \tag{8-90}$$

where

$$K = K_1 K_2$$

C. Assumption of the Most Abundant Surface Intermediate

Boudart [29, 31, 32] introduced the assumption of the *most abundant surface intermediate* (*masi*). By this assumption, the sites occupied by all species except the most abundant surface intermediate (masi) are regarded as negligible relative to those filled by the most abundant intermediate and to those which are empty. For example, if AS is the most abundant surface intermediate in the isomerization reaction [Equation (8-37)], then Equation (8-66) reduces to

$$C_T = C_S + C_{AS} \tag{8-91}$$

Boudart [29, 31, 32] demonstrated that the same form of the rate expression for the isomerization reaction [see Equations (8-38) through (8-40)] is obtained for four different mechanisms when the assumption of the most abundant surface intermediate is utilized along with other assumptions enumerated next.

Consider first the mechanism which is based on the following assumptions:

1. All of the reaction steps are irreversible:

$$A(g) + S \xrightarrow{k_1} AS$$

$$AS \xrightarrow{k_2} BS$$

$$BS \xrightarrow{k_3} B(g) + S$$

2. The most abundant surface intermediate is BS; that is,

$$C_{AS} \cong 0 \quad \text{relative to} \quad C_S \quad \text{and} \quad C_{BS}$$

3. The concentrations of adsorbed A and B may be approximated by means of the pseudo steady-state assumption; that is, $r_{AS} \cong 0$ and $r_{BS} \cong 0$.

From the first assumption, the following rates expression may be stated:

$$\begin{aligned}
r_A &= k_1 p_A C_S \\
r_{AS} &= k_1 p_A C_S - k_2 C_{AS} \\
r_{BS} &= k_2 C_{AS} - k_3 C_{BS}
\end{aligned} \tag{8-92}$$

From the second assumption,

$$C_T = C_S + C_{BS} \tag{8-93}$$

From the third assumption,

$$C_{AS} = \frac{k_1}{k_2} p_A C_S \tag{8-94}$$

$$C_{BS} = \frac{k_2}{k_3} C_{AS} = \frac{k_1}{k_3} p_A C_S \tag{8-95}$$

Elimination of C_{BS} from Equations (8-92), (8-93), and (8-95), followed by rearrangement, yields

$$r_A = \frac{\bar{k} p_A}{1 + \bar{K} p_A} \tag{8-96}$$

where

$$\bar{k} = k_1 C_T$$

$$\bar{K} = \frac{k_1}{k_3}$$

TABLE 8-2

SIMPLIFYING ASSUMPTIONS LEAD TO THE SAME FORM OF THE RATE EXPRESSION
FOR DIFFERENT MECHANISMS

(Taken from M. Boudart, *AIChE J.*, *18*:No. 3, 465 (1972). Used with permission of the American Institute of Chemical Engineers.)

Mechanism	Assumptions*	Meaning of $\bar{k}$ in Equation (8-101)	Meaning of $\bar{K}$ in Equation (8-101)
$A + S \rightleftarrows AS$ (1) $AS \rightarrow BS$ (2) $BS \rightleftarrows B + S$ (3)	$r_1 \cong 0$ $r_3 \cong 0$	$k_2 K_1 C_T$	K_1
$A + S \rightleftarrows AS$ (1) $AS \rightleftarrows BS$ (2) $BS \rightarrow B + S$ (3)	$r_1 \cong 0$ $r_{BS} \cong 0$	$\dfrac{C_T K_1 k_2 k_3}{k_2' + k_3}$	K_1
$A + S \rightleftarrows AS$ (1) $AS \rightarrow BS$ (2) $BS \rightarrow B + S$ (3)	$r_1 \cong 0$ $r_{BS} \cong 0$	$k_2 K_1 C_T$	K_1

*For each mechanism, AS is assumed to be the most abundant surface intermediate.

Three other mechanisms presented by Boudart [29] which give a rate expression of the same form as Equation (8-96) are presented in Table 8-2. The meanings of $\bar{k}$ and $\bar{K}$ are seen to be different for each mechanism and to have entirely different kinetic interpretations, which give rise to the well-known ambiguities of simplified kinetic models. On the other hand, these simplified kinetic models have the advantage that they may be represented by a relatively simple rate expression, such as the one given by Equation (8-96).

The next example is based on a mechanism proposed by Boudart [29] for the catalytic synthesis of ammonia. Recently, Kellogg developed a process for the catalytic synthesis of ammonia on a large scale, producing 1000 tons per day and more. This process makes use of the horizontal reactor pictured in Figure 8-4. A flow diagram of this reactor is shown in Figure 8-5. Exchenbremer and Wagner [33] have presented additional details pertaining to this process.

During operation, the reactants are charged to the converter through the inlet nozzle (Figure 8-5) and directed toward the interchanger via the annulus formed between the cartridge and pressure shell. The reactants flow through the interchanger (shell side) and exit via the gas transfer line to the first bed and are distributed over the catalyst. The reactants flow downward through the catalyst bed, then enter the space below each bed. Quench is distributed over the length of the bed and mixes with the reactants exiting the catalyst bed. The mixture then flows up through the passage provided between adjacent beds and is charged to the next bed of catalyst. Flow is continuous in a similar manner successively through each of the catalyst beds.

Figure 8-4. The M. W. Kellogg horizontal reactor for the synthesis of ammonia (Courtesy of the M. W. Kellogg Company).

The effluent from the last bed is directed toward and flows through the tube side of the interchanger, giving up heat to the reactants being charged to the first bed. The cooled product gases leave the reactor through the outlet nozzle.

EXAMPLE 8-2

For the synthesis of ammonia, $N_2 + 3H_2 \rightleftarrows 2NH_3$ over an iron catalyst, develop the rate expression for the following mechanism:

$$H_2(g) + 2S \rightleftarrows 2HS \tag{1}$$

$$N_2(g) + 2S \rightarrow 2NS \tag{2}$$

$$NS + 3HS \rightleftarrows NH_3S + 3S \tag{3}$$

$$NH_3S \rightleftarrows NH_3(g) + S \tag{4}$$

on the basis of the following assumptions:

1. The most abundant surface intermediate is NS; that is, $C_T = C_S + C_{NS}$.
2. Steps 1, 3, and 4 are in equilibrium and step 2 is the rate determining step.
3. Suppose that the conditions at which the reaction $N_2 + 3H_2 \rightleftarrows 2NH_3$ is carried out are such that the reverse rate is negligible.

(This mechanism and associated assumptions were proposed by Boudart [29].)

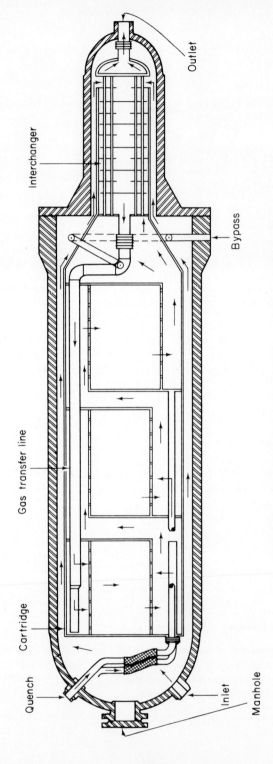

Figure 8-5. Flow diagram of the M. W. Kellogg horizontal reactor for the synthesis of ammonia (Courtesy of the M. W. Kellogg Company).

Outlet

Interchanger

Bypass

Gas transfer line

Cartridge

Quench

Inlet

Manhole

SOLUTION

By assumption (2),

$$r = k_2 p_{N_2} C_S^2 \tag{A}$$

$$K_1 = \frac{C_{HS}^2}{p_{H_2} C_S^2} \tag{B}$$

$$K_3 = \frac{C_S^3 C_{NH_3 S}}{C_{NS} C_{HS}^3} \tag{C}$$

$$K_4 = \frac{C_S p_{NH_3}}{C_{NH_3 S}} \tag{D}$$

When Equations (B), (C), and (D) are solved simultaneously for C_{NS}, the following expression is obtained:

$$C_{NS} = \frac{p_{NH_3} C_S}{K p_{H_2}^{1.5}} \tag{E}$$

where

$$K = K_1^{1.5} K_3 K_4$$

Elimination of C_S from Equations (A), (E), and the expression given for C_T in the first assumption yields the final form of the rate expression:

$$r = k_2 p_{N_2} \left[\frac{C_T}{1 + p_{NH_3}/K p_{H_2}^{1.5}} \right]^2 \tag{F}$$

(In commercial practice, the reactors are operated at conditions such that the reverse reaction cannot be neglected.)

EXAMPLE 8-3

Formaldehyde may be produced by the partial oxidation of methanol over a ferromolybdenum catalyst,

$$CH_3OH + \tfrac{1}{2}O_2 = CH_2O + H_2O$$

This reaction and compounds may be represented symbolically as follows:

$$A + \tfrac{1}{2}B \rightleftarrows C + D$$

Develop the rate expression for the following mechanism and assumptions, which were proposed by Evmenenko and Gorokhoustskii [34]:

$$A + S_1 \rightleftarrows AS_1 \tag{1}$$

$$B + S_2 \rightleftarrows BS_2 \tag{2}$$

$$AS_1 + BS_2 \rightarrow C(g) + DS_1 + S_2 \tag{3}$$

$$DS_1 \rightleftarrows D(g) + S_1 \tag{4}$$

where S_1 = catalyst sites occupied by water and methanol

                      S_2 = catalyst sites occupied by O_2

Assumptions:

1. Sites S_1 and S_2 are of a different type:

$$C_{T_1} = C_{AS_1} + C_{DS_1} + C_{S_1}$$
$$C_{T_2} = C_{BS_2} + C_{S_2}$$

2. Steps 1, 2, and 4 are in dynamic equilibrium and step 3 is the rate controlling step.

SOLUTION

Since steps 1, 2, and 4 are in dynamic equilibrium (by the second assumption), it follows that

$$C_{AS_1} = K_1 p_A C_{S_1} \tag{A}$$

$$C_{BS_2} = K_2 p_B C_{S_2} \tag{B}$$

$$C_{DS_1} = \frac{p_D C_{S_1}}{K_4} \tag{C}$$

Since step 3 is the rate controlling step,

$$r = k_3 C_{AS_1} C_{BS_2} \tag{D}$$

Equations (A) through (D) may be combined with the definitions of C_{T_1} and C_{T_2} in the first assumption, giving the final result:

$$r = \frac{k_3 K_1 K_2 C_{T_1} C_{T_2} p_A p_B}{[1 + K_1 p_A + (p_D/K_4)][1 + K_2 p_B]}$$

D. Rate Expressions for Nonuniform Surfaces

In the preceding developments, the assumption of ideal surfaces or uniform surfaces was used. This assumption means that all sites are kinetically and thermodynamically equal and there is no interaction between sites. Temkin [35] and subsequently Boudart [29, 31] considered the case of noninteracting adsorbed species on sites which differed kinetically and thermodynamically. For such a catalyst surface, the rate of reaction could be expected to differ for each site having a different activity.

In this case, the total rate of reaction over a unit mass of surface which contains C_T (moles of sites) is given by

$$r_A = \int_0^{C_T} r \, ds \tag{8-97}$$

where s = moles of sites having the same activity

        r = the point value of the rate of reaction, moles A reacting per unit time per mole of sites

Attempts to integrate Equation (8-97) on the basis of assumed distributions of site activities have been made by Temkin [35] and others; the resulting expressions for r_A are of the same general form as the expressions developed previously on the basis of uniform site activities.

Lumpkin et al. [36] pointed out that the determination of the rate constants in the preceding rate expressions for catalytic reactions by use of a series of runs made under isothermal conditions at several different temperatures can lead to rate expressions which do not accurately describe the behavior of reactors which are operated adiabatically or nonisothermally. Eleven different rate expressions for the hydrogenation of a codimer which had been curve fitted by different workers on the basis of data collected from isothermally operated reactors were found to predict significantly different conversions for adiabatically operated reactors. In view of this result, the authors made the recommendation that, in addition to data collected at isothermal operation, additional data at adiabatic or nonisothermal operation should be collected and used in the determination of the rate constants.

III. DESIGN OF A FIXED-BED CATALYTIC REACTOR WHEN THE RATE CONTROLLING STEP IS ADSORPTION, DESORPTION, OR THE SURFACE REACTION

If the rates of adsorption, reaction, and desorption are slow relative to the other steps listed at the beginning of this chapter, then the partial pressures appearing in the rate expressions developed in Part II may be set equal to the corresponding partial pressures evaluated at the bulk conditions of the reactant stream.

Consider a fixed-bed catalytic reactor of the type sketched in Figure 8-6. Suppose that the reactant stream which passes through the bed is perfectly mixed in the radial direction and that no mixing occurs in the direction of flow (plug flow). In the description of a catalytic reactor, the mass of catalyst is commonly used as a variable instead of the length or volume of the reactor. The particular choice of variable to be used is arbitrary, since

$$V = Sz = \frac{W}{\rho_b} \tag{8-98}$$

where ρ_b is the bulk density of the catalyst bed (the mass of catalyst per unit volume of the reactor).

A material balance on component A over the element of bed from W_j to $W_j + \Delta W$ for the reactor shown in Figure 8-6 at steady-state operation is given by

$$\left[n_A|_{W_j} - n_A|_{W_j + \Delta W} \right] - \int_{W_j}^{W_j + \Delta W} r_A \, dW = 0 \tag{8-99}$$

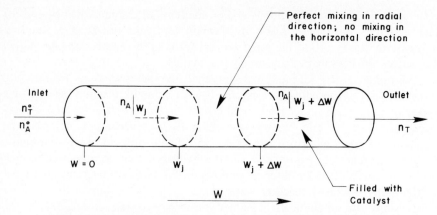

Figure 8-6. Fixed-bed catalytic reactor (W = mass of catalyst contained by the reactor from the inlet up to any point in the reactor).

where r_A is defined in terms of the disappearance of A (moles of A disappearing per unit mass of catalyst per unit time). Application of the mean-value theorems in the same manner as demonstrated for Equation (2-3) followed by the limiting process wherein ΔW is allowed to go to zero leads to the result

$$r_A = -\frac{dn_A}{dW} \tag{8-100}$$

For the case where the reactor is operated isothermally, the pressure drop is negligible, and the gas phase obeys the perfect gas law, the calculation of the mass of catalyst bed required to effect a given conversion is as follows. Suppose that the reaction to be carried out [Equation (8-37)] has a rate expression given by Equation (8-73). Let the conversion be defined in the usual way.

$$x = \frac{n_A^0 - n_A}{n_A^0} \tag{8-101}$$

Then

$$-\frac{dn_A}{dW} = n_A^0 \frac{dx}{dW} \tag{8-102}$$

Since $p_A = (1 - x)P$ and $p_B = xP$ when the feed is composed of pure $A(n_T^0 = n_A^0)$, Equations (8-73) and (8-100) may be restated in terms of the conversion x to give

$$\frac{dx}{dW} = \frac{\alpha(1 - \beta x)}{\gamma + \delta x} \tag{8-103}$$

where
$$\alpha = \frac{C_T k}{n_A^0}, \qquad \beta = 1 + \frac{1}{K}$$

$$\gamma = K_1 + \frac{1}{P}, \qquad \delta = K_3^{-1} - K$$

For an isothermal reactor in which the pressure drop is negligible, the mass of catalyst required to achieve a specified conversion may be obtained by solving Equation (8-103). Separation of the variables in this equation followed by integration yields

$$W = \frac{\delta}{\alpha\beta} \left[\left(\frac{\gamma\beta + \delta}{\beta\delta} \right) \left(\log_e \frac{1}{1 - \beta x} \right) - x \right] \qquad (8\text{-}104)$$

The rate expressions for cases where adsorption, desorption, or the surface reaction is the rate controlling step of the reactions, such as the bimolecular reaction given by Equation (8-79), are integrated in a manner analogous to that demonstrated here. To account for the variation of the reaction temperature and the pressure throughout a flow reactor, energy balances and a modified Fanning equation are needed. The design of such reactors is analogous to that demonstrated in Chapter 7 for thermal reactors.

The design of catalytic reactors is further complicated by the sequence of mass transfer steps (listed at the beginning of this chapter) which occur in series with the catalytic reaction. The mass transfer effects are taken into account in the design of reactors, as demonstrated in Chapter 9.

PROBLEMS

8-1. Verify the result given by Equation (8-78).

8-2. (a) Obtain the rate expression where the rate of adsorption of A is the rate controlling step of the mechanism given by Equations (8-80) through (8-84).

Answer: $r = \dfrac{k_1 C_T [p_A - (p_C p_D / KK_1 p_B)]}{1 + (p_D p_C / K p_B) + p_B K_2 + (p_C / K_4) + (p_D / K_5)}$

where $K = K_2 K_3 K_4 K_5$.

(b) Repeat (a) for the case where the rate of desorption of C is the rate controlling step.

Answer: $r = \dfrac{k_4 C_T [(K p_A p_B / p_D) - (p_C / K_4)]}{[1 + (K p_A p_B / p_D) + p_B K_2 + p_A K_1 + (p_D / K_5)]}$

where $K = K_1 K_2 K_3 K_5$.

8-3. Suppose that the reaction

$$A(g) + B(g) \rightleftarrows C(g)$$

is carried out in the presence of a catalyst, and that the surface reaction mechanism consists of the reaction of A from the gas phase with adsorbed B to produce adsorbed C; that is,

$$A(g) + S \underset{k_1'}{\overset{k_1}{\rightleftarrows}} AS \tag{1}$$

$$B(g) + S \underset{k_2'}{\overset{k_2}{\rightleftarrows}} BS \tag{2}$$

$$A(g) + BS \underset{k_3'}{\overset{k_3}{\rightleftarrows}} CS \tag{3}$$

$$CS \underset{k_4'}{\overset{k_4}{\rightleftarrows}} C(g) + S \tag{4}$$

Develop the rate expressions for r for the cases where steps (2), (3), and (4), respectively, are the rate controlling steps. (The reaction of gas phase molecules with adsorbed molecules is frequently called the Rideal mechanism.)

8-4. Butenes may be dehydrogenated with oxygen to butadiene over a bismuth molybdate catalyst [37]. The reaction is first order with respect to butene and zero order with respect to oxygen. The mechanism is

(1) $O_2(g) + 2S \rightleftarrows 2OS$
(2) $CH_2 = CH - CH_2 - CH_3(g) + OS \rightarrow (C_4H_7)S + OHS$
(3) $(C_4H_7)S + OS \rightarrow CH_2 = CH - CH = CH_2(g) + OHS + S$
(4) $2OHS \rightarrow H_2O(g) + OS + S$

Reactions (2) through (4) are assumed to be irreversible, and the second reaction is assumed to be the rate controlling step. If adsorbed oxygen is the most abundant surface intermediate and reaction (1) is in dynamic equilibrium, show that

$$r_2 = \frac{C_T k \left(\sqrt{K_1 p_{O_2}} \right) p_B}{1 + \sqrt{K_1 p_{O_2}}}$$

which reduces to $r_2 = kC_T p_B$ if $\sqrt{K_1 p_{O_2}} \gg 1$; p_B = partial pressure of butene.

8-5. Initial rate data were used by Ahuja and Mathur [38] to study the oxidation of methane, $CH_4 + 2O_2 = CO_2 + 2H_2O$ over palladium at $300°$ to $340°C$ and pressures of 1.8 to 9.7 atm. The data were correlated by use of the following rate expression:

$$r = \frac{k \left(p_{CH_4} p_{O_2}^2 - p_{H_2O}^2 p_{CO_2}/K \right)}{\left[1 + K_{CH_4} p_{CH_4} + K_{O_2} p_{O_2} + K_{CO_2} p_{CO_2} + K_{H_2O} p_{H_2O} \right]^3}$$

Propose a mechanism that is in agreement with this rate expression and which makes use of the following assumptions:

1. Rate controlling step is the surface reaction

$$2O_2 S + CH_4 S \rightleftarrows 2H_2OS + CO_2 S$$

2. The remaining steps of the mechanism are in equilibrium.

8-6. The dehydrogenation of *sec*-butyl alcohol to methyl ethyl ketone was studied by Thaller and Thodos [39] in a differential flow reactor. The catalyst was brass stock. The temperature and pressure range were 550° to 700°F and 0 to 14 atm. For the temperature range of 550° to 600°F, a dual site mechanism was proposed with the surface reaction as the rate controlling step.

$$A(g) + S \rightleftarrows AS$$
$$AS + S \rightleftarrows BS + CS$$
$$BS \rightleftarrows B(g) + S$$
$$CS \rightleftarrows C(g) + S$$

The stoichiometry is given by

$$CH_3CH_2CHOHCH_3 \rightleftarrows CH_3COC_2H_5 + H_2$$

where
$$A = \text{alcohol}$$
$$B = \text{ketone}$$
$$C = \text{molecular hydrogen}$$

For temperatures equal to and greater than 600°F, the experimental results could be explained by assuming that the rate controlling step was the desorption of hydrogen.

(a) Derive the rate equation for the case where the surface reaction is the rate controlling step.

(b) Derive the rate equation for the case where the desorption of hydrogen is the rate controlling step.

8-7. In a study [40] of the vapor phase oxidation of naphthalene, benzene, and anthracene over vanadium pentoxide at temperatures of 270° to 377°C, the following rate equation was obtained:

$$r_A = \frac{C_T k_2 k_1 p_{O_2} p_A}{\beta k_1 p_A + k_2 p_{O_2}} = \frac{1}{(1/C_T k_1 p_A) + (\beta/C_T k_2 p_{O_2})}$$

where β is equal to the soichiometric moles of oxygen required per mole of the aromatic. Use the following system of reactions plus the pseudo steady-state assumption that $dC_{S_2}/dt \cong 0$ to obtain the preceding expression for r_A.

$$A(g) + S_1 \xrightarrow{k_1} R(g) + S_2, \qquad r_1 = k_1 p_A C_{S_1}$$

$$\beta O_2(g) + S_2 \xrightarrow{k_2} S_1, \qquad r_2 = k_2 p_{O_2} C_{S_2}$$

where $r_1 = (1/\beta)r_2$, and r_A, r_1, r_2 have the units of moles/(time)(vol)

$A = $ benzene, naphthalene, or anthracene

$R = $ oxidized aromatic which obtains the oxygen from the catalyst site, S_1

$S_2 = $ reduced catalyst site

$S_1 = $ oxidized catalyst site

(a) By use of the pseudo steady-state assumption, derive the rate equation.

(b) For an isothermal plug-flow reactor, obtain the following expression for the V/n_T^0. (In the integration, neglect the variation of the change in the total moles with reaction.)

$$\frac{V}{n_T^0} = \frac{-1}{\bar{k}_1} \log_e \frac{p_A}{p_A^0} - \frac{1}{\bar{k}_2} \log_e \frac{p_{O_2}}{p_{O_2}^0}$$

where $\bar{k}_1 = C_T k_1 P$ and $\bar{k}_2 = C_T k_2 P$.

(c) Suggest a form for plotting the data for the purpose of estimating $\bar{k}_1$ and $\bar{k}_2$.

8-8. In the oxidation of ethylene over a silver catalyst to ethylene oxide [37], the selectivity can be approximately fitted by a triangular system of first-order reactions:

$$C_2H_4 \xrightarrow{k_1} C_2H_4O \qquad \frac{k_2}{k_1} = 0.08$$
$$ {}_{k_3}\searrow \swarrow{}_{k_2} \qquad\qquad \frac{k_3}{k_1} = 0.4$$
$$ CO_2$$

For ethylene conversions of 20%, 60%, and 80% in an isothermal, plug-flow reactor, calculate the moles of CO_2 and the moles of ethylene oxide produced per mole of ethylene fed to a plug-flow reactor which is operated isothermally. Assume that the volumetric flow rate is constant throughout the reactor. Also, calculate the moles of ethylene oxide produced per mole of ethylene reacted.

Answer: At $x = 0.2$, 0.6, and 0.8 mol CO_2 per mole C_2H_4 fed = 0.393, 0.318, and 0.113. Moles of C_2H_4O/C_2H_4 fed = 0.142, 0.416, and 0.539.

8-9. A mechanism for the dehydrogenation of butane, $C_4H_{10} \rightleftarrows C_4H_8 + H_2$, over chromia ($Cr_2O_3$) on alumina was postulated by Happel and Atkins [41] as follows:

$$C_4H_{10} + S \rightleftarrows C_4H_{10}S \tag{1}$$

$$C_4H_{10}S + S \rightleftarrows C_4H_9S + HS \tag{2}$$

$$C_4H_9S + S \rightleftarrows C_4H_8S + HS \tag{3}$$

$$C_4H_8S \rightleftarrows C_4H_8 + S \tag{4}$$

$$2HS \rightleftarrows H_2 + 2S \tag{5}$$

(a) On the basis of the following assumptions, develop the corresponding rate expression. Reaction (1) is rate controlling and reactions (2) through (5) are in equilibrium. Assume adsorbed butene is the most abundant surface intermediate.

(b) Repeat part (a) for the case where reaction (2) is the rate controlling step, reactions (1), (3), (4), and (5) are in equilibrium, and butene is the most abundant surface intermediate.

(c) Examine the effect of pressure on the initial rate of reaction for the rate expressions obtained in parts (a) and (b).

8-10. Methanol may be synthesized over a ZnO catalyst in the temperature range of 300° to 400°C at pressures of 200 to 300 atm [42]. The stoichiometry is

$$CO + 2H_2 \rightleftarrows CH_3OH$$

Develop the rate expression for the following mechanism. Let $CO = A$, $H_2 = B$, $CH_3OH = C$; then

$$A(g) + S \rightleftarrows AS \tag{1}$$

$$B(g) + S \rightleftarrows BS \tag{2}$$

$$AS + 2BS \rightleftarrows CS + 2S \tag{3}$$

$$CS \rightleftarrows C(g) + S \tag{4}$$

Assume that reaction (3) is the rate controlling step and that reactions (1), (2), and (4) are in equilibrium.

8-11. The oxidation of n-butane to maleic anhydride over a vanadium phosphorous oxide catalyst was investigated by Centi et al. [43] in a tubular reactor containing the catalyst in the form of stacked pellets. The catalyst had an atomic ratio of phosphorus to vanadium of 1.02, and the oxidation state of vanadium after the catalyst had been activated was 96% V^{4+} and 4% V^{5+}. The x-ray diffraction pattern for the patented β-phase corresponded to a nonstoichiometric vanadium (IV) pyrophosphate. The cylindrical pellets were 2 mm long, 4 mm in diameter, with a hole 2 mm in diameter drilled in the axial direction. The surface area of the catalyst was 31 m^2/g. Residence time distribution experiments at velocities of 0.2 and 0.7 m/s indicated plug-flow behavior through the reactor. The reactions occurring in the reactor could be represented as follows:

$$n\text{-butane} \rightarrow \text{maleic anhydride} \tag{1}$$

$$n\text{-butane} \rightarrow CO \text{ and } CO_2 \tag{2}$$

$$\text{maleic anhydride} \rightarrow CO \text{ and } CO_2 \tag{3}$$

Oxygen is consumed and water is produced in these reactions. A negligible amount of acetic acid was detected in the reactor effluent.

The following rate expressions for this system were developed:

$$r_1 = \frac{k_1 K_B C_B C_{O_2}^\alpha}{1 + K_B C_B}$$

$$r_2 = r_{CO_2} = k_2 C_{O_2}^\beta$$

$$r_3 = \frac{k_3 C_{ma} C_{O_2}^\gamma}{C_B^\delta}$$

where C_B and C_{ma} denote the concentrations of n-butane and maleic anhydride, respectively. The following values of the rate constants, activation energies, and orders were determined:

Reaction	1	2	3
$k_j \times 10^7$ at 320°C	4.62	4.364	0.6606
E_j (kJ mol^{-1})	45.1	110.0	57.4

Values of other constants determined were as follows:

$$K_B = 2616, \ \alpha = \beta = 0.2298, \ \gamma = 0.6345, \ \text{and } \delta = 1.151.$$

The units of the rate constants are consistent with the exponents on the concentrations. The rates are defined in terms of the units of mol/(s g catalyst). The rates for reactions (1) and (3) are defined in terms of the appearance and disappearance of maleic anhydride, respectively, and for reaction (2) in terms of the appearance of carbon dioxide. Thus, r_B, the net rate of disappearance of n-butane, is given by

$$r_B = r_1 + \tfrac{1}{4} r_2$$

and the net rate of appearance of maleic anhydride by

$$r_{ma} = r_1 - r_3$$

On the basis of a feed concentrations of 4.527×10^{-4} mol/liter of n-butane and 6.255×10^{-3} mol/liter of oxygen, with the remainder of the feed being nitrogen, a reaction temperature of 320°C and a reactor pressure of 1 atm, calculate

(a) The conversion as a function of the grams of catalyst per mole of feed per second (W/n_T^0) up to a conversion of 99.9%.

(b) The selectivity of maleic anhydride as a function of butane conversion, where the selectivity is defined as the moles of maleic anhydride produced per mole of butane consumed.

(c) The yield of maleic anhydride as a function of butane conversion, where the yield is defined as the moles of maleic anhydride produced per mole of butane fed to the reactor.

8-12. Kohler et al. [44] investigated the hydrogenolysis of dimethyl succinate (dimethyl butanoate) for the production of 1,4-butanediol, which is used in the production of polybutylene, an engineering plastic. Although several copper-based catalysts were studied throughout the course of the investigation, only the data for the Raney Cu catalyst is used in this problem. The Raney Cu catalyst was prepared by leaching the Al from an alloy of 50% (wt) Cu and 50% (wt) Al with a 20% excess NaOH solution at 323 K. The hydrogenolysis reactions are as follows:

$$\underset{\text{dimethyl succinate}}{H_3COOC(CH_2)_2COOCH_3} + 2H_2 \rightarrow \underset{\substack{\text{methyl} \\ \text{4-hydroxybutanoate}}}{H_3COOC(CH_2)_3OH} + \underset{\text{methanol}}{CH_3OH} \quad (1)$$

$$A \qquad\qquad + 2B \rightarrow \qquad C \qquad + \quad D$$

$$\underset{}{H_3COOC(CH_2)_3OH} + 2H_2 \rightarrow \underset{\text{1,4-butanediol}}{HO(CH_2)_4OH} + CH_3OH \quad (2)$$

$$C \qquad\quad + 2B \rightarrow \qquad E \quad + \quad D$$

At the reaction pressure and temperature of 500 kPa and 543 K, it was found that the rates for reactions (1) and (2) could be represented as follows:

$$r_1 = k_1 p_A^{1/2} p_B \tag{3}$$

$$r_2 = k_2 p_C^{1/2} p_B \tag{4}$$

where r_1 and r_2 are the rates of disappearance of dimethyl succinate and methyl 4-hydroxybutanoate by reactions (1) and (2), respectively, in g mol/(s cm^3). These power law rate expressions and the results of several specially designed experiments led Kohler et al. to postulate the following reaction mechanism:

$$A + 2S \rightleftarrows A_1 S + A_2 S \tag{a}$$

$$B + S \rightleftarrows BS \tag{b}$$

$$A_1 S + BS \rightarrow C + 2S \tag{c}$$

$$C + 2S \rightleftarrows C_1 S + C_2 S \tag{d}$$

$$C_1 S + BS \rightarrow E + 2S \tag{e}$$

$$A_2 S + BS \rightarrow D \tag{f}$$

$$C_2 S + BS \rightarrow D \tag{g}$$

It was further postulated that reactions (c) and (e) are the rate controlling steps with reactions (a), (b), and (d) in dynamic equilibrium, and with reactions (f) and (g) very fast and kinetically insignificant. On the basis of the preceding mechanism and these postulates, the following rate expressions were obtained:

$$r_3 = \frac{k_3 (K_1 p_A)^{1/2} K_2 p_B}{\left[1 + (K_1 p_A)^{1/2} + K_2 p_B + (K_4 p_C)^{1/2}\right]^2} \tag{5}$$

$$r_5 = \frac{k_4 (K_4 p_C)^{1/2} K_2 p_B}{\left[1 + (K_1 p_A)^{1/2} + K_2 p_B + (K_4 p_C)^{1/2}\right]^2} \tag{6}$$

The following rate and equilibrium constants for these rate equations were reported at 543 K:

$$k_3 = 6 \times 10^{-5} \text{ mol}/(\text{s cm}^3), \qquad K_1 = 1.6 \times 10^{-3} \text{ Pa}^{-1}$$

$$k_4 = 4.8 \times 10^{-5} \text{ mol}/(\text{s cm}^3), \qquad K_2 = 7.2 \times 10^{-6} \text{ Pa}^{-1}$$

$$K_4 = 2.4 \times 10^{-3} \text{ Pa}^{-1}$$

(a) To derive the rate expression given by Equations (5) and (6) from the preceding postulates, what additional assumptions must be made regarding the adsorbed species A_1, A_2, C_1, and C_2? Reduce Equations (5) and (6) to the power law model given by Equations (3) and (4).

(b) For a ratio of 9 mol of H_2 per mole of dimethyl succinate, calculate and plot the variation of conversion of dimethyl succinate with respect to space time τ over the range $0 < \tau < 10$ s ($\tau = V/v_T^0 = 1/S_v$) for a plug-flow reactor at a temperature of 543 K and a pressure of 500 kPa.

(c) For the same conditions as in part (b), calculate and plot the selectivity of the monoester and of 1,4-butanediol with the change in conversion. Selectivities are defined as

$$\frac{\text{moles of } i \text{ produced}}{\text{moles of dimethyl succinate reacted}}$$

8-13. The rate equations developed in this chapter are frequently referred to as the Langmuir–Henselwood–Hougen–Watson (LHHW) rate equations. The equations are of the general form

$$r_i(\text{rate}) = \frac{(\text{kinetic parameter})(\text{potential})}{(\text{resistance})}$$

(a) For Equation (8-85), identify the potential term or driving force, the resistance, and the kinetic parameter. Also show that the equation can be put into the following form:

$$Y = \alpha + \beta p_A + \gamma p_B + \delta p_C + \mu p_D$$

where

$$Y = \left[\frac{p_A p_B - (p_C p_D / K)}{r} \right]^{1/2}$$

Identify and define the kinetic parameters represented by α, β, γ, δ, and μ.

(b) For Equation (8-85) and the reaction (8-79), show that the parameters δ and μ cannot be determined independent of α, β, and γ if experiments are not conducted with C and D in the feed.

(c) Show that γ cannot be determined independently if the ratio of the partial pressures of A and B in the feed are not varied independently.

8-14. Cavalli et al. [45] investigated the kinetics of the ammonoxidation of toluene to benzonitrile over a vanadium-titanium oxide catalyst (stacked pellet) in a plug-flow reactor. The reactions are

$$C_7H_8 + NH_3 + 1.5O_2 \rightarrow C_7H_5N + 3H_2O \tag{1}$$

$$C_7H_8 + 9O_2 \rightarrow 7CO_2 + 4H_2O \tag{2}$$

$$2NH_3 + 1.5O_2 \rightarrow N_2 + 3H_2O \tag{3}$$

For the first reaction, the following reaction mechanism was proposed:

$$C_7H_8 + S_1 \rightleftarrows C_7H_8S_1 \quad \text{(d.e.)} \tag{4}$$

$$C_7H_8S_1 + nO_2 \rightarrow AS_1 \quad \text{(r.c.)} \tag{5}$$

$$NH_3 + S_1 \rightleftarrows NH_3S_1 \quad \text{(d.e.)} \tag{6}$$

$$AS_1 + NH_3S_1 + m_1O_2 \rightleftarrows C_7H_5NS_1 + S_1 \quad \text{(d.e.)} \tag{7}$$

$$C_7H_5NS_1 \rightleftarrows C_7H_5N + S_1 \quad \text{(d.e.)} \tag{8}$$

where d.e. and r.c. denote "dynamic equilibrium" and "rate controlling," respectively, and n and m_1 are stoichiometric numbers.

On the basis of the assumptions (1) that the concentrations of A^* and $C_7H_5HS_1$ were small, (2) that water did not affect the surface equilibrium, and (3) that oxygen was fed in such an excess that p_{O_2} could be regarded as a constant, the following expression for the rate expression for the disappearance of C_7H_8 by reaction (1) was obtained:

$$r_1 = \frac{k p_{C_7H_8}}{1 + K_{C_7H_8} p_{C_7H_8} + K_{NH_3} p_{NH_3}} \tag{9}$$

where r_1 has the units of g mol/(s m^3).

For reaction (2), the proposed mechanism consisted of Equations (4) to (8), plus the following rate controlling step:

$$AS_1 + m_2O_2 \rightarrow 7CO_2 + S_1 \quad \text{(r.c.)} \tag{10}$$

where m_2 is a stoichiometric number. The corresponding rate expression for disappearance of C_7H_8 by the second reaction was obtained:

$$r_2 = \frac{k' p_{C_7H_8}}{1 + K_{C_7H_8} p_{C_7H_8} + K_{NH_3} p_{NH_3}} \tag{11}$$

where the units for r_2 are the same as those given for r_1.

For reaction (3), the following mechanism involving sites of the type, S_2, was postulated:

$$NH_3 + S_2 \rightleftarrows NH_3 S_2 \quad \text{(d.e.)} \tag{12}$$

$$NH_3 S_2 + m_3 O_2 \rightarrow N_2 S_2 \quad \text{(r.c.)} \tag{13}$$

$$N_2 S_2 \rightleftarrows N_2 + S_2 \quad \text{(d.e.)} \tag{14}$$

And the corresponding rate expression,

$$r_3 = \frac{k'' p_{NH_3}}{1 + K'_{NH_3} p_{NH_3}} \tag{15}$$

was obtained for the disappearance of NH_3. The symbol m_3 denotes a stoichiometric number, and the units for r_3 are the same as those given for r_1. The following kinetic and equilibrium constants were determined.

Kinetic constants

Rate Constants at 310°C, mol/(s atm m^3)	E_A, kcal/mol
$k = 482.2$	11.1
$k' = 358.5$	15.3
$k'' = 22.1$	10.6

Equilibrium constants (at 310°C)

	$\Delta H°$ (kcal/mole)	$\Delta S°$ (kcal/mol k)
$K_{C_7H_8} = 350$ atm^{-1}	-14.7	-13.6
$K_{NH_3} = 37.6$ atm^{-1}	-10.9	-11.5
$K'_{NH_3} = 5.15$ atm^{-1}	-12.6	-18.3

(a) On the basis of the assumptions enumerated previously, derive the expressions given by Equations (9), (11), and (15) for r_1, r_2, and r_3, respectively.

(b) Calculate the conversions and selectivities of toluene and ammonia to benzonitrile for a space time of 5 s (based on a feed temperature of 318°C) and the following operating conditions. The reactor pressure is 1 atm, the temperature profile is linear from 318° to 340°C, and the partial pressures in atmospheres of toluene, oxygen, ammonia, and helium in the feed are 0.025, 0.15, 0.20, and 0.625, respectively.

(c) Compare your answers from part (b) with the calculated selectivities and conversions for isothermal operation at 318°C and 340°C.

8-15. Hydrogen may be produced from methanol and water for use in fuel cells by use of a low-temperature methanol synthesis catalyst. Amphlett et al. [46] proposed the following model, rate equations, and mechanism for reforming methanol over a Girdler G66B catalyst (CuO–ZnO), which had a BET surface area of 24 m^2/g.

$$CH_3OH \rightleftarrows CO + 2H_2 \tag{1}$$

$$A \rightleftarrows B + 2C$$

$$CO + H_2O \rightleftarrows CO_2 + H_2 \tag{2}$$

$$B + D \rightleftarrows E + C$$

Reaction (2) was assumed to be in dynamic equilibrium at each point in the reactor. The feed consisted of water and methanol ratios of 0.67 to 1.5. W/n_A^0 was varied from 300 to 1500 kg s/g mol.

Temperature was varied from 150° to 250°C and all experiments were conducted at 1 atm pressure. The rate equation for methanol conversion is as follows:

$$r_A = \frac{k_1\left[C_A - \left(C_B C_C^2/K_1\right)\right]}{1 + k_2 C_B}$$

$$k_1 = 7.036(10^6)\, e^{-11,568/T}$$

$$K_1 = 4.275(10^{14})\, e^{-11,160/T}$$

$$k_2 = 1(10^4)\, e^{-3,038/T}$$

where T is in Kelvin.

$$r_A = \text{g mol of methanol reacted}/(s\ kg\ of\ catalyst)$$

$$C_i = \text{concentration of } i \text{ in g mol/m}^3$$

For a $\frac{3}{4}$-in. 316 stainless steel, schedule 40 pipe, 14 mesh Girdler G66B catalyst, a feed temperature of 200°C and pressure of 1 atm, and an equimolar feed of methanol and steam, calculate:

(a) Equilibrium constant for the water gas shift reaction at 200°C.

(b) The feed rate of A required for 7 g of catalyst, isothermal operation, and a methanol conversion of 95%.

Answer: (a) $K = 231.5$, (b) $n_A^0 = 5.84 \times 10^{-4}$ mol/s.

8-16. The formation of formaldehyde by the oxidation of methanol was investigated by Mann and Hahn [47]. The data given in the accompanying table were obtained when the oxidation of methanol was carried out in a plug-flow reactor in the presence of a manganese dioxide–molybdenum trioxide catalyst at 1 atm. Mann and Hahn proposed the reaction mechanism

$$CH_3OH(g) + S_1 \rightarrow HCHO(g) + H_2O(g) + S_2 \tag{1}$$

$$\tfrac{1}{2}O_2(g) + S_2 \rightarrow S_1 \tag{2}$$

which they used to deduce the following rate expression:

$$r_A = \frac{k_1 p_M}{1 + k_1 p_M / \left(2 k_2 p_{O_2}^{1/2}\right)}$$

where r_A is the rate of oxidation of methanol in gmol/(hg catalyst) and p_M and p_{O_2} are in atm. For runs 302 through 307, shown in the table, the selectivity for the production of formaldehyde was 100%.

(a) On the basis of the data presented, evaluate k_1 and k_2 at 365°C.

(b) The following expressions were reported by Mann and Hahn:

$$\log k_1 = 3.432 - \frac{3.81 \times 10^3}{T}$$

$$\log k_2 = -7.508 + \frac{3.23 \times 10^3}{T}$$

where T is in K. Calculate the percentage difference of the k values found in part (a) from those given by the above equations.

EFFECT OF VARIABLES ON CONVERSION, RATE OF FORMATION, AND SELECTIVITY FOR
FORMALDEHYDE PRODUCTION IN METHANOL OXIDATION [47]

Run No.	Temp. °C	W / n_T^o	Feed, g mol / h		
			CH_3OH	O_2	N_2
302	365	2.5	0.030	0.1580	0.5930
303	365	5.0	0.030	0.1580	0.5930
304	365	8.8	0.030	0.1580	0.5930
305	365	13.3	0.030	0.1580	0.5930
306	365	16.3	0.030	0.1580	0.5930
307	365	22.0	0.030	0.1580	0.5930

Run No.	Temp. °C	Analysis of Products, g mol / h					Conversion (x)	Yield (y)
		$HCHO$	CH_3OH	H_2O	O_2	N_2		
302	365	0.0032	0.0269	0.0032	0.1560	0.5930	0.105	0.105
303	365	0.0063	0.0237	0.0063	0.1540	0.5930	0.210	0.210
304	365	0.0108	0.0192	0.0108	0.1520	0.5930	0.360	0.360
305	365	0.0154	0.0146	0.0154	0.1500	0.5930	0.513	0.513
306	365	0.0180	0.0120	0.0180	0.1490	0.5930	0.600	0.600
307	365	0.0219	0.0081	0.0219	0.1470	0.5930	0.730	0.730

n_T^o = molar flow rate, g mol/h
W = grams of catalyst
x = moles of CH_3OH reacted per mole CH_3OH fed
y = moles of $HCHO$ produced per mole CH_3OH fed

Answers: (a) $k_1 = 0.12$ mol/(h g atm), $k_2 = 0.00304$ mol/(h g atm$^{1/2}$).

NOTATION

C_{AS} = moles of A adsorbed per unit time per unit mass of catalyst

C_T = total moles of sites per unit mass of catalyst

E_j = heat of adsorption of layer j of adsorbed molecules [see Equation (8-17)]

p_A = partial pressure of component A in the interior of the pore of a catalyst at any given point in a reactor

P_0 = vapor pressure of a gas being adsorbed at the temperature T of the adsorption

r_A = rate of reaction of component A, moles of A disappearing per unit time per unit mass of catalyst

s_j = surface area covered by j layers of adsorbed molecules [see Equation (8-16)]

v = volume (at 0°C and 1 atm) of a gas adsorbed per gram of adsorbent at the temperature T and pressure P of the adsorption experiment

v_m = volume (at 0°C and 1 atm) per gram of adsorbent when the adsorbent is covered with a unimolecular layer

REFERENCES

1. Frost, A. A., and R. G. Pearson, *Kinetics and Mechanisms*, John Wiley & Sons, Inc., New York (1953).

2. Holland, C. D., and P. G. Murdoch, *AIChE J.*, *3*:386 (1957). C. D. Holland, Ph.D. dissertation, Texas A & M University (May 1953).

3. Brunauer, S., *The Adsorption of Gases and Vapors*, Princeton University Press, Princeton, N.J. (1943).

4. Langmuir, I., *J. Am. Chem. Soc.*, *35*:105 (1913).

5. _____, *J. Am. Chem. Soc.*, *37*:1139 (1915).

6. _____, *J. Am. Chem. Soc.*, *40*:1361 (1918).

7. Beeck, O., *Rev. Mod. Phys.*, *17*:61 (1945).

8. Halsey, G. D., and H. S. Taylor, *J. Chem. Phys.*, *15*:624 (1947).

9. Sips, R., *J. Chem. Phys.*, *16*:490 (1948).

10. Graham, D., *AIChE Symposium Series*, *55*: No. 24 (1927).

11. Volmer, M., *A. Phys. Chem.*, *115*:253 (1925).

12. Fowler, R. H., *Proc. Cambridge Phil. Soc.*, *31*:260 (1935).

13. Loeb, L. B., *Kinetic Theory of Gases*, McGraw-Hill Book Company, New York (1927).

14. Brunauer, S., P. H. Emmett, and E. Teller, *J. Am. Chem. Soc.*, *60*:309 (1938).

15. Livingston, H. K., *J. Colloid Sci.*, *4*:447 (1949).

16. Anderson, R. J., *Experimental Methods in Catalytic Research*, Academic Press, New York (1968).

17. Innes, W. B., *Experimental Methods in Catalytic Research*, Ch. 2 (editor, R. J. Anderson), Academic Press, New York (1968).

18. Smith, J. M., *Chemical Engineering Kinetics*, 2d ed., McGraw-Hill Book Company, New York (1970).

19. Gonzalez, A. J., and C. D. Holland, *AIChE J. 16*:No. 5, 718 (1970) and 17: No. 2, 470 (1971).

20. Griffith, D. P., *The Mechanics of Contact Catalysis*, Oxford University Press, New York (1946).

21. Taylor, H. S., *Proc. Roy. Soc., A108*:105 (1925).

22. _____, and S. C. Liang, *J. Am. Chem. Soc., 69*:1306 (1947).

23. _____, *Trans. Faraday Soc.*, Conference on Catalysis, 3 (April 1950).

24. Roberts, J. K., *Proc. Roy. Soc., A152*:445 (1935).

25. Eucken, A., *Trans. Faraday Soc.*, Conference on Catalysis, 128 (April 1950).

26. Hougen, O. A., and K. M. Watson, *Chem. Proc. Princ.*, Part III, John Wiley & Sons, Inc., New York (1948).

27. Burwell, R. L., Jr., *Chem. Eng. News*, p. 59 (August 22, 1966).

28. Walas, S. M., *Reaction Kinetics for Chemical Engineers*, McGraw-Hill Book Company, New York (1959).

29. Boudart, M., *AIChE J., 18: No.* 3, 465 (1972).

30. Krylov, Oleg V., *Catalysis by Non-Metals*, Academic Press, New York and London (1970).

31. Boudart, M., *Kinetics of Chemical Processes*, Prentice-Hall Inc., Englewood Cliffs, N.J. (1968).

32. _____, *AIChE J., 2*:62 (1956).

33. Exchenbremer, G. P., and G. A. Wagner, III, *Chem. Eng. Prog.,* 68: *No.* 1, 62 (1972).

34. Evmenenko, N. P., and Ya. B. Gorokhoustskii, *Kinetics and Catalysis, 10*:1071 (1969). Translated from *Kinet. Katal.*, 10: *No.* 6, 1299–1304 (1969).

35. Temkin, M. I., *Kinet. Katal., 3*:509 (1962).

36. Lumpkin, R. E., Jr., W. D. Smith, Jr., and J. M. Douglas, *I & EC Fund., 8: No.* 3, 407 (1969).

37. Voge, Hervey, H., and C. R. Adams, *Advances in Catalysis, 17*:151 (1967).

38. Ahuja, O. P., and G. P. Mathur, *Canadian J. Chem. Eng., 45*:367 (1967).

39. Thaller, L. H., and George Thodos, *AIChE J., 6: No.* 3, 369 (1960).

40. Mars, P., and D. W. Van Krevelen, Conference on Oxidation Processes, *Chem. Eng. Sci.*, Supplement *3*:41 (1954).

41. Happel, John, and R. S. Atkins, *I & EC Fund., 9: No.* 1, 11 (1970).

42. Natta, G., P. Pino, G. Mazzanti, and I. Pasquon, *Chimica e Indistria., 35*:705 (1953).

43. Centi, G., G. Fornasari, and F. Trifiro, *Ind. Eng. Chem. Prod. Res., 24*:32 (1985).

44. Kohler, M. A., M. S. Wainwright, D. L. Trimm, and N. W. Cant, *Ind. Eng. Chem. Res., 26*:652 (1987).

45. Cavalli, P., F. Cavani, I. Manenti, F. Trifiro, M. El-Sawi, *Ind. Eng. Chem. Res.*, *26*:804 (1987).

46. Amphlett, J. C., M. J. Evans, R. F. Mann and R. D. Weir, *Canadian J. Chem. Eng.*, *63*:605 (1985).

47. Mann, R. S., and K. W. Hahn, *J. Catalysis*, *15*:329 (1969).

Mass and Heat Transfer in the Design of Catalytic Reactors

9

In Part I of this chapter, the effect of pore diffusion on the rate of reaction is considered, and in Part II the treatment of the rate of mass transfer from the bulk conditions of the phase outside of the catalyst to the entrance of the pores is presented. In Part III, the design equations for a fixed-bed catalytic reactor are developed.

I. EFFECT OF PORE DIFFUSION ON THE RATE OF REACTION

The relatively large surface areas of commercial catalysts suggest that most of the catalytic surface is internal area rather than external area. Then one would expect to find most of the catalytic sites (the sites which promote a given reaction) in the interior of the catalyst. Thus, if the rate of a given reaction depends on the partial pressure of A above the catalytic surface, then the rate of reaction may be affected by the net rate of diffusion into the interior of the catalyst, since the partial pressure A in an element of pore volume depends on the net rate at which A is supplied to the volume by diffusion and the net rate at which A disappears from the volume by reaction. A formulation of the rate expressions follows.

A. Fundamental Rate Expressions for the Rate of Reaction and the Rate of Pore Diffusion

Suppose that component A in the gas phase reacts by a single reaction in the presence of a solid catalyst with an internal pore structure. Furthermore, suppose that the rate at which this reaction proceeds may be represented by a unidirectional nth-order rate expression. Since the number of active sites which are responsible for the activity of the catalyst is proportional to the surface area, a measurable quantity, the defining expression for the rate of reaction for a catalytic reaction is based on a unit of surface area. Then, for the nth-order unidirectional reaction under consideration,

$$r_{sa}\left[\frac{\text{moles of } A \text{ reacted}}{(\text{time})(\text{unit of surface area})}\right] = k_{sa}\, p_A^n \tag{9-1}$$

The partial pressure of a component, say A, at any point within the catalyst is denoted by p_A, and the partial pressure of A at the bulk conditions of the fluid outside the catalyst is denoted by p_{Ab}.

In the analysis which follows, a material balance is made on component A in an element of pore volume. To make this balance, an expression for the rate of reaction which is based on a unit of pore volume is needed; that is, an expression of the form

$$r_{\text{pore}}\left[\frac{\text{moles of } A \text{ reacted}}{(\text{time})(\text{unit of pore volume})}\right] = k_{\text{pore}}\, p_A^n \tag{9-2}$$

The rate constants k_{sa} and k_{pore} may be related through the appropriate use of the following mass densities and the surface area. Let

$$\frac{1}{\rho_p} = \text{volume of pellet per unit mass of pellet}$$

$$\frac{1}{\rho_s} = \text{volume of the solid pellet material per unit mass of the solid}$$

$$S_a = \text{surface area per unit mass of pellet}$$

$$r_{\text{pore}} = r_{sa}\left[\frac{\text{moles of } A \text{ reacted}}{(\text{time})(\text{unit of surface area})}\right] S_a\left(\frac{\text{surface area}}{\text{unit mass}}\right)$$

$$\cdot \frac{1(\text{mass catalyst})}{(1/\rho_p - 1/\rho_s)(\text{volume of pores})} = \left[\frac{S_a}{(1/\rho_p - 1/\rho_s)}\right] r_{sa}$$

Let the volume of pores per unit volume of the pellet be denoted by ε (the volume fraction of voids). Then

$$\varepsilon = \frac{1/\rho_p - 1/\rho_s}{1/\rho_p} \tag{9-3}$$

Thus

$$r_{\text{pore}} = \frac{S_a \rho_p}{\varepsilon} r_{sa} \tag{9-4}$$

and

$$k_{\text{pore}} = \frac{S_a \rho_p k_{sa}}{\varepsilon} \tag{9-5}$$

Next suppose that each unit of catalytic surface in the catalyst bed of a reactor is equally accessible to each reactant molecule A. Let the rate of reaction based on a unit mass of such a catalyst bed be denoted by r and the corresponding rate constant by k; that is, let the intrinsic rate be represented by

$$r \left[\frac{\text{moles of } A \text{ reacted}}{(\text{time})(\text{unit mass of catalyst bed})} \right] = kp_A^n \qquad (9\text{-}6)$$

The rates r and r_{sa} and the associated rate constants are related as follows:

$$r = S_a r_{sa} \qquad (9\text{-}7)$$

and

$$k = S_a k_{sa} \qquad (9\text{-}8)$$

Also, Equations (9-4), (9-5), (9-7), and (9-8) may be used to show that

$$r_{\text{pore}} = \frac{\rho_p}{\varepsilon} r \qquad (9\text{-}9)$$

$$k_{\text{pore}} = \frac{\rho_p}{\varepsilon} k \qquad (9\text{-}10)$$

In the analysis of the effect of pore diffusion on the rate of reaction, it is customary to represent the diffusion of a mole of A through a pore by the following expression. More precisely, the moles of A diffusing in the positive direction of z per unit of cross-sectional area of pore (which is perpendicular to the direction of transfer) is given by

$$\mathscr{I} \left[\frac{\text{moles of } A \text{ transferred}}{(\text{time})(\text{area})} \right] = -\mathscr{D} \frac{dp_A}{dz} \qquad (9\text{-}11)$$

where $\mathscr{D} = $ the diffusion coefficient defined by Equation (9-11) for the diffu-
           sion of A through a pore in which a catalytic reaction occurs

Procedures for estimating effective diffusion coefficients have been given by Satterfield and Sherwood [1].

B. Material Balance on Reactant A in a Catalyst Pore

In 1939, Thiele [2] initiated the study of the effect of pore diffusion on the rate at which a catalytic reaction occurs. Following Thiele, suppose that the actual internal pore structure of a catalyst can be represented by cylindrical pores of length $2L$ and cross-sectional area S as shown in Figure 9-1. Further suppose that the single reaction given by Equation (9-2) occurs at steady-state operation, and that the only mechanism for the transport of A through the pore is by the diffusion mechanism described by Equation (9-11). Again, following Thiele, suppose that all the pores have open ends and that A enters each end at equal rates. For this model, the material balance on A in the

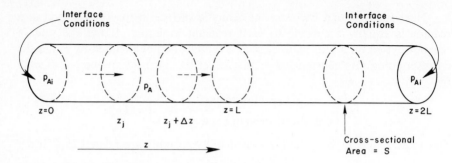

Figure 9-1. Model of an average pore of a catalyst.

element of pore volume $S\Delta z$ (see Figure 9-1) is given by

$$S\mathscr{I}|_{z_j} - S\mathscr{I}|_{z_j + \Delta z} - \int_{z_j}^{z_j + \Delta z} \left(\frac{\rho_p}{\varepsilon}\right) rS\,dz = 0 \tag{9-12}$$

From this integral-difference equation, the following differential equation is obtained as described in Chapter 1.

$$\frac{-d\mathscr{I}}{dz} - \left(\frac{\rho_p}{\varepsilon}\right) r = 0 \tag{9-13}$$

By substituting the definitions of $\mathscr{I}$ and r into Equation (9-13) and by making a change in variables, the following equivalent form of Equation (9-13) is obtained:

$$\frac{d^2\psi}{dZ^2} - L^2\alpha^2\psi^n = 0 \tag{9-14}$$

where $\qquad \alpha = \sqrt{\dfrac{\rho_p k p_{Ai}^{n-1}}{\varepsilon_{\mathscr{D}}}}, \qquad$ for an nth-order reaction

$$\psi = \frac{p_A}{p_{Ai}}$$

$$Z = \frac{z}{L}$$

The product αL is commonly called the Thiele modulus. For the boundary conditions

$$\psi = 1, \qquad \text{at} \quad Z = 0$$
$$\frac{d\psi}{dZ} = 0, \qquad \text{at} \quad Z = 1 \tag{9-15}$$

and for a first-order reaction ($n = 1$), the solution of Equation (9-14) is given by

$$\frac{p_A}{p_{Ai}} = \frac{\cosh(1 - Z)\alpha L}{\cosh \alpha L} \tag{9-16}$$

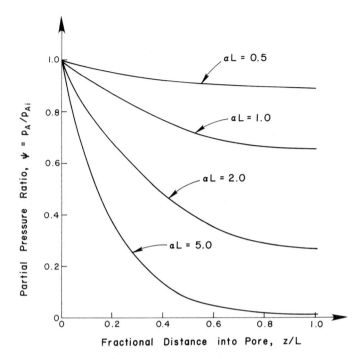

Figure 9-2. Partial pressure of the reactant as a function of the distance into the pore at different values of the Thiele modulus, αL.

Graphs of this function are displayed in Figure 9-2. For an actual pellet, Aris [3] suggested that the characteristic length of a pore, L, be defined as the ratio of the volume of a pellet to its external surface area. For spherical pellets, $L = R/3$. (Upon comparison of the Thiele modulus as presented here with the one presented by others [1, 2, 3, 5, 8, 9, 10], one finds the porosity of the pellet ε does not appear in their expression. This difference is a consequence of the way in which the effective diffusivities are defined. Satterfield and Sherwood [1] base $\mathscr{J}$ and $\mathscr{D}$ on the cross-sectional area of the porous mass, whereas $\mathscr{J}$ and $\mathscr{D}$ in Equation (9-11) are based on the cross-sectional area of the volume of the porous mass. Hence $\mathscr{D}$ presented by others is equal to $\varepsilon\mathscr{D}$ as used herein. The formula presented herein is preferred because it shows the effect of the void volume of the catalyst pellet on the Thiele modulus.)

C. Effectiveness Factor

The effectiveness factor η is defined as follows:

$$\eta = \frac{\text{actual moles of } A \text{ reacting in the pore per unit time}}{\text{moles of } A \text{ which would have reacted in the pore if } p_{Ai} \text{ were equal to } p_A \text{ for all } z} \quad (9\text{-}17)$$

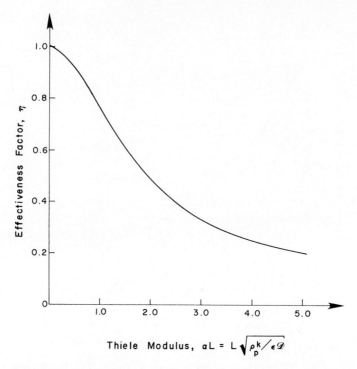

Figure 9-3. Variation of the Thiele modulus for a first-order reaction in a cylindrical pore.

Since the partial pressure of A is symmetrical with respect to the center L of the pore, it follows that the moles of A reacted over $2L$ are twice that reacted over the volume from $z = 0$ to $z = L$. Thus

$$\eta = \frac{2\int_0^L (\rho_p/\varepsilon) rS\, dz}{2\int_0^L (\rho_p/\varepsilon) r|_{p_{Ai}} S\, dz} = \frac{\int_0^1 r\, dZ}{r|_{p_{Ai}}} \qquad (9\text{-}18)$$

where $Z = z/L$. After the rate of reaction (for $n = 1$) has been expressed in terms of A, the integral in Equation (9-18) may be evaluated to give

$$\eta = \frac{\tanh \alpha L}{\alpha L} \qquad (9\text{-}19)$$

A graph of the effectiveness factor η as a function of αL is given in Figure 9-3.

This expression for η may be used to compute the mean value of the rate of reaction within the pore by use of Equation (9-21), which is developed as follows. Let the integral appearing in Equation (9-18) be replaced by its equivalent as given by the mean-value theorem of integral calculus. Then

$$\eta = \frac{\int_0^1 r\, dZ}{r\big|_{p_{Ai}}} = \frac{(r)_m}{r\big|_{p_{Ai}}} = \frac{r_A}{r\big|_{p_{Ai}}} \tag{9-20}$$

where r_A is used to denote the mean value of r over the length of pore L. Thus

$$r_A = \eta r\big|_{p_{Ai}} = \left(\frac{\tanh \alpha L}{\alpha L}\right) r\big|_{p_{Ai}} \tag{9-21}$$

If pore diffusion were the only mechanism that limited the accessibility of the catalyst surface to the reacting molecules, then the experimentally observed rate at any point in the catalyst bed would be equal to r_A.

For values of $\alpha L \geq 5$, the resistance to pore diffusion is large in the sense that pore diffusion has a pronounced effect on the rate of reaction. Since

$$\tanh \alpha L = \frac{e^{\alpha L} - e^{-\alpha L}}{e^{\alpha L} + e^{-\alpha L}} = \frac{1 - e^{-2\alpha L}}{1 + e^{-2\alpha L}}$$

it is evident that, for $\alpha L \geq 5$,

$$\tanh \alpha L \cong 1$$

Thus, for $\alpha L \geq 5$, the expression for the effectiveness factor given by Equation (9-19) reduces to

$$\eta \cong \frac{1}{\alpha L} \tag{9-22}$$

For very small values of αL, the resistance to pore diffusion is said to be "small" in the sense that the rate of reaction becomes independent of pore diffusion as αL approaches zero. For, by l'Hôpital's rule,

$$\lim_{\alpha L \to 0} \eta = \lim_{\alpha L \to 0} \frac{1 - e^{-2\alpha L}}{\alpha L + \alpha L e^{-2\alpha L}}$$

$$= \lim_{\alpha L \to 0} \frac{2e^{-2\alpha L}}{1 - 2\alpha L e^{-2\alpha L} + e^{-2\alpha L}} = 1 \tag{9-23}$$

In the region where the resistance to pore diffusion is large, the rate becomes inversely proportional to the Thiele modulus. The failure to account for pore diffusion may cause a reaction to appear to have a fractional order, as illustrated by the following example.

EXAMPLE 9-1

Suppose that the Thiele model adequately describes an nth-order catalytic reaction which occurs in the presence of a catalyst surface. Thus the expression for the mean rate of reaction which would be exhibited by the reactant in the presence of a catalyst pellet is given by

$$r_A = \eta k p_{Ai}^n$$

where n is the order of the reaction when the catalytic surface is equally accessible to all molecules of A. For the case where the resistance to pore diffusion is large, what are the apparent orders of r_A for different values of n?

SOLUTION

When η in the expression for r_A is replaced by its equivalent as given by Equation (9-22) and the expression for α given by Equation (9-14) is used, the following result is obtained:

$$r_A = \frac{1}{L}\left(\sqrt{\frac{\varepsilon \mathscr{D}}{\rho_p k p_{Ai}^{n-1}}}\right) k p_{Ai}^n = \frac{1}{L}\left(\sqrt{\frac{\varepsilon \mathscr{D} k}{\rho_p}}\right) p_{Ai}^{(n+1)/2}$$

For $n = 0$, the apparent order is $\frac{1}{2}$; for $n = 1$, the apparent order is unity; and for $n = 2$, the apparent order is $\frac{3}{2}$.

The preceding development was based on the assumption of constant temperature throughout a given catalyst pellet. Naturally, the question arises as to what happens when the temperature of the pellet is not uniform throughout. This is a more difficult problem in that it requires simultaneous solution of both heat and material balances. Numerical solutions have been obtained for first- and second-order reactions [4]. Exothermic reactions may have effectiveness factors η greater than unity, and endothermic reactions always have effectiveness factors less than unity [5].

D. Conversion in a Plug-Flow Reactor in Which Pore Diffusion Occurs

Suppose that the rate of transfer of reactant A from the bulk phase outside of the catalyst to the entrance of the pore is exceedingly fast relative to the diffusion of A through the pores. Then the bulk partial pressure of A in the fluid flowing past a catalyst pellet is approximately equal to the partial pressure at the interface.

$$p_{Ab} \cong p_{Ai} \tag{9-24}$$

For the first-order reaction, say,

$$A \rightarrow R \tag{9-25}$$

The expression following Equation (9-14) for α reduces to

$$\alpha = \sqrt{\frac{\rho_p k}{\varepsilon \mathscr{D}}} \tag{9-26}$$

and thus becomes independent of the partial pressure p_{Ai} which varies throughout the length of the catalyst bed. At the inlet to the pores at a given W of the reactor shown in Figure 8-6, the rate of reaction as defined by Equation (9-6) becomes

$$r|_{p_{Ai}} = kp_{Ai} = kp_{Ab} \tag{9-27}$$

Thus, by Equation (9-21), the mean rate of reaction r_A at any W becomes

$$r_A = \eta r|_{p_{Ai}} = \eta kp_{Ab} = \eta k \left(\frac{n_A}{n_T} \right) P \tag{9-28}$$

where n_A is the bulk value of the molar flow rate of A at a given W. The material-balance expression for r_A is given by Equation (8-100):

$$r_A = -\frac{dn_A}{dW} \tag{8-100}$$

Suppose that the fixed-bed catalytic reactor shown in Figure 8-6 is used to carry out the catalytic reaction given by Equation (9-25). The feed consists of pure A ($n_T^0 = n_A^0$), the reactor is operated isothermally with a negligible pressure drop, and the plug-flow assumption is valid. Thus k, n_T, and P in Equation (9-28) are independent of W, and η may likewise be regarded as essentially constant. [For any reaction for which $n \neq 1$, the effectiveness factor η depends on composition; see Equation (9-14)]. Elimination of r_A from Equations (8-100) and (9-28), followed by integration, gives the following expression for n_A at any W:

$$n_A = n_A^0 \exp \left[-\frac{\eta kPW}{n_T^0} \right] \tag{9-29}$$

The preceding relationships suggest the use of the experimental tests to determine if pore diffusion is significant in a given catalytic reaction. One obvious test is to determine the rates of reaction exhibited by a catalyst in two different pellet sizes. If the model constituted an exact description of the process, then, by Equation (9-22), one would expect the rates to be inversely proportional to the characteristic length of the pores of the pellets in systems in which the resistance to diffusion is large. If the rates of reaction are unaffected by changing the length of the catalyst pores, then pore diffusion is fast and not a significant resistance. For nonisothermal exothermic reactions, Peterson [5] suggests the use of three different pellet sizes in order to provide meaningful results in the region where maximum values of η were predicted by the solutions of his model. Boudart [6] questioned the validity of tests of this type for pore diffusion in the zeolites. Such substances have micropores with diameters of the order of 10 ångstroms. For such substances, Boudart [6] suggested the use of several catalysts with differing amounts of active material. If the rate, based on the surface area of active material, is constant, then pore diffusion is fast [6]. For more advanced treatments of this topic, see References [1, 5, 7–10].

EXAMPLE 9-2

Suppose that the Thiele model is obeyed by an actual catalyst which is to be tested in a fixed-bed catalytic reactor. Further suppose that the first-order reaction given by Equation (9-25) is carried out at the operating conditions preceding Equation (9-29) plus the condition that the feed consists of pure A. Two tests are run at the same operating conditions and the same mass of catalyst. In the first test, pellets of length L_1 are used, and in the second test the length of the pellets is doubled ($L_2 = 2L_1$).

(a) If a conversion of $x_1 = 0.3$ (moles of A converted per mole of A fed) is obtained for the first test, find the conversion x_2 to be expected from the second test for the case where resistance to pore diffusion is large ($\alpha L_2 \geq 5$).

(b) Suppose that a third test is run in which the pellets of length L_2 are used ($L_2 = 2L_1$) and the ratio W/n_A^0 is adjusted such that a conversion of $x_3 = x_1 = 0.3$ is obtained. Find the value of $(W/n_A^0)_3$ relative to $(W/n_A^0)_1$ required to achieve the same conversion obtained in the first test. Again, assume that the tests are carried out in the region in which the resistance to pore diffusion is large.

SOLUTION

(a) On the basis of the data given for the first and second tests, Equation (9-29) may be stated as follows for the first test:

$$-\log_e (1 - x_1) = \frac{\eta_1 kPW}{n_A^0}$$

and for the second test

$$-\log_e (1 - x_2) = \frac{\eta_2 kPW}{n_A^0}$$

These two equations are readily solved for x_2 to give

$$x_2 = 1 - (1 - x_1)^{\eta_2/\eta_1}$$

When resistance to pore diffusion is large,

$$\frac{\eta_2}{\eta_1} = \frac{\alpha L_1}{\alpha L_2} = \frac{L_1}{L_2} = \frac{1}{2}$$

Thus

$$x_2 = 1 - \sqrt{0.7} = 0.163$$

(b) In this case, Equation (9-29) takes the following form for the first experiment:

$$-\log_e (1 - x_1) = \eta_1 kP \left(\frac{W}{n_A^0} \right)_1$$

and for the third experiment

$$-\log_e (1 - x_1) = \eta_2 kP \left(\frac{W}{n_A^0} \right)_3$$

Since $\eta_2/\eta_1 = \frac{1}{2}$ [see part (a)], it follows that

$$\left(\frac{W}{n_A^0} \right)_3 = 2 \left(\frac{W}{n_A^0} \right)_1$$

Thus, to achieve the same conversion in the third test as the first, the mass of catalyst must be doubled, or the flow rate of the feed must be reduced by one-half.

II. MASS TRANSFER (FILM DIFFUSION) FROM BULK FLUID TO CATALYST SURFACE

The mechanism of mass transfer includes the transport of the reactant molecules from the flowing stream to the interface at the surface of the catalyst and the transport of the product molecules from the interface to the flowing stream. The mass transfer process includes steps 1 and 7 of the catalytic process described at the beginning of Chapter 8. If mass transfer is the rate controlling step, then the interior surface of the catalyst is not being effectively used.

A. Fundamental Rate Equations

The rate of mass transfer of component A from the bulk conditions in the flowing stream to the interface is commonly represented as follows:

$$r_{mA} = k_A (p_{Ab} - p_{Ai}) \qquad (9\text{-}30)$$

where $k_A = k_G a_v \phi / \rho_b$; k_G = coefficient of mass transfer, moles per unit time per unit pressure per unit of surface area; a_v = interfacial area for mass transfer per unit volume of the empty reactor bed; and ρ_b = bulk density of the catalyst bed, mass of catalyst per unit volume of bed. The quantity ϕ is the shape factor. For spheres $\phi = 1$, for cylinders $\phi = 0.91$, and for flakes $\phi = 0.81$; for irregular granules, $\phi = 0.90$ (assumed) as proposed by Yoshida et al. [11].

r_{mA} = rate of mass transfer of component A from the flowing stream to the surface of the catalyst, moles of A transferred per unit time per unit mass of catalyst

The rate of mass transfer r_{mA} and the rate of reaction r_A have the same units. Furthermore, since the mass transfer step is in series with the reaction step, it follows that at steady-state operation

$$r_{mA} = r_A \tag{9-31}$$

The validity of this relationship may be shown by use of material balances [see Equations (9-72) through (9-75)].

1. Effect of Mass Transfer and Pore Diffusion on the Rate Expression for the Rate of Reaction. These two mechanisms of mass transfer and pore diffusion constitute two rate processes in series. To demonstrate the effect of these two rate processes on the rate of reaction in the pore of a catalyst pellet, consider the simple first-order unidirectional reaction $A \rightarrow R$, given by Equation (9-25). The mean rate throughout the pore is given by rearranging Equation (9-20) to the form

$$r_A = \eta k p_{Ai} \tag{9-32}$$

By solving Equations (9-30) through (9-32), it is found that

$$p_{Ai} = \frac{p_{Ab}}{1 + (\eta k / k_A)} \tag{9-33}$$

Consequently, Equation (9-32) may be stated in terms of the bulk partial pressure p_{Ab} and the resistances to mass transfer and pore diffusion as follows:

$$r_A = \frac{p_{Ab}}{(1/k_A) + (1/\eta k)} \tag{9-34}$$

The resistance to mass transfer is $1/k_A$ and the resistance to chemical reaction as corrected for pore diffusion is represented by $1/\eta k$. If $\eta k \gg k_A$, then Equation (9-34) reduces to

$$r_A = k_A p_{Ab} \tag{9-35}$$

and mass transfer is said to be the rate controlling step. Since k_A depends on the Reynolds number of the flowing stream, it follows that the rate of reaction is dependent on the velocity of the flowing stream.

If $k_A \gg \eta k$, then Equation (9-34) reduces to Equation (9-28):

$$r_A = \eta k p_{Ab}$$

and the rate of reaction (corrected for pore diffusion) is said to be the rate controlling step. The behavior to be expected of a reactor in which the rate of reaction (corrected for pore diffusion) is the rate controlling step has already been demonstrated by Example 9-2.

Consider now the case where the rate of mass transfer is the rate controlling step and the rate of reaction is given by Equation (9-35). Suppose that the reaction given by Equation (9-25) with the rate expression given by Equation (9-35) is carried out isothermally with a negligible pressure drop. Further suppose that the plug-flow assumption is valid. Elimination of r_A from

Equations (8-100) and (9-35) followed by rearrangement and integration gives

$$n_A = n_A^0 \exp\left[-\frac{k_A P W}{n_T^0}\right] \tag{9-36}$$

The rather unusual behavior which may be exhibited by a reactor in which mass transfer is the rate controlling step results from the fact that the mass transfer rate constant k_A is also a function of the total flow rate n_T. The following equation for computing k_G is based on the correlation of Yoshida et al. [11] for $50 < N_{Re} < 1000$:

$$k_G = 0.57\left(\frac{G}{p_f M_m}\right)(N_{Re})^{-0.41}(S_c)^{-2/3} \tag{9-37}$$

where D_{mA} = effective diffusivity of component A in a multicomponent gaseous mixture (see Equations (18) and (19) of Reference [11])

$G = \rho v_T/S$, mass flow rate per unit of cross-sectional area

M_m = molecular weight of the mixture

$N_{Re} = G/a_v \phi \mu$, Reynolds number

$N_{Sc} = \mu/\rho D_{mA}$, Schmidt number

$p_{fA} = P - p_A\left[\dfrac{a + b - c - d}{a}\right]$, for the case where the reaction $aA + bB = cC + dD$ occurs

ρ = mass density of the gaseous mixture

Since $k_A = k_G a_v \phi/\rho_b$ [see Equation (9-30)], it follows that

$$k_A = \left(\frac{0.57 a_v \phi}{\rho_b}\right)\left(\frac{G}{p_{fA} M_m}\right)(N_{Re})^{-0.41}(N_{Sc})^{-2/3} \tag{9-38}$$

EXAMPLE 9-3

Suppose that Equation (9-36) describes the reaction to be studied in a catalytic reactor. Suppose that two test runs are carried out at the same conditions except for the mass of the catalyst and the total flow rate. Also, for convenience, suppose that the feed consists of the pure reactant A ($n_A^0 = n_T^0$). The mass of catalyst used in the second run is twice that of the first ($W_2 = 2W_1$). The flow rate of the second run is twice that of the first ($n_{A,2}^0 = 2n_{A,1}^0$). If 20% of A is converted in the first run, calculate the conversion which may be expected in the second run. Assume that the variations of all quantities in Equation (9-38) except the total flow rate may be neglected; that is, assume

$$k_A \cong C(n_T)^{0.59} \tag{A}$$

SOLUTION

Equation (9-36) may be rearranged and stated in the following form for each run.

$$-\log_e (1 - x_1) = k_{A,1} P \frac{W_1}{n^0_{A,1}} \tag{B}$$

$$-\log_e (1 - x_2) = k_{A,2} P \frac{W_2}{n^0_{A,2}} \tag{C}$$

Since $W_1/n^0_{A,1} = W_2/n^0_{A,2}$ and since k_A is given by Equation (A), Equations (B) and (C) may be restated in the form

$$\log_e (1 - x_2) = \left(\frac{n^0_{A,2}}{n^0_{A,1}}\right)^{0.59} [\log_e (1 - x_1)] = 2^{0.59} \log_e (1 - 0.2)$$

Thus

$$x_2 = 1 - [0.8]^{1.505} = 0.2853$$

If mass transfer had been very fast relative to the rate of reaction as corrected for pore diffusion, then the same conversion would have been obtained for the second test as was obtained for the first.

Many other tests may be devised [8, 9] which make use of the same principles involved in Example 9-3. Boudart [6] urges the use of caution in the application of these tests since at low Reynolds numbers, which are often obtained in laboratory reactors, the coefficients of heat and mass transfer are fairly insensitive to changes in the velocities. The effect of mass transfer on the kinetic behavior of a reaction system is illustrated by the following experimental example.

2. Analysis of Data for Propylene Oxidation. Propylene was oxidized over a copper oxide catalyst in a fixed-bed reactor by Billingsley and Holland [12]. A sketch of the reactor used in this investigation is shown in Figure 9-4. It was found, as discussed later, that the stoichiometry of this oxidation could be represented by the equations

$$C_3H_6 + O_2 \rightleftarrows CH_2CHCHO + H_2O \tag{9-39}$$

$$C_3H_6 + 4.5O_2 \rightleftarrows 3CO_2 + 3H_2O \tag{9-40}$$

For simplicity, these compounds are denoted as follows:

$$A + B \rightleftarrows C + D \tag{9-41}$$

$$A + 4.5B \rightleftarrows 3E + 3D \tag{9-42}$$

These reactions may not go precisely as written; the mechanisms by which these reactions proceed may differ appreciably from these simple equations. These two equations are to be regarded as representing the stoichiometric relationships between two competing mechanisms.

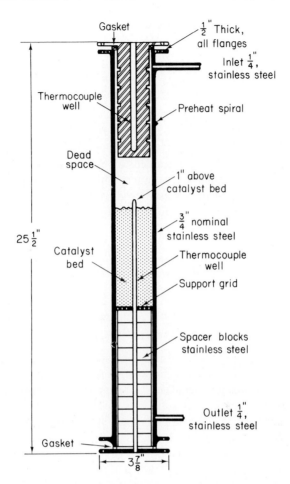

Figure 9-4. Sketch of the experimental reactor used in the oxidation of propylene by Billingsley. [Taken from D. S. Billingsley and C. D. Holland, *I & EC Fundamentals*, 2:252 (1963). Used with permission of the American Chemical Society.]

For reactants A and B, the rates r_A and r_B are defined in terms of the disappearance of A and B. Similarly, for products C, D, and E, the rates r_C, r_D, and r_E are defined in terms of the appearance of C, D, and E. Furthermore, for convenience, rates r_1 and r_2 are introduced. These are the rates at which the reactions given by Equations (9-39) and (9-40) [or Equations (9-41) and (9-42)] proceed:

$$r_1 = r_C \tag{9-43}$$

$$r_2 = \frac{r_E}{3} \tag{9-44}$$

Also, for these two stoichiometric reactions, it is readily shown that the rates are related by three independent equations:

$$r_A = r_C + \frac{r_E}{3} \tag{9-45}$$

$$r_B = r_C + \frac{4.5}{3}r_E \tag{9-46}$$

$$r_D = r_C + r_E \tag{9-47}$$

At steady state, the rate of mass transfer of each reactant from the flowing stream to the interface is equal to the rate at which it is consumed by reaction. Likewise, the rate of production of each product by reaction is equal to its rate of mass transfer from the interface to the flowing fluid. Thus, in terms of mass transfer,

$$r_A = k_A(p_{Ab} - p_{Ai}) \tag{9-48}$$

$$r_B = k_B(p_{Bb} - p_{Bi}) \tag{9-49}$$

$$r_C = k_C(p_{Ci} - p_{Cb}) \tag{9-50}$$

$$r_D = k_D(p_{Di} - p_{Db}) \tag{9-51}$$

$$r_E = k_E(p_{Ei} - p_{Eb}) \tag{9-52}$$

where, for convenience, the rates of mass transfer $r_A, r_B, \ldots, r_E$ have the same set of units as those of catalytic rates of reaction, (moles)/(time)(mass of catalyst). The mass transfer coefficients are defined in the same manner as k_A in Equation (9-30).

When the expressions given by Equations (9-48) through (9-52) are used to eliminate r_A, r_B, r_C, r_D, and r_E from Equations (9-45) through (9-47), the following set of simultaneous equations is obtained:

$$k_A(p_{Ab} - p_{Ai}) = k_C(p_{Ci} - p_{Cb}) + \frac{k_E}{3}(p_{Ei} - p_{Eb}) \tag{9-53}$$

$$k_B(p_{Bb} - p_{Bi}) = k_c(p_{Ci} - p_{Cb}) + \frac{4.5k_E}{3}(p_{Ei} - p_{Eb}) \tag{9-54}$$

$$k_D(p_{Di} - p_{Db}) = k_C(p_{Ci} - p_{Cb}) + k_E(p_{Ei} - p_{Eb}) \tag{9-55}$$

If it is supposed that the rate constants and partial pressures at the bulk conditions are known, then Equations (9-53) through (9-55) constitute three equations in the five unknown interfacial partial pressures, p_{Ai}, p_{Bi}, p_{Ci}, p_{Di}, and p_{Ei}. Thus, to solve these equations, two additional relationships are required, which could consist of two of the postulates that follow:

1. One could assume that the diffusion to and from the interior of the catalyst offers negligible resistance ($p_{Ai} \cong p_A, \ldots, p_{Ei} \cong p_E$). Then two independent rate expressions could be postulated for adsorption, desorption, and surface reactions.

2. One could again assume that the diffusion to and from the interior of the catalyst offers negligible resistance, and that all the adsorption, desorption, and surface reactions are in dynamic equilibrium.

3. One could postulate that the mass transfer of reactants to the interface are the rate controlling steps of the entire process. This assumption gives $p_{Ai} \cong p_{Bi} \cong 0$.

On the basis of the experimental results obtained by Billingsley and Holland [12], some of these postulates were discarded and others were introduced, as described later.

In the experimental investigation of Billingsley and Holland [12], the reactor shown in Figure 9-4 was operated isothermally as a differential reactor. A *differential reactor* is one in which the conversion per pass is small enough so that rate of reaction may be approximated with good accuracy as a constant over the reactor. Thus Equation (8-100) may be rearranged and integrated to give

$$r_j \cong - \frac{n_j - n_j^0}{W} = - \frac{\Delta n_j}{W} \tag{9-56}$$

where r_j is defined in terms of disappearance. For a product component, where r_j is defined in terms of appearance,

$$r_j \cong \frac{\Delta n_j}{W} \tag{9-57}$$

In attempting to correlate the experimental results, the linear relationship shown in Figure 9-5 was discovered. The equation of the best straight line through the points is

$$\frac{r_1}{r_2} = 0.72 \frac{p_{Ab}}{p_{Bb}} \tag{9-58}$$

This expression was taken as an additional relationship. By use of Equations (9-43), (9-44), (9-50), and (9-52), it is possible to restate Equation (9-58) in terms of the same variables appearing in Equations (9-53) through (9-55).

$$\frac{3k_C(p_{Ci} - p_{Cb})}{k_E(p_{Ei} - p_{Eb})} = 0.72 \frac{p_{Ab}}{p_{Bb}} \tag{9-59}$$

Thus Equations (9-53) through (9-55) plus Equation (9-59) constitute four equations in five unknowns. The further postulate was made that the rate of mass transfer of B was rate controlling; that is,

$$p_{Bi} \cong 0 \tag{9-60}$$

which reduces the number of unknowns to four. This postulate ($p_{Bi} = 0$) led to expressions for r_1 and r_2 that fit the experimental results better than the rate expressions based on several other postulates.

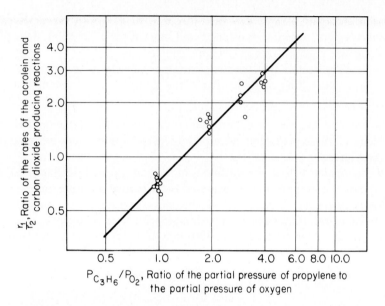

r_1/r_2, Ratio of the rates of the acrolein and carbon dioxide producing reactions

$P_{C_3H_6}/P_{O_2}$, Ratio of the partial pressure of propylene to the partial pressure of oxygen

Figure 9-5. Ratio of rates of reactions producing acrolein and CO_2 varied linearly with ratio of partial pressures. [Taken from D. S. Billingsley and C. D. Holland, *I & EC Fundamentals*, 2:232 (1963). Used with permission of the American Chemical Society.]

The stoichiometry was established by use of the same general approach presented by Woodham and Holland [13] and described in Chapter 4. For any given feed composition, the overall mechanism was found to be independent of the residence time (or W/n_T^0) as shown in Figures 4-8 and 4-9. If secondary reactions, such as the further oxidation of acrolein to CO_2 or the polymerization of acrolein to the dimer were significant, the curves in Figures 4-8 and 4-9 would have had positive slopes. The horizontal lines shown in these figures, plus the fact that other possible products did not appear in the effluent in significant amounts, suggest the reactions are adequately described by the stoichiometric relationships given by Equations (9-39) and (9-40).

The postulates represented by Equations (9-58) and (9-60) are readily utilized to obtain expressions for r_1 and r_2 as follows. Equations (9-43), (9-44), and (9-49) may be combined to give

$$r_B = r_1 + \frac{4.5}{3} 3r_2 \qquad (9\text{-}61)$$

By use of Equations (9-49), (9-58), and (9-60), it is possible to reduce Equation (9-61) to the form

$$k_B p_{Bb} = \left(0.72 \frac{p_{Ab}}{p_{Bb}} + 4.5\right) r_2 \qquad (9\text{-}62)$$

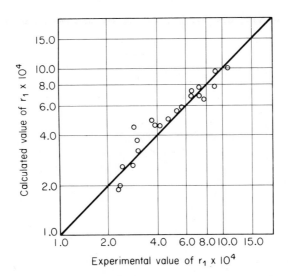

Figure 9-6. Correlation of rate of acrolein producing reaction. [Taken from D. S. Billingsley and C. D. Holland, *I & EC Fundamentals*, 2:232 (1963). Used with permission of the American Chemical Society.]

Thus

$$r_2 = \frac{k_B p_{Bb}^2}{0.72 p_{Ab} + 4.5 p_{Bb}} \tag{9-63}$$

When this result is combined with Equation (9-58), the following expression is obtained:

$$r_1 = k_B \left(\frac{0.72 p_{Ab} p_{Bb}}{0.72 p_{Ab} + 4.5 p_{Bb}} \right) \tag{9-64}$$

The value of the mass transfer coefficient k_B was computed by use of Equation (9-64) and correlated as a function of the total molar flow rate. This result,

$$k_B = 0.94 \times 10^{-3} \left(n_T^0 \right)^{0.555} \tag{9-65}$$

was used to compute the calculated values of r_1 and r_2 shown in Figures 9-6 and 9-7. Comparison of these curves with those in Figures 4-8 and 4-9 suggests that the accuracy of the correlation is about as good as the experimental accuracy.

The mass transfer coefficient k_B is a function of the Reynolds and Schmidt numbers. Of the quantities that appear in these numbers, only the total flow rate was varied to any appreciable extent. Because of this, k_B was correlated as a function of the single variable, total flow rate. The exponent 0.555 on the total flow rate is in general agreement with exponents given by Equation (9-38) and used in Example 9-3. This agreement lends support to the postulates represented by Equations (9-58) and (9-60).

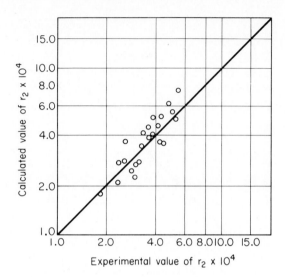

Figure 9-7. Correlation of rate of CO_2 producing reaction. [Taken from D. S. Billingsley and C. D. Holland, *I & EC Fundamentals*, 2:232 (1963). Used with permission of the American Chemical Society.]

III. DEVELOPMENT OF THE MATERIAL AND ENERGY BALANCES FOR A FIXED-BED CATALYTIC REACTOR AT STEADY-STATE OPERATION

To illustrate the development of the material- and energy-balance equations required to describe catalytic reactors, a reactor having a fixed-catalyst bed, such as the one shown in Figure 8-6, was selected. A heterogeneous system is considered in which a reactant in the gas phase is passed through a bed of solid catalyst particles. The development of the equations for this reactor is based on the following set of simplifying assumptions:

1. The gas phase is perfectly mixed in the radial direction and no mixing occurs in the direction of flow of the gas; that is, plug flow is assumed.
2. The rate of reaction can be represented by a mean rate for a catalyst particle.
3. The catalyst particles and the enclosed gas phase are at the same temperature throughout the interior of each particle.
4. The rate of heat transfer along the walls of the reactor in the direction of flow of the gas is negligible.
5. The rate of heat transfer by mixing or eddy diffusion in the direction of flow of the gas is negligible.

These assumptions are the ones commonly made and are usually valid. If the diameter and length of a reactor are of the same order of magnitude, however, the flow pattern may be more nearly approximated by the assumption of a perfectly mixed flow reactor, rather than the plug-flow assumption stated here. Even though the ratio of the diameter to the length of a reactor is small, the temperature gradients in the radial direction may not be negligible for reactors having relatively large diameters.

The second assumption may be approximated by use of the mean value of the rate of reaction within the interior of the catalyst as discussed in the previous section, or if the resistance to pore diffusion is negligible relative to the resistances for all other steps, then the rate of reaction is the same at all interior points. The equations are developed for the general case of a reaction whose stoichiometry is represented by

$$aA + bB = cC + dD$$

A. Material Balances

In all the material and energy balances which follow, a fixed-catalytic bed in a single tube is considered. The rates of mass transfer of the reactants from the bulk phase to the interior of the catalyst are defined as follows:

$$r_{mA} = k_A(p_{Ab} - p_{Ai})$$
$$r_{mB} = k_B(p_{Bb} - p_{Bi}) \tag{9-66}$$

and, for the products, the rate of mass transfer from the interior of the catalyst to the bulk phase are defined as follows:

$$r_{mC} = k_C(p_{Cb} - p_{Ci})$$
$$r_{mD} = k_D(p_{Db} - p_{Di}) \tag{9-67}$$

In the interest of generalization, the rates for all components are stated in terms of disappearance from the gas phase. Thus, when components C and D are transferred from the catalyst to the gas phase, the rates r_{mC} and r_{mD} as defined by Equation (9-67) are negative.

1. Material Balances on the Reactants and Products in the Gas Phase outside the Catalyst. A schematic diagram of the streams involved is shown in Figure 9-8. A material balance on component A over the element of volume ΔV is given by

$$n_A|_V - n_A|_{V+\Delta V} - \int_V^{V+\Delta V} r_{mA}\rho_b \, dV = 0 \tag{9-68}$$

For product C, the balance is given by

$$n_C|_V - n_C|_{V+\Delta V} - \int_V^{V+\Delta V} r_{mC}\rho_b \, dV = 0 \tag{9-69}$$

The corresponding differential equations listed next are obtained in the usual way by use of the mean-value theorems, followed by the limiting process

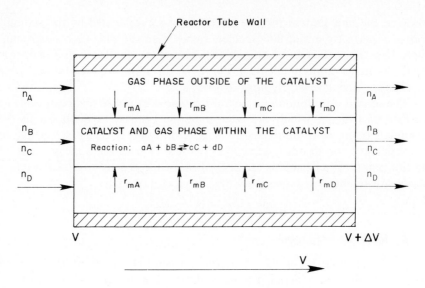

Figure 9-8. Sketch of the reactor used in making the material balances.

wherein ΔV is allowed to go to zero:

$$\frac{dn_A}{dV} + r_{mA}\rho_b = 0$$

$$\frac{dn_B}{dV} + r_{mB}\rho_b = 0$$

$$\frac{dn_C}{dV} + r_{mC}\rho_b = 0 \tag{9-70}$$

$$\frac{dn_D}{dV} + r_{mD}\rho_b = 0$$

Note also that since the mass of catalyst in the bed is related to the volume of the bed ($W = \rho_b V$) these equations may be restated in terms of the variable W rather than V in a manner analogous to that which follows for component A.

$$\frac{dn_A}{dW} + r_{mA} = 0 \tag{9-71}$$

2. Material Balance on the Gas Phase in the Interior of the Catalyst Pellets. The equality of the rate of mass transfer and the rate of reaction as given by Equation (9-31) for component A is shown to be true by making the following material balance. Consider component A which is being transferred from the bulk gas phase to the gas phase in the interior of the catalyst at the rate r_{mA}. In the interior of the catalyst, component A is being consumed by reaction at the mean rate r_A. Then, over the element of volume ΔV (see Figure

9-8), the material balance on A is given by

$$\int_{V}^{V+\Delta V} r_{mA}\rho_b \, dV - \int_{V}^{V+\Delta V} r_A\rho_b \, dV = 0 \tag{9-72}$$

Since r_{mA} and r_A are regarded as continuous functions, Equation (9-72) may be restated in the form

$$\int_{V}^{V+\Delta V} (r_{mA}\rho_b - r_A\rho_b) \, dV = 0 \tag{9-73}$$

Since Equation (9-73) applies for all values of V greater than zero and all values of $V + \Delta V$ less than the total volume of the reactor, it follows that the integrand must be equal to zero for values of V in the interval $(0 < V < V_{\text{total}})$. Thus

$$r_{mA} - r_A = 0 \tag{9-74}$$

When the remaining equations are treated in a similar manner, the following relationships are obtained:

$$r_{mB} = r_B, \qquad r_{mC} = r_C, \qquad r_{mD} = r_D \tag{9-75}$$

Also the stoichiometry of the reaction requires that the rates of reaction be related by

$$r_B = \frac{b}{a}r_A, \qquad r_C = -\frac{c}{a}r_A, \qquad r_D = -\frac{d}{a}r_A \tag{9-76}$$

B. Energy Balances

A schematic diagram of the stream flows involved in these balances is shown in Figure 9-9. The rate of heat transfer through the wall to the gas phase outside the catalyst is denoted by q_c, and the rate of heat transfer from the gas phase to the catalyst is denoted by q_g. The flow rate of gas into the element of volume ΔV is denoted by n_T, and the total heat content of this gas stream is denoted by H. The rate of mass transfer of each component to (or from) the catalyst phase from (or to) the gas phase outside the catalyst is shown in Figure 9-9.

For definiteness, suppose that heat is being transferred across the wall and the gas film to the gas phase in the catalyst bed at the rate q_c (Btu per unit time per unit length) at all points along the boundary. If a heat-transfer medium is used, then q_c at any V (that is, at any point along the boundary) is given by

$$q_c = U_c a_c (T_c - T) \tag{9-77}$$

where T_c is the temperature of the heat-transfer medium and T is the temperature of the gas phase. The heat-transfer area per unit length of reactor is denoted by a_c. (The overall heat-transfer coefficients U_c and a_c are consistent in that if U_c is based on the interior surface area of the tube then a_c is the interior surface area.)

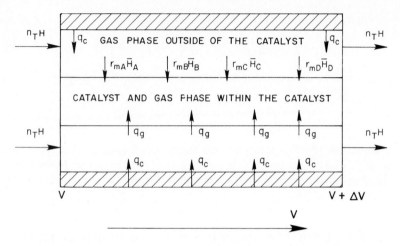

Figure 9-9. Sketch used in making the energy balances on the gas and the catalyst phases.

The heat of heat transfer from the gas phase to the catalyst phase is given by

$$q_g = U_g a_g (T - T_s) \qquad (9\text{-}78)$$

where a_g is the exterior surface area of the catalyst particles per unit length of the reactor bed which is effective in heat transfer, and T_s is the temperature of the catalyst particle and the enclosed gas.

In Figure 9-9, it is implied that the enthalpy transferred from the bulk gas phase to the gas phase within the catalyst is given by $r_{mi}\overline{H}_i$ for each component of the mixture. Taking this product equal to the enthalpy transferred for each component amounts to assigning to the partial molar enthalpy $\overline{H}_i$ the same algebraic properties possessed by H_i, the enthalpy of pure component i in an ideal solution. In some instances $\overline{H}_i$ and H_i have the same algebraic properties in multiplication and in other instances they do not. For example, they have the same algebraic properties when $n_T H$ and $C_T H$ are expanded by Euler's theorem at constant temperature and pressure, as shown in Chapter 6.

$$n_T H = \sum_{i=1}^{c} n_i \frac{\partial (n_T H)}{\partial n_i} = \sum_{i=1}^{c} n_i \overline{H}_{n_i} = \sum_{i=1}^{c} n_i \overline{H}_i$$

$$C_T H = \sum_{i=1}^{c} C_i \frac{\partial (C_T H)}{\partial C_i} = \sum_{i=1}^{c} C_i \overline{H}_{C_i} = \sum_{i=1}^{c} C_i \overline{H}_i$$

The first expression is based on the fact that $n_T H$ is homogeneous of degree 1 in the $\{n_i\}$ and the second on the fact that $C_T H$ is homogeneous of degree 1 in the $\{C_i\}$. Upon comparison of the preceding sums with the sum $\sum_{i=1}^{c} r_{mi}\overline{H}_i$, one might be inclined to infer that the total enthalpy transferred is homogeneous in the $\{r_{mi}\}$; however, this is not the case.

To show that the enthalpy transferred from the bulk gas phase to the gas phase within the catalyst is given by $\sum_{i=1}^{c} r_{mi}\overline{H}_i$, consider the general case of the plug flow of a reacting stream through a tubular reactor. Then, at constant temperature and pressure, the total change in $n_T H$ with respect to reactor volume is given by the chain rule:

$$\left(\frac{d(n_T H)}{dV}\right)_{P,T} = \frac{\partial(n_T H)}{\partial n_1}\frac{dn_1}{dV} + \frac{\partial(n_T H)}{\partial n_2}\frac{dn_2}{dV} + \cdots + \frac{\partial(n_T H)}{\partial n_c}\frac{dn_c}{dV}$$

and, by Equations (9-70) and (9-75), it follows upon substitution and rearrangement that

$$-\left[\frac{d(n_T H)}{dV}\right]_{P,T} = \rho_b \sum_{i=1}^{c} r_i \overline{H}_i = \rho_b \sum_{i=1}^{c} r_{mi} \overline{H}_i \qquad (9\text{-}79)$$

Let the net rate of disappearance of enthalpy from the bulk gas phase at constant temperature and pressure be denoted by R_H and defined by

$$R_H = \frac{\text{enthalpy disappearing from the bulk gas phase by reaction}}{(\text{time})(\text{reactor volume})} \qquad (9\text{-}80)$$

Then it is readily shown by an enthalpy balance at constant temperature and pressure that R_H is equal to the left side of Equation (9-79), and thus

$$R_H = \rho_b \sum_{i=1}^{c} r_{mi}\overline{H}_i \qquad (9\text{-}81)$$

Observe that if the reaction occurs in the gas phase within the catalyst (or more precisely on the catalyst and associated gas phase) the net rate of disappearance of enthalpy from the bulk gas phase is equal to the net rate of appearance of enthalpy in the gas phase within the catalyst; that is,

$$R_H = \frac{\text{enthalpy appearing in the gas phase within the catalyst}}{(\text{time})(\text{reactor volume})} \qquad (9\text{-}82)$$

This result is seen to be in agreement with the diagram shown in Figure 9-9.

Two independent energy balances may be stated for the system shown in Figure 9-9: (1) a balance enclosing the gas phase outside of the catalyst and (2) the gas phase within it. Alternately, a third balance enclosing the gas phase plus the catalyst could be used in lieu of either of the two listed. A formulation of the independent energy balances follows.

1. Energy Balance on the Gas Phase outside of the Catalyst Particles. On the basis of the assumptions stated at the beginning of this section and by use of the schematic diagram presented in Figure 9-9, an energy balance on the gas phase on the outside of the catalyst pellets in the element of volume from V to $V + \Delta V$ is given by

$$n_T H|_V - n_T H|_{V+\Delta V} + \int_V^{V+\Delta V} \frac{q_c - q_g}{S} \, dV - \int_V^{V+\Delta V} R_H \, dV = 0 \quad (9\text{-}83)$$

The differential equation corresponding to Equation (9-83) is

$$-\frac{d(n_T H)}{dV} + \frac{q_c - q_g}{S} - \rho_b \sum_{i=1}^c r_{mi} \overline{H}_i = 0 \quad (9\text{-}84)$$

where R_H has been replaced by its equivalent as given by Equations (9-81) and (9-82). Expansion of $d(n_T H)/dV$ by the chain rule, followed by the evaluation of the individual terms by use of the virtual values of the partial molar enthalpies as shown in Chapter 6, yields

$$-\frac{d(n_T H)}{dV} = -n_T C_p \frac{dT}{dV} - n_T \frac{\partial \Omega}{\partial P} \frac{dP}{dV} + \rho_b \sum_{i=1}^c r_{mi} \overline{H}_i \quad (9\text{-}85)$$

where

$$C_p = \sum_{i=1}^c \frac{n_i}{n_T} \left(C_{pi}^0 + \frac{\partial \Omega}{\partial T} \right)$$

Elimination of $d(n_T H)/dV$ from Equations (9-84) and (9-85) yields the result

$$-n_T C_p \frac{dT}{dV} - n_T \frac{\partial \Omega}{\partial P} \frac{dP}{dV} + \frac{q_c - q_g}{S} = 0 \quad (9\text{-}86)$$

2. Energy Balance Enclosing the Catalyst and the Gas Phase within the Catalyst. As seen from Figure 9-9, the energy balance enclosing the catalyst and the gas phase within it from V to $V + \Delta V$ is given by

$$\int_V^{V+\Delta V} \left(R_H + \frac{q_g}{S} \right) dV = 0 \quad (9\text{-}87)$$

Since Equation (9-87) is valid for all choices of V and $V + \Delta V$ ($0 < V < V + \Delta V < V_T$, where V_T is the total volume of the reactor), it follows that

$$R_H + \frac{q_g}{S} = 0 \quad (9\text{-}88)$$

or

$$\rho_b \sum_{i=1}^c r_{mi} \overline{H}_i + \frac{q_g}{S} = 0 \quad (9\text{-}89)$$

Since

$$\rho_b \sum_{i=1}^c r_{mi} \overline{H}_i = \rho_b \sum_{i=1}^c r_i \overline{H}_i = \rho_b \sum_{i=1}^c r_i \left[H_{fi}^0 + \frac{\partial(n_T \Omega)}{\partial n_i} \right]$$

it follows that, for the single reaction $aA + bB = cC + dD$,

$$\rho_b \sum_{i=1}^{c} r_{mi} \overline{H}_i = -\rho_b r_A \left[\frac{c}{a} H_{fC}^{\circ} + \frac{d}{a} H_{fD}^{\circ} - H_{fA}^{\circ} - \frac{b}{a} H_{fB}^{\circ} \right.$$

$$\left. + \frac{c}{a} \frac{\partial (n_T \Omega)}{\partial n_C} + \frac{d}{a} \frac{\partial (n_T \Omega)}{\partial n_D} - \frac{\partial (n_T \Omega)}{\partial n_A} - \frac{b}{a} \frac{\partial (n_T \Omega)}{\partial n_B} \right]$$

$$= -\rho_b r_A \Delta H_r \tag{9-90}$$

where $\quad \Delta H_r = \Delta H_r^{\circ} + \dfrac{c}{a} \dfrac{\partial (n_T \Omega)}{\partial n_C} + \dfrac{d}{a} \dfrac{\partial (n_T \Omega)}{\partial n_D} - \dfrac{\partial (n_T \Omega)}{\partial n_A} - \dfrac{b}{a} \dfrac{\partial (n_T \Omega)}{\partial n_B}$

$\Delta H_r^{\circ} = \dfrac{c}{a} H_{fC}^{\circ} - \dfrac{d}{a} H_{fD}^{\circ} - H_{fA}^{\circ} - \dfrac{b}{a} H_{fB}^{\circ}$, the heat of reaction of the perfect gas mixture at the temperature T of the reacting mixture

Thus, Equation (9-90) may be stated in the form

$$-\rho_b r_A \Delta H_r + \frac{q_g}{S} = 0 \tag{9-91}$$

When the effect of heat transfer in the axial direction by conduction $(Q = -kdT / dz)$ is taken into account, Equation (9-87) becomes

$$SQ|_V - SQ|_{V+\Delta V} + \int_V^{V+\Delta V} \left(R_H + \frac{q_g}{S} \right) dV = 0 \tag{9-92}$$

and thus Equation (9-88) becomes

$$-S \frac{dQ}{dV} + R_H + \frac{q_g}{S} = 0 \tag{9-93}$$

Instead of Equation (9-91), the following expression is obtained when the effect of conduction in the axial direction is included.

$$S^2 k \frac{d^2 T}{dV^2} - \rho_b r_A \Delta H_r + \frac{q_g}{S} = 0 \tag{9-94}$$

3. Energy Balance on the Heat-Transfer Medium. As in the formulation of the energy balance on the gas phase outside of the catalyst phase, suppose that heat is transferred through the walls of the reactor at the rate q_c from the heat-transfer medium as depicted in Figure 9-10. This heat-transfer process gives rise to an additional relationship: an energy balance on the heat-transfer medium. Suppose that the heat-transfer medium is passed through the jacket of the reactor in the same direction as the reacting stream and that the outside of the jacket is perfectly insulated. An energy balance on an element of fluid in

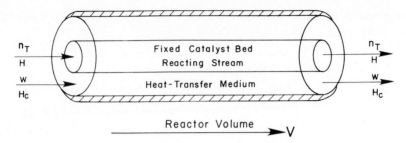

Figure 9-10. Sketch used in making the energy balance on the heat-transfer medium.

the jacket is given by

$$wH_c|_V - wH_c|_{V+\Delta V} - \int_V^{V+\Delta V} \frac{q_c}{S} \, dV = 0 \tag{9-95}$$

The corresponding differential equation is

$$wC_{pc} \frac{dT_c}{dV} + \frac{q_c}{S} = 0 \tag{9-96}$$

4. Pressure Drop in Fixed-Bed Catalytic Reactors. Equations for describing the flow of fluids through a fixed-catalyst bed are to be found in the literature on the flow of fluids through porous media. The correlation presented is based on the one proposed by Ergun [14].

$$-\frac{dP}{dz} = \frac{fG^2}{g_c D_p \rho_f} \tag{9-97}$$

where $\qquad f = \left(\dfrac{1 - \varepsilon_T}{\varepsilon_T^3}\right)\left[1.75 + \dfrac{150(1 - \varepsilon_T)}{N_{\mathrm{Re}}}\right], \qquad \dfrac{N_{\mathrm{Re}}}{1 - \varepsilon_T} < 500$

$$N_{\mathrm{Re}} = \frac{GD_p}{\mu}$$

$$a_v = \frac{\text{surface area of catalyst particle}}{\text{volume of the catalyst particle}}$$

$$D_p = \frac{6}{a_v}$$

G = mass flow rate per unit of cross-sectional area

S = cross-sectional area of the bed

v_T = volumetric flow rate of the fluid

ε_T = total volume fraction of voids in the catalyst bed

μ = viscosity of the fluid in consistent units

ρ_f = mass density of the fluid

For spherical particles, Hicks [15] recommended the following formula for the friction factor f:

$$f = 6.8 \frac{(1 - \varepsilon_T)^{1.2}}{\varepsilon_T^3} N_{Re}^{-0.2}, \qquad 300 < \frac{N_{Re}}{1 - \varepsilon_T} < 60,000$$

For gases described by $Pv_T = zn_T RT$, Equation (9-93) becomes

$$-P \frac{dP}{dz} = \frac{fTG^2}{g_c D_p \rho_f^0} \beta \left(\frac{n_T}{n_T^0} \right)$$

where $\beta = (z/z^0)(P^0/T^0)$, and the superscript 0 denotes that the quantity is at the feed conditions.

5. Summary of the Equations Describing the Fixed-Bed Catalytic Reactor Shown in Figures 9-8 through 9-10. The equations presented for the description of the steady-state behavior of the reactor shown in Figure 9-10 are perhaps best summarized by demonstrating how they might be applied to a design problem.

EXAMPLE 9-4

Suppose that the reaction having the stoichiometry

$$aA + bB \rightarrow cC + dD$$

has been carried out in an experimental reactor. The experimental reactor was filled with the commercial catalyst of interest. A series of tests demonstrated that the rate of reaction was first order with respect to A.

$$r_A = k_{exp} p_{Ab}$$

For a given set of design conditions for a commercial reactor, it is desired to determine the reactor volume required to achieve a specified conversion of the reactant A. It may be supposed that all the diameters and wall thicknesses of the reactor have been fixed as well as the feed rate n_T^0 and the flow rate w of the heat-transfer medium. The feed composition is fixed as well as its inlet temperature. Likewise the composition of the nonreactive heat-transfer medium and its inlet temperature are fixed. All the physical properties of the streams and the catalyst bed are known as well as the overall heat-transfer coefficients. Also, the enthalpy of the reacting stream is independent of pressure, $\partial \Omega / \partial P = 0$. The problem thus reduces to finding the values of n_A, n_B, n_C, n_D, T, T_s, T_c, and P for each choice of the volume V.

Formulate a set of functional expressions which may be used to solve this design problem numerically.

SOLUTION

For a given choice of n_A at any $V > 0$, the corresponding values of n_B, n_C, and n_D may be found by use of the following expressions which follow immediately from the stoichiometry of the reaction.

$$n_B = n_B^0 - \frac{b}{a}n_A^0 x = n_B^0 - \frac{b}{a}\left(n_A^0 - n_A\right)$$

$$n_C = n_C^0 + \frac{c}{a}n_A^0 x = n_C^0 + \frac{c}{a}\left(n_A^0 - n_A\right) \tag{A}$$

$$n_D = n_D^0 + \frac{d}{a}n_A^0 x = n_D^0 + \frac{d}{a}\left(n_A^0 - n_A\right)$$

Of the set of flow rates, the flow rate n_A will be regarded as an independent variable and the remaining flow rates will be regarded as dependent variables. For each independently selected element of volume ΔV, there exist five independent equations which must be satisfied over this element of volume: Equations (9-86), (9-91), (9-96), and (9-97) plus the differential equation for the rate of reaction. The last equation is obtained by eliminating r_A from Equations (9-71), (9-74), and from the rate expression given in the statement of the problem.

$$-\frac{1}{\rho_b}\frac{dn_A}{dV} = k_{\exp}p_{Ab} \tag{B}$$

In the numerical solution of these five equations over each element of volume, the problem to be solved consists of finding the values of five variables, n_A, T, T_c, T_s, and P at the end of each element of volume. Of the many numerical methods which could be used to solve this set of equations [16], the method demonstrated in Chapter 7 will be employed in the formulation of the problem. The implicit method is used to transform the set of differential equations into a corresponding set of algebraic equations. The algebraic equations as obtained are stated in functional notation and solved for the values of n_A, T, T_c, T_s, and P at the end of each element of volume by use of the Newton–Raphson method. The functions to be solved are as follows:

$$F_1 = n_A - n_A|_① + \left\{\mu\rho_b k_{\exp}\left(\frac{n_A}{n_T}\right)P + (1-\mu)\left[\rho_b k_{\exp}\left(\frac{n_A}{n_T}\right)P\right]\Big|_①\right\}\Delta V$$

$$F_2 = -\left\{\mu n_T C_p + (1-\mu)\left[n_T C_p\right]|_①\right\}\left(T - T|_①\right)$$
$$\quad + \left\{\mu(q_c - q_g) + (1-\mu)[q_c - q_g]|_①\right\}\frac{\Delta V}{S}$$

$$F_3 = -\rho_b k_{\exp}\left(\frac{n_A}{n_T}\right)P\Delta H_r + \frac{q_g}{S}$$

$$F_4 = w\left\{\mu C_{pc} + (1-\mu)C_{pc}|_①\right\}\left(T_c - T_c|_①\right)$$
$$\quad + \left\{\mu q_c + (1-\mu)q_c|_①\right\}\frac{\Delta V}{S}$$

$$F_5 = (P - P_1) + \left\{\mu\Phi + (1-\mu)\Phi|_①\right\}\frac{\Delta V}{S}$$

where Φ is used to denote the right side of Equation (9-97) and μ is the weight factor which may be assigned any arbitrary value lying between zero and one. (For $\mu = 0$, the implicit method reduces to Euler's one-point predictor, and for $\mu = 1/2$, the implicit method reduces to the trapezoidal rule. Equations such as those given are likely to be stiff, and a value of μ lying between $1/2$ and 1 should be used.) The symbol ① is used to denote the fact that the values of the variables so denoted are to be evaluated at the beginning of the increment of volume. The absence of ① denotes the fact that the variables are to be evaluated at the end of the increment of volume.

For a given increment of volume ΔV, the correct set of values of the variables n_A, T, T_c, T_s, and P at the end of the increment of volume ΔV is found by successive application of the Newton–Raphson equations until the set of values of the variables has been found that makes $F_1 = F_2 = F_3 = F_4 = F_5 = 0$, simultaneously. For the system under consideration, the Newton–Raphson equations may be represented as follows:

$$
\begin{bmatrix}
\dfrac{\partial F_1}{\partial n_A} & \dfrac{\partial F_1}{\partial T} & \dfrac{\partial F_1}{\partial T_c} & \dfrac{\partial F_1}{\partial T_s} & \dfrac{\partial F_1}{\partial P} \\
\vdots & \vdots & \vdots & \vdots & \vdots \\
\dfrac{\partial F_5}{\partial n_A} & \dfrac{\partial F_5}{\partial T} & \dfrac{\partial F_5}{\partial T_c} & \dfrac{\partial F_5}{\partial T_s} & \dfrac{\partial F_5}{\partial P}
\end{bmatrix}
\begin{bmatrix}
\Delta n_A \\ \Delta T \\ \Delta T_c \\ \Delta T_s \\ \Delta P
\end{bmatrix}
= -
\begin{bmatrix}
F_1 \\ F_2 \\ F_3 \\ F_4 \\ F_5
\end{bmatrix}
$$

PROBLEMS

9-1. Show that the general solution to the differential equation given by Equation (9-14),

$$
Z = \int \frac{d\psi}{\sqrt{(2\alpha^2 L^2/n + 1)\,\psi^{n+1} + C_1}} + C_2
$$

may be obtained by use of the following relationship:

$$
\frac{d^2\psi}{dZ^2} = \frac{d(d\psi/dZ)}{dZ} = \frac{d(d\psi/dZ)}{d\psi}\frac{d\psi}{dZ} = \frac{1}{2}\frac{d(d\psi/dZ)^2}{d\psi}
$$

Hint: Use this relationship and the differential equation [Equation (9-14)] to show that

$$
\left(\frac{d\psi}{dZ}\right)^2 = \frac{2\alpha^2 L^2}{n+1}\psi^{n+1} + C_1
$$

9-2. (a) By use of the boundary condition that at $z = L$, $dp_A/dz = 0$, and the differential equation given by Equation (9-13), show that the moles of A reacting per unit time in the pore volume is equal to rate of transfer of A by diffusion at the entrance to pore $z = 0$; that is, show that

$$
\mathcal{I}|_{z=0} = -\mathscr{D}\left(\frac{dp_A}{dz}\right)\Bigg|_{z=0} = \int_0^L \left(\frac{\rho_p r}{\varepsilon}\right) dz
$$

(b) For the case of an nth-order reaction, show that, when the result given in part (a) is used to evaluate the effectiveness factor defined by Equation (9-17), the following result is obtained:

$$
\eta = \frac{-1}{(\alpha L)^2}\left(\frac{d\psi}{dZ}\right)\Bigg|_{Z=0}
$$

(c) On the basis of the boundary condition stated in part (a) and the additional boundary condition that $\psi = 0$ at $Z = 1$ and the result of Problem 9-1, show that

$$\frac{d\psi}{dZ} = -\sqrt{\frac{2\alpha^2 L^2 \psi^{n+1}}{n+1}}$$

Then use this result to evaluate the expression given in part (b) for η when the reaction is zero, first, or second order.

(d) Show that the additional boundary condition introduced in part (c) implies that αL is large.

9-3. In a manner analogous to that shown for Equation (9-29), obtain the corresponding integral expression for the general case of the reaction $A \rightarrow R$ which has a rate of order n ($n \neq 1$) when carried out in the presence of a catalytic surface which is not equally accessible to all reactant molecules.

Answer:

$$\frac{1}{n_A^{(n-1)/2}} - \frac{1}{\left(n_A^0\right)^{(n-1)/2}} = \frac{1}{L}\left(\sqrt{\frac{\varepsilon \mathscr{D} k}{\rho_p}}\right)\left(\frac{P}{n_T^0}\right)^{(n+1)/2}\left[\frac{(n-1)}{2}\right]\left[\frac{2}{n+1}\right]^{1/2} W$$

9-4. Suppose that the experimental evidence for the reaction $A \rightarrow B + 2C$ shows that the rate of mass transfer from the bulk conditions of the gas phase to the catalyst is the rate controlling step. The rate of reaction is given by $r_A = k_A p_{Ab}$ where $k_A = (k_G a_v \phi)/\rho_b = 0.048$ g mol/h/atm/gram of catalyst bed. It is desired to produce 100 g mol of B per minute. Calculate the mass of catalyst required to obtain 90% conversion of A in a plug-flow reactor. The feed contains 50% A and 50% inerts (on a mole basis). Also, calculate the molar flow rate of A in the feed which would be required to achieve the specified production rate of B. The reactor is to be operated isothermally at a pressure of 5 atm.
Answer: 206 kg

9-5. Maleic anhydride (MA) is widely used in the synthesis of a number of valuable commercial products because it has the potential of reacting as an anhydride, a dicarboxylic acid, an ethylene comonomer, a dienophile, or a combination of the functionalities. In Europe, MA is produced primarily from benzene, while in the United States it is made predominately from n-butane. Denka Chemical Corporation licenses a process and catalyst for butane oxidation to maleic anhydride. The major MA producer in the United States is Monsanto, which uses its own technology. Wohlfahrt et al. [17] have reported the stoichiometry and kinetics for a V_2O_5/P_2O_5 catalyst. The stoichiometric reactions are as follows:

$$C_4H_{10} + 3.5O_2 = C_4H_2O_3 + 4H_2O \tag{1}$$

$$C_4H_{10} + 4.5O_2 = 4CO + 5H_2O \tag{2}$$

$$C_4H_{10} + 6.5O_2 = 4CO_2 + 5H_2O \tag{3}$$

$$C_4H_2O_3 + O_2 = 4CO + H_2O \tag{4}$$

$$C_4H_2O_3 + 3O_2 = 4CO_2 + H_2O \tag{5}$$

Maleic anhydride has the following structural representation:

These stoichiometric equations are represented kinetically by the following sequence:

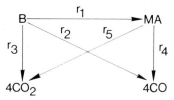

The rates of disappearance of butane (B) in kg mol/(s kg catalyst) for reactions 1, 2, and 3 are denoted by r_1, r_2, and r_3, respectively, and the rates of disappearance of MA by reactions 4 and 5 by r_4 and r_5. The corresponding rate expressions are:

$$r_1 = \frac{k_1 C_{O_2}^n b_1 C_B}{1 + b_1 C_B}$$

$$r_2 = \frac{k_2 C_{O_2}^n b_2 C_B}{1 + b_2 C_B}$$

$$r_3 = \frac{k_3 C_{O_2}^n b_2 C_B}{1 + b_2 C_B}$$

$$r_4 = \frac{k_4 C_{O_2}^n C_{MA}}{1 + b_1 C_B}$$

$$r_5 = \frac{k_5 C_{O_2}^n C_{MA}}{1 + b_1 C_B}$$

where

$$b_1 = 429 \text{ m}^3/\text{kg mol}, \qquad b_2 = 2024 \text{ m}^3/\text{kg mol}$$

$$n = 0.282$$

$$k_j = \exp\left[\log_e k_{0j} - \frac{E_j}{RT}\right] (\text{kg mol/m}^3)^{1-n} (1/\text{s})$$

Reaction	1	2	3	4	5
$\log_e k_{0j}$	2.901	−3.433	−3.938	2.265	3.284
E/R (K)	10,672	7012	7012	7012	7012
$-\Delta H_r$ (kJ/kg mol)	1,260,600	1,526,200	2,658,600	265,600	1,398,000

For this problem, assume the heat of reaction ΔH_{rj} is constant and that the mean heat capacity of the reaction gas mixture is 32 J/(g mole K).

On the basis of the preceding data of Wohlfahrt et al. and other data presented later, it is desired to predict the behavior of a pilot plant reactor. The reactor is 3.66 m long and 0.021 m in diameter (i.d.). The catalyst section of the tubes is 3.2 m, and each tube contains 0.001128 m^3 of catalyst. The bulk density of the catalyst is 1000 kg/m^3. The catalyst is in the form of cylindrical pellets (4.8 mm in diameter with a 1.6-mm hole drilled through each pellet along the axis), and it has a density of 2000 kg/m^3. Heat generated by the exothermic reactions is removed from the shell side by a molten-salt heat-transfer medium whose temperature may be regarded as approximately constant at 405°C.

The reactor feed is 1.7% (mole) *n*-butane in air. The inlet reactor pressure is 20 psig, the pressure drop is 9 psig, and the space velocity (based on 0°C and 1 atm) is 2500 h^{-1}. The heat transfer coefficient is 0.15 kJ/(s m^2 K). The feed temperature is 400°C. The viscosity of the reaction mixture may be regarded as constant at 0.032 cp (= 0.00032 p = 0.00032 g/cm s = 3.2×10^{-5} kg/m s).

For these conditions and the rate equations given previously, calculate temperature, concentration, butane conversion, yield of MA, selectivity of MA, and the pressure drop as a function of the reactor length.

$$\text{Yield} = \frac{\text{g mol MA produced}}{\text{g mol butane fed}}$$

$$\text{Selectivity} = \frac{\text{g mol MA produced}}{\text{g mol butane consumed}}$$

9-6. Monti et al. [18] studied the vapor phase hydrogenolysis of methyl formate over copper-on-silica catalysts in an external recycle reactor. Recycle rates were greater than 20 times the feed rate to the reactor. The reactor consisted of an 0.8-cm (i.d.) stainless steel U-tube containing 3 to 4 cm^3 (about 1 to 1.5 g) of catalyst and immersed in a stirred oil bath. The recycle loop was 0.64-cm (o.d.) stainless steel tubing and was heated to 50°C to prevent condensation. A stainless steel bellows pump was used in the recycle loop. Feed rates to the reactor were varied over the range of 100 to 400 cm^3/min, and the recycle loop flow rate was 8000 cm^3/min.

The catalyst was prepared by using an ion-exchange technique, which appeared to result in one copper atom per silanol site. Silanol site density was reported as 3.2×10^{18} sites (surface silonol groups) per square meter. Rates were based on the surface area of the copper as determined by the decomposition of N_2O.

The decomposition reactions are hydrogenolysis,

$$\underset{\text{methyl formate}}{CH_3OCHO} + 2H_2 \rightleftarrows \underset{\text{methanol}}{2CH_3OH} \tag{1}$$

$$A + 2B \rightleftarrows 2C$$

and decarbonylation,

$$CH_3OCHO \rightleftarrows CH_3OH + CO$$

$$A \rightleftarrows C + D$$

The rate expressions were found to be

$$r_{A1} = 3.6(10^9) e^{-14,072/T} p_A^{0.39} p_D^{-0.17}$$

$$r_{A2} = 2.2(10^9) e^{-13,471/T} p_A$$

where $\quad r_{Ai} = \dfrac{\text{g mol methyl formate consumed by reaction } i}{\text{min m}^2 \text{ of Cu surface}}$

p_i = partial pressure of i in bars

T = temperature in Kelvin

Catalyst properties are as follows:

Property	CuOX 50	Cu 200	Cu 300
Total surface area, m²/g	45	190	247
Copper content, % (wt)	1.65	6.6	8.2
Copper surface area, m²/g	1.4	5.0	6.7
Average pore radius, nm	25	13	10
Bulk density, g/cm³	0.375	0.333	0.333

The catalyst particle diameter varied between 350 to 500 μm.

The feed is an 8:1 molar ratio of hydrogen to methyl formate, and the reactor temperature and pressure are 187°C and 1 bar.

(a) For a feed rate of 423 cm³/min measured at 0°C and 1.03 bar, 1.2 g of catalyst, and average particle size of 400 μm, calculate the mass transfer coefficient k_A and r_{AM}/k_A for 61% conversion and 96% selectivity. Is the resistance to mass transfer negligible?

(b) The rate equations were developed assuming that resistance to pore diffusion was negligible. Was this a valid assumption? For this calculation, assume the rate of the second reaction is negligible relative to the first and that the reactor contents are 0.97% CO, 24.25% methyl formate, and 72.75% hydrogen.

(c) In the absence of any diffusional resistances, calculate the methyl formate conversion, methanol selectivity, and the yield for each catalyst for a feed rate of 350 cm³/min (measured at 0°C and 1.013 bar) and 1.2 g of catalyst for each catalyst.

(d) Calculate the superficial velocity and pressure drop across the catalyst bed for each catalyst for the condition of part (c).

9-7. The isomerization of cyclopropane over a Li–Y zeolite catalyst was studied by Schobert and Ma [19]. The following results and data were reported for the operating conditions of 250°C and 1 atm. Film diffusion resistance was negligible. The catalyst pellets were formed from zeolite crystals and held together with a binder such as alumina. The crystals contain micropores with pore openings of 0.8 nm. Diffusion occurs through the macropores in the pellet to the surface of the crystals and then into the crystals. The Thiele modulus ϕ_μ for the crystals is defined as follows:

$$\phi_\mu = R_\mu \sqrt{\frac{k_S K_A}{D_\mu}}$$

and the effectiveness factor η_μ for a spherical particle is

$$\eta_\mu = \frac{3}{\phi_\mu} \left[\frac{1}{\tanh \phi_\mu} - \frac{1}{\phi_\mu} \right]$$

(Definitions and numerical values used in the definitions are given later.) For the macropores, the Thiele modulus is defined by

$$\phi_m = R_m \sqrt{\frac{k_S K_A}{D_m} \frac{\varepsilon_\mu (1 - \varepsilon_m)}{\varepsilon_m}} \, \eta_\mu$$

where the effectiveness factor for the macropores is given by

$$\eta_m = \frac{3}{\phi_m} \left(\frac{1}{\tanh \phi_m} - \frac{1}{\phi_m} \right)$$

The rate of reaction for the crystal is

$$r_\mu = k_S C_{AS} = k_S K_A C_{A\mu}$$

where r_μ has the units of g mol/(s volume of crystal), C_{AS} is the concentration of A on the interior surface of the crystal, and $C_{A\mu}$ is the concentration of A in the micropores. The rate of reaction for the pellets is given by

$$r_{Ap} = \eta_\mu \eta_m \varepsilon_\mu (1 - \varepsilon_m) k_S K_A C_{Ab}$$

where r_{Ap} has the units of g mol/(s volume of pellet) and C_{Ab} is the concentration of A in the bulk fluid. A material balance on an element of the reactor reduces to

$$-\frac{dn_A}{dV} = \frac{1 - \varepsilon_R}{\varepsilon_R} \eta_\mu \eta_m \varepsilon_\mu (1 - \varepsilon_m) k_S K_A C_{Ab}$$

Definitions and numerical values of the parameters appearing in the preceding expressions are as follows:

$k_S = 3.7 \times 10^2$ (s^{-1}); first-order reaction rate constant
$K_A = 291$; adsorption equilibrium constant, which is dimensionless
$D_\mu = 2.65 \times 10^{-16}$ (m^2/s); intracrystalline effective diffusivity
$D_m = 3.83 \times 10^{-8}$ (m^2/s); the macropore effective diffusivity
$R_\mu = 25$ μm; radius of the crystal
$R_p = 3$ mm; radius of the pellet compound of crystals bound together by a binder
$\varepsilon_\mu = 0.48$; porosity of the crystals
$\varepsilon_m = 0.385$; porosity of the pellet
$\varepsilon_R = 0.4$; void fraction between the pellets in the reactor
Catalyst surface area = 443 m^2/g
Pellet density = 1120 kg/m^3

(a) Based on the data and definitions given, evaluate the effectiveness, η_μ.

(b) Evaluate the macropore effectiveness factor, η_m.

(c) Beginning with a material balance on the crystal, the pellets, and an element of the reactor, develop the corresponding differential equations and reduce them to dimensionless form to obtain the definitions for ϕ_μ, ϕ_m, and the equations presented.

(d) On the basis of the data presented, and the following operating conditions, calculate the reactor volume required to achieve a conversion of 40%. The tubular reactor is to be operated isothermally and the plug-flow assumption is valid. The reactor is packed with spherical pellets which are 6 mm in diameter. The reaction temperature is 250°C, the reactor pressure is 1 atm, the feed is pure A, and the feed rate is 10 g mol/h.

NOTATION

a_c = heat-transfer area per unit length of the reactor; based on the diameter consistent with the overall heat-transfer coefficient U_c

a_g = heat-transfer area per unit length of the reactor; based on the diameter consistent with the overall heat-transfer coefficient U_g

a_v = surface area of a catalyst particle per unit volume of the catalyst particle

D_p = $6/a_v$, mean particle diameter

$\mathscr{D}$ = diffusion coefficient; defined by Equation (9-11)

H_{fi}° = enthalpy of pure component i at a given temperature T in the perfect gas state at 1 atm above its elements in their standard states at some arbitrary datum temperature, cal/g mol

$\overline{H}_i$ = partial molar enthalpy of component i; partial derivative of the function $n_T H$ with respect to n_i with all other n_i's and P and T held fixed, cal/g mol

$\mathscr{J}$ = moles of a component transferred in the positive direction of z per unit time per unit of area which is available for diffusion and which is perpendicular to the direction of transfer [see Equation (9-11)]

k = a reaction rate constant; defined by Equation (9-6)

k_A = mass transfer coefficient; defined by Equation (9-30)

k_{pore} = reaction rate constant; defined by Equation (9-2)

k_{sa} = reaction rate constant; defined by Equation (9-1)

L = one-half of the length of a cylindrical pore

p_A = partial pressure of component A at any given location in a cylindrical pore

p_{Ai} = partial pressure of component A at the inlet of a pore

p_{Ab} = partial pressure of component A at the bulk conditions of the flowing stream

P = total pressure

q_c = rate of heat transfer [see Equation (9-77)]

q_g = rate of heat transfer [see Equation (9-78)]

r = rate of reaction; defined by Equation (9-6)

r_A = mean rate of reaction over the length of pore L; defined by Equation (9-21)

r_{pore} = rate of reaction; defined by Equation (9-2)

r_{sa} = rate of reaction; defined by Equation (9-1)

r_{mA} = rate of mass transfer of component A from the gas phase to the interface at the entrance of the pore [see Equation (9-30)]

S = cross-sectional area of the fixed-catalyst bed

S_a = surface area of a catalyst per unit mass

T = temperature of the gas phase

T_c = temperature of the heat-transfer medium

T_s = temperature of the catalyst phase

U_c = overall heat-transfer coefficient; defined by Equation (9-77)

U_g = overall heat-transfer coefficient; defined by Equation (9-78)

W = mass of catalyst

z = length of pore; $z = 0$ at the inlet, $z = L$ at the center, and $z = 2L$ at the opposite end

$Z = z/L$, a dimensionless length

Greek Letters

α = a parameter; defined following Equation (9-14); the product αL is called the Thiele modulus

ε = volume of pore per unit volume of a pellet

ε_T = fraction of voids [see Equation (9-93)]

η = effectiveness factor; defined by Equation (9-17)

μ = viscosity [see Equation (9-93)]; also used as a weight factor in Example 9-4

ρ = mass density of the gas phase [see Equation (9-93)]

ρ_b = bulk density of the catalyst bed

ρ_p = mass density of a pellet

ρ_s = mass density of the solid catalyst material

ψ = ratio of partial pressure, p_A/p_{Ai}

REFERENCES

1. Satterfield, C. N., and T. K. Sherwood, *The Role of Diffusion in Catalysts*, Addison-Wesley Publishing Company, Reading, Mass. (1963); *Mass Transfer in Catalysis*, MIT Press, Cambridge, Mass. (1970).

2. Thiele, E. W., *Ind. Eng. Chem.*, *31*:916 (1939).

3. Aris, R., *Chem. Eng. Sci.*, 6:262 (1957).

4. Tinkler, J. D., and A. B Metzner, *IE & C*, *53*:663 (1961).

5. Peterson, E. E., *Chemical Reaction Analysis*, Prentice-Hall, Inc., Englewood Cliffs, N.J. (1965).

6. Boudart, M., *AIChE J.*, *18*: No. 3, 465 (1972).

7. Wheeler, A., in *Catalysis*, Vol. 2, P. H. Emmett, editor, Van Nostrand Reinhold, New York (1955).

8. Levenspiel, O., *Chemical Reaction Engineering*, John Wiley & Sons, Inc., New York (1972).

9. Hougen, O. A., and K. M. Watson, *Chemical Process Principles*, Part III, John Wiley & Sons, Inc., New York (1948).

10. Carberry, J. J., *AIChE J.*, *7*:350 (1961).

11. Yoshida, F., D. Ramajwami, and O. A. Hougen, *AIChE J.*, *8*:5–11 (1962).

12. Billingsley, D. S., and C. D. Holland, *I & EC Fund.*, *2*:252 (1963).

13. Woodham, J. F., and C. D. Holland, *I & EC Fund.*, *52*:985 (1960).

14. Ergun, S., *Chem. Eng. Prog.*, *48*:89 (1952).

15. Hicks, R. E., *Ind. Eng. Chem.*, *40*:500 (1970).

16. Carnahan, B., H. A. Luther, and J. O. Wilkes, *Applied Numerical Methods*, John Wiley & Sons, Inc., New York (1969).

17. Wohlfahrt, K., H. Hofmann, and K. Dialer, *Chem.-Ing.-Tech.*, *52*, No. 10:1811 (1980).

18. Monti, D. M., M. S. Wainwright, and D. L. Trimm, *Ind. Eng. Chem. Prod. Res. Dev.*, *24*:397 (1985).

19. M. A. Schobert and Y. A. Ma, *J. Catalysis*, *70*:111 (1981).

Section III:
Advanced Topics

Polymerization   10
Reactions

Polymerization reactions are a combination of series and simultaneous reactions. These reactions are involved in the production of plastics, synthetic rubber, paint resins, and waxes. The reactions by which these macromolecules are produced are called *addition* and *condensation* reactions [1].

Addition reactions are further classified according to two types: *free radical addition* and *stepwise addition*. In free radical addition, the long-chain polymers are produced almost instantaneously at low conversions, whereas in stepwise addition long-chain polymers are only produced at high conversions. Monomers such as ethylene, propylene, vinyl chloride, styrene, butadiene, isoprene, and methylmethacrylate polymerize by addition reactions. Cationic, anionic, and coordination polymerizations are also examples of addition polymerizations. Examples of products formed by condensation reactions are the polyurethanes, polyureas, polyphenylesters, polycarbonates, and phenolic resins.

In Part I of this chapter, addition reactions which are initiated by free radicals are treated. Although only the one class of polymerizations is treated, a number of other polymerizations such as ionic, cationic, and coordination polymerizations may be treated in an analogous manner. All these polymerizations have in common the fact that the pseudo steady-state assumption may be used in the analysis of their corresponding mechanisms. In Part II, stepwise addition and condensation polymerizations and copolymerizations are treated.

In the design of polymerization reactors, appropriate provisions must be made for mixing and removing the heat of reaction from the reacting mixture. Heats of polymerizations commonly range from 13 to 21 kcal/g mol [2]. An

Figure 10-1. Continuously stirred reactor used for the batch production of polystyrene (Courtesy of Cosden Oil & Chemical Company).

industrial reactor used in the production of polystyrene is shown in Figure 10-1.

I. POLYMERIZATIONS BY FREE-RADICAL ADDITION REACTIONS

The common addition monomers such as ethylene ($CH_2{=}CH_2$), vinyl chloride ($CH_2{=}CHCl$), vinyl acetate ($CH_3COOCH{=}CH_2$), and styrene ($CH_2{=}CHC_6H_5$) may be represented by the vinyl group $CH_2{=}CHX$, where X represents any one of the preceding attached groups. The monomer methyl methacrylate [$CH_2{=}C(CH_3)COOCH_3$] may be represented as $CH_2{=}C(CH_3)X$.

Most of the commercially significant addition polymerizations are carried out in the presence of a catalyst. The two types of catalysts commonly employed may be classified as (1) free radical-type catalysts and (2) ionic-type catalysts. These catalysts are also called initiators. For free radical-type catalysts, the role of the initiator is to produce molecules with unpaired electrons or free radicals.

Typical initiators are benzoyl peroxide and azobisisobutyronitrile (AIBN). For benzoyl peroxide, the decomposition reactions are as follows:

$$\langle\bigcirc\rangle-\overset{\overset{O}{\|}}{C}-O:O-\overset{\overset{O}{\|}}{C}-\langle\bigcirc\rangle \longrightarrow 2\langle\bigcirc\rangle-\overset{\overset{O}{\|}}{C}-O\cdot$$

$$2\langle\bigcirc\rangle-\overset{\overset{O}{\|}}{C}-O\cdot \longrightarrow 2\langle\bigcirc\rangle\cdot + 2CO_2$$

where the dot $\cdot$ represents a free electron and two dots $:$ or $-$ represents a pair of shared electrons or a single bond. A double bond or two pairs of shared electrons is represented by $=$ or $::$. Also, the aromatic ring which is commonly symbolized by $\bigcirc$ is represented herein by the symbol $\langle\bigcirc\rangle$, which has been suggested by several recent authors.

The decomposition of azobisisobutyronitrile is commonly represented by

$$(CH_3)_2-C:N=N:C-(CH_3)_2 \longrightarrow 2(CH_3)_2\,C\cdot + N_2$$

with pendant $\underset{N}{\overset{C}{\underset{\|\|\|}{|}}}$ groups on each carbon.

The decomposition of the initiator I to produce the free radicals ϕ is commonly represented as follows:

$$I \longrightarrow 2\phi$$

The free radical ϕ then reacts with a monomer to produce another free radical. For example, the benzoyl free radical may react with the monomer $CH_2{=}CHX$ to produce another free radical:

$$\langle\bigcirc\rangle\cdot + \underset{\underset{H}{|}}{\overset{\overset{H}{|}}{C}}:\underset{\underset{X}{|}}{\overset{\overset{H}{|}}{C}} \longrightarrow \langle\bigcirc\rangle\cdot\cdot\underset{\underset{H}{|}}{\overset{\overset{H}{|}}{C}}\cdot\cdot\underset{\underset{X}{|}}{\overset{\overset{H}{|}}{C}}\cdot$$

This reaction is called the *initiation reaction*. In the preceding reaction, the free radical is seen to add to the monomer through the formation of a single bond with the monomer, which is made possible by the redistribution of electrons. This initiation of free radicals is represented by

$$\phi + M \longrightarrow R_1$$

where R_1 is the free radical formed and M is the monomer. A monomer may now add to the free radical R_1. For example,

$$\langle\bigcirc\rangle-\underset{\underset{H}{|}}{\overset{\overset{H}{|}}{C}}-\underset{\underset{X}{|}}{\overset{\overset{H}{|}}{C}}\cdot + CH_2{=}CH_2X \longrightarrow \langle\bigcirc\rangle-\underset{\underset{H}{|}}{\overset{\overset{H}{|}}{C}}-\underset{\underset{H}{|}}{\overset{\overset{H}{|}}{C}}-\underset{\underset{H}{|}}{\overset{\overset{H}{|}}{C}}-\underset{\underset{H}{|}}{\overset{\overset{H}{|}}{C}}\cdot$$

This reaction is called the *propagation reaction*, and it is represented symbolically by

$$R_1 + M \longrightarrow R_2$$

Thus, in general,

$$R_j + M \longrightarrow R_{j+1}$$

where R_j consists of the free radical ϕ with j attached monomer units. For example,

$$\langle \bigcirc \rangle - (CH_2CHX)_{j-1} - CH_2CHX\cdot$$

The propagation reactions continue until a termination reaction occurs.

The combination of any two free radicals, say R_j and R_i (where j and i represent any two arbitrary number of monomer units), serves to terminate the propagation step. For example,

$$\langle \bigcirc \rangle - (CH_2CHX)_{j-1} - CH_2CHX\cdot \ + \ \langle \bigcirc \rangle - (CH_2CHX)_{i-1}$$

$$- CH_2CHX\cdot \longrightarrow \langle \bigcirc \rangle - (CH_2CHX)_{j-1}CH_2CHX$$

$$- CHXCH_2(CHXCH_2)_{i-1} - \langle \bigcirc \rangle$$

Symbolically,

$$R_j + R_i \longrightarrow P_{j+i}$$

where the dead polymer is represented by the symbol P. The propagation reactions may also be terminated by a disproportionation reaction in which a hydrogen atom is transferred from one radical to another. A terminal double bond is formed in one radical and a saturated single bond in the other. For example,

$$\langle \bigcirc \rangle - (CH_2CHX)_{j-1} - CH_2CHX\cdot \ + \ \langle \bigcirc \rangle - (CH_2CHX)_{i-1}$$

$$- CH_2CHX\cdot \longrightarrow \langle \bigcirc \rangle - (CH_2CHX)_{j-1}CH=CHX$$

$$+ \langle \bigcirc \rangle - (CH_2CHX)_{i-1} - CH_2CH_2X$$

or symbolically

$$R_j + R_i \longrightarrow P_j + P_i$$

Normally, the monomer M adds to the free radical R_j. However, the monomer may terminate the polymerization by reacting as follows:

$$\langle\!\!\langle O \rangle\!\!\rangle-(CH_2CHX)_{j-1}-CH_2CHX\cdot \ + \ CH_2{=}CHX$$

$$\longrightarrow \langle\!\!\langle O \rangle\!\!\rangle-(CH_2CHX)_{j-1}-CH{=}CHX \ + \ CH_3CHX\cdot$$

The reactivity of $CH_3CHX\cdot$ is customarily assumed to be the same as that of the free radical $\langle\!\!\langle O \rangle\!\!\rangle-CH_2CH_2\cdot$ (called R_1). Symbolically, this termination reaction is represented as follows:

$$R_j + M \longrightarrow P_j + R_1$$

The propagation reactions may also be terminated by the action of the free radical R_j with the solvent S. For example, suppose that the polymerization considered previously is carried out in the solvent toluene. In this case, the following termination reaction may occur:

$$\langle\!\!\langle O \rangle\!\!\rangle-(CH_2CHX)_{j-1}-CH_2CHX\cdot \ + \ \langle\!\!\langle O \rangle\!\!\rangle-CH_3$$

$$\longrightarrow \langle\!\!\langle O \rangle\!\!\rangle-(CH_2CHX)_{j-1}-CH_2CHXH \ + \ \langle\!\!\langle O \rangle\!\!\rangle-CH_2\cdot$$

Again, it is customary to assume that the free-radical form of toluene has the same reactivity as that of R_1 for the polymerization. This reaction may then be represented by

$$R_j + S \longrightarrow P_j + R_1$$

Termination of the propagation step may be promoted by the addition of a chain transfer agent. For the preceding polymerization, carbon tetrachloride acts as a chain transfer agent as follows:

$$\langle\!\!\langle O \rangle\!\!\rangle-(CH_2CHX)_{j-1}-CH_2CHX\cdot \ + \ CCl_4$$

$$\longrightarrow \langle\!\!\langle O \rangle\!\!\rangle-(CH_2CHX)_{j-1}-CH_2CHXCl \ + \ CCl_3\cdot$$

Again, it is customary to assume that the free radical $CCl_3\cdot$ has the same reactivity as R_1 for the preceding polymerization. Symbolic representation of the termination of the propagation step by use of a chain transfer agent C is

$$R_j + C \longrightarrow P_j + R_1$$

In our example, "head-to-tail" addition of the monomer was assumed, which produces a chain of the form

$$\cdots -\overset{\overset{\displaystyle H}{|}}{\underset{\underset{\displaystyle H}{|}}{C}} - \overset{\overset{\displaystyle H}{|}}{\underset{\underset{\displaystyle X}{|}}{C}} - \overset{\overset{\displaystyle H}{|}}{\underset{\underset{\displaystyle H}{|}}{C}} - \overset{\overset{\displaystyle H}{|}}{\underset{\underset{\displaystyle X}{|}}{C}} - \cdots \qquad \text{(head-to-tail)}$$

"Head-to-head" addition is also possible, and it produces a chain of the form

$$\cdots -\overset{\overset{\displaystyle H}{|}}{\underset{\underset{\displaystyle H}{|}}{C}} - \overset{\overset{\displaystyle H}{|}}{\underset{\underset{\displaystyle X}{|}}{C}} - \overset{\overset{\displaystyle H}{|}}{\underset{\underset{\displaystyle X}{|}}{C}} - \overset{\overset{\displaystyle H}{|}}{\underset{\underset{\displaystyle H}{|}}{C}} - \cdots \qquad \text{(head-to-head)}$$

"Head-to-tail" addition is generally the most probable [3, 4].

A. Characterization of Mixtures of Polymer Molecules

The word *polymer* is used herein to refer to a single long-chain molecule as well as a collection of molecules having any arbitrary distribution of chain lengths. Definitions of the quantities commonly used in the characterization of these polymers follow.

1. Weight Fraction. The weight fraction W_j of polymer molecules having chains composed of j monomer units in a mixture of polymer molecules having all possible chain lengths is defined by

$$W_j = \frac{\text{mass of molecules composed of } j \text{ monomer units}}{\text{total mass of polymer}}$$

The weight fraction is calculated as follows:

$$W_j = \frac{M_j P_j}{\sum M_j P_j} \qquad (10\text{-}1)$$

where P_j = concentration of polymer P_j in moles per unit volume

M_j = molecular weight of the polymer P_j

$\sum M_j P_j$ = sum of the products $M_j P_j$ over all possible chain lengths, j $(j = 1, 2, \ldots)$

A molecule composed of j monomer units has a molecular weight equal to j

times the molecular weight of the monomer,

$$M_j = jM_M \tag{10-2}$$

where M_M = molecular weight of the monomer

[*Note*: Equation (10-2) neglects the contribution of the free radical $R\cdot$ to the molecular weight of the polymer. Instead of Equation (10-2), the correct molecular weight of the polymer is given by $M_j = jM_M + M_R$, where M_R is the molecular weight of the free radical $R\cdot$.]

When M_j in Equation (10-1) is replaced by its equivalent as given by Equation (10-2), the following expression is obtained:

$$W_j = \frac{jP_j}{\Sigma jP_j} = \frac{jP_j}{\Gamma_1(P_j)} \tag{10-3}$$

where $\Gamma_1(P_j) = \sum_{j=1}^{\infty} jP_j$,   the first moment of the P_j's

Since the mass of the monomer which has reacted at any time must be in the form of polymer, it follows that

$$M_M(M^0 - M) = \Sigma M_j P_j = M_M \Sigma jP_j \tag{10-4}$$

and that

$$M^0 - M = \Gamma_1(P_j) \tag{10-5}$$

where M^0 and M represent the initial and final concentrations of the monomer.

2. Molecular Weight Distribution. As polymerization reactions proceed, polymer molecules of different lengths are produced. The plot of the weight fraction of molecules of the same length versus the length is called a *molecular weight distribution* (see Figure 10-2). The plot of the mole fraction of molecules of the same length versus the length is called the *molecular number* or the *number fraction distribution* (see Figure 10-3). The mole fraction x_j is defined in the usual way:

$$x_j = \frac{P_j}{\Sigma P_j} \tag{10-6}$$

3. Number Average Molecular Weight. The number average molecular weight $\overline{M}_n$ is defined by

$$\overline{M}_n = \frac{\Sigma P_j M_j}{\Sigma P_j} \tag{10-7}$$

Since $M_j = jM_M$ [Equation (10-2)], $\overline{M}_n$ may be stated in the form

$$\overline{M}_n = M_M \frac{\Sigma jP_j}{\Sigma P_j} = M_M \frac{\Gamma_1(P_j)}{\Gamma_0(P_j)} \tag{10-8}$$

where $\Gamma_0(P_j) = \sum_{j=1}^{\infty} P_j$,   the total concentration of all polymers

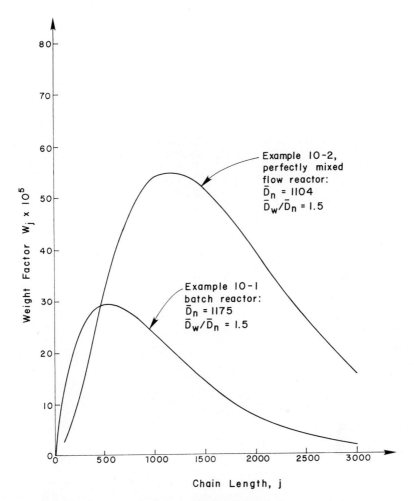

Figure 10-2. Molecular weight distributions computed for the polymerizations of styrene in a batch reactor (Example 10-1) and in a perfectly mixed reactor (Example 10-2).

4. Weight Average Molecular Weight. This average molecular weight is defined by

$$\overline{M}_W = \frac{\Sigma W_j M_j}{\Sigma W_j} \tag{10-9}$$

Elimination of M_j and W_j from Equation (10-9) by use of Equations (10-2) and (10-3), respectively, gives

$$\overline{M}_W = M_M \frac{\Sigma j^2 P_j}{\Sigma j P_j} = M_M \frac{\Gamma_2(P_j)}{\Gamma_1(P_j)} \tag{10-10}$$

where $\Gamma_2(P_j)$ is the second moment of the P_j's.

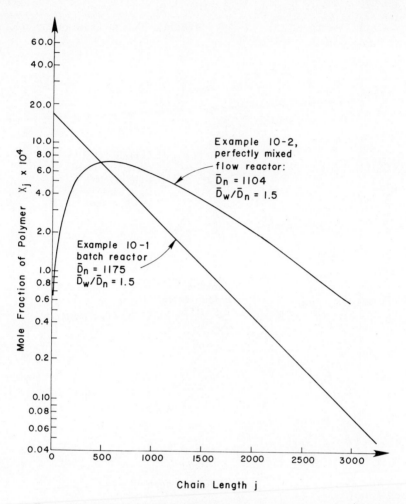

Figure 10-3. Mole fraction distributions for the polymerization of styrene in a batch and in a perfectly mixed reactor.

5. Number Average Degree of Polymerization. This quantity is denoted by the symbol $\overline{D}_n$ and defined by

$$\overline{D}_n = \frac{\sum j P_j}{\sum P_j} = \frac{\Gamma_1(P_j)}{\Gamma_0(P_j)} \tag{10-11}$$

A comparison of Equations (10-8) and (10-11) shows that

$$\overline{D}_n = \frac{\overline{M}_n}{M_M} \tag{10-12}$$

6. Weight Average Degree of Polymerization. This measure of the extent of the polymerization is defined by

$$\overline{D}_W = \frac{\sum j^2 P_j}{\sum j P_j} = \frac{\Gamma_2(P_j)}{\Gamma_1(P_j)} \tag{10-13}$$

From the definition of $\overline{M}_W$ [Equation (10-10)], it is seen that

$$\overline{D}_W = \frac{\overline{M}_W}{M_M} \tag{10-14}$$

7. Dispersion Index. This quantity, also called the *heterogeneity index* by some authors, is defined as the ratio of $\overline{D}_W/\overline{D}_n$. By use of the preceding relationships, the dispersion index may be stated in the form

$$\frac{\overline{D}_W}{\overline{D}_n} = \frac{\overline{M}_W}{\overline{M}_n} = \frac{\left(\sum j^2 P_j\right)\left(\sum P_j\right)}{\left(\sum j P_j\right)^2} \tag{10-15}$$

When Equation (10-15) is restated in terms of the mole fractions [Equation (10-6)], the following expression is obtained for the dispersion index:

$$\frac{\overline{D}_W}{\overline{D}_n} = \frac{\sum j^2 x_j}{\left(\sum j x_j\right)^2} \tag{10-16}$$

The significance of the dispersion index is perhaps best demonstrated by the consideration of some numerical examples which present limiting cases. First, if all the polymer molecules are of the same length, then Equation (10-16) reduces to

$$\frac{\overline{D}_W}{\overline{D}_n} = \frac{j^2\left(\sum x_j\right)}{j^2\left(\sum x_j\right)^2} = 1 \tag{10-17}$$

Next, suppose that a mixture is composed of 1 molecule of 1000 monomer lengths and 19 molecules of 1 monomer length. (*Note*: A polymer of one monomer unit consists of one monomer and one initiator molecule.) For this case, Equation (10-16) becomes

$$\frac{\overline{D}_W}{\overline{D}_n} = \frac{(1000)^2\left(\frac{1}{20}\right) + (1)^2\left(\frac{19}{20}\right)}{\left[(1000)\left(\frac{1}{20}\right) + (1)\left(\frac{19}{20}\right)\right]^2}$$

$$= 20\left[\frac{(1000)^2 + 19}{(1000 + 19)^2}\right] = 19.26$$

or

$$\frac{\overline{D}_W}{\overline{D}_n} \cong \frac{1}{x_{1000}} = 20$$

Thus the dispersion index varies from unity for polymers composed of molecules of equal lengths and increases to numbers greater than unity as the range of molecular weights of the constituents is increased and the relative number of molecules having the higher molecular weights is decreased. The precise distribution of molecular weights is lost in the value given by the dispersion index because, for any value of the dispersion index greater than unity, there are infinitely many distributions that have the same index. Thus, to characterize a mixture of polymers, several different types of experimental determinations are needed.

In particular, the weight average molecular weight may be measured by light scattering techniques [3, 4]; the number average molecular weight may be found by use of vapor phase osmometry [3, 4] or by membrane osmometry [3, 4]. The molecular weight distribution of a polymer such as the ones which are plotted in Figures 10-1 and 10-2 may be found by use of gel permeation chromatography [4]. The distribution so obtained may be used to compute the number and weight average molecular weights of the polymer. The monomer concentration, temperature, and pressure are other variables which are commonly measured in experimental investigations.

B. A Mechanism for Free-Radical Polymerization

Polymerization reactions are normally composed of initiation, propagation, termination, transfer, and branching reactions. The following general mechanism is applicable to the polymerization of styrene, vinyl chloride, methylmethacrylate, and acrylamides [3, 4, 5, 6]. The important termination and transfer reactions for each monomer are shown in Table 10-1.

Decomposition:
$$I \xrightarrow{k_0} 2\phi \qquad (10\text{-}18)$$

Initiation:
$$\phi + M \xrightarrow{k_1} R_1 \qquad (10\text{-}19)$$

Propagation:
$$R_j + M \xrightarrow{k_p} R_{j+1} \qquad (10\text{-}20)$$

Termination by addition (or combination):
$$R_j + R_i \xrightarrow{k_c} P_{j+i} \qquad (10\text{-}21)$$

Termination by disproportionation:
$$R_j + R_i \xrightarrow{k_d} P_j + P_i \qquad (10\text{-}22)$$

Termination by monomer transfer:
$$R_j + M \xrightarrow{k_{tM}} P_j + R_1 \qquad (10\text{-}23)$$

Termination by solvent transfer:
$$R_j + S \xrightarrow{k_{tS}} P_j + R_1 \qquad (10\text{-}24)$$

Termination by chain transfer:
$$R_j + C \xrightarrow{k_{tC}} P_j + R_1 \qquad (10\text{-}25)$$

TABLE 10-1
TERMINATION AND TRANSFER REACTIONS IN FREE RADICAL POLYMERIZATION
OF VINYL MONOMERS*

Monomer	Termination	Transfer Reaction	Heats of Polymerization, kcal / g mol
Styrene	Combination	Solvent and monomer transfer	16.7
Vinyl chloride	Disproportion	Monomer transfer	22.7
Methylmethacrylate	Disproportion	None of major importance	13.2
Acrylamide	Disproportion	Monomer transfer	19.8

*Prepared from information presented in [2] and [6].

where $j = 1, 2, \ldots$, and $i = 1, 2, \ldots$. The solvent is denoted by S and the chain transfer agent by C.

The first step in the analysis of the mechanism given by Equations (10-18) through (10-25) is the formulation of the rate expressions. The decomposition of the initiator I to form free radicals ϕ is represented by Equation (10-18). The rate of decomposition of the initiator r_0 is generally first order,

$$r_0 = k_0 I \qquad (10\text{-}26)$$

where I denotes the concentration of the initiator. A rate constant k_0 may be determined independently of the given polymerization for which it is to be used. For most initiators in common use, the rate constants k_0 have been determined [3, 4, 5].

The rate of initiation r_i is defined as the rate of disappearance of free radicals ϕ or the rate of formation of free radicals R_1, and

$$r_i = k_i \phi M \qquad (10\text{-}27)$$

where ϕ and M denote the concentrations of the free radicals and the monomers, respectively. It is customary to let f denote the efficiency of the utilization of the free radicals ϕ. The value of f for many polymer initiator systems has been determined by analyzing the polymer to determine the number of initiator fragments associated with the polymer molecules. The net rate of disappearance of free radicals ϕ is given by

$$r_\phi = k_i \phi M - 2 f k_0 I \qquad (10\text{-}28)$$

Consider next the propagation reactions represented by Equation (10-20). To produce polymers having high molecular weights, it is necessary that the rate for the propagation reactions be greater than those for all other reactions of the mechanism. An assumption commonly made is that the rate constant k_p is independent of the chain length. The rate of disappearance of free radicals R_j by propagation is given by

or

$$\begin{aligned}
(r_p)_{R_j} &= k_p M R_j - k_p M R_{j-1} \\
(r_p)_{R_j} &= k_p M (R_j - R_{j-1}), \qquad (j = 2, 3, \ldots)
\end{aligned} \qquad (10\text{-}29)$$

The *rate of propagation*, a term in common usage, means the rate of disappearance of the monomer by all possible propagation reactions,

$$(r_p)_M = k_p M \sum_{j=1}^{\infty} R_j \tag{10-30}$$

All termination reactions of a given type, such as termination by combination [Equation (10-21)], are assumed to have the same rate constants. For example, consider the case where the dead polymer P_6 is formed by combination. The distinguishable reactions by which P_6 may be formed are as follows:

$$R_1 + R_5 \rightarrow P_6, \qquad r_5 = r_1 = k_c R_1 R_5$$
$$R_2 + R_4 \rightarrow P_6, \qquad r_4 = r_2 = k_c R_2 R_4$$
$$R_3 + R_3 \rightarrow P_6, \qquad r_3 = k_c R_3 R_3$$

where r_1, r_2, r_3, r_4, and r_5 are the rates of disappearance of the free radicals R_1, $R_2, \ldots,$ and R_5. The net rate of appearance of P_6 by the combination mechanism is given by

$$(r_c)_{P_6} = r_1 + r_2 + \tfrac{1}{2} r_3$$

The factor of $\tfrac{1}{2}$ appears in this expression because the rate constant k_c is defined with respect to the disappearance of the free radicals (two molecules of R_3 disappear by the third reaction for each molecule of P_6 formed). Since the reactions $R_5 + R_1 \rightarrow P_6$ and $R_4 + R_2 \rightarrow P_6$ are indistinguishable from the first two, it follows that the rate of appearance of the polymer P_6 by combination may be represented as follows:

$$(r_c)_{P_6} = \frac{k_c}{2} R_1 R_5 + \frac{k_c}{2} R_5 R_1 + \frac{k_c}{2} R_2 R_4 + \frac{k_c}{2} R_4 R_2 + \frac{k_c}{2} R_3 R_3$$

or

$$(r_c)_{P_6} = \frac{k_c}{2} \sum_{i=1}^{5} R_{6-i} R_i$$

Then for the rate of appearance of any polymer P_j,

$$(r_c)_{P_j} = \frac{k_c}{2} \sum_{i=1}^{j-1} R_{j-i} R_i \tag{10-31}$$

The rate of disappearance of the free radical R_j by all possible termination reactions by the combination mechanism is given by

$$(r_c)_{R_j} = k_c R_j \sum_{i=1}^{\infty} R_i \tag{10-32}$$

The rate of production of the polymer P_j by the disproportionation mechanism [Equation (10-22)] is given by

$$(r_d)_{P_j} = k_d R_j \sum_{i=1}^{\infty} R_i \tag{10-33}$$

For the disproportionation reaction, the rate of appearance of polymers P_j is equal to the rate of disappearance of free radicals R_j; that is,

$$(r_d)_{P_j} = (r_d)_{R_j}$$

The rate of disappearance of free radicals R_j by each type of transfer reaction is represented as follows. For monomer transfer [Equation (10-23)],

$$(r_{tM})_{R_j} = k_{tM}MR_j \tag{10-34}$$

For solvent transfer [Equation (10-24)],

$$(r_{tS})_{R_j} = k_{tS}SR_j \tag{10-35}$$

For chain transfer [Equation (10-25)],

$$(r_{tC})_{R_j} = k_{tC}CR_j \tag{10-36}$$

The rates of appearance of polymers P_j by each of these transfer reactions are equal to the rates of disappearance of the free radicals R_j by each of these reactions; that is,

$$(r_{tM})_{P_j} = (r_{tM})_{R_j}, \qquad (r_{tS})_{P_j} = (r_{tS})_{R_j}, \qquad (r_{tC})_{P_j} = (r_{tC})_{R_j} \tag{10-37}$$

The monomer M disappears by the initiation reaction, the propagation reactions, and the termination reaction by monomer transfer. Thus the total rate of disappearance of the monomer M is given by

$$r_M = r_i + (r_p)_M + \sum_{j=1}^{\infty} (r_{tM})_{R_j} \tag{10-38}$$

and by use of Equations (10-30) and (10-34) it follows that

$$r_M = r_i + (k_p + k_{tM})M \sum_{j=1}^{\infty} R_j \tag{10-39}$$

C. Analysis of the Proposed Mechanism for Free Radical Polymerization

Because the concentrations of the free radicals R_j ($j = 1, 2, \ldots$) present at any time during a polymerization process are generally too small to be conveniently measured, it is desirable to solve for the R_j's in terms of other variables which may be followed experimentally.

1. Statement of the Initiation Rate r_i in Terms of the Concentration of the Initiator I. The concentration ϕ of free radicals is generally small because they are consumed by the initiation reaction almost as fast as they are formed by the decomposition of the initiator I. Thus, if the pseudo steady-state assumption is applied to the free radicals ϕ, then $r_\phi = 0$, and Equation (10-28) may be solved for ϕ to give

$$\phi = \frac{2fk_0I}{k_iM} \tag{10-40}$$

Elimination of ϕ from Equations (10-27) and (10-40) gives

$$r_i = 2fk_0I \tag{10-41}$$

Since k_0 is generally known for most initiators and f is known for a given initiator and polymerization [4], Equation (10-41) may be used to evaluate r_i for any known concentration of the initiator.

2. Statement of the Total Concentration of the Free Radicals R_j in Terms of r_i. The total concentration $\Gamma_0(R_j)$ of the R_j's is defined as follows:

$$\Gamma_0(R_j) = \sum_{j=1}^{\infty} R_j \tag{10-42}$$

Since the concentrations of the R_j's are not easily determined and r_i can be evaluated, it is fortunate that the total concentration of the R_j's may be expressed in terms of r_i. The desired result may be obtained by making use of the fact that the free radicals R_j are consumed by reaction at approximately the same rate at which they are formed (the pseudo steady-state assumption): that is,

$$\sum_{j=1}^{\infty} r_{R_j} = 0 \tag{10-43}$$

where r_{R_j} is the net rate of disappearance of free radicals R_j by all steps of the mechanism given by Equations (10-18) through (10-25). The net rate of disappearance of R_1 is given by

$$r_{R_1} = k_pR_1M + k_cR_1\Gamma_0 + k_dR_1\Gamma_0 - r_i$$
$$- k_{tM}M \sum_{j=2}^{\infty} R_j - k_{tC}C \sum_{j=2}^{\infty} R_j - k_{tS}S \sum_{j=2}^{\infty} R_j \tag{10-44}$$

(In this equation and in the remainder of the development the argument R_j of the functions Γ_0, Γ_1, and Γ_2 is omitted whenever possible.) For $j \geq 2$, the net rate of disappearance of R_j is given by

$$r_{R_j} = k_pM(R_j - R_{j-1}) + (k_c + k_d)R_j\Gamma_0$$
$$+ k_{tM}MR_j + k_{tC}SR_j + k_{tS}CR_j, \qquad j \geq 2 \tag{10-45}$$

After Equation (10-45) has been summed over all $j \geq 2$ and the result so obtained has been added to Equation (10-44), the following result is obtained:

$$\sum_{j=1}^{\infty} r_{R_j} = (k_c + k_d)\Gamma_0^2 - r_i \tag{10-46}$$

When the pseudo steady-state assumption [Equation (10-43)] is imposed upon Equation (10-46), the resulting expression may be solved for Γ_0 to give

$$\Gamma_0(R_j) = \left(\frac{r_i}{k_c + k_d}\right)^{1/2} \tag{10-47}$$

Since r_i may be evaluated by use of Equation (10-41), Equation (10-47) may be used to evaluate the total concentration of the R_j's.

3. Statement of r_M in Terms of r_i and M. When the expression given by Equation (10-47) for Γ_0 is substituted into Equation (10-39), the following expression is obtained for r_M:

$$r_M = r_i + \left(k_p + k_{tM}\right)M\left(\frac{r_i}{k_c + k_d}\right)^{1/2} \tag{10-48}$$

For a high-molecular-weight polymer to be formed, the rate of free radical initiation [Equation (10-9)] and the rate of monomer transfer [Equation (10-23)] must be small relative to the rate of propagation [Equation (10-20)]. These conditions imply that $k_{tM} \ll k_p$, and that r_i is very small, relative to the second term. Thus, the expression given by Equation (10-48) reduces to

$$r_M = k_p M\left(\frac{r_i}{k_c + k_d}\right)^{1/2} \tag{10-49}$$

When $[r_i/(k_c + k_d)]^{1/2}$ is replaced by its equivalent as given by Equation (10-47), the following expression for r_M is obtained:

$$r_M = k_p M \Gamma_0(R_j) = k_p M \sum_{j=1}^{\infty} R_j \tag{10-50}$$

In the developments which follow, the expression given by Equation (10-50) is a more convenient form to use than the one given by Equation (10-49).

In addition to the concentrations of I and M, the concentrations of dead polymers P_j ($j = 1, 2, \ldots$), the solvent S, and the chain transfer agent C may be followed experimentally. Also, on the basis of the proposed mechanism, the concentrations of the solvent S and the chain transfer agent C may be expressed relative to the concentration of the monomer M. Let r_{tS} denote the rate of disappearance of the solvent S by the solvent transfer reaction. Then, by Equation (10-35),

$$r_{tS} = \sum_{j=1}^{\infty} (r_{tS})_{R_j} = k_{tS} S \sum_{j=1}^{\infty} R_j \tag{10-51}$$

Similarly, let r_{tC} denote the rate of disappearance of the chain transfer agent C by the chain transfer reaction. Then, by Equation (10-36),

$$r_{tC} = \sum_{j=1}^{\infty} (r_{tC})_{R_j} = k_{tC} C \sum_{j=1}^{\infty} R_j \tag{10-52}$$

Thus, by use of the expression given by Equation (10-50) for r_{tM}, it follows that

$$\frac{r_{tS}}{r_M} = \frac{k_{tS}S}{k_p M} \tag{10-53}$$

and

$$\frac{r_{tC}}{r_M} = \frac{k_{tC}C}{k_p M} \tag{10-54}$$

For a given type of reactor, the concentrations of S and C may be expressed in terms of M, as shown in the following section on reactor design.

Unfortunately, most rate expressions contain the concentrations of the free radicals R_j. Since these concentrations are generally very small, they are not customarily measured. Thus the major effort in the analysis of the mechanism consists of expressing the R_j's in terms of r_i, I, M, C, and S. The first step in the analysis of the reaction mechanism is the development of expressions for the instantaneous fractional degree of polymerization. These expressions, which are valid for any type of reactor, may be used with the material-balance equations in the analysis of various types of reactors.

4. Instantaneous Fractional Degree of Polymerization. This new variable is denoted by $\bar{d}_n$ and is defined as the total moles of polymer produced per mole of monomer reacted at any instant at any position in any given type of reactor. Thus

$$\bar{d}_n = \frac{\sum_{j=1}^{\infty} r_{P_j}}{r_M} \tag{10-55}$$

where r_{P_j} is the rate of appearance of dead polymer P_j by the termination and transfer steps of the mechanism [Equations (10-21) through (10-25)]. (The reciprocal of the function $\bar{d}_n$ is sometimes called the *instantaneous number degree of polymerization* because it may be used to compute $\bar{D}_n$ for a given type of reactor.)

5. Elimination of the Free Radical Concentrations R_j from the Expression for $\bar{d}_n$. The rate of appearance of polymers P_1 by all possible reactions is given by

$$r_{P_1} = k_d R_1 \Gamma_0(R_j) + k_{tM} R_1 M + k_{tS} R_1 S + k_{tC} R_1 C \tag{10-56}$$

and, for $j \geq 2$, the rate of appearance of polymer P_j by all possible reactions is given by

$$r_{P_j} = \frac{k_c}{2} \sum_{i=1}^{j-1} R_{j-i} R_i + k_d R_j \Gamma_0(R_j) + k_{tM} R_j M + k_{tS} R_j S + k_{tC} R_j C \tag{10-57}$$

When the expression given by Equation (10-57) is summed over all j ($j = 2, 3, \dots$) and added to the expression given by Equation (10-56) for r_{P1}, the following result is obtained upon rearrangement:

$$\sum_{j=1}^{\infty} r_{P_j} = \frac{k_c}{2} \sum_{j=2}^{\infty} \sum_{i=1}^{j-1} R_{j-i} R_i + k_d \left[\Gamma_0(R_j) \right]^2$$

$$+ \left[k_{tM} M + k_{tS} S + k_{tC} C \right] \Gamma_0(R_j) \qquad (10\text{-}58)$$

The double summation in Equation (10-58) may be expressed as follows [7]:

$$\sum_{j=2}^{\infty} \sum_{i=1}^{j-1} R_{j-i} R_i = \left(\sum_{j=1}^{\infty} R_j \right)^2 = \left[\Gamma_0(R_j) \right]^2 \qquad (10\text{-}59)$$

Thus

$$\sum_{j=1}^{\infty} r_{P_j} = \left(\frac{k_c}{2} + k_d \right) \left[\Gamma_0(R_j) \right]^2 + \left[k_{tM} M + k_{tS} S + k_{tC} C \right] \Gamma_0(R_j) \quad (10\text{-}60)$$

By use of this expression for the sum of the r_{P_j}'s and the expression given by Equation (10-50) for r_{tM}, the following expression is obtained for $\bar{d}_n$ for the formation of high-molecular-weight polymers:

$$\bar{d}_n = \left[\frac{(k_c/2) + k_d}{k_p} \right] \left[\frac{\Gamma_0(R_j)}{M} \right] + T \qquad (10\text{-}61)$$

where
$$T = \frac{k_{tM}}{k_p} + \frac{k_{tC}}{k_p} \left(\frac{C}{M} \right) + \frac{k_{tS}}{k_p} \left(\frac{S}{M} \right)$$

The first term on the right side of Equation (10-61) represents the contribution of the terminations by combination and disproportionation, and the second term represents the contributions of terminations by monomer, solvent, and chain transfer to $\bar{d}_n$.

6. Instantaneous Weight Fraction. The instantaneous weight fraction w_j for polymer P_j is defined as the weight of polymer P_j produced per unit weight of monomer M reacted at any instant in any given type of reactor,

$$w_j = \frac{M_j r_{P_j}}{M_M r_M} \qquad (10\text{-}62)$$

Since $M_j = j M_M$ [Equation (10-2)], this definition may be stated in the alternate form

$$w_j = \frac{j r_{P_j}}{r_M} \qquad (10\text{-}63)$$

The instantaneous weight fraction is given this name because it may be used to compute the weight fraction in any type of reactor.

7. Instantaneous Weight Degree of Polymerization. The instantaneous weight degree of polymerization $\bar{d}_W$ is defined as follows:

$$\bar{d}_W = \sum j w_j \tag{10-64}$$

Elimination of w_j from this expression by use of Equation (10-62) gives

$$\bar{d}_W = \frac{\sum_{j=1}^{\infty} j^2 r_{P_j}}{r_M} \tag{10-65}$$

The function $\bar{d}_W$ is useful for computing the weight average degree of polymerization, $\bar{D}_W$, for any given type of reactor, as demonstrated in a subsequent section.

8. Elimination of the Free Radical Concentrations R_j from the Expression for $\bar{d}_w$. The summation in Equation (10-65) may be evaluated by multiplying the expression given by Equation (10-57) for r_{P_j} by j^2. After this expression has been summed over all $j \geq 2$, the result so obtained may be added to the expression given by Equation (10-56) for r_{P_1} to give

$$\sum_{j=1}^{\infty} j^2 r_{P_j} = \frac{k_c}{2} \sum_{j=2}^{\infty} j^2 \sum_{i=1}^{j-1} R_{j-i} R_i$$

$$+ \left[k_d \Gamma_0(R_j) + k_{tM} M + k_{tS} S + k_{tC} C \right] \Gamma_2(R_j) \tag{10-66}$$

The double summation may be expressed [7] as

$$\sum_{j=2}^{\infty} j^2 \sum_{i=1}^{j-1} R_{j-i} R_i = 2 \left(\sum_{j=1}^{\infty} j R_j \right)^2 + 2 \left(\sum_{j=1}^{\infty} R_j \right) \left(\sum_{j=1}^{\infty} j^2 R_j \right) \tag{10-67}$$

and thus Equation (10-66) may be stated in the form

$$\sum_{j=1}^{\infty} j^2 r_{P_j} = k_c \left[\Gamma_1(R_j) \right]^2 + \left[(k_c + k_d) \Gamma_0(R_j) + k_{tM} M \right.$$

$$\left. + k_{tS} S + k_{tC} C \right] \Gamma_2(R_j) \tag{10-68}$$

For the proposed mechanism, the first and second moments, $\Gamma_1(R_j)$ and $\Gamma_2(R_j)$, may be expressed in terms of $\Gamma_0(R_j)$ or r_i, M, C, S, and the rate constants. The first step in the evaluation of $\Gamma_1(R_j)$ and $\Gamma_2(R_j)$ is to obtain an expression for R_j. On the basis of the pseudo steady-state assumption, $r_{R_j} = 0$, Equations (10-44) and (10-45) may be solved for R_1 and R_j ($j \geq 2$), respectively. After r_{R1} has been set equal to zero, Equation (10-44) may be solved for R_1 to give

$$R_1 = \gamma \psi \tag{10-69}$$

where
$$\gamma = \frac{r_i}{k_p M} + T \Gamma_0(R_j)$$

$$\psi = \left[1 + T + \left(\frac{k_c + k_d}{k_p M} \right) \Gamma_0(R_j) \right]^{-1}$$

On the basis of the assumption that $r_{R_j} = 0$ for $j \geq 2$, Equation (10-45) may be solved for R_j to give

$$R_j = \psi R_{j-1}, \qquad j \geq 2 \tag{10-70}$$

Then

$$R_j = \psi^{j-1} R_1 = \gamma \psi^j \tag{10-71}$$

Thus

$$\Gamma_1(R_j) = \sum_{j=1}^{\infty} j R_j = \gamma \sum_{j=1}^{\infty} j \psi^j = \gamma \psi \sum_{j=1}^{\infty} j \psi^{j-1}$$

and since $\psi < 1$, it follows [7] that

$$\Gamma_1(R_j) = \frac{\gamma \psi}{(1 - \psi)^2} \tag{10-72}$$

Similarly,

$$\Gamma_2(R_j) = \sum_{j=1}^{\infty} j^2 R_j = \frac{(1 + \psi) \gamma \psi}{(1 - \psi)^3} \tag{10-73}$$

Now observe that Equation (10-68) may be restated in terms of ψ as follows;

$$\sum_{j=1}^{\infty} j^2 r_{P_j} = k_c \left[\Gamma_1(R_j) \right]^2 + k_p M \left(\frac{1 - \psi}{\psi} \right) \Gamma_2(R_j) \tag{10-74}$$

The desired expression for $\bar{d}_W$ may now be obtained by use of Equations (10-50) and (10-72) through (10-74):

$$\bar{d}_W = \frac{\sum_{j=1}^{\infty} j^2 r_{P_j}}{r_M} = \frac{\gamma}{(1 - \psi)^2 \Gamma_0(R_j)} \left[1 + \psi + \left(\frac{\psi}{1 - \psi} \right)^2 \frac{\gamma k_c}{k_p M} \right] \tag{10-75}$$

9. Elimination of the Free Radical Concentrations from the Expression for the Instantaneous Weight Fraction. The expression given by Equation (10-57) for r_{P_j} ($j \geq 2$) may be restated in terms of the termination function T [see Equation (10-61)] as follows:

$$r_{P_j} = \frac{k_c}{2} \sum_{i=1}^{j-1} R_{j-i} R_i + k_d R_j \Gamma_0(R_j) + k_p M T R_j \tag{10-76}$$

The summation may be expanded to give

$$\sum_{i=1}^{j-1} R_{j-i}R_i = R_{j-1}R_1 + R_{j-2}R_2 + \cdots + R_1R_{j-1} \tag{10-77}$$

When each R_j appearing in this series is replaced by its equivalent as given by Equation (10-71), the following result is obtained:

$$\sum_{i=1}^{j-1} R_{j-i}R_i = \gamma^2\left[\psi^{j-1}\psi^1 + \psi^{j-2}\psi^2 + \cdots + \psi^1\psi^{j-1}\right] = \gamma^2(j-1)\psi^j \tag{10-78}$$

When R_j in the second term of Equation (10-76) is replaced by its equivalent $\gamma\psi^j$ [Equation (10-71)] and the summation is replaced by the result given by Equation (10-78), the following expression is obtained for r_{P_j} ($j \geq 2$):

$$r_{P_j} = \gamma\psi^j\left[\frac{k_c\gamma(j-1)}{2} + k_d\Gamma_0(R_j) + k_pMT\right] \tag{10-79}$$

Thus, for $j \geq 2$, the expression for the instantaneous weight fraction becomes

$$w_j = \frac{jr_{P_j}}{r_M} = j\gamma\psi^j\left[\frac{k_c\gamma(j-1)}{2k_pM\Gamma_0(R_j)} + \frac{k_d}{k_pM} + \frac{T}{\Gamma_0(R_j)}\right] \tag{10-80}$$

where r_M is given by Equation (10-50).

When $j = 1$, Equation (10-80) gives the correct formula for w_1. This statement is readily confirmed by commencing with the expressions for r_M and r_{P_1} and eliminating R_1 by use of its equivalent given by Equation (10-69).

D. Design of Reactors for Carrying Out Free Radical Polymerizations

The design of reactors for carrying out polymerizations follows an approach similar to that demonstrated for thermal and catalytic reactors. The reaction time for a batch reactor or the volume of a flow reactor must be selected such that the specified number average or weight average degree of polymerization [$\overline{D}_n$ and $\overline{D}_W$, see Equations (10-12) and (10-14)] is satisfied. These quantities as well as the molecular weight distributions [see Equations (10-1) and (10-6)] may be used in the quality control of polymers. All these functions may be found by use of the results of the various types of chemical and physical analyses described at the end of the introduction. From such analyses, the total concentration and the moments of the concentrations of the dead polymers [$\Gamma_0(P_j)$, $\Gamma_1(P_j)$, and $\Gamma_2(P_j)$] may be evaluated, as well as other functions such as $\overline{D}_n$, $\overline{D}_W$, $\overline{M}_n$, $\overline{M}_W$, and W_j.

On the basis of the proposed mechanism, the selectivities defined previously, and the material-balance equation for a given type of reactor, the monomer and initiator conversion, molecular weights, and molecular weight distribution may be predicted as demonstrated next for the three types of reactors: batch, plug-flow, and perfectly mixed flow reactors. Because of the similarity of the design of polymerization reactors to the design of thermal and catalytic reactors, an abbreviated treatment of the design of polymerization reactors is presented. In particular, it is supposed that the reactors are operated isothermally and that the mass density of the reacting mixture is independent of the extent of polymerization.

1. Batch Reactors. The design equations for a perfectly mixed batch reactor are developed as follows. As shown in Chapter 1, the limit of a material balance on the monomer in the reactor as the time period over which the balance is made is allowed to go to zero is given by

$$r_M = -\frac{dM}{dt} \tag{10-81}$$

where it has been supposed that the volume of the reacting mixture remains constant throughout the course of the polymerization. Elimination of r_M from Equations (10-50) and (10-81) gives

$$-\frac{dM}{dt} = k_p M \left(\frac{r_i}{k_c + k_d} \right)^{1/2} \tag{10-82}$$

By continuously feeding a small amount of initiator I to the reactor and by operating the reactor isothermally, the rate r_i [see Equation (10-41)] may be held constant throughout the polymerization. For the case where r_i is held fixed and the reactor is operated isothermally, Equation (10-82) may be integrated and rearranged to give

$$M = M^0 \exp \left[-k_p \left(\frac{r_i}{k_c + k_d} \right)^{1/2} t \right] \tag{10-83}$$

Since r_i may be evaluated by use of Equation (10-41), the concentration of the monomer M at any time t may be computed by use of Equation (10-83). After the concentration M at time t has been computed, the corresponding concentrations of the solvent S and the chain transfer agent C may be computed by relationships which are developed as follows. First observe that Equation (10-53) for r_{tS}/r_M may be restated as follows:

$$\frac{dS}{dM} = \frac{k_{tS}S}{k_p M} \tag{10-84}$$

Separation of the variables in this equation followed by integration and rearrangement gives

$$S = S^0 \left(\frac{M}{M^0} \right)^{k_{tS}/k_p} \tag{10-85}$$

where S^0 is the concentration of the solvent at time $t = 0$. Similarly, by restating Equation (10-54) for r_{tC}/r_M in the form

$$\frac{dC}{dM} = \frac{k_{tC}C}{k_p M} \tag{10-86}$$

and integrating, it is found that

$$C = C^0 \left(\frac{M}{M^0} \right)^{k_{tC}/k_p} \tag{10-87}$$

where C^0 is the concentration of the chain transfer agent at time $t = 0$.

On the basis of the proposed mechanism and the resulting expression given by Equation (10-61) for the instantaneous number fraction of polymerization, an expression for computing the number average degree of polymerization [defined by Equation (10-11)] may be developed as follows. First observe that the definition of the instantaneous fractional degree of polymerization [Equation (10-55)] may be restated as follows for a batch reactor operated at constant volume

$$\bar{d}_n = \frac{\sum_{j=1}^{\infty} r_{P_j}}{r_M} = \frac{\sum_{j=1}^{\infty} \left(dP_j/dt \right)}{-\left(dM/dt \right)} \tag{10-88}$$

Separation of the variables followed by integration gives

$$\int_0^t \sum_{j=1}^{\infty} \frac{dP_j}{dt} dt = -\int_0^t \bar{d}_n \frac{dM}{dt} dt \tag{10-89}$$

For $P_j = 0$ for all j at $t = 0$, this equation may be integrated to give

$$\sum_{j=1}^{\infty} P_j = -\int_{M^0}^{M} \bar{d}_n \, dM \tag{10-90}$$

By use of this result, Equations (10-5) and (10-11), the following expression, which may be used to compute $\bar{D}_n$, is obtained:

$$\bar{D}_n = -\frac{M^0 - M}{\int_{M^0}^{M} \bar{d}_n \, dM} \tag{10-91}$$

where $\bar{d}_n$ and M are given by Equations (10-61) and (10-83), respectively. [Note that when the proper set of rate constants has been selected, the value of $\bar{D}_n$ computed by use of Equation (10-91) must be equal to the value found by use of the experimentally determined moments: $\bar{D}_n = \Gamma_1(P_j)/\Gamma_0(P_j)$.]

Similarly, the weight average degree of polymerization may be computed on the basis of the proposed mechanism by use of an expression which is developed as follows. By commencing with the following form of the defining equation for the instantaneous weight degree of polymerization for a batch reactor,

$$\bar{d}_W = \frac{\sum_{j=1}^{\infty} j^2 (dP_j/dt)}{-(dM/dt)} \tag{10-92}$$

and integrating as before, it is found that

$$\sum_{j=1}^{\infty} j^2 P_j = -\int_{M^0}^{M} \bar{d}_W \, dM \tag{10-93}$$

By use of the definition of the weight average degree of polymerization $\bar{D}_W$ [Equation (10-13)] and Equations (10-5) and (10-93), the following expression for computing $\bar{D}_W$ on the basis of the proposed mechanism is obtained:

$$\bar{D}_W = -\frac{\int_{M^0}^{M} \bar{d}_W \, dM}{M^0 - M} \tag{10-94}$$

where M is given by Equation (10-83).

Similarly, an expression for the weight fraction [defined by Equation (10-3)] may be obtained by commencing with the instantaneous weight fraction [Equation (10-63)] and following a procedure similar to that shown for $\bar{d}_n$. The final result is

$$W_j = -\frac{\int_{M^0}^{M} w_j \, dM}{M^0 - M} \tag{10-95}$$

where the expression for w_j is given by Equation (10-80) and the expression for M is given by Equation (10-83).

EXAMPLE 10-1

Styrene is to be polymerized at 60°C with the free radical initiator, benzoyl peroxide, in a batch reactor. The initial concentration of the styrene is 8.35 g mol/liter, and the concentration of the initiator is to be maintained constant at 0.04 g mol/liter throughout the polymerization. (To achieve this condition, the initiator is fed continuously to the reactor throughout the course of the reaction.) In this example, it is supposed that the significant reactions of the mechanism proposed are initiation, propagation, and termination by addition (or combination). The rate constants at the temperature of the polymerization, 60°C, are as follows [3]:

$$k_0 = 3.2 \times 10^{-6} \text{ s}^{-1}$$

$$f = 0.6$$

$$k_p = 176 \text{ liters/g mol s}$$

$$k_c = 3.6 \times 10^7 \text{ liters/g mol s}$$

(a) Calculate the rate at which the initiator must be fed to the reactor in order for the concentration of the initiator to remain constant throughout the polymerization.

The volume of the reactor filled by the reacting mixture is 3760 liters, and the variation of the volume of the reacting mixture with time may be neglected.

(b) For a reaction time of 180 min, compute the following: (1) the percentage of the styrene polymerized, (2) the number average molecular weight, (3) the weight average molecular weight, (4) the molecular weight distribution, and (5) the mole fraction and molar distributions.

SOLUTION

(a) As illustrated in Chapter 2, the limit of a material balance on the initiator entering the perfectly mixed reactor as the time period over which the balance is made is allowed to go to zero is given by

$$n_I^0 - k_0 IV = \frac{d(IV)}{dt}$$

where n_I^0 is the feed rate of the initiator to the reactor at any time t. Since the concentration I of the initiator and the volume V are constant, $d(VI)/dt = 0$, and thus

$$n_I^0 = k_0 IV$$
$$= (3.2 \times 10^{-6})(0.04)(3760) = 4.813 \times 10^{-4} \text{ g mol/s}$$

(b) (1) The styrene conversion is calculated by use of Equations (10-41) and (10-83). By Equation (10-41),

$$r_i = 2fk_0 I = (2)(0.6)(3.2)(10^{-6})(0.04) = 1.536 \times 10^{-7} \text{ g mol/s}$$

From Equation (10-83),

$$M = M^0 \exp\left[-k_p\left(\sqrt{r_i/k_c}\right)t\right]$$
$$= 8.35 \exp\left[-176\left(\sqrt{(1.536 \times 10^{-7})/(3.6 \times 10^7)}\right)10,800\right]$$
$$= 7.375 \text{ g mol/liter}$$

And the percentage of the styrene converted is

$$\frac{8.35 - 7.375}{8.35} \times 100 = 11.68\%$$

(2) The number average molecular weight is computed by use of Equations (10-47), (10-61), and (10-91). The total concentration of polymers is given by Equation (10-47):

$$\Gamma_0(R_j) = \left(\frac{r_i}{k_c}\right)^{1/2} = \left(\frac{1.536 \times 10^{-7}}{3.6 \times 10^7}\right)^{1/2} = 6.532 \times 10^{-8} \text{ g mol/liter}$$

By Equation (10-61),

$$\bar{d}_n = \frac{k_c \Gamma_0}{2k_p M} = \frac{(3.6 \times 10^7)(6.532 \times 10^{-8})}{(2)(176) M}$$
$$= \frac{6.68 \times 10^{-3}}{M}$$

After this expression for $\bar{d}_n$ has been substituted into Equation (10-91), the number average degree of polymerization may be computed as follows:

$$\bar{D}_n = -\frac{M^0 - M}{\int_{M^\circ}^M \bar{d}_n\, dM} = -\frac{M^0 - M}{\int_{M^\circ}^M 6.68 \times 10^{-3}\, \dfrac{dM}{M}}$$

Thus

$$\bar{D}_n = -\left(\frac{8.35 - 7.375}{6.68 \times 10^{-3} \log_e (7.375)/(8.35)}\right)$$

$$= 1175.5$$

The number average molecular weight of the polystyrene may now be computed by use of Equation (10-12). Since the molecular weight of styrene $M_M = 104$,

$$\bar{M}_n = \bar{D}_n M_M = (1175.5)(104) = 122{,}252$$

(3) To compute the weight average molecular weight, the weight average degree of polymerization is first computed by use of Equation (10-94). To evaluate $\bar{D}_W$ by this equation, it is necessary to obtain an expression for $\bar{d}_W$ by use of Equation (10-75). On the basis of the assumptions given in the statement of the problem, the expression for ψ given after Equation (10-69) reduces to

$$\psi = \frac{1}{1 + (k_c \Gamma_0 / k_p M)} = \frac{1}{1 + 2\bar{d}_n}$$

$$= \frac{1}{1 + 0.01336/M}$$

The expression for γ, given after Equation (10-69), reduces to

$$\gamma = \frac{r_i}{k_p M} = \frac{1.536 \times 10^{-7}}{176 M} = \frac{8.727 \times 10^{-10}}{M}$$

By use of these values for ψ and γ, the given values of k_c and k_p, and the computed value of Γ_0, the instantaneous weight degree of polymerization $\bar{d}_W$ by Equation (10-75) may be expressed in terms of M. The following approximations are valid in the evaluation of $\bar{d}_W$ because M has values that lie between 7.375 and 8.35. First

$$1 - \psi = \frac{0.01336}{M + 0.01336} \cong \frac{0.01336}{M}$$

and

$$\psi = \frac{M}{M + 0.01336} \cong 1$$

Then, by use of Equation (10-75),

$$\bar{d}_W = \frac{(8.727 \times 10^{-10})\, M}{(1.336 \times 10^{-2})^2 (6.532 \times 10^{-8})}\left[2 + \frac{(8.727 \times 10^{-10})(3.6 \times 10^7)}{(1.336 \times 10^{-2})^2 (1.76 \times 10^2)}\right]$$

$$= 224.56 M$$

This result may now be used to compute the weight average degree of polymerization by use of Equation (10-94):

$$\bar{D}_W = -\frac{\int_{M^0}^M \bar{d}_W \, dM}{M^0 - M} = \frac{224.56 \int_M^{M_0} dM}{M^0 - M}$$

$$= 224.56 \frac{M^0 + M}{2} = 224.56 \frac{8.35 + 7.375}{2}$$

$$= 1765.6$$

By use of this value of $\bar{D}_W$, the weight average molecular weight may be computed by use of Equation (10-14):

$$\bar{M}_W = \bar{D}_W M_M = (1765.6)(104) = 183.622$$

(4) For the polymerization under consideration, the expression given by Equation (10-80) for the instantaneous weight fraction reduces to

$$w_j = \frac{\gamma^2 \psi^j}{M} \left[\frac{j(j-1)k_c}{2k_p \Gamma_0(R_j)} \right]$$

For convenience, let the expressions obtained previously for ψ and γ be restated as follows:

$$\psi = \frac{1}{1 + a/M}, \qquad \gamma = \frac{b}{M}$$

Then

$$w_j = \frac{1}{M^3(1 + a/M)^j} \left[\frac{j(j-1)b^2 k_c}{2k_p \Gamma_0(R_j)} \right]$$

Substitution of this expression for $\bar{d}_W$ into Equation (10-95) followed by the change in variable

$$u = \frac{1}{M}$$

gives

$$W_j = \frac{j(j-1)b^2 k_c}{a[(1/u^0) - (1/u)]k_p \Gamma_0(R_j)} \left[\int_{u^0}^u \frac{au \, du}{(1 + au)^j} \right]$$

For $j = 1$, $W_j = 0$. Since the integral in the expression may be stated in the form

$$\int_{u^0}^u \frac{au \, du}{(1 + au)^j} = \int_{u^0}^u \frac{du}{(1 + au)^{j-1}} - \int_{u^0}^u \frac{du}{(1 + au)^j}$$

it is readily shown that, for $j = 2$,

$$\int_{u^0}^u \frac{au \, du}{(1 + au)^2} = \frac{1}{a} \left\{ \left[\log_e (1 + au) + \frac{1}{(1 + au)} \right] \Big|_{u^0}^u \right\}$$

and, for $j > 2$,

$$\int_{u^0}^{u} \frac{au \, du}{(1 + au)^j} = \frac{1}{a} \left\{ \left[\frac{1}{(j-1)(1+au)^{j-1}} - \frac{1}{(j-2)(1+au)^{j-2}} \right] \Big|_{u^0}^{u} \right\}$$

A graph of the values of W_j versus j, which were computed by use of this expression, is presented in Figure 10-2.

(5) After W_j has been computed, the corresponding mole fractions x_j may be found as follows:

$$x_j = W_j \left(\frac{\overline{D}_n}{j} \right)$$

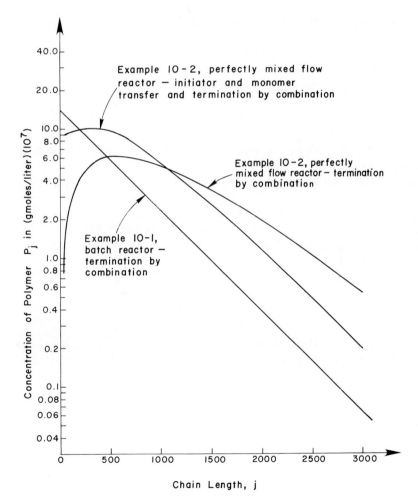

Figure 10-4. Molar distributions for the polymerization of styrene in batch and perfectly mixed flow reactors.

This expression is shown to reduce to the definition of x_j given by Equation (10-6) by use of Equations (10-3) and (10-11). A graph of x_j versus chain length j is shown in Figure 10-3.

Similarly, after W_j has been computed, the corresponding molar concentrations of polymers P_j of chain length j are given by

$$P_j = \frac{W_j(M^0 - M)}{j}$$

A plot of the results so obtained is shown in Figure 10-4. A variety of other mechanisms and corresponding molecular weight distributions have been presented by Peebles [7].

2. Plug-Flow Reactors. As demonstrated in Chapter 2, the design equations for plug-flow reactors may be obtained from those for isothermally operated batch reactors provided that the following conditions are satisfied: (1) the plug-flow reactor is operated isothermally at steady state, (2) the flow rate through the reactor satisfies the condition for plug-flow (see Chapter 2), and (3) the density or volumetric flow rate of the reacting mixture remains constant throughout the reactor. To obtain the design equations for plug-flow reactors from those stated previously, concentrations and time in the equations for the batch reactors are replaced by their equivalents, molar flow rates divided by the total volumetric flow rate and plug-flow reactor volume divided by the total volumetric flow rate, respectively.

$$M = \frac{n_M}{v_T}, \qquad t = \frac{V}{v_T} \tag{10-96}$$

where n_M = flow rate of the monomer in moles of monomer per unit time

       v_T = total volumetric flow rate of the reacting mixture in volume units per unit time

      V = volume of the plug-flow reactor

3. Perfectly Mixed Flow Reactors. For a perfectly mixed flow reactor at steady-state operation, a component material balance on the monomer yields

$$r_M = \frac{n_M^0 - n_M}{V} \tag{10-97}$$

When the density of the reacting mixture is constant, Equation (10-97) may be expressed as follows:

$$r_M = \frac{\left(n_M^0/v_T\right) - \left(n_M/v_T\right)}{V/v_T} = \frac{M^0 - M}{\theta} \tag{10-98}$$

where M^0 is the concentration of the monomer entering the reactor, M is the concentration of the monomer in the stream leaving the reactor, and θ is the residence time. By eliminating r_M from Equations (10-48) and (10-98) and solving the expression so obtained for M, the following result for computing M is obtained:

$$M = \frac{M^0}{1 + k_p[r_i/(k_c + k_d)]^{1/2}\theta} \qquad (10\text{-}99)$$

By beginning with Equations (10-53) and (10-54), the following expressions for the concentrations of the solvent S and the chain transfer agent C are obtained in a manner analogous to that demonstrated for M:

$$S = \frac{S^0}{1 + (k_{tS}/k_p)[(M^0/M) - 1]} \qquad (10\text{-}100)$$

and

$$C = \frac{C^0}{1 + (k_{tC}/k_p)[(M^0/M) - 1]} \qquad (10\text{-}101)$$

By use of the component material balances for a perfectly mixed flow reactor, it is possible to restate the defining equation for the instantaneous fractional degree of polymerization in the following form:

$$\bar{d}_n = \frac{\sum_{j=1}^{\infty} P_j}{M^0 - M} \qquad (10\text{-}102)$$

By use of this result as well as the one given by Equation (10-5), the definition of $\bar{D}_n$ given by Equation (10-11) may be restated as follows:

$$\bar{D}_n = \frac{\sum_{j=1}^{\infty} jP_j}{\sum_{j=1}^{\infty} P_j} = \frac{M^0 - M}{(M_0 - M)\bar{d}_n} = \bar{d}_n^{-1} \qquad (10\text{-}103)$$

By use of an approach analogous to that demonstrated for $\bar{d}_n$ and $\bar{D}_n$, the following expressions are obtained for the instantaneous weight degree of polymerization and the weight average degree of polymerization:

$$\bar{d}_W = \frac{\sum_{j=1}^{\infty} j^2 r_{P_j}}{r_M} = \frac{\sum_{j=1}^{\infty} j^2 P_j}{M^0 - M} \qquad (10\text{-}104)$$

and

$$\bar{D}_W = \frac{\sum_{j=1}^{\infty} j^2 P_j}{\sum_{j=1}^{\infty} jP_j} = \frac{(M^0 - M)\bar{d}_W}{M^0 - M} = \bar{d}_W \qquad (10\text{-}105)$$

Similarly, it can be shown that the weight fraction and the instantaneous weight fraction are equal when the reaction is carried out in a perfectly mixed flow reactor at steady state; that is,

$$W_j = w_j \qquad (10\text{-}106)$$

EXAMPLE 10-2

   Styrene is to be polymerized at 60°C in a perfectly mixed flow reactor. The styrene concentration in the feed is 8.35 g mol/liter. The concentration of benzoyl peroxide in the effluent is to be maintained at 0.04 g mol/liter. The volume of the reactor filled by the reacting mixture at any time is 3760 liters. The rate constants and mechanism of Example 10-1 are applicable in parts (a), (b), and (c).

(a) Calculate the residence time required to obtain 11.67% conversion of the styrene.
(b) Calculate the volumetric flow rate of the feed and the concentration of initiator, benzoyl peroxide, in the feed.
(c) Calculate the following: (1) the number average molecular weight, (2) the weight average molecular weight, (3) the molecular weight distribution, and (4) the mole fraction and molar distributions.
(d) Calculate the molar distribution of the styrene polymer for the case where termination by monomer transfer and termination by chain transfer with the initiator are also considered. The rate constants for these two transfer mechanisms are as follows:

$$k_{tM} = 105.6 \times 10^{-4} \text{ liter/g mol s}$$

$$k_{tC} = 9.68 \text{ liters/g mol s}$$

SOLUTION

(a) The residence time θ may be found solving Equation (10-99) for θ.

$$\theta = \frac{M^0 - M}{k_p M [r_i / k_a]^{1/2}}$$

The value of r_i was computed in Example 10-1(b), and the values for the remaining variables appearing in this expression were given in the statement of the example. Thus

$$\theta = \frac{8.35 - 7.375}{(176)(7.375)\left[(1.536 \times 10^{-7})/(3.6 \times 10^7)\right]^{1/2}}$$

$$= 1.15 \times 10^4 \text{ s} = 3.19 \text{ h}$$

(b) $v_T = V/\theta = 3760/(1.15 \times 10^4) = 0.327$ liter/s. Since

$$r_0 = \frac{I^0 - I}{\theta}, \qquad I^0 = I + r_0 \theta$$

and thus

$$I^0 = 0.04 + (1.28 \times 10^{-7})(1.15 \times 10^4)$$

$$= 0.0415 \text{ g mol/liter}$$

(c) (1) The number average degree of polymerization may be found by use of Equation (10-103):

$$\overline{D}_n = \overline{d}_n^{-1}$$

As demonstrated in Example 10-1, the expression given by Equation (10-61) for $\bar{d}_n$ may be used to compute $\bar{D}_n$ as follows:

$$\bar{D}_n = \bar{d}_n^{-1} = \left(\frac{6.68 \times 10^{-3}}{7.375} \right)^{-1} = 1.104 \times 10^3$$

The number average molecular weight follows from Equation (10-12),

$$\bar{M}_n = \bar{D}_n M_M = (1.104 \times 10^3)(104) = 114,816$$

(2) The weight average degree of polymerization is given by [Equation (10-105)]

$$\bar{D}_W = \bar{d}_W$$

and by use of Equation (10-75), as demonstrated in Example 10-1,

$$\bar{D}_W = \bar{d}_W = 224.56M = 1.656 \times 10^3$$

Then, by Equation (10-14),

$$\bar{M}_W = \bar{D}_W M_M = (1.652 \times 10^3)(104) = 1.718 \times 10^5$$

(3) By Equation (10-106),

$$W_j = w_j$$

The expression for w_j given in Part (b4) of Example 10-1 is applicable. In this case $M = 7.375$ and the mass fraction W_j of polymers of length j may be computed as follows:

$$W_j = w_j = \frac{j(j-1)b^2 k_c}{M^3(1 + a/M)^j \left[2k_p \Gamma_0(R_j) \right]}$$

$$= \frac{j(j-1)[8.727 \times 10^{-10}]^2 (3.6 \times 10^7)}{(7.375)^3 \left[1 + \dfrac{0.01336}{7.375} \right]^j (2)(176)(6.532 \times 10^{-8})}$$

A graph of the results given by this expression is presented in Figure 10-2.
(4) After the weight fraction W_j has been computed, the corresponding values of x_j and P_j are computed as described in Problem 10-1. These results are plotted in Figures 10-2 and 10-3, respectively.
(d) The two additional reactions to be considered are as follows:

$$R_j + M \overset{k_{tM}}{\to} P_j + R_1$$

$$R_j + I \overset{k_{tC}}{\to} P_j + R_1$$

Since some additional initiator is required for the termination reactions, and since it has been specified that the outlet concentration of I is to be 0.04 g mol/liter, it is necessary to recompute the concentration of the initiator in the feed. In this case,

$$\frac{I^0 - I}{\theta} = r_0 + k_{tC} I \Gamma_0(R_j)$$

Then, from the results of parts (b) and (c) of Example 10-1,

$$I^0 = I + \theta\left[r_0 + k_{tC}I\Gamma_0(R_j)\right]$$

$$= 0.04 + 1.15 \times 10^4\left[1.28 \times 10^{-7} + (9.68)(0.04)(6.532 \times 10^{-8})\right]$$

$$= 0.04 + 0.00176 = 0.04176$$

Thus, the initiator transfer reaction has only a small effect on the feed concentration of the initiator.

In the calculation of the number average degree of polymerization, the expression for $\bar{d}_n$ now becomes

$$\bar{d}_n = \frac{k_c\Gamma_0}{2k_pM} + T$$

where
$$T = \frac{k_{tM}}{k_p} + \frac{k_{tC}}{k_p}\left(\frac{I}{M}\right) = \frac{105.6 \times 10^{-4}}{176} + \left(\frac{9.68}{176}\right)\left(\frac{0.04}{7.375}\right)$$

$$= 0.358 \times 10^{-3}$$

By use of the expression obtained in part (b2) of Example 10-1,

$$\bar{d}_n = \frac{6.68 \times 10^{-3}}{7.375} + 0.358 \times 10^{-3} = 1.264 \times 10^{-3}$$

Then

$$\bar{D}_n = \bar{d}_n^{-1} = \frac{1}{1.264 \times 10^{-3}} = 791$$

and
$$\bar{M}_n = \bar{D}_nM_M = (791)(104) = 8.23 \times 10^4$$

Note that the effect of the two transfer reactions is to reduce the number average molecular weight by 28.3%.

In the weight average degree of polymerization, the term $T\Gamma_0(R_j)$ must be added to the value of γ found in part (c3) of Example 10-1. That is, in this case,

$$\gamma = \frac{r_i}{k_pM} + T\Gamma_0(R_j)$$

$$= \frac{8.727 \times 10^{-10}}{7.375} + (0.358 \times 10^{-3})(6.532 \times 10^{-8})$$

$$= 1.417 \times 10^{-10}$$

In this case the expression for ψ becomes

$$\psi = \frac{1}{1 + T + (k_c\Gamma_0/k_pM)} = \frac{1}{1.000358 + (0.01336/7.375)}$$

$$= 0.99783$$

By use of these values for γ and ψ as well as other computed results, the expression given by Equation (10-75) for $\bar{d}_W$ may be evaluated to give

$$\bar{d}_W = 1303$$

From Equation (10-105),

$$\bar{d}_W = \bar{D}_W = 1303$$

and by Equation (10-14),

$$\bar{M}_W = \bar{D}_W M_M = (1311)(104) = 1.355 \times 10^5$$

Note that the transfer reactions have reduced the weight average molecular weight found in part (c2) by 20.7%.

The distribution curve of weight fraction may be computed by use of the following expression, which is obtained by use of Equations (10-80) and (10-106):

$$W_j = w_j = \frac{j\gamma\psi^j}{\Gamma_0(R_j)} \left[\frac{k_c\gamma(j-1)}{2k_p M} + T \right]$$

The values for γ, ψ, and T are those computed previously for the case where the additional termination reactions are considered. The resulting distribution computed by this expression for W_j is represented in Figure 10-2.

After the weight fraction W_j has been computed, the corresponding values of P_j are computed as described in Problem 10-1. These results are plotted in Figure 10-4.

4. The Gel or Trommsdorff Effect. The names *gel* or *Trommsdorff effect* [5, 8, 9] are used to describe the tendency of reactions in which high-molecular-weight polymers are formed to become "diffusion controlled." Because of the size of the relatively large polymer molecules, their kinetic velocities are far less than those of the monomers. Thus the chance of collision between two large polymer molecules followed by a termination reaction [Equations (10-21) and (10-22)] is much less than the chance of collision between a monomer and a polymer followed by propagation reaction [Equation (10-20)]. The net result of this effect on the rate of reaction of the monomer may be seen by examining the expression obtained by elimination of r_i from Equations (10-41) and (10-49).

$$r_M = \left(\frac{k_p\sqrt{2k_0 f}}{\sqrt{k_c + k_d}} \right) MI^{1/2} = k_{ob}MI^{1/2} \tag{10-107}$$

where k_{ob} is the observed rate constant. When the polymerization becomes diffusion controlled, the rate constants for termination, k_c and k_d, decrease, and consequently the observed rate constant k_{ob} increases. Consequently, the tendency to produce higher-molecular-weight polymers by the propagation reaction rather than termination of the polymerization [Equations (10-21) and (10-22)] is increased.

Hamielec [6] observed that the onset of diffusion control begins with a viscosity of 10 centipoise in the polymerization of styrene. Hui and Hamielec [9] and Biesenberger et al. [8] have treated the diffusion effect in more detail.

Sacks et al. [10, 11], Beckman [12], and Lynn and Huff [13] have considered the cases of optimal temperature profiles of batch reactors, problems of designing large-scale reactor vessels, and deviations from plug flow, respectively.

II. STEPWISE ADDITION AND CONDENSATION POLYMERIZATIONS AND COPOLYMERIZATIONS

First the chemistry of these types of polymerizations is presented and then stepwise addition polymerizations are treated quantitatively.

A. The Chemistry of Selected Polymerizations

1. Stepwise Addition Polymerizations. Stepwise addition polymerizations are characteristic of the *living polymers*, as coined by Szwarc [14]. Two basic reactions occur, which are

$$I + M \xrightarrow{k_i} R_1 \qquad \text{(initiation)} \qquad (10\text{-}108)$$

$$R_j + M \xrightarrow{k_p} R_{j+1} \qquad \text{(propagation)} \qquad (10\text{-}109)$$

For this simple case, no transfer or termination reactions occur. The polymer species R_j has a long lifetime, and the reaction must be terminated by an outside agent such as water.

Cationic, anionic, and coordinated polymerizations are examples of stepwise addition polymerizations. An example of the use of an ionic type of catalyst is the stepwise polymerization of styrene in a solvent by use of butyllithium. The initiation step is

where the abbreviated form

This initiation reaction may be represented symbolically by Equation (10-108).

The propagation step then begins by the reaction of R_1 with M to give R_2; that is,

$$
\left[\begin{array}{c} \underset{\displaystyle\bigcirc}{\overset{\displaystyle H\ \ H}{Bu-\underset{\displaystyle H}{\overset{}{C}}-\overset{}{C}\!:}} \end{array} \right]^{-} [Li]^{+} + \underset{\displaystyle\bigcirc}{\overset{\displaystyle H\ \ H}{\underset{\displaystyle H}{\overset{}{C}}\!:\ \ :\overset{}{C}}} \longrightarrow \left[\begin{array}{c} \underset{\displaystyle\bigcirc}{Bu(CH_2CH)_2} \end{array} \right]^{-} [Li]^{+}
$$

This reaction is represented symbolically by Equation (10-109). Stepwise addition polymerizations do not have a termination step, and water or some other impurity must be injected into the reacting mass in order to stop the polymerization reaction [14]. In ionic polymerizations, the length of the polymer molecules is dependent on the initiator concentration and the monomer conversion. To produce long-chain polymers, low initiator concentrations of the order of 10^{-6} mol/liter and high monomer conversions are required.

2. Condensation Polymerizations. Examples of products formed by condensation reactions are the polyurethanes, polyureas, polyphenylesters, polycarbonates, and phenolic resins. An example of condensation polymerizations is the formation of nylon by the reaction of diamino-hexane and adipic acid [3]. The first step is

$$
\underset{\text{Diaminohexane}}{\overset{\displaystyle H\ \ H\ \ H\ \ H\ \ H\ \ H\ \ H\ \ H}{\underset{\displaystyle H\ \ H\ \ H\ \ H\ \ H\ \ H\ \ H\ \ \boxed{H}}{N-\overset{}{C}-\overset{}{C}-\overset{}{C}-\overset{}{C}-\overset{}{C}-\overset{}{C}-N}}} \quad + \quad \underset{\text{Adipic acid}}{\overset{\displaystyle O\ \ H\ \ H\ \ H\ \ H\ \ O}{\underset{\displaystyle \ \ \ H\ \ H\ \ H\ \ H\ \ OH}{C-\overset{}{C}-\overset{}{C}-\overset{}{C}-\overset{}{C}-\overset{}{C}}}}
$$

$$
\longrightarrow \underset{\displaystyle H}{\overset{\displaystyle H}{N}}-(CH_2)_6-\underset{}{\overset{\displaystyle H\ \ O}{N-\overset{\|}{C}}}(CH_2)_4-\underset{\displaystyle OH}{\overset{\displaystyle O}{\overset{\|}{C}}} + H_2O
$$

Additional reactions occur at both ends of this polymer molecule to form long-chain polymers. A characteristic of condensation reactions is the production of small molecules such as H_2O, NH_3, HCHO, and NaCl. Usually the small molecules must be removed during the polymerization in order to produce high-molecular-weight molecules.

3. Copolymerization. A copolymer is a polymer chain composed of two monomers, which may be represented as monomers A and B. Possible copolymer chains are [15] as follows:

A—B—A—B—A—B   (alternating copolymer)

$\left.\begin{array}{l} A—A—B—A—B—B—A—B—A—B—A—A \\ A—A—A—B—A—B—B—B—A \end{array}\right\}$ (random copolymers)

A—A—A—A—A—A—A—A—B—B—B—B—B—B   (block copolymer)

Examples of commercially important copolymers are styrene–butadiene, vinylidene chloride–vinyl chloride, ethylene–vinyl acetate, and butadiene–acrylonitrile.

B. Analysis of Stepwise Addition Polymerizations

The reaction mechanism considered consists of the initiation reaction [Equation (10-108)] and the propagation reactions [Equation (10-109)]. The rate of disappearance of the initiator I is represented by

$$r_i = k_i IM \tag{10-110}$$

Again the assumption made in the analysis of the propagation reactions in part I is employed; that is, the rate constant k_p is independent of the chain length. Let the rate of appearance of the active species R_j be denoted by r_{R_j}. Then, for $j = 1$,

$$r_{R_1} = k_i IM - k_p MR_1 \tag{10-111}$$

and, for $j \geq 2$,

$$r_{R_j} = k_p M(R_{j-1} - R_j), \qquad j = 2, 3, \ldots \tag{10-112}$$

The rate of propagation or the rate of disappearance of the monomer M by all propagation reactions is given by

$$(r_p)_M = k_p M \sum_{j=1}^{\infty} R_j \tag{10-113}$$

Let the total rate of disappearance of the monomer be denoted by r_M. Then

$$r_M = r_i + (r_p)_M \tag{10-114}$$

The summation of the r_{R_j}'s over all j ($j = 1, 2, \ldots$) is given by

$$\sum_{j=1}^{\infty} r_{R_j} = k_i IM - k_p MR_1 + k_p M \left[\sum_{j=2}^{\infty} R_{j-1} - \sum_{j=2}^{\infty} R_j \right]$$

$$= k_i IM + k_p M \left[\sum_{j=2}^{\infty} R_{j-1} - \sum_{j=1}^{\infty} R_j \right]$$

$$= k_i IM + k_p M \left[\sum_{k=1}^{\infty} R_k - \sum_{j=1}^{\infty} R_j \right]$$

$$= k_i IM = r_i \tag{10-115}$$

The total polymer concentration, $\sum_{j=1}^{\infty} R_j$, which appears in Equation (10-113), may be evaluated by making a material balance on the initiator. In particular, since each polymer molecule R_j has associated with it one molecule of initiator I, it follows that

$$I^0 - I = \sum_{j=1}^{\infty} R_j = \Gamma_0(R_j) \tag{10-116}$$

Similarly, by making a mass balance on the monomer, the first moment of the R_j's may be evaluated. That is, since the monomer M must be in either the monomer form or in the polymer form, it follows that

$$M_M(M^0 - M) = \sum_{j=1}^{\infty} M_j R_j = M_M \sum_{j=1}^{\infty} j R_j$$

and thus

$$M^0 - M = \sum_{j=1}^{\infty} j R_j = \Gamma_1(R_j) \tag{10-117}$$

1. Definitions of the Instantaneous Fractional Degree and The Weight Degree of Polymerization. For stepwise addition polymerizations, the instantaneous fractional degree of polymerization, $\bar{d}_n$, is defined by

$$\bar{d}_n = \frac{\sum_{j=1}^{\infty} r_{R_j}}{r_M} \tag{10-118}$$

Then

$$\bar{d}_n = \frac{r_i}{r_i + (r_p)_M} = \frac{1}{1 + (k_p/k_i)(I^0/I - 1)} \tag{10-119}$$

The instantaneous weight degree of polymerization is defined by

$$\bar{d}_W = \frac{\sum_{j=1}^{\infty} j^2 r_{R_j}}{r_M} \tag{10-120}$$

In this case,

$$\sum_{j=1}^{\infty} j^2 r_{R_j} = k_i I M + k_p M \sum_{j=1}^{\infty} \left[(j+1)^2 - j^2 \right] R_j$$

$$= k_i I M + k_p M \sum_{j=1}^{\infty} (2j+1) R_j$$

$$= k_1 I M + 2 k_p M (M^0 - M) + k_p M (I^0 - I) \tag{10-121}$$

Since $r_M = r_i + (r_p)_M$, it follows that

$$\bar{d}_W = 1 + \frac{2 k_p (M^0 - M)}{k_i I + k_p (I^0 - I)} \tag{10-122}$$

C. Polymerization in Batch Reactors

Throughout the following developments, it is supposed that the batch reactor operates isothermally and that the volume of the reacting mixture remains constant with respect to time.

1. Number and Weight Average Degrees of Polymerization. Although the instantaneous degrees of polymerization $\bar{d}_n$ and $\bar{d}_W$ found previously may be used to obtain expressions for the number average degree of polymerization, $\bar{D}_n$, and the weight average degree of polymerization, $\bar{D}_W$, the following approach is simpler for the treatment of batch reactors. The first step of this approach is to obtain expressions for the amounts of the initiator I and monomer M as functions of a new variable τ which is related to time. The second step consists of the development of an expression for computing the time t corresponding to each value of τ.

Elimination of r_i from the component material balance

$$r_i = -\frac{dI}{dt}$$

and Equation (10-110) gives

$$\frac{-dI}{dt} = k_i I M \tag{10-123}$$

Let the new variable τ be defined as follows;

or
$$\tau = \int_0^t M \, dt \tag{10-124}$$

$$d\tau = M \, dt$$

After this change in variable has been made, Equation (10-123) reduces to

$$\frac{-dI}{d\tau} = k_i I \tag{10-125}$$

Separation of variables followed by integration and rearrangement yields

$$I = I^0 e^{-k_i \tau} \tag{10-126}$$

The expression obtained for I as a function of τ may now be used to express the amount of monomer as a function of τ. Equations (10-81), (10-110), (10-113), and (10-114) may be combined to give

$$\frac{-dM}{dt} = k_i I M + k_p M \sum_{j=1}^{\infty} R_j \tag{10-127}$$

After each member of this equation has been divided by M and the sum of the R_j's has been eliminated by use of Equation (10-116), one obtains

$$\frac{-dM}{d\tau} = k_i I + k_p (I^0 - I) \tag{10-128}$$

Elimination of I by use of Equation (10-126) gives

$$\frac{-dM}{d\tau} = k_i I^0 e^{-k_i \tau} + k_p I^0 (1 - e^{-k_i \tau}) \qquad (10\text{-}129)$$

Multiplication of each term of this equation by $d\tau$, followed by termwise integration and rearrangement, yields

$$(M^0 - M) = I^0 (1 - e^{-k_i \tau}) \left(1 - \frac{k_p}{k_i}\right) + k_p I^0 \tau \qquad (10\text{-}130)$$

A one-to-one correspondence between τ and t may now be obtained by first eliminating M from Equations (10-124) and (10-130). The resulting expression may be integrated to give

$$t = \int_0^\tau \frac{d\tau}{M^0 - I^0 \left[(1 - k_p/k_i)(1 - e^{-k_i \tau}) + k_p \tau\right]} \qquad (10\text{-}131)$$

By use of this equation to obtain the t for any given τ, the corresponding values of I and M at time t may be found by use of Equations (10-126) and (10-130), respectively.

Since the number average degree of polymerization as given by Equation (10-11) is equal to $\Gamma_1(R_j)$ divided by $\Gamma_0(R_j)$, and since these moments are given by Equations (10-116) and (10-117), respectively, it follows that

$$\overline{D}_n = \frac{\Gamma_1(R_j)}{\Gamma_0(R_j)} = \frac{M^0 - M}{I^0 - I} \qquad (10\text{-}132)$$

It should be remarked that this expression is valid for any type of reactor. For the case of a batch reactor, I and M in Equation (10-132) may be stated in terms of τ by use of Equations (10-126) and (10-130), respectively, to give

$$\overline{D}_n = \left(1 - \frac{k_p}{k_i}\right) + \frac{k_p \tau}{1 - e^{-k_i \tau}} \qquad (10\text{-}133)$$

and, for any τ, the corresponding value of t is given by Equation (10-131).

To express the weight average degree of polymerization $\overline{D}_W$ [$\overline{D}_W = \Gamma_2(R_j)/\Gamma_1(R_j)$] in terms of τ requires the development of an expression for the second moment of the R_j's in terms of τ. Such an expression may be obtained by commencing with Equation (10-121) and the material balances

$$r_{R_j} = \frac{dR_j}{dt}, \qquad j = 1, 2, \ldots \qquad (10\text{-}134)$$

Elimination of the r_{R_j}'s in Equation (10-121) by use of Equation (10-134) gives

$$\sum_{j=2}^{\infty} j^2 \frac{dR_j}{dt} = \left[k_i IM + 2k_p M(M^0 - M) + k_p M(I^0 - I)\right] \qquad (10\text{-}135)$$

After each member of this equation has been divided by M, and then I and $(M^0 - M)$ are replaced by their equivalents as given by Equations (10-126) and (10-130), the resulting expression may be integrated to give the desired result:

$$\sum_{j=1}^{\infty} j^2 R_j = I^0 \left\{ \left(1 - \frac{k_p}{k_i} \right) \left(1 - \frac{2k_p}{k_i} \right) (1 - e^{-k_i \tau}) \right.$$

$$\left. + k_p \tau \left[(1 + k_p \tau) + 2 \left(1 - \frac{k_p}{k_i} \right) \right] \right\} \qquad (10\text{-}136)$$

Then, by use of this result and the results given by Equations (10-117) and (10-130), the following expression is obtained for $\bar{D}_W$:

$$\bar{D}_W = \frac{\left(1 - \dfrac{k_p}{k_i} \right) \left(1 - \dfrac{2k_p}{k_i} \right) (1 - e^{-k_i \tau}) + k_p \tau \left[1 + k_p \tau + 2 \left(1 - \dfrac{k_p}{k_i} \right) \right]}{\left(1 - \dfrac{k_p}{k_i} \right) (1 - e^{-k_i \tau}) + k_p \tau}$$

$$(10\text{-}137)$$

where the time t corresponding to a given τ is found by use of Equation (10-131).

2. Distribution of the Polymers R_j. To compute the weight fraction or mole fraction of R_j in the mixture, expressions are needed for computing the R_j's ($j = 1, 2, \ldots$). The desired expressions for the R_j's may be obtained by the following procedure.

Elimination of r_{R_j} from Equations (10-111) and (10-112) by use of the corresponding material balances,

$$r_{R_j} = \frac{dR_j}{dt}, \qquad j = 1, 2, \ldots$$

gives

$$\frac{dR_1}{d\tau} = k_i I - k_p R_1 \qquad (10\text{-}138)$$

and

$$\frac{dR_j}{d\tau} = k_p (R_{j-1} - R_j) \qquad (10\text{-}139)$$

after the change of variable $d\tau = M\, dt$ has been made. Since I is known as a function of τ [Equation (10-126)], Equation (10-138) may be rearranged to the form

$$\frac{dR_1}{d\tau} + k_p R_1 = k_i I^0 e^{-k_i \tau} \qquad (10\text{-}140)$$

The solution to this differential equation is well known [see Equations (3-57) and (3-58)]. For the boundary condition $R_1 = 0$ at $\tau = 0$, the solution of Equation (10-140) is given by

$$R_1 = \left(\frac{k_i I^0}{k_p - k_i}\right)(e^{-k_i \tau} - e^{-k_p \tau}) \tag{10-141}$$

where $k_p \neq k_i$. By use of this expression and Equation (10-112) for $j = 2$, the following equation in R_2 and τ is obtained:

$$\frac{dR_2}{d\tau} + k_p R_2 = k_p \left(\frac{k_i I^0}{k_p - k_i}\right)(e^{-k_i \tau} - e^{-k_p \tau}) \tag{10-142}$$

The solution of Equation (10-142) is of the same general form as the one for Equation (10-140), and for $R_2 = 0$ at $\tau = 0$,

$$R_2 = \frac{k_i I^0 k_p}{(k_p - k_i)^2}(e^{-k_i \tau} - e^{-k_p \tau}) - \frac{k_p k_i I^0 \tau e^{-k_p \tau}}{k_p - k_i} \tag{10-143}$$

By continuation of this procedure, the following expression is obtained:

$$R_j = \frac{(k_i I^0)k_p^{j-1}e^{-k_i \tau}}{(k_p - k_i)^j}\left[1 - e^{-z}\left(\sum_{n=0}^{j-1}\frac{z^n}{n!}\right)\right] \tag{10-144}$$

where

$$z = (k_p - k_i)\tau$$

The incomplete gamma function

$$\Gamma(z, n) = \int_0^z \frac{x^n e^{-x} \, dx}{n!} \tag{10-145}$$

has been tabulated. By successive integration of the right side of Equation (10-145) by parts, the following equivalent form of the incomplete gamma function is obtained:

$$\Gamma(z, n) = 1 - e^{-z}\left(\sum_{k=0}^{n}\frac{z^k}{k!}\right) \tag{10-146}$$

Thus Equation (10-144) reduces to

$$R_j = \frac{k_i I^0 k_p^{j-1} e^{-k_i \tau}}{(k_p - k_i)^j}[\Gamma(z, j - 1)] \tag{10-147}$$

where the time t corresponding to a given τ is found by use of Equation (10-131). [Zeman and Amundson [15] proposed numerical procedures for the calculation of the R_j's for more complex mechanisms and more general operating conditions than those used to obtain Equation (10-147).]

After the set of R_j's has been computed, the weight fractions and other distribution functions may be computed. For example, the weight fraction W_j of polymer R_j may be computed by use of the following formula:

$$W_j = \frac{jR_j}{M^0 - M} \tag{10-148}$$

where $M^0 - M$ is given by Equation (10-130) and the time t corresponding to each τ is given by Equation (10-131). The relationship given by Equation (10-148) follows immediately from Equations (10-3) and (10-117).

EXAMPLE 10-3

Suppose that, for the reactions given by Equations (10-108) and (10-109), $k_i =$ 0.032 (liter/g mol min) and $k_p = 1.2$ (liters/g mol min). (a) For the initial concentrations of $M^0 = 1$ (g mol/liter) and $I^0 = 0.01$ (g mol/liter), calculate the times required to achieve 50% and 85% conversion of the monomer. (b) At 50% and 85% conversion of the monomer, calculate the corresponding initiator conversions. (c) At 50% and 85% conversions of the monomer, calculate the corresponding values for the number average degree of polymerization, the weight average degree of polymerization, and the dispersion index ($\overline{D}_W/\overline{D}_n$). Also plot the corresponding distributions on a molar and a weight fraction basis.

SOLUTION

(a) To find the time t required to achieve a given conversion of the monomer requires a graphical integration of Equation (10-131). Computed values of the integrand of Equation (10-131) for different choices of τ are listed next and plotted in Figure 10-5.

τ	$1/M$	τ	$1/M$
5	1.006	60	1.691
10	1.020	70	2.057
15	1.043	80	2.654
20	1.072	90	3.780
25	1.110	100	6.661
30	1.156	105	10.83
40	1.276	110	29.24
50	1.446		

That $1/M$ is equal to the integrand of Equation (10-131) follows immediately from Equations (10-124) and (10-130).

At 50% conversion, $M = M^0 - 0.5M^0$ and $1/M = 2.0$. From Figure 10-5, $\tau = 67.5$ at $1/M = 2.0$. At 85% conversion, $1/M = 6.67$ and $\tau = 100$. Then at 50% conversion of the monomer, the value of t is found by graphical evaluation of the

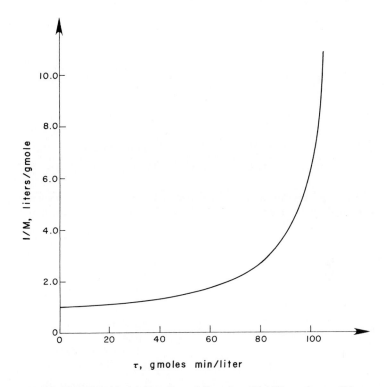

Figure 10-5. Graphical integration of Equation (10-131) at the conditions stated in Example 10-3.

following integral through the use of Figure 10-5; that is, at 50% conversion,

$$t = \int_{\tau=0}^{\tau=67.5} \frac{d\tau}{M} \cong 87 \text{ min}$$

Similarly, at 85% conversion of the monomer,

$$t = \int_{\tau=0}^{\tau=100} \frac{d\tau}{M} = 200 \text{ min}$$

(b) At 50% conversion of the monomer, the corresponding conversion of the initiator may be computed by rearranging Equation (10-126) to the form

$$\frac{I^0 - I}{I^0} = 1 - e^{-k_i\tau} = 1 - e^{-0.032 \times 67.5}$$

$$= 0.885 \quad \text{or} \quad 88.5\%$$

Similarly, at 85% conversion of the monomer, the corresponding conversion of the initiator is computed as follows:

$$\frac{I^0 - I}{I^0} = 1 - e^{-0.032 \times 100} = 0.959 \quad \text{or} \quad 95.9\%$$

The number average degree of polymerization at each of the two conversions may now be computed by use of Equation (10-132). At 50% conversion of the monomer,

$$\overline{D}_n = \frac{M^0\left[(M^0 - M)/M^0\right]}{I^0\left[(I^0 - I)/I^0\right]} = \frac{(1)(0.5)}{(0.01)(0.885)} = 56.5$$

At 85% conversion of the monomer,

$$\overline{D}_n = \frac{M^0\left[(M^0 - M)/M^0\right]}{I^0\left[(I^0 - I)/I^0\right]} = \frac{(1)(0.85)}{(0.01)(0.959)} = 88.6$$

The weight average degree of polymerization at each of the conversions may be computed by use of Equation (10-137). At 50% conversion of the monomer, the

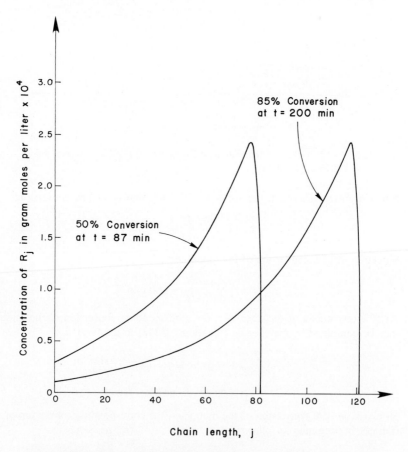

Figure 10-6. Molar concentrations of polymers R_j for Example 10-3 at 50% and 85% conversion of the monomer.

factors in Equation (10-137) take on the following values:

$$\left(1 - \frac{k_p}{k_i}\right)\left(1 - \frac{2k_p}{k_i}\right)\left(1 - e^{-k_i\tau}\right)$$

$$= \left(1 - \frac{1.2}{0.032}\right)\left(1 - \frac{2 \times 1.2}{0.032}\right)\left(1 - e^{-0.032 \times 67.5}\right) = 2389.5$$

$$k_p\tau\left[1 + k_p\tau + 2\left(1 - \frac{k_p}{k_i}\right)\right]$$

$$= 1.2 \times 67.5\left[1 + 1.2 \times 67.5 + 2\left(1 - \frac{1.2}{0.032}\right)\right] = 729$$

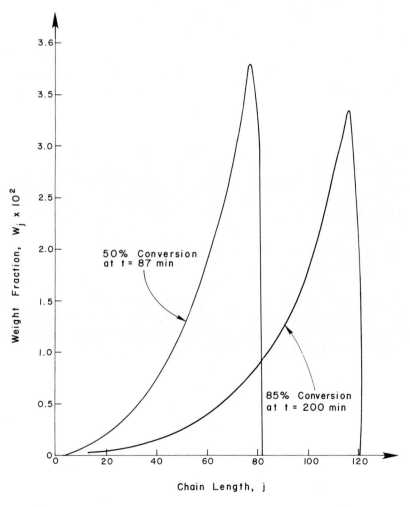

Figure 10-7. Molecular weight distributions for Example 10-3 at 50% and 85% conversion of the monomer.

and

$$\left(1 - \frac{k_p}{k_i}\right)(1 - e^{-k_i\tau}) + k_p\tau$$

$$= \left(1 - \frac{1.2}{0.032}\right)(1 - e^{-0.032 \times 67.5}) + 1.2 \times 67.5 = 48.71$$

After these values have been substituted into Equation (10-137), it is found that the weight average degree of polymerization has the following value:

$$\overline{D}_W = 64.02$$

At 50% conversion, the dispersion index may now be calculated:

$$\frac{\overline{D}_W}{\overline{D}_n} = \frac{64.02}{56.5} = 1.134$$

At 85% conversion of the monomer, the value $\overline{D}_W$ is computed in a manner analogous to that demonstrated for 50% conversion. The result so obtained is $\overline{D}_W = 98.2$. Thus the dispersion index at 85% conversion is given by

$$\frac{\overline{D}_W}{\overline{D}_n} = \frac{98.2}{88.6} = 1.11$$

(c) By use of these results and Equation (10-144) [or Equation (10-147)], the distributions shown in Figure 10-6 are obtained. (Alternately, the procedure proposed by Zeman et al. [15] may be used.) After each R_j has been computed, the corresponding W_j is found by use of Equation (10-148). The results so obtained are shown in Figure 10-7.

D. Polymerization in Plug-Flow and Perfectly Mixed Flow Reactors

1. Plug-Flow Reactors. The equations describing plug-flow reactors at isothermal operation and constant volumetric flow rate v_T throughout the reactor are of the same form as those stated for batch reactors and may be obtained from them by making the following substitutions:

$$t = \frac{V}{v_T}, \qquad M = \frac{n_M}{v_T}$$

$$I = \frac{n_I}{v_T}, \qquad R_j = \frac{n_{R_j}}{v_T}$$

2. Perfectly Mixed Flow Reactors. In the following development, the reactor is assumed to be operated at steady state, and it is further supposed that the volumetric flow rate v_T is independent of the amount of polymerization which has taken place. Material balances on the initiator and monomer are as follows:

$$r_i = \frac{n_I^0/v_T - n_I/v_T}{V/v_T} = \frac{I^0 - I}{\theta} \tag{10-149}$$

$$r_M = \frac{n_M^0/v_T - n_M/v_T}{V/v_T} = \frac{M^0 - M}{\theta} \tag{10-150}$$

Elimination of r_i from Equations (10-110) and (10-149) gives

$$I^0 - I = k_i I M \theta \tag{10-151}$$

Similarly, elimination of r_M from Equations (10-150) and (10-114) gives

$$M^0 - M = k_i I M \theta + k_p M (I^0 - I) \theta \tag{10-152}$$

where $(r_p)_M$ has been replaced by its equivalent as given by Equations (10-113) and (10-116).

Equations (10-151) and (10-152) are seen to involve the three variables I, M, and θ. To obtain the value of M for a given value of I, the following expression, which was obtained by eliminating θ from Equations (10-151) and (10-152), may be used.

$$M = M^0 - (I^0 - I)\left[1 + \frac{k_p(I^0 - I)}{k_i I}\right] \tag{10-153}$$

After M has been computed for a given I by use of this equation, then the corresponding value of θ may be found by use of these values of M and I and Equation (10-151). Alternately, Equation (10-153) may be used to eliminate M from either Equation (10-151) or Equation (10-152) to give an explicit expression for I as a function of θ.

3. Number Average Degree and Weight Average Degree of Polymerization. The number average degree of polymerization is given by Equation (10-132). Replacing $M^0 - M$ in this expression for $\bar{D}_n$ gives

$$\bar{D}_n = \frac{M^0 - M}{I^0 - I} = \frac{k_i I M \theta + k_p M (I^0 - I) \theta}{k_i I M} = 1 + \frac{k_p}{k_i}\left(\frac{I^0 - I}{I}\right) \tag{10-154}$$

By comparison of Equations (10-154) and (10-119), it is seen that

$$\bar{D}_n = \frac{1}{\bar{d}_n} \tag{10-155}$$

For a perfectly mixed flow reactor in which the volumetric flow rate is constant, the r_{R_j}'s and r_M may be eliminated from the defining equation for

$\bar{d}_W$ by use of the component material balances

$$r_M = \frac{M^0 - M}{\theta} \tag{10-156}$$

$$r_{R_j} = \frac{R_j}{\theta}, \qquad j = 1, 2, \ldots \tag{10-157}$$

to give
$$\bar{d}_W = \frac{\sum_{j=1}^{\infty} j^2 r_{R_j}}{r_M} = \frac{\sum_{j=1}^{\infty} j^2 R_j}{M^0 - M}$$

Since $\bar{D}_W = \Gamma_2(R_j)/\Gamma_1(R_j)$ and since $\Gamma_1(R_j) = M^0 - M$, it follows from Equation (10-122) that

$$\bar{D}_W = \bar{d}_W = 1 + \frac{2k_p(M^0 - M)/k_i I}{1 + k_p(I^0 - I)/k_i I} \tag{10-158}$$

Elimination of $k_p(I^0 - I)/k_i I$ and $(M^0 - M)$ by use of the relationship given by Equation (10-153) gives

$$\bar{D}_W = 2\bar{D}_n - 1 \tag{10-159}$$

For large values of $\bar{D}_n$, Equation (10-159) reduces approximately to

$$\frac{\bar{D}_W}{\bar{D}_n} \cong 2 \tag{10-160}$$

4. Distribution of the Polymers R_j. On the basis of the assumptions of steady-state operation and constant density of the reacting mixture, the material balances on the polymers $R_1, R_2, \ldots,$ may be represented by Equation (10-157), where it has been supposed that the feed to the perfectly mixed reactor contains no polymers. Elimination of r_{R_j} $(j = 1, 2, \ldots)$ from Equations (10-111), (10-112), and (10-157) gives

$$\frac{R_1}{\theta} = k_i I M - k_p R_1 M \tag{10-161}$$

$$\frac{R_j}{\theta} = k_p M(R_{j-1} - R_j), \qquad j = 2, 3, \ldots \tag{10-162}$$

Equation (10-161) may be solved for R_1 and Equation (10-162) for R_j to give

$$R_1 = \frac{k_i I M \theta}{1 + k_p \theta M} \tag{10-163}$$

and
$$R_j = \frac{k_p \theta M R_{j-1}}{1 + k_p \theta M} \tag{10-164}$$

For $j = 2$, Equation (10-164) contains R_1, which may be eliminated by use of

Equation (10-163) to give

$$R_2 = \frac{(k_i IM\theta)(k_p\theta M)}{(1 + k_p\theta M)^2} \qquad (10\text{-}165)$$

Continuation of this procedure leads to

$$R_j = k_i IM\theta \frac{(k_p\theta M)^{j-1}}{(1 + k_p\theta M)^j} \qquad (10\text{-}166)$$

or

$$R_j = R_1 \left[\frac{k_p\theta M}{1 + k_p\theta M} \right]^{j-1} \qquad (10\text{-}167)$$

For a given polymerization, θ is fixed. The concentrations of M and I in the effluent from the reactor may be computed by use of Equations (10-151) through (10-153). Then R_1 and R_j ($j = 2, 3, \ldots$) may be computed by use of Equations (10-163) and (10-167), respectively. The weight fraction of polymer R_j may then be computed by use of the equation

$$W_j = \frac{jR_j}{M^0 - M} \qquad (10\text{-}168)$$

Procedures for calculating molecular weight distributions for living polymer systems with reversible termination and complex initiation rate expressions have been presented by Edgar et al. [16], Landon and Anthony [17], Porter et al. [18], Maggott et al. [19], Ahmad [20], Ahmad et al. [21, 22] and Treybig and Anthony [23].

EXAMPLE 10-4

The reactions given by Equations (10-108) and (10-109) with the rate constants given in Example 10-3 are to be carried out in a perfectly mixed flow reactor. The initial concentrations of the monomer and initiator are $M^0 = 1.0$ g mol/liter, and I^0 is equal to 0.01 mol/liter. (a) For a 75% conversion of the initiator, compute the corresponding conversion of the monomer. (b) Calculate the residence time $\theta = V/v_T$ required to obtain a 75% conversion of the initiator. (c) At a conversion of 75% of the initiator, compute the number and weight average degrees of polymerization. Also plot the corresponding distributions on a molar and a weight fraction basis.

SOLUTION

(a) Since it is given that $I = 0.25I^0$, the fraction of monomer converted may be found by use of Equation (10-153) after it has been rearranged to the following form:

$$\frac{M^0 - M}{M^0} = \frac{I^0 - I}{M^0}\left[1 + \frac{k_p(I^0 - I)}{k_i I}\right]$$

$$= \left[\frac{(0.01)(0.75)}{1.0}\right]\left[1 + \frac{(1.2)(0.01)(0.75)}{(0.032)(0.25)(0.01)}\right]$$

Thus

$$\frac{M^0 - M}{M^0} = 0.8512$$

and $M = 0.1488$ g mol/liter

(b) The residence time required to achieve a 75% conversion of the initiator is found by use of the result of part (a) and Equation (10-151):

$$\theta = \frac{I^0 - I}{k_i IM} = \frac{(0.01)(0.75)}{(0.032)(0.25)(0.01)(0.1488)}$$

$$= 630.04 \text{ min} = 10.5 \text{ h}$$

(c) The number average degree of polymerization may be found by use of Equation

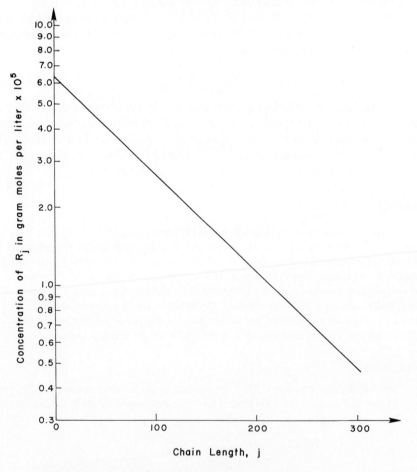

Figure 10-8. Molar concentrations of polymers R_j for Example 10-4 at 75% conversion of the initiator.

(10-132):

$$\overline{D}_n = \frac{M^0 - M}{I^0 - I} = \frac{0.8512}{(0.75)(0.01)} = 113.5$$

The weight average degree of polymerization may be computed by use of Equation (10-159):

$$\overline{D}_W = 2\overline{D}_n - 1 = 226$$

Then the dispersion index is given by

$$\frac{\overline{D}_W}{\overline{D}_n} = \frac{226}{113.5} = 1.991$$

The distribution of polymers may now be computed by use of Equations (10-163) and (10-167). By use of Equation (10-163),

$$R_1 = \frac{k_i IM\theta}{1 + k_p\theta M} = \frac{(0.032)(0.25)(0.01)(0.1488)(630.04)}{1 + (1.2)(630.04)(0.1488)}$$

$$= 6.608(10^{-5}) \text{ g mol/liter}$$

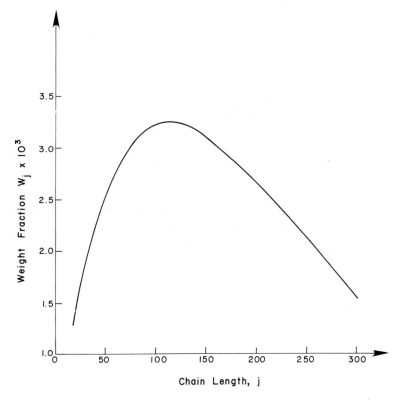

Figure 10-9. Molecular weight distribution for Example 10-4 at 75% conversion of the initiator.

For each chain length j, the corresponding polymer concentration R_j is found by use of Equation (10-167) as follows:

$$R_j = R_1\left[\frac{k_p\theta M}{1 + k_p\theta M}\right]^{j-1} = (6.608)(10^{-5})\left[\frac{(1.2)(630.04)(0.1488)}{1 + (1.2)(630.04)(0.1488)}\right]^{j-1}$$

Thus

$$R_j = (6.608)(10^{-5})[0.99189]^{j-1}$$

and

$$W_j = \frac{jR_j}{M^0 - M} = \frac{jR_j}{0.8512}$$

Plots of the results obtained by use of these equations for R_j and W_j are shown in Figures 10-8 and 10-9, respectively.

PROBLEMS

10-1. By use of the relationships given in the text, show that

(a) $\overline{M}_n = M_M\Gamma_1(x_j)$

(b) $\overline{M}_W = M_M\dfrac{\Gamma_2(x_j)}{\Gamma_1(x_j)}$

(c) $\overline{D}_n = \Gamma_1(x_j)$

(d) $\overline{D}_W = \dfrac{\Gamma_2(x_j)}{\Gamma_1(x_j)}$

10-2. By use of the definition of x_j, W_j, and $\overline{D}_n$, show that

(a) $x_j = \dfrac{W_j\overline{D}_n}{j}$

(b) $P_j = \dfrac{W_j(M^0 - M)}{j}$

10-3. Begin with Equation (10-145) and produce the result given by Equation (10-146).

10-4. Begin with the expression given by Equation (10-158) for $\overline{D}_W$ and obtain the expression given by Equation (10-159) for $\overline{D}_W$ in terms of $\overline{D}_n$.

NOTATION

C = chain transfer agent; also used to denote the concentration of the chain transfer agent; C^0 is the initial concentration of the chain transfer agent

$\bar{d}_n$ = instantaneous fractional degree of polymerization; total moles of polymer produced per mole of monomer reacted at any instant at any position in any given type of reactor; defined by Equation (10-55)

$\overline{D}_n$ = number average degree of polymerization; defined by Equation (10-11)

$\bar{d}_W$ = instantaneous weight degree of polymerization; defined by Equation (10-64)

$\overline{D}_W$ = weight average degree of polymerization; defined by Equation (10-13)

I = initiator; also used to denote the concentration of the initiator

j = number of monomer molecules in a given polymer P_j

M = monomer; also used to denote the concentration of the monomers, moles per unit volume; M^0 is the initial concentration of monomers

M_j = molecular weight of a polymer composed of j monomer molecules

$\overline{M}_n$ = number average molecular weight; defined by Equation (10-7)

$\overline{M}_W$ = weight average molecular weight; defined by Equation (10-10)

P_j = polymer composed of j monomer molecules; also used to denote the concentration of such polymers

$(r_c)_{P_j}$ = rate of appearance of polymer P_j by addition (or combination)

$(r_c)_{R_j}$ = rate of disappearance of free radicals or active species R_j by addition (or combination)

r_{tC} = rate of disappearance of the chain transfer agent C by the chain transfer reaction

$(r_{tC})_{R_j}$ = rate of disappearance of free radicals or active species R_j by chain transfer

$(r_d)_{P_j}$ = rate of appearance of polymer P_j by the disproportionation reaction

$(r_d)_{R_j}$ = rate of disappearance of free radicals or active species R_j by the disproportionation reaction

r_i = rate of disappearance of free radicals ϕ, moles/(unit time) (unit volume) [see Equation (10-27)]

R_j = free radical or active species of j monomer molecules; also used to denote the concentration of these free radicals, moles/unit volumes

$(r_{tM})_{R_j}$ = rate of disappearance of free radicals or active species R_j by monomer transfer

r_{P_j} = net rate of appearance of polymers P_j

$(r_p)_M$ = rate of disappearance of the monomer M by propagation [see Equation (10-30)]

$(r_p)_{R_j}$ = rate of disappearance of free radicals or active species R_j by propagation [see Equation (10-29)]

r_{R_j} = net rate of disappearance of free radicals R_j by all steps of the mechanism

r_{tS} = rate of disappearance of the solvent S by the solvent transfer reaction

$(r_{tS})_{R_j}$ = rate of disappearance of free radicals R_j by solvent transfer

r_ϕ = net rate of disappearance of free radicals

S = solvent; also used to denote the concentration of the solvent

W_j = weight fraction of polymer molecules having chains composed of j monomer units in a mixture of polymer molecules having all possible chain lengths

w_j = instantaneous weight fraction for polymer P_j; defined by Equation (10-62)

x_j = mole fraction of polymer P_j

Greek Letters

$\Gamma_0(P_j)$ = total concentration of all polymers of all possible chain lengths

$\Gamma_1(P_j)$ = first moment of the P_j; defined following Equation (10-3)

$\Gamma_2(P_j)$ = second moment of the P_j's [see Equation (10-10)]

ϕ = free radical produced by the initiator I; also used to denote the concentration of these free radicals, moles/unit volume

REFERENCES

1. Carothers, W. H., *J. Am. Chem. Soc.*, *51*:2548 (1929).

2. Platzer, N., IUPAC International Symposium on Macromolecular Chemistry, pp. 155–184, Budapest, Hungary (1969).

3. Flory, P. J., *Principles of Polymer Chemistry*, Cornell University Press, Ithaca, N.Y. (1953).

4. Rodriquez, F., *Principles of Polymer Systems*, McGraw-Hill Book Company, New York (1970).

5. Bamford, C. H., W. G. Barb, A. D. Jenkins, and P. F. Onyon, *The Kinetics of Vinyl Polymerization by Radical Mechanisms*, Butterworth & Co. Publishers Limited, London (1958).

6. Hamielec, A. E., Symposium on Polymer Reaction Engineering, Laval University, Quebec City, Quebec, Canada (June 4–7, 1972).

7. Peebles, L. H., Jr., *Molecular Weight Distributions in Polymers*, Wiley/Interscience Publisher, New York (1971).

8. Biesenberger, J. A., M. Sacks, and I. Dubdevani, *Proceedings of the Symposium on Polymer Reaction Engineering*, Laval University, Quebec City, Quebec, Canada (June 5–7, 1972).

9. Hui, A. W., and A. E. Hamielec, *J. Applied Polymer Sci.*, Part C., *No. 25*, 167 (1968).

10. Sacks, M. E., Soo-il Lee, and J. A. Biesenberger, *Chem. Eng. Sci.*, *28*:2281–2289 (1972).

11. Sacks, M. E., Soo-il Lee, and J. A. Biesenberger, *Chem. Eng. Sci.*, *28*:241–257 (1973).

12. Beckman, G., *Chem. Tech.*, p. 304 (May 1973).

13. Lynn, S., and J. E. Huff, *AIChE J.*, *17*: No. 2, 475 (1971).

14. Szwarc, M., *Carbanions, Living Polymers, and Electron Transfer Processes*, Wiley/Interscience Publishers, New York (1968).

15. Zeman, R. J., and N. R. Amundson, *AIChE J.*, *9*:297 (1968).

16. Edgar, T. D., S. Hasan, and R. G. Anthony, *Chem. Eng. Sci.*, *25*:1463–1423 (1970).

17. Landon, T. R., and R. G. Anthony, *AIChE J.*, *18*:843 (1972).

18. Porter, R. E., A. Ahmad, and R. G. Anthony, *J. Applied Polymer Sci. 18*, 1805–1819 (1974).

19. Maggott, R. J., A. Ahmad, and R. G. Anthony, *J. Applied Polymer Sci. 19*, 165 (1975).

20. Ahmad, A., and R. G. Anthony, *J. Applied Polymer Sci. 21*, 1401–1408 (1977).

21. Ahmad, A., M. N. Treybig, and R. G. Anthony, *J. Applied Polymer Sci. 21*, 2021–2028 (1977).

22. Ahmad, A., M. N. Treybig, and R. G. Anthony, *J. Applied Polymer Sci. 22*, 314–333 (1978).

23. Treybig, M. N., and R. G. Anthony, *ACS Symposium Series, 104*, 295 (1979).

Using Tracer   11
Techniques
to Determine
Mixing Effects

Tracer techniques may be used to determine the effect of mixing on the extent of reaction. The fundamental principles involved in the analysis of mixing effects by use of tracer techniques are closely related to those found in other fields such as the kinetic theory of gases, statistics, and mechanics. In Part I, the fundamental principles involved in the analysis of linear systems by use of tracer techniques is presented. In Part II, the use of experimental techniques in the determination of mixing characteristics of systems is presented. In particular, the use of results obtained from tracer experiments to determine the dispersion coefficient, as well as the number of perfect mixers required to describe a given system, is presented.

Because of mixing in flow reactors, some of the elements of fluid remain in the reactor longer than others, with the consequence that the extent of the reaction may vary from element to element depending on its residence time or age. As pointed out by Foss et al. [1], the concept of residence time distributions dates back to Stewart [2], who in 1894 determined flows through small blood vessels by measuring the time required for an injected salt tracer to flow from one point to another. In 1935, MacMullin and Weber [3] investigated the distribution of residence times of particles flowing through a series of stirred reactors. Gilliland and Mason [4, 5] used the diffusion model as well as the concept of residence times to describe backmixing in beds of fluidized solids. Later Danckwerts [6, 7], followed by Zwietering [8], introduced F and C diagrams for representing the output responses to step inputs and impulse inputs, respectively. A further summary of recent contributors is presented by Levenspiel [9].

I. FUNDAMENTAL PRINCIPLES INVOLVED IN TRACER TECHNOLOGY

The relationship used in the kinetic theory of gases for expressing the probability that a molecule has a velocity lying between u and $u + du$ is of the same form as that used to express the probability that an element of fluid has a residence time lying between t and $t + dt$. The development of the expression for the probability that an element of fluid has a residence time in a system between t and $t + dt$ follows.

Consider the system in which a pure solvent is flowing through a horizontal tube. The system is at steady state and the volumetric flow rate of the solvent is v_T. Furthermore, suppose that between times $t = t_0$ and $t_0 + \beta$, a small pulse consisting of A moles of tracer is introduced uniformly over the time period β. Let the inlet molar flow rate of the tracer be denoted by $n^0(t)$ and the output flow rate by $n(t)$. Then the introduction of the pulse of tracer is represented by the equations

$$n^0(t) = 0, \qquad 0 < t_0$$

$$n^0(t) = \frac{A}{\beta}, \qquad t_0 \le t \le t + \beta \qquad (11\text{-}1)$$

$$n^0(t) = 0, \qquad t > t_0$$

It is further supposed that the amount A of tracer added is so small that the total volumetric flow rate v_T is unaltered by the introduction of the pulse of tracer. Typical input $n^0(t)$ and output $n(t)$ flow rates of the tracer are shown in Figure 11-1.

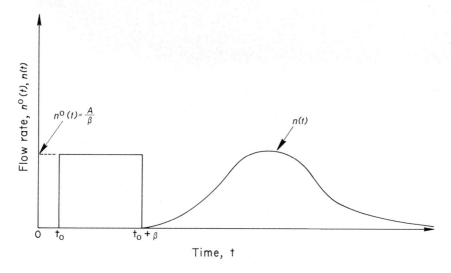

Figure 11-1. Introduction of A moles of tracer into a solvent.

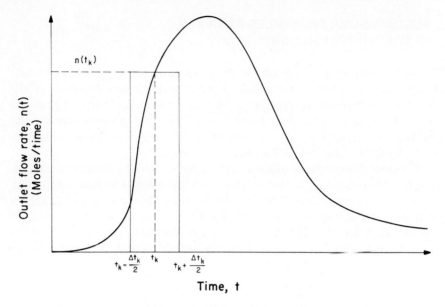

Figure 11-2. Distribution of residence times.

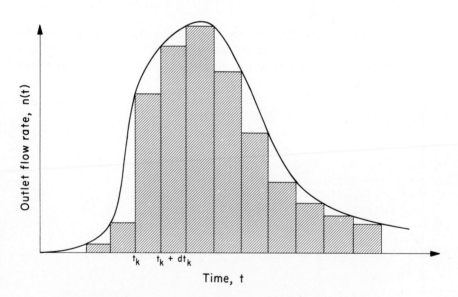

Figure 11-3. The molecules represented by the approximate area $n(t_k)\,dt_k$ have residence times between t_k and $t_k + dt_k$.

The total input of the pulse over the time period β is represented by

$$\int_0^\infty n^0(t) \, dt = \int_{t_0}^{t_0+\beta} \left(\frac{A}{\beta}\right) dt = \frac{A}{\beta}(t_0 + \beta - t_0) = A \qquad (11\text{-}2)$$

Since all the tracer introduced eventually leaves the system, it follows that

$$\int_0^\infty n(t) \, dt = A \qquad (11\text{-}3)$$

The area $n(t_k) \cdot \Delta t_k$ of the rectangle shown in Figure 11-2 is approximately equal to the moles of tracer that remain in the tube for a period of time lying between $t_k - \Delta t_k/2$ and $t_k + \Delta t_k/2$, or on the average the moles of tracer $n(t_k) \cdot \Delta t_k$ remain in the tube a period of time t_k. The moles $n(t_k) \cdot \Delta t_k$ are also said to have the age t_k. The language used in these statements may be placed in a one-to-one correspondence with that employed in the classical kinetic theory of gases in the following manner. Since the elements will be involved eventually in a limit, the precise construction of the rectangles is arbitrary and that shown in Figure 11-3 may be used. Also, since t is an independent variable, dt_k can be made any size or $dt_k = \Delta t_k$. Thus, the moles of tracer represented by the area $n(t_k) \Delta t_k$ [or $n(t_k) \, dt_k$] are those having a residence time between t_k and $t_k + \Delta t_k$ (or $t_k + dt_k$).

Then the fraction of the tracer input having a residence time between t_k and $t_k + dt_k$ is given by

$$\begin{array}{c}\text{Fraction of tracer} \\ \text{leaving between} \\ t_k \text{ and } t_k + dt_k\end{array} = \frac{n(t_k) \, dt_k}{A} = E(t_k) \, dt_k \qquad (11\text{-}4)$$

The fraction of the tracer leaving between times t_k and $t_k + dt_k$ is also equal to the probability P_{t_k} of its leaving at time t_k; that is,

$$P_{t_k} = E(t_k) \, dt_k \qquad (11\text{-}5)$$

Recall that A is equal to the total moles (or molecules) of tracer. Molecules may leave with residence times lying between t_k and $t_k + \Delta t_k$ (regarded as successes) or with residence times between t_i and $t_i + \Delta t_i$ (regarded as failures). Thus A also represents the number of successes plus the number of failures for molecules to leave. That is,

$$A = n(t_k) \, \Delta t_k + \sum_{i=1, \neq k}^{n} n(t_i) \, \Delta t_i$$

| Number of successes (molecules having residence times between t_k and $t_k + \Delta t_k$) | Number of failures (molecules leaving which do not have residence times between t_k and $t_k + \Delta t_k$) | (11-6) |

Then, in the limit as Δt_k and Δt_i go to zero and as n increases without bound,

Equation (11-6) reduces to Equation (11-3). Division of each member of Equation (11-3) by A gives

$$\int_0^\infty \frac{n(t)}{A} \, dt = \int_0^\infty E(t) \, dt = 1 \tag{11-7}$$

From an experimental point of view, it is convenient to restate the equations describing the pulse of tracer in terms of concentrations. Since $v_T C(t) = n(t)$, where v_T is the volumetric flow rate of the solvent, it follows that

$$\int_0^\infty \frac{n(t) \, dt}{A} = \int_0^\infty \frac{v_T C(t) \, dt}{A} = \int_0^\infty E(t) \, dt = 1 \tag{11-8}$$

The function $E(t)$ is called the *residence time distribution function*.

A. Residence Time Distribution Function for a Perfect Mixer

Consider the case of a single perfect mixer having a holdup V at steady state with an inlet and outlet volumetric flow rate v_T. Also suppose that the input to the mixer contains a small amount of tracer of concentration C^0 and molar flow rate n^0. The corresponding steady-state holdup of the tracer is N^0 moles or concentration $C^0 = N^0/V$. Suppose that at $t = 0-$ the tracer input to the feed stream is cut off, with the total volumetric flow rate remaining constant at v_T. Note that the preceding formulation is equivalent to injecting A moles of tracer into the holdup in the reactor at time $t = 0-$. Thus $A = N^0$ in Equation (11-9), which becomes

$$\int_0^\infty E(t) \, dt = \int_0^\infty \frac{C(t) \, dt}{(V/v_T)(N^0/V)} = \int_0^\infty \frac{C(t) \, dt}{\bar{t} C^0} = 1 \tag{11-9}$$

and thus

$$E(t) = \frac{C(t)}{\bar{t} C^0} \tag{11-10}$$

where $\bar{t} = V/v_T$, the residence time for the perfect mixer. An expression for $C(t)/C^0$ is found by solving the following differential equation, which represents a material balance on the tracer:

$$-n = \frac{dN}{dt}, \qquad t \geq 0 \tag{11-11}$$

Thus

$$-v_T C = V \frac{d(N/V)}{dt} = V \frac{dC}{dt} \tag{11-12}$$

and integration gives

$$\frac{C(t)}{C^0} = e^{-tv_T/V} = e^{-t/\bar{t}} \tag{11-13}$$

The distribution function $E(t)$ for the perfect mixer is given by

$$E(t) = \frac{C(t)}{C^0\bar{t}} = \frac{e^{-t/\bar{t}}}{\bar{t}} \tag{11-14}$$

Thus Equation (11-8) reduces to

$$\int_0^\infty \frac{e^{-t/\bar{t}}}{\bar{t}}\, dt = 1 \tag{11-15}$$

for a perfect mixer, or

$$\int_0^\infty e^{-t/\bar{t}}\, dt = \bar{t} \tag{11-16}$$

Equation (11-14) was restated in the form given by Equation (11-15) because in this form it corresponds to the Laplace transform with s of the Laplace transform being equal to $1/\bar{t}$. Thus the corresponding transforms with respect to $e^{-t/\bar{t}}$ are obtained from those given for the Laplace transforms by setting $s = 1/\bar{t}$. Laplace transforms are useful for solving linear systems of ordinary and partial differential equations. Before pursuing this subject further, linear systems are defined.

B. Linear Systems

Linear systems may be classified as those described by equations in which each dependent variable (concentration, molar flow rate) and its derivatives and integrals are of the first power and are combined in a linear equation. Such terms as $(dC_A/dt)^2$, C_A^2, and $C_A C_B$ are not permitted. It is important to note that linearity does not require that the coefficients of an equation be constants; they may depend on the independent variable t, but not on the dependent variable C_A. The analysis of linear systems rests on their unique properties, such as superposition.

C. Application of the Perfect Mixer Transform to a Batch Reactor

The *perfect mixer transform* of the function $F(t)$ is defined by

$$L\{F(t)\} = f(\bar{t}) = \int_0^\infty F(t)e^{-t/\bar{t}}\, dt \tag{11-17}$$

It will now be shown that when the differential equation describing a batch reactor is solved by use of the perfect mixer transform, the solution so obtained corresponds to the steady-state concentration produced by a perfect mixer.

Consider the differential equation for the batch reactor

$$-\frac{dC_A}{dt} - kC_A = 0$$

or

$$-\frac{d(C_A - C_A^0)}{dt} - k(C_A - C_A^0) - kC_A^0 = 0 \tag{11-18}$$

(In the interest of simplicity, the subscript c on the rate constant k_c has been omitted in this chapter.) Let $F(t) = C_A - C_A^0$, and observe that $F(0) = 0$. Multiplication of each member of Equation (11-18) by $e^{-t/\bar{t}}$, followed by integration from $t = 0$ to $t = \infty$, yields

$$-\int_0^\infty F'(t)e^{-t/\bar{t}}\, dt - k\int_0^\infty F(t)e^{-t/\bar{t}}\, dt - kC_A^0\int_0^\infty e^{-t/\bar{t}}\, dt = 0 \tag{11-19}$$

Since

$$\int_0^\infty F'(t)e^{-t/\bar{t}}\, dt = \frac{1}{\bar{t}}\int_0^\infty F(t)e^{-t/\bar{t}}\, dt$$

Equation (11-19) may be rearranged to give

$$\int_0^\infty (C_A - C_A^0)e^{-t/\bar{t}}\, dt = -\frac{kC_A^0\bar{t}}{(1/\bar{t}) + k}$$

where $F(t)$ has been replaced by its equivalent $C_A - C_A^0$. Thus

$$\int_0^\infty C_A e^{-t/\bar{t}}\, dt = \bar{t}C_A^0 - \frac{kC_A^0\bar{t}}{(1/\bar{t}) + k} = \frac{\bar{t}C_A^0}{1 + k\bar{t}} \tag{11-20}$$

The right side of Equation (11-20) is recognized as $\bar{t}$ times the steady state outlet concentration of a perfectly mixed reactor having a volume V, a volumetric flow rate v_T, and an inlet concentration C_A^0. This equivalence follows immediately from the material balance

$$v_T(C_A^0 - (C_A)_{\text{out}}) - Vk(C_A)_{\text{out}} = 0$$

Thus

$$(C_A)_{\text{out}} = \frac{C_A^0}{1 + k\bar{t}} \tag{11-21}$$

and substitution of this result in the right side of Equation (11-20) yields

$$\int_0^\infty C_A e^{-t/\bar{t}}\, dt = \bar{t}(C_A)_{\text{out}} \tag{11-22}$$

Application of the *generalized mean-value theorem** to the integral on the right side of Equation (11-22) yields

$$\int_0^\infty C_A e^{-t/\bar{t}}\, dt = \bar{C}_A \int_0^\infty e^{-t/\bar{t}}\, dt = \bar{C}_A \bar{t}$$

or

$$\bar{C}_A = \int_0^\infty C_A \frac{e^{-t/\bar{t}}}{\bar{t}}\, dt \qquad (11\text{-}23)$$

Thus for the case of a first-order reaction, it follows from Equations (11-22) and (11-23) that

$$(C_A)_{\text{out}} = \bar{C}_A \qquad (11\text{-}24)$$

In the following section, a physical interpretation is given to Equation (11-23).

D. Physical Interpretation of the Perfect Mixer Transformation

To deduce a physical interpretation of the integral given by Equation (11-23), it is helpful to restate this equation in the form:

$$\frac{\bar{C}_A}{C_A^0} = \int_0^\infty \frac{C_A(t)}{C_A^0} \frac{e^{-t/\bar{t}}}{\bar{t}}\, dt$$

$$= \int_0^\infty \left[\frac{C_A(t)}{C_A^0} \right] [E(t)\, dt] \qquad (11\text{-}25)$$

The two terms enclosed by brackets in Equation (11-25) may be regarded as the product of two probabilities. As will be recalled, $P_t = E(t)\, dt$ is the probability that a molecule will have a residence time between t and $t + dt$, while $C_A(t)/C_A^0 = N_A(t)/N_A^0$ is the fraction of unreacted molecules which have a residence time lying between t and $t + dt$ or the probability that a molecule having residence time t will be unreacted. Since these two events may be regarded as independent, it follows that the product of their respective probabilities gives the probability of finding an unreacted molecule having a residence time between t and $t + dt$. Thus the sum of these products over all possible times is given by the infinite integral, Equation (11-25).

EXAMPLE 11-1

Suppose the first-order reaction $A \rightarrow$ products is carried out in a batch reactor for which $C_A(t) = C_A^0 e^{-kt}$. Show that the outlet concentration found by carrying out the

---

*If $f(x)$ and $p(x)$ are continuous functions in x ($a \le x \le b$), and $p(x) \ge 0$, then

$$\int_a^b f(x) p(x)\, dx = f(\xi) \int_a^b p(x)\, dx$$

where $a \le \xi \le b$.

integration of Equation (11-25) is equal to that for a perfectly mixed flow reactor at steady state.

SOLUTION

For the first-order reaction, Equation (11-25) becomes

$$\frac{\bar{C}_A}{C_A^0} = \int_0^\infty \frac{C_A(t)e^{-t/\bar{t}}}{C_A^0\bar{t}}\,dt = \int_0^\infty e^{-(1+k\bar{t})(t/\bar{t})}\,d(t/\bar{t})$$

$$= \frac{1}{1 + k\bar{t}}$$

E. Second- and Higher-Order Reactions

Regardless of the order of a reaction, the integral given by the right side of Equation (11-25) may be formed and given the physical interpretation outlined previously and the generalized mean value theorem applied to give

$$\frac{\bar{C}_A}{C_A^0} = \int_0^\infty \frac{C_A(t)}{C_A^0}E(t)\,dt \tag{11-26}$$

If the reaction is first order (or zero order), the mean value $\bar{C}_A/C_A^0$ can be identified with the output $(C_A)_{\text{out}}/C_A^0$ of a reactor, such as the perfectly mixed reactor of Example 11-1. If the reaction is second order (or higher), the mean value $\bar{C}_A/C_A^0$ cannot be identified with the output $(C_A)_{\text{out}}/C_A^0$, as demonstrated in Example 11-2.

EXAMPLE 11-2

For the case where the second-order reaction $2A \rightarrow$ products is carried out in a batch reactor at constant volume, one obtains

$$\frac{C_A(t)}{C_A^0} = \frac{1}{1 + kC_A^0 t}$$

Compare the mean value of $\bar{C}_A/C_A^0$ found by substitution of this expression for $C_A(t)/C_A^0$ for a batch reactor into Equation (11-25) with the outlet concentration of a perfectly mixed flow reactor at steady state.

SOLUTION

$$\frac{\bar{C}_A}{C_A^0} = \int_0^\infty \left(\frac{1}{1 + kC_A^0 t}\right)\frac{e^{-t/\bar{t}}}{\bar{t}}\,dt = \alpha e^\alpha \int_\alpha^\infty \frac{e^{-x}}{x}\,dx \tag{A}$$

where

$$\alpha = 1/(kC_A^0\bar{t})$$

$$x = \alpha + t/\bar{t}$$

The integral on the right side of equation (A) is recognized as the exponential integral function, which is tabulated in most standard mathematical handbooks.

When a second-order reaction is carried out at steady state in a perfectly mixed flow reactor, the outlet concentration is found by use of the material balance

$$v_T\left[C_A^0 - (C_A)_{out}\right] - Vk(C_A)_{out}^2 = 0$$

or

$$\frac{(C_A)_{out}}{C_A^0} = \frac{-1 + \sqrt{1 + 4C_A^0 k\bar{t}}}{2C_A^0 k\bar{t}} \tag{B}$$

The expressions given for the mean value $\bar{C}_A/C_A^0$ and outlet value $(C_A)_{out}/C_A^0$ are not equal in this case because the mean value or its corresponding perfect mixer transform [see Equations (11-16) and (11-17)] is not a solution to the nonlinear differential equation describing the occurrence of a second-order reaction in a batch reactor.

EXAMPLE 11-3

Demonstrate that a solution is not given by the perfect mixer transformation of the differential equation for a second-order reaction carried out in a batch reactor.

SOLUTION

The differential equation to be solved is

$$-\frac{dC_A}{dt} - kC_A^2 = 0 \tag{A}$$

Let

$$F(t) = C_A - C_A^0$$

and rearrange Equation (A) to

$$F'(t) + k[F(t)]^2 + 2kC_A^0 F(t) + k(C_A^0)^2 = 0 \tag{B}$$

Multiplication by $e^{-t/\bar{t}}$, followed by the integration of each term from $t = 0$ to $t = \infty$, yields upon rearrangement

$$\left(\frac{1}{\bar{t}} + 2kC_A^0\right)f(\bar{t}) + k\int_0^\infty [F(t)]^2 e^{-t/\bar{t}} dt + k(C_A^0)^2\bar{t} = 0 \tag{C}$$

where $f(\bar{t}) = \int_0^\infty F(t)e^{-t/\bar{t}} dt$, the perfect mixer transform. Observe that analogous to Laplace transforms, the integral,

$$\int_0^\infty [F(t)]^2 e^{-t/\bar{t}} dt$$

cannot be expressed in terms of the perfect mixer transform $f(\bar{t})$. Thus the perfect mixer transformation does not yield a solution to Equation (A).

In summary, the mean concentration given in Equation (11-25) (or by the equivalent transformation technique) is equal to the outlet steady-state concentration of a reactor described by the mixing distribution function $E(t)$ only for the case of first-order and zero-order reactions (see Problem 11-1). Although mixing distribution functions and the corresponding transformations may be applied to other types of distributions, this approach is not generally recommended because of the labor involved in the development of the transforms for the various types of distribution functions other than the perfect mixer.

The introduction of the perfect mixer transform made it possible to demonstrate that the mean concentration given by Equation (11-25) can be identified with the output of an actual system only for the case of a linear system. If the system is not linear, the perfect mixer transformation no longer yields a solution to the nonlinear differential equation, as demonstrated by Example 11-3. The lack of agreement between $\bar{C}_A$ given by Equation (A) and $(C_A)_{\text{out}}$, as given by Equation (B) in Example 11-2, has been interpreted by others in terms of macromixing effects.

Since the mean value of the concentration given by Equation (11-25) is equal to the outlet concentration only for the case of first- and zero-order isothermal reactions, it is not very useful in the modeling of reactors. The recommended approach is to find a mixing model which is in agreement with the experimentally determined residence time distribution function, $E(t)$. Then the actual reactor system is described by the formulation of the material and energy balances required to describe the model when a reaction is carried out. However, considerable care should be exercised in the use of this approach, as illustrated by the following example.

EXAMPLE 11-4

The residence time distribution function is the same for the following two types of reactor configurations. Consider the case of two reactors, a plug-flow reactor and a perfectly mixed reactor connected in series and operating isothermally. In case I, the sequence is the plug-flow reactor followed by the perfect mixer. In case II, the sequence is a perfect mixer followed by a plug-flow reactor.

For the second-order reaction $2A \rightarrow 2B$ and for $kC_A^0\theta = 5$, show that the conversion depends on the sequencing of the reactors; that is,

(a) Find the conversion for case I.
(b) Find the conversion for case II.

SOLUTION

(a) Case I, plug-flow reactor followed by a perfectly mixed reactor: For the plug-flow reactor,

$$\frac{C_{A1}}{C_A^0} = \frac{1}{1 + k_1 C_A^0 \theta_1} \tag{A}$$

and for the perfectly mixed reactor,

$$\frac{C_{A2}}{C_A^0} = \frac{-1 + \sqrt{1 + 4k_2 C_{A1}\theta_2}}{2k_2 C_A^0 \theta_2} \tag{B}$$

Since $k_1 = k_2$ and $\theta_1 = \theta_2$, C_{A1} may be eliminated from these equations to give

$$\frac{C_{A2}}{C_A^0} = \frac{-1 + \sqrt{1 + 4kC_A^0\theta\left(\dfrac{1}{1 + kC_A^0\theta}\right)}}{2kC_A^0\theta} \tag{C}$$

or

$$\frac{C_{A2}}{C_A^0} = \frac{-1 + \left[\dfrac{(1 + 5kC_A^0\theta)}{(1 + kC_A^0\theta)}\right]^{1/2}}{2kC_A^0\theta}$$

Then, for $kC_A^0\theta = 5$

$$\frac{C_{A2}}{C_A^0} = \frac{-1 + \sqrt{\dfrac{26}{6}}}{10} = \frac{1.08}{10} = 0.108$$

and thus the conversion

$$x = 1 - C_{A2}/C_A^0 = 1 - 0.108 = 0.892$$

for case I.

(b) Case II, perfectly mixed flow reactor followed by a plug-flow reactor: For the perfectly mixed flow reactor,

$$\frac{C_{A1}}{C_A^0} = \frac{-1 + \sqrt{1 + 4k_1 C_A^0 \theta_1}}{2k_1 C_A^0 \theta_1} \tag{D}$$

and for the plug-flow reactor,

$$C_{A2} = \frac{C_{A1}}{1 + k_2 C_{A1}\theta_2} \tag{E}$$

Since $k_1 = k_2$ and $\theta_1 = \theta_2$, C_{A1} may be eliminated from Equations (D) and (E) to give

$$\frac{C_{A2}}{C_A^0} = \frac{-1 + \sqrt{1 + 4k_c C_A^0\theta}}{kC_A^0\theta\left[1 + \sqrt{1 + 4k_c C_A^0\theta}\right]}$$

For $kC_A^0\theta = 5$, one obtains

$$\frac{C_{A2}}{C_A^0} = 0.128$$

and thus for case II,

$$x = 0.872$$

In cases I and II % Diff. $= \dfrac{0.892 - 0.872}{0.872} \times 100 = 2.3\%$

Hence, the order of mixing can affect the results even though the residence time distribution is the same. The deviation increases for smaller values of $kC_A^0\theta$.

The next section deals with the development of more elaborate models which may be used to account for the mixing resulting from turbulent fluid motion and diffusion.

II. DETERMINATION OF THE DISPERSION COEFFICIENTS AND NUMBER OF PERFECTLY MIXED TANKS REQUIRED TO DESCRIBE A REACTOR SYSTEM

In this section a further description of the experimental techniques and mathematical tools which may be used in the characterization of the mixing in reactors resulting from turbulent motion and diffusion is given. The results so obtained, such as the dispersion coefficients, are then used in the modeling of a reactor, as demonstrated in Chapters 2 and 6.

A. Experimental Determination of Dispersion Coefficients by Use of Tracer Techniques

To demonstrate the use of the tracer technique for the determination of dispersion coefficients, the experiments performed by Holland, Welch, et al. [10, 11, 12] are described. As shown in Figure 11-4, a salt tracer was injected at the inlet of a plate of a distillation column having ballast plates. The concentration of the salt pulse in the direction of flow was followed by use of conductivity cells. Typical of the output responses are those shown in Figure 11-5.

B. The Dispersion Model

The mixing on the plate of a distillation column results primarily from eddy diffusion caused by the turbulent motion of the fluid. The contribution of molecular diffusion is usually relatively small. As shown in Chapters 2 and 6,

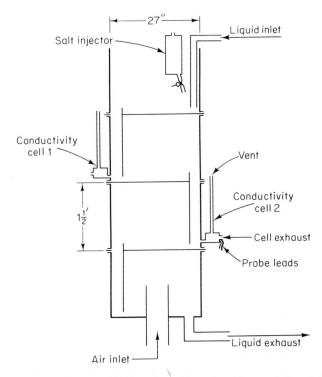

Figure 11-4. Sketch of distillation column (Reproduced by permission of the *AIChE.*).

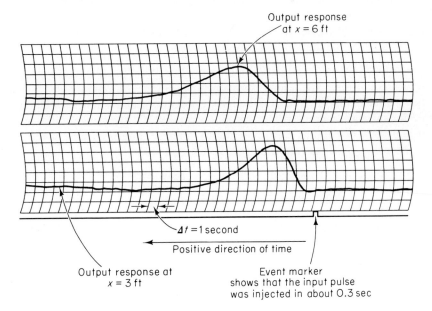

Figure 11-5. Typical output responses.

the combined effects of eddy and molecular diffusion are represented by Fick's law with an appropriate coefficient D, which is referred to as the dispersion coefficient. By use of this modified version of Fick's law and a material balance on the tracer, the following partial differential equation is obtained:

$$D\frac{\partial^2 C}{\partial z^2} - u\frac{\partial C}{\partial z} = \frac{\partial C}{\partial t} \tag{11-27}$$

where D = dispersion coefficient (also called eddy diffusivity), a constant with the dimensions of cm^2/s

t = time, s

u = mean linear velocity of the solvent flowing across the plate, cm/s

z = distance, cm; measured from the point of introduction of the pulse of tracer in the direction of flow of the liquid

C = concentration of the tracer, mol/cm^3

It should be noted that if the dispersion coefficient D and the velocity u are independent of concentration C, the model given by Equation (11-27) is classified as a linear system.

Equation (11-27) is based on the following assumptions:

1. The only concentration gradient existing on the plate is the one in the z direction, which implies perfect mixing in the vertical and horizontal directions that are perpendicular to the direction of flow.
2. The departure of the outer boundaries of the flow path from parallel lines is negligible.
3. The depth of liquid on the plate is taken to be the same for all z, which amounts to the assumption of a closed system.

Taken together, assumptions 2 and 3 imply that the cross-sectional area of the liquid holdup on the plate is independent of z. Equation (11-27), these assumptions, and the following boundary conditions and solutions are applicable for tubular reactors with or without a catalyst. Of the several different sets of boundary conditions corresponding to different interpretations of the input of the pulse, only the following set is presented [10].

$$
\begin{array}{lll}
(1) & C(z,0) = 0 & (z > 0) \\
(2) & \lim_{z \to \infty} C(z,t) = 0 & (t \geq 0) \\
(3) & C(0,T) = 0 & (t < t_0) \\
(4) & C(0,t) = \dfrac{A}{v_T \beta} & (t_0 \leq t \leq t_0 + \beta) \\
(5) & C(0,t) = 0 & (t > t_0 + \beta)
\end{array} \tag{11-28}
$$

As shown in reference [10], the solution which satisfies Equation (11-27) and this set of boundary conditions is as follows:

$$C(z, t) = \frac{zA}{2v_T \pi Dt^3} \exp\left[-\frac{(z - ut)^2}{4Dt}\right] \tag{11-29}$$

This result may be restated in the following dimensionless form,

$$\Phi(N, \tau) = \frac{N}{\pi \tau^3} \exp\left[-\frac{N(1 - \tau)^2}{\tau}\right] \tag{11-30}$$

by making the following changes in variables:

$$\bar{t} = \frac{V}{v_T} = \frac{z}{u} \text{ (s)} = \frac{\text{length of flow path}}{\text{mean velocity of solvent}} = \text{residence time}$$

$$N = \frac{uz}{4D} = \frac{N_{Pe}}{4} = \frac{\text{Peclet number}}{4}$$

$$\tau = t/\bar{t}$$

$$\Phi(N, \tau) = \frac{\bar{t}C(z, t)}{A/v_T}$$

A graph of $\Phi(N, \tau)$ for different choices of the Peclet number is shown in Figure 11-6.

When counterdiffusion at the inlet weir is permitted as proposed by Levenspiel and Smith [13], the following solution is obtained:

$$C(z, t) = \frac{A/2S}{\pi Dt} \exp\left[-\frac{(z - ut)^2}{4Dt}\right] \tag{11-31}$$

which is readily restated in the dimensionless form:

$$\Phi(z, \tau) = \frac{N}{\pi \tau} \exp\left[-\frac{N(1 - \tau)^2}{\tau}\right] \tag{11-32}$$

In the course of the investigation by Welch [11], solvent flow rates ranging from 26 to 120 gal/min and air rates ranging from 530 to 980 ft³/min (at 32°F and 1 atm) were employed. Solvents having surface tensions ranging from 22 to 72.7 dyn/cm, viscosities from 0.83 to 6.7 centipoise and densities 0.961 to 1.122 g/cm³ were employed.

Within the limits of experimental error and within the range of variables investigated ($4.4 \leq N_{Pe} \leq 27$), Welch found that N/z was independent of the solvent and air rates as well as the physical properties (surface tension, viscosity, density) of the solvent. On the basis of all the data for each location,

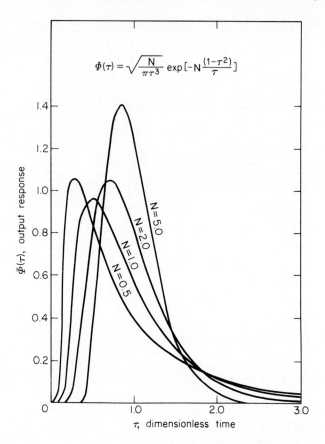

Figure 11-6. Variation of the output response of the diffusion model for different choices of the Peclet number ($N = N_{Pe}/4$).

it was found that

$$\frac{N}{z} = 0.85, \quad \text{for } z = 3 \text{ ft}, \qquad \frac{N}{z} = 0.87, \quad \text{for } z = 6 \text{ ft} \qquad (11\text{-}33)$$

The value of N in each instance was that which minimized the sum of the squares of the deviations [Φ (experimental) $-$ Φ (calculated)]. The standard deviations at $z = 3$ ft and $z = 6$ ft were ± 0.05 and ± 0.01, respectively. Comparison of the experimental and calculated values of Φ are presented in Figures 11-7 and 11-8.

The expression for N/z may be rearranged to give the final result:

$$D \left(\text{ft}^2/\text{s}\right) = \frac{z}{4N} u = 0.29u \text{ (ft/s)} \qquad (11\text{-}34)$$

An alternate method for determination of the Peclet number and other parameters is to evaluate the moments numerically by use of the output data.

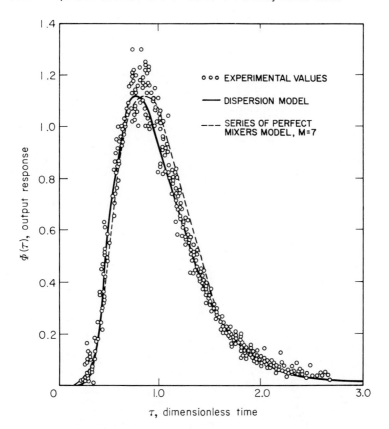

Figure 11-7. Predicted output and experimental responses for the dispersion model and pool model at $z = 3$ ft (Reproduced by permission of the *AIChE J.*).

As shown in Figure 11-9, the moment arm of the area $v_T C(t_k) \Delta t_k$ with respect to the $v_T C(t)$ axis is t_k, and the first moment of this area with respect to the $v_T C(t)$ axis is $t_k v_T C(t_k) \Delta t_k$. Let the t axis be subdivided into intervals of size Δt_i. Then the moment of each rectangle with base Δt_i and height $v_T C(t_i)$ is $t_i v_T C(t_i) \Delta t_i$. By definition of the definite integral, it follows that over the interval from $t = 0$ to $t = t_M$

$$\lim_{\substack{n \to \infty \\ \Delta t_i \to 0}} \sum_{i=1}^{n} t_i C(t_i) \Delta t_i = \int_0^{t_M} t v_T C(t) \, dt \qquad (11\text{-}35)$$

Then, as t_M is allowed to increase without bound, the infinite integral used to compute the first moment is obtained; that is,

$$\lim_{t_M \to \infty} \int_0^{t_M} t v_T C(t) \, dt = \int_0^{\infty} t v_T C(t) \, dt \qquad (11\text{-}36)$$

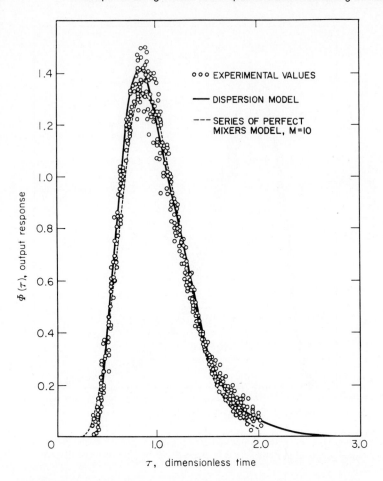

Figure 11-8. Predicted output and experimental responses for the dispersion model and pool model at $z = 6$ ft (Reproduced by permission of the *AIChE J.*).

By use of the *generalized mean-value theorem of integral calculus*, the following expression is obtained for the mean time $\bar{t}$, the moment arm which when used with the total area under the curve gives the same moment as the one computed by use of Equation (11-36):

$$\int_0^\infty t v_T C(t)\, dt = \bar{t} \int_0^\infty v_T C(t)\, dt \qquad (11\text{-}37)$$

or

$$\bar{t} = \frac{\displaystyle\int_0^\infty t C(t)\, dt}{\displaystyle\int_0^\infty C(t)\, dt} \qquad (11\text{-}38)$$

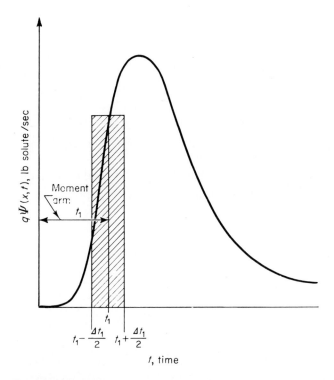

Figure 11-9. Moment arm of the area $v_T C(t_k)\,\Delta t_k$ with respect to the $v_T C(t)$ axis is t_k.

Since the moment arm of an area is equal to the age of the particles represented by the given area, the quantity $\bar{t}$ represents the mean time that the complete collection of particles making up the output response will stay in the system. The integral in the numerator of Equation (11-38) is called the first moment and the one in the denominator the zero moment and denoted by M_1 and M_0, respectively. Expressions for the second moment with respect to the $C(t)$ axis as well as the centroidal axis $\bar{t}$ are formulated in an analogous manner, and the corresponding expressions obtained for the mean square time $\overline{t^2}$ and variance $[\sigma_t^2 = \overline{(t - \bar{t})^2}]$ are given by

$$\overline{t^2} = \frac{\displaystyle\int_0^\infty t^2 C(z, t)\, dt}{\displaystyle\int_0^\infty C(z, t)\, dt}, \qquad \sigma_t^2 = \frac{\displaystyle\int_0^\infty (t - \bar{t})^2 C(z, t)\, dt}{\displaystyle\int_0^\infty C(z, t)\, dt} \qquad (11\text{-}39)$$

where $(t - \bar{t})$ is the moment arm with respect to the centroidal axis $\bar{t}$. As shown elsewhere [10], the moments are related to the parameters of the system as follows:

$$M_0 = \int_0^\infty C(z, t)\,dt = \frac{A}{v_T}$$

$$M_1 = \int_0^\infty tC(z, t)\,dt = \left(\frac{A}{v_T}\right)\frac{z}{u} \tag{11-40}$$

$$M_2 = \int_0^\infty t^2 C(z, t)\,dt = \frac{A}{v_T}\left(\frac{z}{u}\right)^2\left(1 + \frac{1}{2N}\right)$$

By use of these relationships and Equations (11-38) and (11-39), the mean time $\bar{t}$, mean square time $\overline{t^2}$, and variance σ_t^2 may be expressed in terms of the parameters of the system as follows:

$$\bar{t} = \frac{z}{u} = \frac{V}{v_T}$$

$$\overline{t^2} = (\bar{t})^2\left(1 + \frac{1}{2N}\right) \tag{11-41}$$

$$\sigma_t^2 = \frac{(\bar{t})^2}{2N}$$

The three moments M_0, M_1, and M_2 are seen to completely describe the dispersion model; that is, all higher-order moments may be expressed in terms of M_0, M_1, and M_2. Although the mixing is well described by the dispersion model and the boundary conditions, it is also equally well described by other sets of boundary conditions [10].

Instead of using a continuous model to describe the mixing, a series of perfect mixers may be used, as shown next.

C. Mixing Model: Series of Perfect Mixers

Figure 11-10 shows the model, which consists of a series of perfect mixers. Foss et al. [1] and Gautreaux and O'Connell [14] used this model to correlate the experimental results of mixing on the plate of a distillation column. The differential equations describing this model are as follows:

$$V\frac{dC_1(t)}{dt} = v_T[C_i(t) - C_1(t)]$$

$$V\frac{dC_m(t)}{dt} = v_T[C_{m-1}(t) - C_m(t)], \qquad m = 2, 3, \ldots, M \tag{11-42}$$

where V is the volumetric holdup of each mixer and $C_i(t)$ is the concentration of the input pulse of tracer. (Note that if the system under consideration has a total volumetric holdup of V_T then, for a model consisting of M perfect

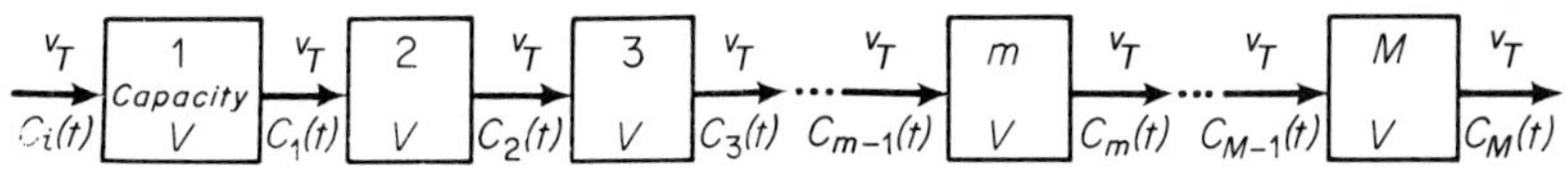

Figure 11-10. Series of perfect mixers.

mixers, the volumetric holdup per mixer is given by $V = V_T/M$.) The initial conditions are as follows:

$$C_i(t) = 0, \qquad t < 0$$

$$C_i(t) = \frac{A}{v_T \beta}, \qquad 0 \leq t \leq \beta \qquad (11\text{-}43)$$

$$C_i(t) = 0, \qquad t > \beta$$

The solution which satisfies both the differential equations and the initial conditions is given by

$$C_M(t) = \left(\frac{A/v_T}{\bar{t}^M}\right) \frac{t^{M-1}e^{-t/\bar{t}}}{(M-1)!} \qquad (11\text{-}44)$$

where
$$\bar{t} = \frac{V}{v_T} = \frac{\text{volumetric holdup of each mixer}}{\text{total volumetric flow rate}}$$

$$= \text{residence time}$$

When stated in dimensionless form, Equation (11-44) becomes

$$\Phi_M(\tau) = \frac{\tau^{M-1}e^{-\tau}}{(M-1)!} \qquad (11\text{-}45)$$

where
$$\tau = \frac{t}{\bar{t}}, \qquad \Phi_M(\tau) = \frac{\bar{t}}{A/v_T}C_M(\tau)$$

A graph of $\phi_M(\tau)$ as a function of the parameter M appears in Figure 11-11. At $z = 3$ ft and $z = 6$ ft, values of $M = 7$ and $M = 10$ gave the best fits of the data. The curve fits for $M = 7$ and $M = 10$ for $z = 3$ ft and $z = 6$ ft, respectively, are shown in Figures 11-7 and 11-8.

An extension of this model whereby a stream is recycled from each mixer to the previous one was originally proposed by Roemer and Durbin [15] and has been used by a number of investigators.

1. Use of the Results of Tracer Experiments in the Modeling of Chemical Reactors. If it is desired to model the mixing effects in a reactor by use of the dispersion model, the dispersion coefficient D should be determined for the reactor as demonstrated previously. The resulting dispersion coefficient so obtained is then used in the model for the reactor.

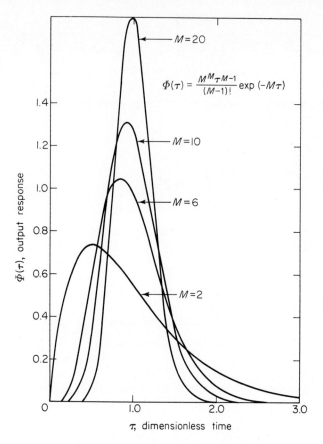

$$\Phi(\tau) = \frac{M^M \tau^{M-1}}{(M-1)!} \exp(-M\tau)$$

Figure 11-11. Profiles of the series of perfect mixers of different values of the parameter M.

Similarly, if the reactor is to be represented by a series of perfect mixers, then the number of mixers required to describe the mixing effects in the reactor is determined by use of tracer experiments as described previously. The number of mixers so obtained is then used in the modeling of the reactor.

2. Use of Residence Time Distribution as a Diagnostic Tool. Tracer response data may be used to determine the extent of bypassing and existence of stagnant regions in a continuous flow reactor [16]. Naor and Shinnar [17] have presented a summary of response curves which might be obtained for several kinds of reactors and suggest the use of the intensity function to gain physical insight into the mixing process. They also present an operational definition of stagnant zones. Wolf and Resnick [18] discuss residence time distribution in real systems. For the cases of plug flow, perfectly mixed flow, plug flow with stagnant zones, perfectly mixed with dead zones, plug flow with bypassing, perfectly mixed with bypassing, perfectly mixed with an error in the mean residence time, plug flow with perfectly mixed, and perfectly mixed with lag in the response, the response curves can be described by

$$\log_e C/C^0 = -\eta\frac{tv_T}{V} + \eta\frac{\epsilon v_T}{V} \tag{11-46}$$

where η is the slope and $\epsilon v_t/V$ is the system phase shift. By comparison of Equation (11-46) with Equation (11-13) after the logarithms of both sides of Equation (11-13) have been taken demonstrates that $\eta = 1$ and $\epsilon = 0$ for the perfectly mixed flow reactor. Wolf and Resnick [18] summarized the meaning of the parameters as follows:

	η	$\epsilon v_T/V$
Perfect mixing	1	0
with plug flow	> 1	> 0
with dead space	> 1	0
with bypassing	< 1	< 0
with error in v_T/V	$\lessgtr 1$	0
with system lag	1	> 0

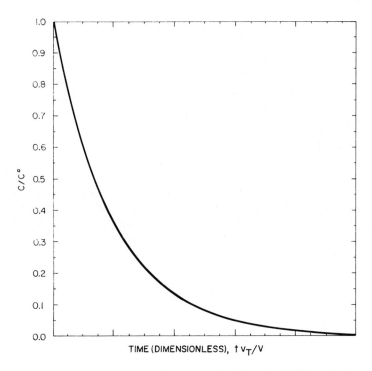

Figure 11-12. Typical response to a unit step change for the Berty reactor using 200 g of chabazite AW500 powder (10 to 100 μm).

TABLE 11-1
RESIDENCE TIME DISTRIBUTION EXPERIMENTAL RESULTS FOR THE BERTY REACTOR
WITH 200 g OF CHABAZITE POWDER OF 30 TO 100 MICROMETER PARTICLE SIZE
200 g OF $\frac{1}{16}$-INCH PELLETS, AND EMPTY

Run	Flow rate,* cm^3 / s	Impeller Speed, rev / s	Temperature, K	Pressure, kN / m^2	η	$\epsilon / \bar{t}$	Catalyst
1	5.108	0.0	769.5	102.4	0.8317	-0.0244	Powder
2	5.108	5.8	769.8	102.4	0.9436	0.0521	Powder
3	5.108	9.9	769.9	102.4	0.9595	0.0510	Powder
4	81.214	9.9	769.4	154.7	0.7930	-0.5177	Powder
5	18.981	5.8	292.0	119.7	0.9571	-0.0601	Powder
6	18.970	9.9	292.0	119.8	1.0028	-0.0004	Powder
7	69.740	5.8	580.8	202.3	0.7822	-0.3848	Powder
8	69.808	9.9	580.8	202.3	0.8408	-0.2855	Powder
9	4.556	11.8	689.0	102.1	0.9936	0.0045	Powder
10	0.644	11.8	681.1	715.5	1.0042	0.0103	Powder
11	4.597	9.9	691.5	102.2	1.0031	0.0612	Pellets
12	81.987	9.9	830.2	257.7	1.0016	-0.0310	Empty
13	83.703	0.0	849.3	257.6	1.0032	-0.0082	Empty

*The flow rate is for the temperature and pressure given in this table.

For kinetic studies, an ideal reactor (plug flow or perfectly mixed flow) should be used. Tracer response data can be used to determine whether the proposed operating conditions in the reactor will yield the ideal reactor.

Vera-Castaneda [19] conducted a series of tracer response studies for a 1.146-liter Berty reactor to determine if it could be operated as a perfectly mixed flow reactor. The Berty reactor is an internal recycle reactor. The experiments were conducted for a range of temperatures, pressures, impeller speeds, and flow rates. The experiments were conducted by feeding nitrogen which contained approximately 0.05 mole fraction of helium. As a steady state was approached, the helium concentration was reduced to zero. A typical response curve is illustrated in Figure 11-12 for run 6 listed in Table 11-1, where the flow rates, impeller speed, temperature, pressure, type of catalyst, and the values of η and $\epsilon v_T/V$ or $\epsilon/\bar{t}$ are presented. Bypassing appeared to have occurred in the experiments conducted at low impeller speeds, high temperatures, and low pressures. Perfect mixing appears to have been obtained for runs 6 and 9 to 13. Perfect mixing with plug flow or system lag appear to have occurred in run 11 when pellets were used. The effect of pressure and temperature can be combined into the one variable, density. Analysis of these data indicates that experiments conducted with impeller speeds greater than 11.4 revolutions per second and gas densities greater than 5×10^{-4} g/cm^3 should result in perfectly mixed reactor conditions. Vera-Castaneda conducted most of his catalyst studies at impeller speeds of 20 to 25 revolutions per second, pressures of 500 to 700 kPa, and temperatures of 650 and 750 K. Examination of the catalyst after a series of experiments indicated no maldistribution of the flow.

PROBLEMS

11-1. (a) Show that for the case of a zero-order reaction the mean value $\bar{C}_A/C_A^0$ found by use of the perfect mixer transformation of the differential equation

$$-\frac{dC_A}{dt} - k = 0 \tag{A}$$

is equal to the mean value obtained by integrating Equation (11-23) with $C_A(t)$ set equal to the solution of Equation (A); that is, $C_A(t) = C_A^0 - kt$.

(b) Show that when the zero-order reaction is carried out in a perfect mixer the expression obtained for the outlet concentration $(C_A)_{\text{out}}$ is the same as that obtained in part (a) for $\bar{C}_A$.

11-2. (a) Develop the mixing model for the case where two perfect mixers of equal volume V are connected in series. At time $t = 0 -$, suppose N_1^0 moles are added to the first mixer to give it a concentration of $C_1^0 = N_1^0/V$. The volumetric flow rate of solvent to and from the system is constant at v_T. Obtain the following differential equations by making a material balance on each of the two mixers:

$$-n_1(t) = \frac{dN_1(t)}{dt}$$

$$n_1(t) - n_2(t) = \frac{dN_2(t)}{dt}$$

where $n_1(t)$ and $n_2(t)$ are the molar flow rates at which the tracer leaves mixers 1 and 2, respectively, at any time t. Show the distribution function for this system is given by

$$E(t) = \frac{t}{\bar{t}} e^{-t/\bar{t}}$$

(b) Show for the case where the first-order reaction $A \rightarrow$ products $(C_A = C_A^0 e^{-kt})$ occurs in this system that the mean value $\bar{C}_{A2}$ given by Equation (11-24) is equal to the steady-state value of the outlet concentration when the reaction is carried out (or modeled) by two perfectly mixed reactors in series.

11-3. Develop the residence time distribution functions for the case where two equal-size reactors are operated isothermally in series in two different ways. Case the I: A plug-flow reactor is followed by a perfectly mixed reactor. Case II: A perfectly mixed reactor is followed by a plug-flow reactor.

11-4. Read points from the response curve in Figure 11-12 and evaluate the parameters in Equation (11-46). Calculate the error in the parameters that is introduced by reading the points from the graph by comparing your answers with the results reported for run 6 in Table 11-1.

NOTATION

A = moles of tracer

$C(t)$ = concentration of tracer

$C_A(t)$ = concentration of solute A; C_A^0 = initial concentration of solute A

$C_M(t)$ = concentration of the tracer in the Mth perfectly mixed pool at time t

D = dispersion coefficient, cm^2

$E(t)$ = normalized part of a probability function

m = counting integer, used to denote the number of a mixed pool

M_n = nth moment for the dispersion model

N_{Pe} = Peclet number

$N = N_{Pe}/4$

$n(t)$ = molar flow rate of solute at time t

$n^0(t)$ = molar flow rate of the tracer

$n_A(t)$ = molar flow rate of solute A

P_t = probability of a molecule or small element of liquid having an age between t and $t + dt$

t = time, residence time, age

$\bar{t}$ = mean time or age that all elements remain in the tube or reactor; also used to denote residence time

$\overline{t^2}$ = mean square time

$(t - \bar{t})^2$ = second moment of an element with respect to the centroidal axis $\bar{t}$

$\overline{(t - \bar{t})^2}$ = mean value of $(t - \bar{t})^2$; also called the variance

u = linear velocity of the solvent

v_T = total volumetric flow rate of the solvent

V = volumetric holdup of a pool of the perfectly mixed pool model

z = length of flow path or reactor

Greek Letters

β = period of time required to inject the tracer into the system

$\sigma_t^2(z)$ = variance

τ = dimensionless time

$\Phi_M(\tau)$ = normalized outlet concentration of tracer from the perfectly mixed model

REFERENCES

1. Foss, A. S., J. A. Gerster, and R. L. Pigford, *AIChE J.*, 4:231 (1958).

2. Stewart, G. H., *Physiology*, 15:1 (1894).

3. MacMullin, R. B., and W. Weber, Jr., *Chem. Eng. Sci.*, 7:187 (1958).

4. Gilliland, E. R., and E. A. Mason, *Ind. Eng. Chem.*, 44:218 (1952).

5. Gilliland, E. R., and E. A. Mason, *Ind. Eng. Chem.*, *41*:1191 (1949).

6. Danckwerts, P. V., *Chem. Eng. Sci.*, *2*:1 (1953).

7. Danckwerts, P. V., *Chem. Eng. Sci.*, *9*:78 (1958).

8. Zwietering, T. N., *Chem. Eng. Sci.*, *11*:1 (1959).

9. Levenspiel, O., *Chemical Reaction Engineering*, 2nd ed., John Wiley & Sons, Inc., New York (1972).

10. Holland, C. D., *Unsteady State Processes with Applications in Multicomponent Distillation*, Prentice-Hall, Inc., Englewood Cliffs, N.J. (1966).

11. Welch, H. E., Ph.D. dissertation, Texas A & M University, College Station, Tex., 1963.

12. Welch, H. E., L. D. Durbin, and C. D. Holland, *AIChE J.*, *10*:373 (1964).

13. Levenspiel, O., and W. K. Smith, *Chem. Eng. Sci.*, *6*:227 (1957).

14. Gautreaux, M. F., and H. E. O'Connell, *Chem. Eng. Progr.* 51 (No. 5):232 (1955).

15. Roemer, M. H., and L. D. Durbin, *Ind. Eng. Chem.*, *6*:120 (1967).

16. Himmelblau, D. M., and K. B. Bischoff, *Process Analysis and Simulation in Deterministic Systems*, John Wiley & Sons Inc., New York (1968).

17. Naor, P., and R. Shinnar, *Ind. Eng. Chem. Fundamentals*, 2(4):278 (1963).

18. Wolf, D., and W. Resnick, *Ind. Eng. Chem. Fundamentals*, 2(4):287 (1963).

19. Vera-Castaneda, E., *Study of the Methanol Conversion to Ethylene and Propylene Using Small Pore Size Zeolites*, Ph.D. thesis, Texas A & M University, College Station, Tex. (May, 1985).

Theory of   **12**
Reaction Rates

The concept of first-order reaction rates was proposed by Wilhelmy in 1850 to explain the experimental observation of the inversion of cane sugar. In the presence of an aqueous solution of a dilute acid, the concentration of the sucrose was found to decrease exponentially with time to form glucose and fructose.

During the 15-year period from 1864 to 1879, Guldberg and Waage proposed and applied the *principle of mass action* to reaction rates and chemical equilibria. If a reaction goes precisely as written, then the law of mass action asserts that the rate is proportional to the product of the concentration of the reactants raised to powers equal to their respective stoichiometric numbers. For example, at high pressures, the reaction

$$2NO + O_2 \rightleftarrows 2NO_2$$

goes as written. By the principle of mass action, the forward rate of a reaction of NO is given by

$$r_f = k_c C_{NO}^2 C_{O_2} \tag{12-1}$$

and the rate of appearance of NO by reverse reaction is

$$r_r = k_c' C_{NO_2} \tag{12-2}$$

Then the net rate of disappearance of NO is given by

$$r_{NO} = r_f - r_r \tag{12-3}$$

518

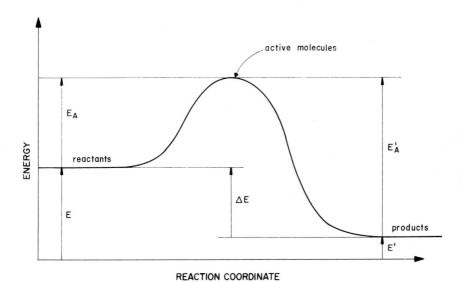

Figure 12-1. Relationship between the heat of reaction at constant volume ΔE and the energies of activation E_A and E_A'.

At equilibrium, $r_{NO} = 0$, and

$$K_c = \frac{k_c}{k_c'} = \frac{C_{NO_2}^2}{C_{NO}^2 C_{O_2}} \tag{12-4}$$

In 1889, Arrhenius proposed that molecules must become activated before they react. To relate the energy of activation to the rate constant, Arrhenius commenced with the Van't Hoff equation.

$$\frac{d \log_e K_c}{dt} = \frac{\Delta E}{RT^2}, \qquad \text{at constant volume} \tag{12-5}$$

where ΔE = change in internal energy when the reaction goes one time;
          $\Delta E = E'$ (products) $- E$ (reactants)

The relationship between the internal energies of the products and reactants is displayed in Figure 12-1. The term *reaction coordinate* is loosely used to mean the energy levels between reactants and products and to demonstrate the relationship

$$E + E_A = E' + E_A' \tag{12-6}$$

$$\Delta E = E' - E = E_A - E_A' \tag{12-7}$$

The internal energy E of the reactants may be regarded as the average energy of all reactant molecules, and E_A the average energy of the *active molecules*. The same interpretations are applicable for E' and E_A'. The active molecules are those that possess enough energy to undergo reaction. As a further interpretation of the reaction process, Arrhenius proposed that in a reacting system an equilibrium exists between the reactant molecules and the active molecules.

Since $K_c = k_c/k_c'$, Equations (12-5) and (12-7) may be combined to give

$$\frac{d \log_e k_c}{dT} - \frac{d \log_e k_c'}{dT} = \frac{E_A}{RT^2} - \frac{E_A'}{RT^2} \tag{12-8}$$

Then, in general,

$$\frac{d \log_e k_c}{dT} = \frac{E_A}{RT^2} + f(T) \tag{12-9}$$

and

$$\frac{d \log_e k_c'}{dT} = \frac{E_A'}{RT^2} + f(T) \tag{12-10}$$

Note that the same function $f(T)$ is required for each equation in order to satisfy the thermodynamic relationship, Equation (12-8), or more precisely Equation (12-5). Integration of Equation (12-9) yields

$$\log_e k_c = \frac{-E_A}{RT} + \int f(T)\, dT + \log_e A \tag{12-11}$$

where $\log_e A$ is the constant of integration. In many cases, experimental results may be described by taking

$$\int f(T)\, dT = 0 \tag{12-12}$$

On the basis of this assumption, Equation (12-11) reduces to the relationship commonly called the Arrhenius equation:

$$k_c = Ae^{-E_A/RT} \tag{12-13}$$

The concept of the activated state carries over both into the *collision theory* for bimolecular reactions and into the *transitions state theory*. These are the two principal theories that have been advanced to explain the rates at which reactions proceed. The collision theory for bimolecular reactions makes use of relationships belonging to the body of knowledge known as the *kinetic theory*.

I. KINETIC THEORY

Since concepts of the kinetic theory of gases are useful not only in the interpretation of reaction rates but also in the development of rate expressions for adsorption, the development of the Maxwellian distribution of velocities as presented follows closely the proof first advanced by Maxwell around 1860. A more general derivation based on the study of equilibrium states existing among the molecules making elastic impacts was presented by Boltzmann toward the end of the 19th century.

In the kinetic theory of gases, the formulas used to express the probability of the occurrence of a given event are of the same form as that developed in Chapter 11 for a molecule or globule of liquid leaving a tube to have a residence time or age lying between t_k and $t_k + dt_{kj}$; that is,

$$P_{t_k} = E(t_k)\, dt_k \tag{12-14}$$

A. Maxwell's Deduction of the Distribution of Velocities

If one measured the speed of each automobile passing a given reference line on a highway, it would be found that the speeds would range from some small value to speeds greater than the maximum speed limit. The average of all observed speeds would probably lie somewhere in the near neighborhood of the maximum speed limit.

A similar though more random behavior is exhibited by a collection of gas molecules in a closed vessel. The distribution of velocities of such a collection of molecules was deduced by Maxwell on the basis of the following postulates.

1. It is postulated that the system is at steady state, the velocity distribution is independent of time, the observed temperature of the mixture is constant, the total pressure is fixed, the mixture is not subject to any type of force field which would promote one direction of motion over another, and the mixture consists of monatomic gas molecules of the same type.
2. Let the scalar value of the velocity vector of a molecule in three-dimensional space be denoted by c. Let the velocity components along the x, y, z axes of a rectangular coordinate system be denoted by u, v, and w, respectively. It is postulated that u, v, and w are mutually independent, and that all directions of the component velocities are equally likely. The velocity c is related to the component velocities as follows:

$$c^2 = u^2 + v^2 + w^2 \tag{12-15}$$

3. It is postulated that the probability that a molecule has a velocity u in the x-direction is a function of u alone, that it has a velocity v in the y-direction is a function of v alone, and that it has a velocity w in the z-direction is a function of w alone. Thus, the probability that the velocity u lies between u and $u + du$ is given by

$$P_u = f(u)\, du \tag{12-16}$$

Similarly, the probability that v lies between v and $v + dv$, and that w lies between w and $w + dw$ is given, respectively, by

$$P_v = f(v)\, dv \tag{12-17}$$

$$P_w = f(w)\, dw \tag{12-18}$$

Since there is no preferred direction in space, the functions $f(u)$, $f(v)$ and $f(w)$ are necessarily of the same form. Consequently,

$$\int_{-\infty}^{\infty} f(\zeta)\, d\zeta = 1 \tag{12-19}$$

where ζ may be set equal to u, v, or w. That is, the probability that a given component velocity has some value between $-\infty$ and $+\infty$ is a certainty.

Since the probabilities P_u, P_v, and P_w are mutually independent, the probability W that a molecule has velocity components simultaneously between u and $u + du$, v and $v + dv$, and w and $w + dw$ is equal to the product $P_u P_v P_w$, or

$$W(u, v, w) = f(u)f(v)f(w) \, du \, dv \, dw \qquad (12\text{-}20)$$

This equation also represents the probability that a particular velocity vector (whose components are u, v, and w) will terminate in the volume $du \, dv \, dw$.

It follows from the basic postulates that all velocity directions are equally probable. Then the probability that a molecule has a velocity c, regardless of direction, is a function of c alone; that is, $F(c)$, and the probability W that c terminates in the volume $du \, dv \, dw$ is given by

$$W = F(c) \, du \, dv \, dw \qquad (12\text{-}21)$$

Comparison of Equations (12-20) and (12-21) shows that

$$f(u)f(v)f(w) = F(c) = F(\sqrt{u^2 + v^2 + w^2}) \qquad (12\text{-}22)$$

Then

$$\log_e f(u) + \log_e f(v) + \log_e f(w) = \log_e F(c) \qquad (12\text{-}23)$$

Since u, v, and w are independent, differentiation of Equation (12-23) with respect to u gives

$$\frac{d \log_e f(u)}{du} = \left[\frac{d \log_e F(c)}{dc} \right] \frac{dc}{du} = \left[\frac{d \log_e F(c)}{dc} \right] \frac{u}{c} \qquad (12\text{-}24)$$

where dc/du was evaluated by use of Equation (12-15). Differentiation of Equation (12-23) with respect to v and w yields expressions of the same form as Equation (12-24). Thus

$$\frac{1}{u} \frac{d \log_e f(u)}{du} = \frac{1}{v} \frac{d \log_e f(v)}{dv} = \frac{1}{w} \frac{d \log_e f(w)}{dw} = \frac{1}{c} \frac{d \log_e F(c)}{dc} \qquad (12\text{-}25)$$

Since u, v, and w are independent, the only way that the equality given by Equation (12-25) can hold is for each term to be equal to a constant, say $-2/\alpha^2$. The negative sign is selected in order that Equation (12-19) be satisfied. Then

$$\int \frac{d \log_e f(u)}{du} \, du = -\frac{2}{\alpha^2} \int u \, du + \log_e A \qquad (12\text{-}26)$$

where $\log_e A$ is the constant of integration. Integration of this expression as well as the corresponding ones for v and w yields

$$f(u) = Ae^{-u^2/\alpha^2}$$

$$f(v) = Ae^{-v^2/\alpha^2} \tag{12-27}$$

$$f(w) = Ae^{-w^2/\alpha^2}$$

Since functions $f(v)$, $f(u)$, and $f(w)$ are of the same form, it follows that the constant of integration $\log_e A$ is the same for each function. Also, in view of Equations (12-15), (12-22), and (12-27), it follows that

$$F(c) = A^3 e^{-c^2/\alpha^2} \tag{12-28}$$

The constant A is evaluated by use of Equations (12-19) and (12-27); that is,

$$\int_{-\infty}^{\infty} Ae^{-u^2/\alpha^2} \, du = 1 \tag{12-29}$$

Integration of this expression gives

$$A = \frac{1}{\alpha\sqrt{\pi}} \tag{12-30}$$

and, consequently, $f(u)$, $f(v)$, and $f(w)$ are of the form

$$f(u) = \left(\frac{1}{\alpha\sqrt{\pi}}\right) e^{-u^2/\alpha^2} \tag{12-31}$$

and

$$F(c) = \frac{1}{\alpha^3 \pi^{3/2}} e^{-c^2/\alpha^2} \tag{12-32}$$

1. Evaluation of α. By use of the result given by Equation (12-32) for the probability that a molecule has a velocity c regardless of direction, it is possible to compute the distribution of velocities in a gaseous mixture containing N_T molecules. That is, all the molecules that have velocity vectors between c and $c + dc$ terminate in a spherical shell described about the origin by the radii c and $c + dc$. Since the volume of this shell is $4\pi c^2 \, dc$, the fraction of the molecules having a velocity between c and $c + dc$ is given by

$$\frac{dN_c}{N_T} = F(c)\left[4\pi c^2 \, dc\right] \tag{12-33}$$

or

$$W = \frac{dN_c}{N_T} = \left[\frac{1}{\alpha^3 \pi^{3/2}}\right] e^{-c^2/\alpha^2} 4\pi c^2 \, dc \tag{12-34}$$

where dN_c = molecules having a velocity between c and $c + dc$

To evaluate α that appears in Equation (12-32), the translational or kinetic energy of a mixture of molecules with the distribution of velocities given by Equation (12-34) is computed as follows. Let $\bar{E}_{tr}$ denote the total kinetic energy per molecule in the mixture. Then

$$N_T \bar{E}_{tr} = \int_0^{N_T} \frac{mc^2}{2} dN_c \tag{12-35}$$

where
$$m = \text{mass per molecule}$$

By the *mean-value theorem of integral calculus*, it follows that

$$\int_0^{N_T} \frac{mc^2}{2} dN_c = \frac{\overline{mc^2}}{2} N_T \tag{12-36}$$

where
$$\overline{c^2} = \text{mean square velocity}$$

$$\sqrt{\overline{c^2}} = \text{root mean square velocity}$$

By rearranging Equation (12-36), the following expression is obtained for $\overline{c^2}$:

$$\overline{c^2} = \int_0^{N_T} c^2 \frac{dN_c}{N_T}$$

Elimination of dN_c/N_T from this expression by use of Equation (12-34) yields

$$\overline{c^2} = \frac{4}{\alpha^3 \sqrt{\pi}} \int_0^\infty e^{-c^2/\alpha^2} c^4 \, dc \tag{12-37}$$

After the integration indicated has been carried out, the following result is obtained:

$$\overline{c^2} = \frac{3\alpha^2}{2} \tag{12-38}$$

In a similar manner, the kinetic energy in each direction is computed. For example, in the x-direction,

$$\overline{u^2} = \frac{1}{\alpha \sqrt{\pi}} \int_{-\infty}^\infty u^2 e^{-u^2/\alpha^2} \, du \tag{12-39}$$

Thus

$$\overline{u^2} = \frac{\alpha^2}{2}, \qquad \overline{v^2} = \frac{\alpha^2}{2}, \qquad \overline{w^2} = \frac{\alpha^2}{2} \tag{12-40}$$

It should be noted at this point that Equations (12-38) and (12-40) imply equipartition of kinetic energy; that is, one-third of the kinetic energy of the mixture is distributed along each axis.

$$\frac{N_T m \overline{u^2}}{2} + \frac{N_T m \overline{v^2}}{2} + \frac{N_T m \overline{w^2}}{2} = \frac{N_T m}{2} \left(\overline{u^2} + \overline{v^2} + \overline{w^2} \right) = \frac{n_T m}{2} \overline{c^2} \tag{12-41}$$

where the final equality follows from Equations (12-38) and (12-40).

   Thus the translational energy (the kinetic energy) may be divided equally among the three directions of motion or freedom. Therefore,

$$\bar{E}_{tr} = 3\varepsilon_{tr} \qquad (12\text{-}42)$$

where ε_{tr} = translational energy per molecule per degree of freedom

From a kinetic interpretation of gas pressure, Joule first showed in 1851 that

$$PV = \tfrac{1}{3}N_T\overline{mc^2} \qquad (12\text{-}43)$$

Since experiments show that $PV = N_T kT$, it follows from Equations (12-35), (12-36), and (12-43) that

$$PV = \frac{2}{3}\left(\frac{N_T\overline{mc^2}}{2}\right) = \frac{2}{3}\left(N_T\bar{E}_{tr}\right) = N_T kT \qquad (12\text{-}44)$$

where k = Boltzmann constant = 1.3805×10^{-16} erg/(K molecule)

Thus

$$\frac{2}{3}\left(N_T 3\varepsilon_{tr}\right) = N_T kT$$

or

$$\varepsilon_{tr} = \frac{kT}{2} \qquad (12\text{-}45)$$

or the translational energy (kinetic energy per degree of freedom) is $kT/2$. Thus

$$\frac{\overline{mc^2}}{2} = \frac{3kT}{2} \qquad (12\text{-}46)$$

Elimination of $\overline{c^2}$ from Equations (12-38) and (12-46) gives

$$\alpha^2 = \frac{2kT}{m} \qquad (12\text{-}47)$$

for a system with three degrees of freedom. Thus Equations (12-31) and (12-34) may be stated in the form

$$f(u) = \left(\frac{m}{2\pi kT}\right)^{1/2} e^{-mu^2/2kT} \qquad (12\text{-}48)$$

and

$$\frac{dN_c}{N_T} = \left(\frac{m}{2\pi kT}\right)^{3/2} e^{-mc^2/2kT} 4\pi c^2\, dc \qquad (12\text{-}49)$$

B. The Average and Most Probable Velocities

The *average velocity*, c_{av}, is defined by

$$c_{av} N_T = \int_0^{N_T} c\, dN_c \qquad (12\text{-}50)$$

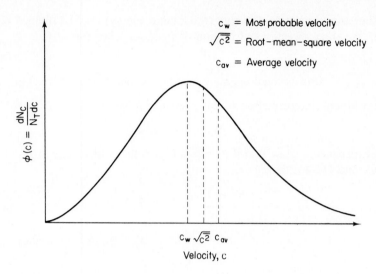

Figure 12-2. Maxwellian distribution velocities c for motion in three dimensions.

This equality is seen to follow immediately from the *mean-value theorem of integral calculus*. Then

$$c_{av} = \int_0^{N_T} c \, \frac{dN_c}{N_T} = 4\pi \left(\frac{m}{2\pi kT} \right)^{3/2} \int_0^\infty e^{-mc^2/2kT} c^3 \, dc \qquad (12\text{-}51)$$

Integration yields

$$c_{av} = \sqrt{\frac{8kT}{\pi m}} \qquad (12\text{-}52)$$

The root mean square velocity, $\sqrt{\overline{c^2}}$, the average velocity, c_{av}, and the most probable velocity, c_w, are compared in the graph shown in Figure 12-2. The most probable velocity is obtained as follows. Let

$$\frac{dN_c}{N_T \, dc} = \phi(c) = \left(\frac{m}{2\pi kT} \right)^{3/2} e^{-mc^2/2kT} 4\pi c^2 \qquad (12\text{-}53)$$

The *most probable velocity* is that value of c possessed by the greatest number of the molecules of a given mixture. At this value of c,

$$\frac{d\phi(c)}{dc} = 0 \qquad (12\text{-}54)$$

When this differentiation is carried out, it is found that the most probable velocity, c_w, is given by

$$c_w = \sqrt{\frac{2kT}{m}} \qquad (12\text{-}55)$$

The *mean relative velocity* $\bar{r}$ for two different types of molecules A and B takes into account the complete distribution of velocities of these molecules in space as well as their difference in mass. The somewhat lengthy derivation of $\bar{r}$ is not presented, but it is generally available [1, 2]. The result is

$$\bar{r} = \sqrt{\frac{8kT}{\pi\mu}} \tag{12-56}$$

where $$\mu = \frac{m_A m_B}{m_A + m_B}$$

m_A, m_B = mass per molecule of A and B, respectively

For two like molecules, $\mu = m/2$ and Equation (12-56) reduces to

$$\bar{r} = \sqrt{2}\, c_{av} \tag{12-57}$$

C. Postulates of the Kinetic Theory of Bimolecular Reactions

The postulates for the kinetic theory of bimolecular reactions are as follows:

1. A collision between the two reacting molecules A and B must occur before they can react.
2. Of the total number of collisions, only those molecules react that have a kinetic energy equal to or greater than ε along the line of centers at the moment of contact, provided that the molecules are properly orientated.

The number of collisions of molecules A and B per second per cubic centimeter with a kinetic energy equal to or greater than ε along the line of centers at the time of collision may be calculated directly by use of the distribution laws as shown by Frost and Pearson [1]. Alternately, the number of two-body collisions may be computed and then the fraction f of the molecules having an energy equal to or greater than ε may be computed. The latter procedure is used. The rate of reaction of A or B is then given by

$$r_A \left(\frac{\text{molecules of } A \text{ reacting}}{\text{s cm}^3} \right) = r_B = pZ_{AB}f \tag{12-58}$$

where Z_{AB} = number of two-body collisions between two unlike molecules A and B per s/cm^3

p = fraction of the molecules that collide that have the proper orientation

D. Number of Two-Body Collisions per Second per Cubic Centimeter

Consider first two unlike molecules A and B with diameters σ_A and σ_B, respectively. A collision is said to have occurred when there is contact between the surface of molecule A and the surface of molecule B. Consider a molecule

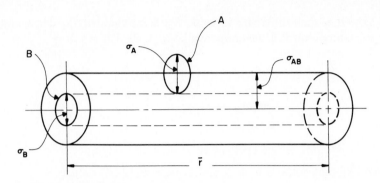

Figure 12-3. Cylinder containing the centers of the molecules that will collide per second.

of type A moving in an arbitrary direction with a mean relative velocity $\bar{r}$ with respect to a molecule of type B. If the centers of the two molecules come within a distance

$$\sigma_{AB} = \frac{\sigma_A + \sigma_B}{2} \qquad (12\text{-}59)$$

of one another, a collision will occur. During a period of 1 s, molecule A will travel a distance $\bar{r}$ relative to molecule B. Then the total number of collisions of 1 molecule of A per second is approximated by the volume swept out by the sphere of influence of radius, σ_{AB}, times the number of molecules of B per cubic centimeter. The volume swept out by the sphere of influence is represented graphically in Figure 12-3. Then

$$\begin{array}{l}\text{Number of collisions which one molecule of} \\ A \text{ experiences per second per cubic centimeter}\end{array} = \pi\sigma_{AB}^2\bar{r}C_B \quad (12\text{-}60)$$

where $\qquad C_B$ = concentration of B, molecules/cm^3

Since each molecule of A experiences the same number of collisions given by Equation (12-60), then the total number of two-body collisions of unlike molecules A and B per second per cubic centimeter is given by

$$Z_{AB} = \pi\sigma_{AB}^2\bar{r}C_A C_B \qquad (12\text{-}61)$$

The number of collisions of like molecules, say A and A, is equal to one-half the number of two-body collisions given by setting $A = B$ in Equation (12-61). Thus, for this case, the number of collisions suffered by molecule A per second per cubic centimeter is given by

$$Z_{2A} = \frac{\pi\sigma_A^2\sqrt{2}\,c_{\text{av}}C_A^2}{2} \qquad (12\text{-}62)$$

However, for every collision that results in a reaction, it is evident that two molecules of A react. Then, for bimolecular reactions of like molecules,

$$r_A = 2\,pfZ_{2A} \qquad (12\text{-}63)$$

E. Fraction of the Colliding Molecules That Have an Energy Equal to or Greater Than the Activation Energy

Let this minimum activation energy be denoted by ε (ergs per molecule) or E (calories per gram mole). The fraction of colliding molecules that has an energy equal to or greater than ε may be approximated by considering the kinetic energy in the plane of reaction as being available for reaction. This calculation is carried out for like molecules. Let u and v be the velocities of the two molecules in the plane of collision. Then

$$\overline{c^2} = \overline{u^2} + \overline{v^2} \tag{12-64}$$

and, for two-dimensional space, Equation (12-43) reduces to

$$\overline{c^2} = \alpha^2 \tag{12-65}$$

For two degrees of freedom,

$$\frac{m\overline{c^2}}{2} = 2\frac{kT}{2} \tag{12-66}$$

Thus

$$\alpha = \sqrt{\frac{2kT}{m}} \tag{12-67}$$

and for two degrees of freedom the function $F(c)$ is given by

$$F(c) = \left(\frac{m}{2\pi kT}\right)e^{-mc^2/2kT} \tag{12-68}$$

Furthermore, all velocity vectors between c and $c + dc$ terminate in the annulus described about an origin by circles of radii c and $c + dc$. Since the area of this annulus is $2\pi c\, dc$, the fraction of molecules having a velocity between c and $c + dc$ is given by

$$\frac{dN_c}{N_T} = \left(\frac{m}{2\pi kT}\right)e^{-mc^2/2kT}2\pi c\, dc \tag{12-69}$$

Since the kinetic energy ε_{tr} per molecule is given by

$$\varepsilon_{tr} = \frac{mc^2}{2} \tag{12-70}$$

it follows that $\qquad\qquad d\varepsilon_{tr} = mc\, dc \tag{12-71}$

Equation (12-69) may be restated in terms of the fraction of molecules having an energy lying between ε_{tr} and $\varepsilon_{tr} + d\varepsilon_{tr}$ by substituting Equations (12-70) and (12-71) into Equation (12-69) to obtain

$$\frac{dN_c}{N_T} = \frac{1}{kT}e^{-\varepsilon_{tr}/kT}\, d\varepsilon_{tr} \tag{12-72}$$

Thus the fraction f of the total molecules that have an energy equal to or greater than ε is given by

$$f = \int_{\varepsilon}^{\infty} e^{-\varepsilon_{tr}/kT} \left(\frac{d\varepsilon_{tr}}{kT} \right) = e^{-\varepsilon/kT} \tag{12-73}$$

If the value of f for unlike molecules is approximated by taking it equal to the value computed for like molecules, then for the reaction

$$A + B \rightarrow \text{products}$$

$$r_A = r_B = pfZ_{AB} = pe^{-\varepsilon/kT}\pi\sigma_{AB}^2\bar{r}C_AC_B$$

$$= p\sigma_{AB}^2\sqrt{\frac{8\pi kT}{\mu}}\ e^{-\varepsilon/kT}C_AC_B \tag{12-74}$$

For the reaction

$$2A \rightarrow \text{products}$$

$$r_A = 2pfZ_{2A} = 4p\sigma_A^2\sqrt{\frac{\pi kT}{m}}\ e^{-\varepsilon/kT}C_A^2 \tag{12-75}$$

Instead of expressing the rate expressions in the units of molecules, cm^3, K, and ergs, the system of units of g mol, cm^3, K, and cal is used. For this case, Equation (12-75) becomes

$$r_A\left(\frac{\text{g mol}}{\text{s cm}^3}\right) = 4p\sigma_A^2\sqrt{\frac{\pi RT}{M}}\ e^{-E/RT}C_A^2 \tag{12-76}$$

where
$$C_A = \text{concentration of } A, \text{g mol/cm}^3$$
$$E = \text{energy of activation, cal/g mol}$$
$$R = \text{gas constant, cal/(g mol)(K)}$$
$$M = \text{molecular weight of } A, \text{g/g mol}$$

Then the rate constant k_c, as predicted by the kinetic theory for the reaction $2A \rightarrow$ products, with the rate $r_A = k_cC_A^2$, is given by

$$k_c = \left(4p\sigma_A^2\sqrt{\frac{\pi RT}{M}}\right)e^{-E/RT} \tag{12-77}$$

In general, the number of collisions far exceeds the number of molecules reacting. For example, Z_{2A} as given by Equation (12-62) is approximately equal to 10^{28} collisions/s/cm^3 for a typical gas at standard conditions. Although the fractions p and f do reduce the number of collisions, the predicted rate constants are inaccurate, unreliable, and must be, therefore, determined experimentally. Equation (12-77) predicts that the variation of k_c with temperature is given by

$$k_c \propto \sqrt{T}\,e^{-E/RT} \tag{12-78}$$

The most significant term of this expression is the exponential $e^{-E/RT}$, which is in agreement with many experimental observations of the variation of k_c with temperature.

Also, the kinetic theory predicts the correct form of the rate expression; that is, for $2A \rightarrow$ products,

$$r_A \propto C_A^2 \tag{12-79}$$

The collision theory can be extended to account for third-order reactions on the basis of trimolecular collisions as demonstrated by Frost and Pearson [1]. However, it remained for Lindermann [3] to explain unimolecular reactions on the basis of the collision theory. The theory developed by Lindermann [3] and Hinshelwood [4] also accounts for the observation that certain reactions exhibit a second-order rate at low pressures and a first-order rate at high pressures. The Lindermann–Hinshelwood theory is demonstrated by an example in Chapter 4.

II. TRANSITION-STATE THEORY

More modern than the kinetic theory of reaction rates is the transition-state theory, also called the activated complex theory, as well as the theory of absolute reaction rates. Laidler [5] credits Mercelin [6] with the original formulation of this theory. The theory was first applied to a specific reaction by Pelzer and Wigner [7] and further developed to its present general form by Eyring and Polanyi [8] and Eyring [9].

The transition-state theory takes into account the internal energy distribution within the molecules as well as the translational energy possessed by the molecules. According to this theory, there is some configuration through which the reacting molecules must pass in going from reactants to products. This particular configuration in space is called the *transition state*, and the molecules in this state are called the *activated complex*.

The postulates of the transition-state theory are as follows:

1. For a chemical reaction to occur, it is necessary for the atoms or molecules to come together and form an activated complex.
2. The activated complexes are in dynamic equilibrium with the reactant molecules.
3. The rate of reaction is equal to the concentration of activated complexes that pass from reactants to products (from left to right) divided by the time required for an activated complex to traverse the distance across the energy barrier in the direction from left to right.

The energies of the reactants, the activated complex, and the products under different conditions are commonly represented by potential energy surfaces. Dynamic equilibrium between reactants and activated complexes is treated by statistical mechanics. Before further consideration of the transition-state theory, the concepts of potential energy surfaces and statistical mechanics are reviewed briefly.

A. Potential Energy Surfaces

The concept of potential energy surfaces is introduced herein by consideration of the case where the atoms X, Y, and Z are involved in the reaction

$$XY + Z = X + YZ \tag{12-80}$$

The potential energy of the system may be represented in three-dimensional space by plotting the potential energy of X, Y, and Z as a function of the $X - Y$ and $Y - Z$ distances. The lowest energy states correspond to the stable molecules XY and YZ. The potential energy corresponding to different $X - Y$ and $Y - Z$ distances may be presented as a contour map of the same form as that which might be used to represent a mountain range. Lines of equal elevation would correspond to lines of equal energy. The two stable states XY and YZ (the reactant and product molecules) have the lowest potential energies of any configuration and may be placed in correspondence with two valleys separated by mountains. All the ancient roads and most of the modern ones cross mountain ranges through passes, gaps, or saddle points. Similarly, most of the reacting molecules follow an energy path from reactants to

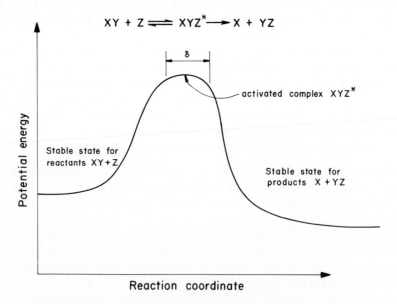

Figure 12-4. Potential energy along the reaction coordinate. [Taken from Kenneth Denbigh, *The Principles of the Chemical Equilibrium*, 3rd ed. (1971). Used with permission of Cambridge University Press.]

products that has a minimum potential energy barrier, which corresponds to a "pass," "gap," or "saddle point" in the potential energy surface. This energy path followed by most of the molecules is called the reaction coordinate. The potential energy along this path is depicted in Figure 12-4.

In the stable states, the molecule XY is far removed from the center of the atom Z, and YZ is far removed from X. As molecule XY and atom Z approach each other, higher energies are required to overcome the van der Waals repulsive force. As XY and Z approach the transition state, the atom Y takes a position about halfway between X and Z. This configuration is called the activated complex and denoted by XYZ^*. The descent into the YZ valley is marked by a breaking away of X from YZ, which is reflected by a decrease in energy.

The concept presented of potential energy surfaces has been simplified considerably. In principle, the potential energy surface can be obtained by quantum mechanical calculations [10]. However, because of computational difficulties, the potential energy surface of even the simplest of systems, H_2, is obtained by means of a semiempirical method such as the one devised by Eyring and Polanyi [8].

B. Relationships from Statistical Mechanics

The development of this subject from first principles has been presented by others [11–17], and certain results of this theory are taken as the starting point in the present analysis. Maxwell–Boltzmann statistics are based on the suppositions that the molecules of a gas are independent (the gas obeys the perfect gas law) and that every molecule is in a discrete energy state. These postulates lead to the partition function Q for a gas mixture containing N molecules of the same type.

$$Q = \frac{1}{N!} f^N$$

$$f = \sum_{i=1}^{n} \omega_i e^{-\varepsilon_i/kT}$$

(12-81)

where n = number of energy levels

ω_i = number of times a given energy state ε_i occurs

ε_i = molecular energy levels

For a mixture containing N_1 molecules of type 1, N_2 of type 2, and N_3 of type 3, the expression for the partition function is given by

$$Q = \frac{1}{N_1! \, N_2! \, N_3!} f_1^{N_1} f_2^{N_2} f_3^{N_3}$$

(12-82)

To evaluate the molecular partition function f, it is necessary to postulate that Schrödinger's wave equation describes the mechanical behavior of matter on the atomic scale. Only the one solution of this equation that corresponds to one degree of freedom of translation (one coordinate) is needed in the subsequent analysis. For a molecule of mass m which is permitted to move a distance δ in the x-direction, the partition function f_x is given by

$$f_x = \frac{\delta}{h}(2\pi mkT)^{1/2} \tag{12-83}$$

where h = Planck's constant = 6.624×10^{-27} (erg s)/molecule

A given molecular energy level, ε_i, is made up of the sum of the translational (ε_t), rotational (ε_r), vibrational (ε_v), and electronic (ε_e) energies; that is,

$$\varepsilon_i = \varepsilon_t + \varepsilon_r + \varepsilon_v + \varepsilon_e \tag{12-84}$$

Then, as shown by Denbigh [11], the molecular partition function f can be stated in terms of the corresponding partition functions

$$f = f_t f_r f_v f_e \tag{12-85}$$

where the individual partition functions are of the same form as that which follows for f_t.

$$f_t = \sum_j \omega_{tj} e^{-\varepsilon_{tj}/kT} \tag{12-86}$$

where ω_{tj} is the number of times the energy state ε_{tj} occurs.

The partition function Q for a gaseous mixture containing N molecules is related to the Helmholtz free energy function $A[A \equiv U(\text{internal energy}) - TS(\text{entropy})]$ as follows:

$$A = -kT \log_e Q \tag{12-87}$$

From classical thermodynamics [1], the following condition must be satisfied at equilibrium:

$$\sum_i \alpha_i \mu_i = 0 \tag{12-88}$$

The α_i's are the stoichiometric coefficients. For example, for the reaction

$$aA + bB \rightleftarrows cC + dD$$
$$\alpha_A = -a, \qquad \alpha_B = -b, \qquad \alpha_C = c, \quad \text{and} \quad \alpha_D = d \tag{12-89}$$

The μ_i's are the partial molar quantities, which may be stated in terms of the Helmholtz free energy A of a mixture as follows:

$$\mu_i = \left(\frac{\partial A}{\partial N_i}\right)_{T, V, N_j} \tag{12-90}$$

The subscript N_j means that, in carrying out the partial differentiation, the moles of all components except component i are held fixed. For the general reaction given by Equation (12-89),

$$A = -kT \log_e \frac{f_A^{N_A} f_B^{N_B} f_C^{N_C} f_D^{N_D}}{N_A! N_B! N_C! N_D!} \tag{12-91}$$

From Equations (12-87) through (12-90) it follows that

$$c \frac{\partial A}{\partial N_C} + d \frac{\partial A}{\partial N_D} - a \frac{\partial A}{\partial N_A} - b \frac{\partial A}{\partial N_B} = 0 \tag{12-92}$$

Since the number of molecules is relatively large, Stirling's approximation for $N!$ is used:

$$N! \cong N^N e^{-N} \sqrt{2\pi N} \tag{12-93}$$

which may be further simplified to

$$N! \cong N^N e^{-N} \tag{12-94}$$

Use of this approximation to restate Equation (12-91) yields

$$A = -kT \left(- \sum_{A,B,C,D} N_i \log_e N_i + \sum_{A,B,C,D} N_i + \sum_{A,B,C,D} N_i \log_e f_i \right) \tag{12-95}$$

Then

$$\left(\frac{\partial A}{\partial N_i} \right)_{T,V,N_j} = -kT(-\log_e N_i + \log_e f_i)$$

$$= -kT \left(-\log_e \frac{N_i}{V} + \log_e \frac{f_i}{V} \right), \qquad i = A, B, C, D \tag{12-96}$$

where V is the total volume of the reacting mixture. Substitution of the result given by Equation (12-96) for $i = A$, B, C, and D into Equation (12-92) gives the following result for a perfect gas mixture:

$$K_c = \frac{(f_C/V)^c (f_D/V)^d}{(f_A/V)^a (f_B/V)^b} \tag{12-97}$$

where

$$K_c = \frac{C_C^c C_D^d}{C_A^a C_B^b}$$

Equation (12-97) requires that each partition function be computed relative to the same zero datum. However, it is customary to compute the partition function of each molecule relative to its own lowest energy level. Relative to any arbitrary zero energy level, let the lowest energy levels of molecules A, B, C, and D be denoted by ε_A, ε_B, ε_C, and ε_D, respectively. Then Equation (12-97) becomes

$$K_c = \frac{\phi_C^c \phi_D^d}{\phi_A^a \phi_B^b} e^{-[c\varepsilon_C + d\varepsilon_D - a\varepsilon_A - b\varepsilon_B]/kT} \tag{12-98}$$

where the partition functions f are related to the partition functions ϕ in the same manner as that which follows for component C:

$$\phi_C = \frac{f_C}{V} e^{-\varepsilon_C/kT} \tag{12-99}$$

Instead of stating the energies on a molecule basis, a mole basis may be employed; that is, let

$$\frac{\Delta E_0}{RT} = \frac{c\varepsilon_C + d\varepsilon_D - a\varepsilon_A - b\varepsilon_B}{kT} \tag{12-100}$$

Then the general result follows immediately from Equation (12-98):

$$K_c = \frac{\phi_C^c \phi_D^d}{\phi_A^a \phi_B^b} e^{-\Delta E_0/RT} \tag{12-101}$$

C. Derivation of the Rate Equation

On the basis of the postulates of the transition-state theory, the forward rate of the reaction given by Equation (12-89) is formulated in the following manner. Since existence of an activated complex is given by the first postulate, Equation (12-89) may be restated as follows:

$$aA + bB \rightleftarrows M \rightleftarrows cC + dD \tag{12-102}$$

In view of the second postulate (the reactants and the activated complex are in equilibrium), it follows from Equations (12-97) and (12-101) that

$$K_c = \frac{C_M}{C_A^a C_B^b} = \frac{\phi_M}{\phi_A^a \phi_B^b} e^{-\Delta E_0/RT} \tag{12-103}$$

The activated complex M has only one degree of translational freedom as it travels the distance δ along the reaction coordinate. Then, by use of Equations (12-83) and (12-85), it follows that

$$\phi_M = \phi_M^\dagger \left(\frac{\delta}{h}\right)(2\pi m_M kT)^{1/2} \tag{12-104}$$

where $\phi_M^\dagger$ denotes the remainder of the partition function ϕ_M after the translational partition function over δ has been factored out. Thus Equation (12-103) may be restated as follows:

$$K_c = \frac{C_M}{C_A^a C_B^b} = \left[\frac{\phi_M^\dagger}{\phi_A^a \phi_B^b}\right]\left(\frac{\delta}{h}\right)(2\pi m_M kT)^{1/2} e^{-\Delta E_0/RT} \tag{12-105}$$

Let

$$K^\dagger = \frac{\phi_M^\dagger e^{-\Delta E_0/RT}}{\phi_A^a \phi_B^b} \tag{12-106}$$

Then

$$K_c = \frac{C_M}{C_A^a C_B^b} = K^\dagger \left(\frac{\delta}{h} \right) (2\pi m_M kT)^{1/2} \tag{12-107}$$

To compute the mean velocity at which activated complexes move to the right along the reaction coordinate and through the distance δ, the velocity distribution for one-dimensional motion [Equation (12-48)] may be employed. Since one-half of the activated complex molecules are free to move to the left and one-half are free to move to the right, it follows that, for those moving to the right,

$$\bar{v} \frac{N_M}{2} = \int_0^{N_M/2} v \, dN_M, \qquad \bar{v} = 2 \int_0^{N_M/2} v \frac{dN_M}{N_M} \tag{12-108}$$

Then from Equation (12-48) it follows that

$$\bar{v} = 2 \left(\frac{m_M}{2\pi kT} \right)^{1/2} \int_0^\infty e^{-m_M v^2/2kT} v \, dv \tag{12-109}$$

where m_M = mass per molecule of the activated complex M

The limits of 0 to ∞ result from the fact that only those velocities having a direction to the right are to be included in the average. Integration of Equation (12-109) yields

$$\bar{v} = \left(\frac{2kT}{\pi m_M} \right)^{1/2} \tag{12-110}$$

The average time required for an activated complex to move to the right through the transition state is given by

$$\frac{\delta}{\bar{v}} = \delta \left(\frac{\pi m_M}{2kT} \right)^{1/2}$$

Then the rate of reaction r in terms of the number of complexes crossing per unit time per unit volume is given by

$$r = \left(\frac{C_M}{2} \right) \left(\frac{\bar{v}}{\delta} \right) = \left(\frac{C_M}{\delta} \right) \left(\frac{kT}{2\pi m_M} \right)^{1/2} \tag{12-111}$$

When C_M is replaced by its equivalent as given by Equation (12-107), the result

$$r = \left(\frac{kT}{h} \right) K^\dagger C_A^a C_B^b \tag{12-112}$$

is obtained. Thus the transition-state theory predicts the correct molecularity $(a + b)$ of the forward reaction. It also accounts for a unimolecular reaction. The quantity $K_c^\dagger$ is not actually an equilibrium constant, but differs from the equilibrium constant K_c by the translational partition function for the activated complex [see Equations (12-103) and (12-106)].

To allow for the probability that not every activated complex reaching the top of the potential energy barrier is converted into a reaction product, Laidler [5] and Eyring [10] employ a transmission coefficient κ in Equation (12-112) to give

$$r = \kappa \frac{kT}{h} K^\dagger C_A^a C_B^b \qquad (12\text{-}113)$$

As suggested by Denbigh [11], the effect of temperature on the rate of reaction may be accounted for by regarding $K^\dagger$ as an equilibrium constant and relating it to a "standard" free energy change $\Delta G^\dagger$.

$$\Delta G^\dagger = -RT \log_e K^\dagger \qquad (12\text{-}114)$$

Then for $\kappa = 1$, the rate constant in the expression $r_A = k_c C_A^a C_B^b$ is given by

$$k_c = \left(\frac{kT}{h}\right) e^{-\Delta G^\dagger/RT} = \left[\frac{kT}{h} e^{\Delta S^\dagger/R}\right] e^{-\Delta H^\dagger/RT} \qquad (12\text{-}115)$$

A similar approach for the treatment of the effect of temperature on k_c is suggested by Laidler [5] and Glasstone et al. [14].

PROBLEMS

12-1. Use the fact that

$$\int_0^\infty e^{-a^2 x^2} \, dx = \frac{\sqrt{\pi}}{2a}$$

to show that Equation (12-29) may be integrated to give Equation (12-30).

12-2. Use the fact that

$$\int_0^\infty x^4 e^{-ax^2} \, dx = \frac{3}{8a^2} \sqrt{\frac{\pi}{a}}$$

to obtain Equation (12-38) from Equation (12-37).

12-3. (a) Use the fact that

$$\int_0^\infty x^2 e^{-ax^2} \, dx = \frac{1}{4a} \sqrt{\frac{\pi}{a}}$$

to obtain the result given by Equation (12-40) from Equation (12-39).
    (b) Show that the result given by Equation (12-52) may be obtained by evaluating the integral appearing in Equation (12-51).

12-4. Begin with Equation (12-53) and the expression for $\phi(c)$, and make use of Equation (12-54) to show that the most probable velocity is given by Equation (12-55).

12-5. Perform the integration indicated in Equation (12-73) and obtain the result shown.

12-6. **(a)** Begin with Equation (12-91) and make use of Equation (12-94) to obtain the expression given by Equation (12-95).

(b) Use the result of part (a) and the relationship given by Equation (12-92) to obtain the expression given by Equation (12-97).

REFERENCES

1. Frost, A. A., and R. G. Pearson, *Kinetics and Mechanisms*, John Wiley & Sons, Inc., New York (1953).

2. Loeb, L. B., *Kinetic Theory of Gases*, McGraw-Hill Book Company, New York (1927).

3. Lindermann, F. A., *Trans. Faraday Soc.*, *17*:598 (1922).

4. Hinshelwood, C. N., *Proc. Roy. Soc.*, *A113*:230 (1927).

5. Laidler, K. J., *Chemical Kinetics*, McGraw-Hill Book Company, New York (1950).

6. Mercelin, A., *Ann. Phys.*, *3*:158 (1915).

7. Pelzer, H., and E. Wigner, *Z. Physikal Chem.*, *B15*:123 (1932).

8. Eyring, H., and M. Polanyi, *Z. Physikal Chem.*, *12B*:279 (1931).

9. Eyring, H., *J. Chem. Phys.*, *3*:107 (1935).

10. Eyring, H., and E. M. Eyring, *Modern Chemical Kinetics*, Van Nostrand Reinhold, New York (1965).

11. Denbigh, Kenneth, *The Principles of Chemical Equilibrium with Applications in Chemistry and Chemical Engineering*, Cambridge University Press, New York (1955).

12. Fowler, R. H., *Statistical Mechanics*, Cambridge University Press, New York (1938).

13. Mayer, J. E., and M. G. Mayer, *Statistical Mechanics*, John Wiley & Sons, Inc., New York (1940).

14. Glasstone, S., K. J. Laidler, and H. Eyring, *Theory of Rate Processes*, McGraw-Hill Book Company, New York (1941).

15. Rushbrook, G. S., *Introduction to Statistical Mechanics*, Oxford University Press, New York (1949).

16. Sommerfeld, Arnold, *Thermodynamics and Statistical Mechanics*, Academic Press, New York (1964).

17. Tolman, R. C., *Principles of Statistical Mechanics*, Oxford University Press, New York (1938).

Index